Heather Mortimer

Contents

OXFORD MEDICAL PUBLICATIONS

Oxford Handbook of
Clinical Specialties

This work is dedicated to

M & V

Oxford Handbook of Clinical Specialties

J. A. B. COLLIER
J. M. LONGMORE
and T. J. HODGETTS

Oxford   New York   Tokyo
OXFORD UNIVERSITY PRESS

Oxford University Press, Walton Street, Oxford OX2 6DP

Orders by telephone (UK) 01865 242913

Oxford   New York
Athens   Auckland   Bangkok   Bombay
Calcutta   Cape Town   Dar es Salamm   Delhi
Florence   Hong Kong   Istanbul   Karachi
Kuala Lumpur   Madras   Madrid   Melbourne
Mexico City   Nairobi   Paris   Singapore
Taipei   Tokyo   Toronto

and associated companies in
Berlin   Ibadan

Oxford is a trade mark of Oxford University Press

Published in the United States
by Oxford University Press Inc., New York

© J. A. B. Collier and J. M. Longmore, 1987, 1989, 1990;
J. A. B. Collier, J. M. Longmore, and J. H. Harvey, 1991;
First published 1987
Second edition 1989
Third edition 1991
Fourth edition 1995
Reprinted (with corrections and updatings) 1995, 1996
Spanish translation 1991
German translation 1991, 1994
Hungarian translation 1992
Polish translation 1993

A catalogue record for this book is available from the British Library.

Library of Congress Cataloging in Publication Data
ISBN 0 19 262537-3

Typeset by AMA Graphics Limited, Preston
Printed in China

Preface to the fourth edition

The text of this edition has been entirely reset, to make way for the present flash of colour. This has given us the opportunity to make many hundreds of small changes to keep this text up to date, and to make its appreciation easier.

This volume sees some major changes since the arrival of the third edition (1991). We are pleased to welcome Dr Tim Hodgetts as an author, and his many contributions are manifest in the following pages. Chief among these is our new chapter on *Pre-hospital immediate care* (p 780-800).

The *Obstetrics* section has been completely rewritten. An early version of this changed chapter appeared in the last printing of this edition, and has been changed again for the current edition. We are very grateful to Oxford University Press for allowing us the latitude to make these changes at reprintings (as well as at the time of a new edition): this enables us to offer our readers not just the *latest* thoughts and refinements on any topic, but also a continuous striving for the *best*.

Other new topics included since the third edition appeared are male infertility (p 59), voiding difficulties (p 71), the essence of reproductive health (p 77), labour policies (p 113), abdominal pain in pregnancy (p 135), epilepsy in pregnancy (p 161), the 'triple blood test' (p 167), is this child seriously ill? (p 175), primary antibody deficiency (p 201), murmurs and heart sounds in children (p 225), child coma scales (p 257), purpura in children (p 275), paediatric cardiorespiratory arrest (p 310-11), how to identify and use the full range of psychiatric services (p 328), selective serotonin reuptake inhibitors (p 341), community care (p 387), seasonal affective disorder (p 404), poverty and mental health (p 406), purchasers and providers (p 466), protocols and guidelines (p 434), the placebo effect (p 435), eye terms (p 475), colour blindness (p 517), allergic conjunctivitis (p 522), photosensitive eruptions (p 599), mastocytosis (p 602), diagnostic peritoneal lavage (p 719), Li-Fraumeni syndrome (p 752), and moyamoya disease (p 754).

As mentioned on p 446, we have now entered the realm of electronic publication, and more recent updatings have been arranged with our partner in this enterprise—Egton Medical Information System (EMIS). We wish to record our thanks to this company for enabling these updates to be made available to readers every two weeks, by modem. These updates are woven into the fabric of the text, but may also be seen in isolation, enabling users to view, for example, all the important updates in *Obstetrics* in the last month, or to see what randomized trials have been included in the text.

Herring
1994

J.A.B.C.
J.M.L.

Preface to the 1st edition

When someone says that he is 'doing obstetrics'—or whatever, this shoul not hide the fact that much more is being done besides, not just a little c each of medicine, psychiatry, gynaecology and paediatrics, but also a goo deal of work to elicit and act upon the patient's unspoken hopes and fear At the operating table he must concentrate minutely on the problem i hand; but later he must operate on other planes too, in social and psycho logical dimensions so as to understand how the patient came to need to b on the operating table, and how this might have been prevented. All th best specialists practice a holistic art, and our aim is to show how specia ism and holism may be successfully interwoven, if not into a fully wate tight garment, then at least into one which keeps out much of th criticism rained upon us by the proponents of alternative medicine.

We hope that by compiling this little volume we may make the arduou task of learning medicine a little less exhausting, so allowing more energ to be spent at the bedside, and on the wards. For a medical student com ing fresh to a specialty the great tomes which mark the road to knowledg can numb the mind after a while, and what started out fresh is in dange of becoming exhausted by its own too much. It is not that we are again the great tomes themselves—we are simply against reading them too muc and too soon. One starts off strong on 'care' and weak on knowledge, an the danger is that this state of affairs becomes reversed. It is easier to lear from books than from patients, yet what our patients teach us may be c more abiding significance: the value of sympathy, the uses of compassio and the limits of our human world. It is at the bedside that we learn ho to be of practical help to people who are numbed by the mysterious disa ters of womb or tomb, for which they are totally unprepared. If this sma book enables those starting to explore the major specialties to learn a they can from their patients, it will have served its purpose—and can the be discarded.

Because of the page-a-subject format, the balance of topics in the follow ing pages may at first strike the reader as being odd in places. However, has been our intention to provide a maximally useful text rather than on which is perfectly balanced in apportioning space according to how com mon a particular topic is—just as the great *Terrestrial Globes* made by Georg Phillips in the 1960s may seem at first to provide an odd balance of plac names, with Alice Springs appearing more prominently than Amsterdam To chart a whole continent, and omit to name a single central location ou of respect for 'balance' is to miss a good opportunity to be useful. Georg Phillips did not miss this opportunity, and neither we hope, have we. It inevitable that some readers will be disappointed that we have left ou their favoured subjects (the Phillips' Globe does not even mentio Oxford!). To these readers we offer over 300 blank pages by way of apolog

Ferring
1987

J.A.B.C
J.M.

Acknowledgements

We thank the British Lending Library, Ms Susan Merriott, and Mrs Thelma Royce for their diligence in supplying us with hard-to-find references—without their tireless and much appreciated help this publication would not have been possible. We also thank Dr S. Mercer for reading and correcting the proofs, and Dr T. Toma (USA/Romanian fax correspondant) for his timely updates. For typographic expertise, we thank Mr Simon Mather, and Mr Andrew Mather.

For their invaluable comments on the text we thank Mr A. Carr, Dr R. Hope, Dr J. Colston, Dr R. Downes, Mr J. Braithwaite, Dr C. Moulton, Dr P. Muthusamy, Ms A. Peattie, Dr B. Phillips, Professor C. Redman, Mr S. Western, Mr T Lavy, Dr M. Too-Chung.

For additional peer reviewing we are especially grateful to Dr T. Toma, Dr . Scott, Dr J. Cox, Dr J. Orrell and Dr H. Thomas.

We thank the following for their suggestions—via our reader's comments card enclosed with this volume): K. Abou-Elhmd; A. Adiele; R. Adley; M. Al-Amin; H. Albrecht; Zulfiqar Ali; V. Atamyan; anonymous readers from Hinckley and District Hospital; N. Balasuriyar, D. Bansevicius; D. Boddie; . Bourke; P. Piotr Brykalski; C. Budd; K. Burn; I. Cardozo; P. Collins; Corcoran; H. Constantinides; J. Crane; D. Dharmi; J. Dart; T. Davies; Eley; J. Fagan; O. Fenton; M. Fry; D. Foss; E. France; J. Gutman; . Hennigan; J. Hill; S. Holliday; R. William Howe; G. Hutchison; M. Ip; Kuber; N. Lees; J. Lehane; J. McFazdean; R. McLaughlin; A. Martin; C. Maytum; E. Miller; D. Moskopp; M. Naraen; K. O'Driscoll; E. Olson; J. Olson; . Payne; A. Peattie; J Rees P. Rees; R. Reynolds; C. Robertson; A. Rodgers; . Pyper; N. Tseraidi; M Tsolaks; M. Turur; D. Warren; E. Wright; P. Zack.

We particularly thank all the staff at Oxford University Press for their imaginative help at all stages of publication.

We thank the following authors, publishers and editors for permission to reproduce illustrative material: R. A. Hope; D. Kinshuck; A. Land; .. O'Driscoll; A. Swain; the *British Journal of Hospital Medicine*; the *British Medical Journal*; *General Practitioner*; the *Journal of Pediatrics*; the Association for Consumer Research; Baillière Tindall; John Wright; Edward Arnold; ange; Churchill Livingstone. We thank Mark Edwards for drawing the illustrations on p 389–97.

Notes

Pronouns For brevity, the pronoun *he* or *she* has been used in places where *he or she* would have been appropriate. Such circumlocutions do not aid the reader in forming a vivid visual impression, which is one of the leading aims of good authorship. Therefore, for balance and fairness, and where sense allows, we have tried alternating *he* with *she*.

Drugs While every effort has been made to check drug doses, it is still possible that errors have been missed. Dosage schedules are frequently revised and new side-effects recognized. Oxford University Press makes no guarantee, express or implied, that drug doses in this book are correct. For these reasons, readers are urged to consult the fortnightly updated electronic form of this text (*Mentor*, from EMIS, see OHCM p 17–18), the *British National Formulary*, or the drug company's printed instructions, before giving any of the drugs recommended in this book.

Except in Chapter 3 (*Paediatrics*), drug doses are for the non-pregnant adult who is not breast feeding.

Most of the time we chug along treating patients quite well, without ever really understanding them. The idea that we should strive to understand and empathize with *all* our patients is ultimately unsustainable. Out-patient clinics and surgeries would grind to a halt, and urgent visits would never get done. It is also possible that to do so would be counter-productive from the patient's point of view. For two human beings to understand each other's inner life is a rare event, and if we offered this understanding to all our patients they might become addicted to us, and be unable to get on with the rest of their lives. Nevertheless it is good practice to have a go, on occasion, at trying to understand some patients. Doing so may entail swallowing an alien world and digesting it rather slowly. Paradoxically, to achieve this, we very often need to keep our mouths *shut*, particularly with those in whom we have reached a therapeutic impasse—for example if the illness is untreatable, or the patient has rejected our treatment, or if the patient seems to be asking or appealing for something more. Eye contact is important here. One of the authors (JML) recalls forever his very first patient—found on a surgical ward recovering from the repair of a perforated duodenal ulcer: a nice simple surgical patient, ideal for beginners. I asked all the questions in the book, and knew all his answers and his physical features: even the colour of his eyes. Luckily, the house officer who was really looking after him did not ask many questions, and knew how to interpret the appeal for help behind those eyes, and in his busy day found space to receive the vital clue beyond my grasp—that my patient was a drug addict and under great stress as he could no longer finance his activity.

So the first step in trying to understand a patient is to sit back and listen. Next, if possible, it is very helpful to see your patient often, to establish rapport, and mutual respect. If the relationship is all one way, with the doctor finding out all about the patient, but revealing nothing of him- or herself, this mutual respect can take a very long time to grow. But beware of sharing too much of your own inner life with your patients: you may overburden them, or put them off. Different patients respond to different approaches. Understanding patients inevitably takes time, and it may be hard in a series of short appointments. A visit to the patient's home may be very revealing, but for many doctors trapped in hospital wards or clinics, this is impossible. But it is usually possible to have a longish, private interview, and take whatever opportunity turns up. We once worked with a consultant who infuriated his junior staff on busy ward rounds by repeatedly selecting what seemed to us the most boring and commonplace medical 'cases' (such as someone with a stroke) and proceeding to draw the curtain around the patient's bed to exclude us, and engage in what seemed like a long chat with the patient, all in very hushed voices, so that we never knew what he said—until Sister told us that he never said anything much and simply received anything that was on the patient's mind. For the most part, he was swallowing their world in silence. We came to realize that there was nothing these patients, robbed as they were of health and wholeness, appreciated more in their entire hospital stay.

Symbols and abbreviations

►	This is important	CVP	Central venous pressure
►►	Don't dawdle! Prompt action saves lives	CVS	Cardiovascular system
		CXR	Chest X-ray
	Conclusion from evidence-based medicine (see p433). Examples appear on pp 102, 186, 624	D	Dimension
		D&C	Dilatation (cervix) & curettage
		dB	Decibel
	More details available in the electronic version of this text (see p 446)	DIC	Disseminated intravascular coagulation
		DIP	Distal interphalangeal
	Conflict (controversial topic)	dl	Decilitre
♂/♀	Male to female ratio	DM	Diabetes mellitus
~	approximately	DoH	Department of Health
−ve	Negative	D&V	Diarrhoea and vomiting
+ve	Positive	DVT	Deep venous thrombosis
↓	Decreased (eg plasma level)	EBM	*Evidence-based Medicine*
↑	Increased (eg plasma level)	ECG	Electrocardiogram
↔	Normal (eg plasma level)	ECT	Electroconvulsive therapy
ac	Ante cibum (before food)	EEG	Electroencephalogram
ACLS	Advanced Cardiac Life Support	ENT	Ear, nose and throat
ACTH	Adrenocorticotrophic hormone	ESR	Erythrocyte sedimentation rate
ADH	Antidiuretic hormone	ET	Endotracheal
AFP	Alpha-fetoprotein	FB	Foreign body
AIDS	Acquired immuno-deficiency syndrome	FBC	Full blood count
		FH	Family history
Alk	Alkaline (Phos = phosphatase)	FHSA	Family Health Service Authority
ANF	Antinuclear factor	FSH	Follicle stimulating hormone
AP	Anteroposterior	GA	General anaesthesia
APH	Antepartum haemorrhage	g	Gram
APLS	Advanced Life Support *Manual* reference (p 171)	GI	Gastrointestinal
		GP	General practitioner
ASO	Antistreptolysin O (titre)	GGT	Gamma glutamyl transpeptidase
ATLS	Advanced Trauma Life Support *Manual* reference	GU	Genitourinary
		h	Hour
AV	Atrioventricular	Hb	Haemoglobin
BMJ	*British Medical Journal*	HBsAg	Hepatitis B surface antigen
BP	Blood pressure	HDL	High density lipoprotein
CCF	Combined (right & left sided) cardiac failure	HIV	Human immuno-deficiency virus
ChVS	Chorionic villus sampling	HRT	Hormone replacement therapy
CI	Contraindications	HVS	High vaginal swab
CMV	Cytomegalovirus	Ib/ibid	Ibidem (in the same place)
CNS	Central nervous system	ICP	Intracranial pressure
CoC	combined oral contraceptive pill	IE	Infective endocarditis
		Ig	Immunoglobulin
CO₂	Carbon dioxide	IHD	Ischaemic heart disease
CPR	Cardiopulmonary resuscitation	IM	Intramuscular
CRP	C-reactive protein	IP	Interphalangeal
CSF	Cerebrospinal fluid		

IPPV	Intermittent +ve pressure ventilation	**P$_a$O$_2$**	Partial pressure of oxygen in arterial blood
iu	International unit	**pc**	Post cibum (after food)
IUCD	Intrauterine contraceptive device	**PCV**	Packed cell volume
IV	Intravenous	**PET**	Pre-eclamptic toxaemia
IVI	Intravenous infusion	**PID**	Pelvic inflammatory disease
IVU	Intravenous urography	**PIP**	Proximal interphalangeal
JRCGP	*Journal of the Royal College of General Practitioners*	**PO**	Per os (by mouth)
		PoP	Progesterone only pill
JVP	Jugular venous pressure	**PPH**	Post-partum haemorrhage
K$^+$	Potassium	**PR**	Per rectum
kg	Kilogram	**PTR**	Prothrombin ratio
kPa	Kilopascal	**PUO**	Pyrexia of unknown origin
l	Litre	**PV**	Per vaginam
LA	Local anaesthesia	**RBC**	Red blood cell
LFT	Liver function test	**RCGP**	Royal College of General Practitioners
LH	Luteinizing hormone		
LMP	First day of the last menstrual period	**RA**	Rheumatoid arthritis
		Rh	Rhesus
LP	Lumbar puncture	**RTA**	Road traffic accident(s)
LVH	Left ventricular hypertrophy	**RUQ**	Right upper quadrant
µg	Micrograms	**RVH**	Right ventricular hypertrophy
MAOI	Monoamine oxidase inhibitor	**SBE**	Infective endocarditis (subacute endocarditis is a subset of this)
MCP	Metacarpophalangeal		
MCV	Mean cell volume		
mg	Milligrams	**SC**	Subcutaneous
min	Minute(s)	**SCBU**	Special care baby unit
ml	Millilitre	**SE**	Side effects
mmHg	Millimetres of mercury	**sec**	Seconds
MRI	Magnetic resonance imaging	**SpO$_2$**	Oximetry estimation of capillary blood oxygen saturation (also SaO$_2$)
MSU	Midstream urine (culture of)		
MTP	Metatarsophalangeal	**STD**	Sexually transmitted disease
NaCl	Sodium chloride	**SVC**	Superior vena cava
NBM	Nil by mouth	**t$^{1/2}$**	Half life
NEJM	*New England Journal of Medicine*	**T°**	Temperature
		T$_3$	Triiodothyronine
NHS	National Health Service	**T$_4$**	Thyroxine
NMR	Nuclear magnetic resonance scan	**TB**	Tuberculosis
		TIA	Transient ischaemic attack
N$_2$O	Nitrous oxide	**TFT**	Thyroid function tests
NSAIDs	Non-steroidal anti-inflammatory drugs	**TPR**	Temperature, pulse, and respirations
O$_2$	Oxygen	**TSH**	Thyroid stimulating hormone
OHCM	*Oxford Handbook of Clinical Medicine* 4th ed, OUP	**u (or U)**	Units
		U&E	Urea and electrolytes
ORh-ve	Blood group O Rhesus negative	**UTI**	Urinary tract infection
		VSD	Ventricular septal defect
PA	Posteroanterior	**WCC**	White blood cell count
PAN	Polyarteritis nodosa	**wt**	Weight
P$_a$CO$_2$	Partial pressure of CO$_2$ in arterial blood	**yrs**	Years (old)

Note: other abbreviations are given in full on the pages where they occur.

Contents

Note: The content of individual chapters
is detailed on each chapter's first page.

1. Gynaecology

Relevant pages in other chapters
Puberty p 200; virilism and hirsutism *OHCM* p 556

History and examination

History Let her tell the story, but remember that she may be reluctant to admit some problems, particularly if you are a man, so make sure to cover them in your questions.

1 *Menstrual history:* ►Note date of last menstrual period (LMP) ie 1st day of bleeding, or menopause. Was the last period normal? Cycles: no. days bleeding/no. days from day 1 of one period to day 1 of next (eg 5/26). Are periods regular? If heavy, are there clots or flooding? How many pads/tampons are needed (this may not be a reliable guide)? Are periods painful? Is there bleeding between periods or since the menopause? Age at menarche?

2 *Obstetric history:* How many children? Antenatal problems, delivery, outcome (and weights of babies) and puerperium of each pregnancy. Terminations and miscarriages—at what stage, why and (terminations) how.

3 *Symptoms:* If she has *pain* what is it like? Uterine pain tends to be colicky and felt in the sacrum and groins. Ovarian pain tends to be felt in the iliac fossa and radiates down front of the thigh to the knee. Ask about *dyspareunia* (painful intercourse). Is it superficial (round the outside) or deep inside? If she has *vaginal discharge* what is it like (amount, colour, smell, itch); when does she get it? Ask about *prolapse* and *incontinence.* When? How bad?

4 *Sex and contraception:* Is she sexually active? Are there physical or emotional problems with sex? What contraception is she using and is she happy with it? What has she tried previously? Has she had problems conceiving? If so, has she had treatment for infertility? What about sexually transmitted diseases? Date and result of last smear?

5 *Other:* General health, smoking. Former gynaecological treatment.

Examination ►Many women find pelvic examination painful and undignified, and particularly embarrassing if you are male. Explain what you are going to do. Be gentle. Use a chaperone, if available.

General: Is she well or ill? Is she shocked? If so, treat it.

Abdomen: Look for tenderness and peritonism. If there is a mass, could it be a pregnancy? Listen for a fetal heart (p 84).

Vaginal examination: (p 6). Use your eyes to inspect the vulva, a *speculum* to examine the vagina and cervix and your *fingers* to assess the uterus and adnexae bimanually. Examination is usually done with the patient on her back or in the left lateral position (preferable for detecting prolapse). *Sims' speculum* has 2 right-angle bends, and is used for inspecting the vaginal walls, eg for prolapse and incontinence. ◻

Cusco's (bivalve) speculum is used for inspecting the cervix with the aid of a light. Insert the speculum closed (warmed under a tap) with the blades parallel to the labia; lubricating jelly makes it easier. When it is in, rotate it and open it. The speculum should achieve its full length before opening and usually the cervix will pop into view. If it doesn't, do a bimanual to check the position of the cervix and try again. Do swabs (p 48) and a cervical smear (p 32) if indicated. Close the speculum gradually as you withdraw it, to avoid trapping the cervix.

Sexual health

Sexual health is the enjoyment of sexual activity of one's choice, without causing or suffering physical or mental harm.[1] Of course there is more to sex than enjoyment. 'Perhaps the sexual life is the great test. If we can survive it with charity to those we love, and affection to those we have betrayed, we needn't worry so much about the good and the bad in us. But jealousy, distrust, cruelty, revenge, recrimination. . . then we fail. The wrong is in that failure even if we are the victims and not the executioners. Virtue is no excuse. . .'[2]

1 P Greenhouse 1995 *BMJ* i 1468   **2** G Greene 1965 *The Comedians*, Penguin, p139, ISBN 0-14-018494-5

Gynaecological anatomy

4 The vulva comprises the entrances to the vagina and urethra, the structures which surround them (clitoris, labia minora and fourchette), and the encircling labia majora and perineum. The hymen, when broken (by tampons, intercourse) leaves tags at the mouth of the vagina.

Look for: Rashes; atrophy; ulcers; lumps (p 28–30); deficient perineum (you can see the back wall of the vagina).

The vagina is a potential space with distensible folded muscular walls. The contents of the rectum, which runs behind the posterior wall, may be palpated through the vagina. The cervix projects into the vault at the top which forms a moat around it, deepest posteriorly, conventionally divided into anterior, posterior and two lateral fornices. From puberty till the menopause lactobacilli in the vagina keep it acid, discouraging infection.

Look for: Inflammation; discharge (p 48); prolapse (p 54).

The cervix is mostly connective tissue. It feels firm, and has a dent in the centre (the opening, or os, of the cervical canal). Mucin-secreting glands on its surface lubricate the vagina. The os is circular in nulliparous women, but is a slit in the parous.

Look for: Pain on moving the cervix (excitation—p 24 & p 50); ectopy; cervicitis and discharge; polyps, carcinoma (p 32).

The uterus has a thick muscular-walled *body* lined internally with columnar epithelium (the endometrium) connected to the cervix or neck. It is supported by the uterosacral ligaments. The peritoneum is draped over the uterus. The valley so formed between it and the rectum is the rectovaginal pouch (of Douglas), and the fold of peritoneum in which the Fallopian tubes lie is known as the broad ligament. The *size* of the uterus is by convention described by comparison with its size at different stages of pregnancy. Since that is variable, estimates are approximate, but the following is a guide: non-pregnant—plum-sized; 6 weeks—egg; 8 weeks—small orange; 10 weeks—large orange; 14 weeks—fills pelvis.

In most women the uterus is *anteverted*, ie its long axis is directed forward and the cervix points backwards. The body then flops forwards on the cervix—*anteflexed*. An anteverted uterus can be palpated between the two hands on bimanual examination (unless the woman is obese or tense).

In 20% it is *retroverted and retroflexed* (p 6).

Look for: Position (important to know for practical procedures); mobility (especially if retroverted); size; tenderness (p 24 & p 50).

Adnexae These are the *Fallopian tubes, ovaries* and associated connective tissue (parametria). They are palpated bimanually in the lateral fornices, and if normal cannot be felt. The ovaries are the size of a large grape and may lie in the rectovaginal pouch.

Look for masses (p 42), tenderness (p 50).

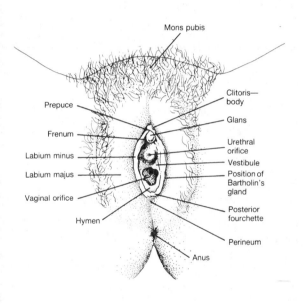

Mons pubis

Prepuce

Frenum

Labium minus

Labium majus

Vaginal orifice

Hymen

Clitoris—
body

Glans

Urethral
orifice

Vestibule

Position of
Bartholin's
gland

Posterior
fourchette

Perineum

Anus

Genital abnormalities

Vagina and uterus These are derived from the Müllerian duct system and formed by fusion of the right and left parts. Different degrees of failure to fuse lead to duplication of any or all parts of the system.

Vaginal septae are quite common (and often missed on examination).

Duplication of the cervix and/or uterus may also be missed, eg until the woman becomes pregnant in the uterus without the IUCD!

A partially divided (*bicornuate*) uterus or a uterus where one side has failed to develop (*unicornuate*) may present as recurrent miscarriage, particularly in the second trimester, or as difficulties in labour. Such abnormalities are diagnosed by *hysterosalpingogram*.

An absent uterus or a rudimentary uterus with absent endometrium are rare. They present with primary amenorrhoea.

An absent or short vagina is uncommon but can be corrected by plastic surgery. The membrane at the mouth of the vagina where the Müllerian and urogenital systems fuse (the hymen) may be imperforate. There is apparent primary amenorrhoea, with a history of monthly abdominal pain and swelling, and the membrane bulging under the pressure of dammed up menstrual blood (haematocolpos). It is relieved by incising the membrane.

▶Renal system abnormalities often coexist with genital ones, so IVU and ultrasound should be performed.

Ovary Thin, rudimentary 'streak' ovaries are found in Turner's syndrome (p 758). Ovaries are absent in testicular feminization syndrome, but primitive testes are present (p 222). Remnants of developmental tissue (eg the Wolffian system) may result in cysts around the ovary and in the broad ligament.

Uterine retroversion

About 20% of women normally have a retroverted retroflexed uterus which is fully mobile. It is difficult to palpate bimanually unless you can push it into anteversion by pressure on the cervix. It causes no problems except (rarely) if it fails to lift out of the pelvis at 12 weeks of pregnancy, presenting with discomfort and retention of urine eg at 14 weeks; catheterization and lying prone may relieve it.

Inflammation in the pelvis (due to infection or endometriosis) can cause adhesions which tether the uterus in a retroverted position. The patient may present with dysmenorrhoea, dyspareunia, or infertility—problems which can only be relieved by treatment of the underlying disease.

Normal menstruation

Puberty is the development of adult sexual characteristics. The sequence: breast buds develop, then growth of pubic hair, then axillary hair. Then menses begin (menarche) from 10½ years onwards (mean ~13yrs; age was falling in the UK, but no longer; it is earlier if short and overweight; investigate if no periods by 16yrs (p 10). Next there is a growth spurt (p 200).

The menstrual cycle The cycle is controlled by the 'hypothalamic-pituitary-ovarian (HPO) axis'. Pulsatile production of gonadotrophin-releasing hormones by the hypothalamus stimulates the pituitary to produce the gonadotrophins: follicle stimulating hormone (FSH) and luteinizing hormone (LH). These stimulate the ovary to produce oestrogen and progesterone. The ovarian hormones modulate the production of gonadotrophins by feeding back on the hypothalamus and pituitary.

Day 1 of the cycle is the first day of menstruation. Cycle lengths vary greatly; only 12% are 28 days. Cycles soon after menarche and before the menopause are most likely to be irregular and anovulatory. In the first 4 days of the cycle, FSH levels are high, stimulating the development of a primary follicle in the ovary. The follicle produces oestrogen, which stimulates the development of a thin, glandular 'proliferative' endometrium and of cervical mucus which is receptive to sperm. The mucus becomes clear and stringy (like raw egg white) and if allowed to dry on a slide produces 'ferning patterns' due to its high salt content. Oestrogen also controls FSH and LH output by negative feedback.

14 days before the onset of menstruation (on the 16th day of the cycle of a 30-day cycle) the oestrogen level becomes high enough to stimulate a surge of LH. This stimulates ovulation. Having released the ovum, the primary follicle then forms a corpus luteum and starts to produce progesterone. Under this influence, the endometrial lining is prepared for implantation: it increases in thickness and glands become convoluted ('secretory phase'). The cervical mucus becomes viscid and hostile to sperm and no longer ferns. If the ovum is not fertilized the corpus luteum breaks down, so hormone levels fall. This causes the spiral arteries in the uterine endothelial lining to constrict and the lining sloughs—hence menstruation.

Menstruation This is the loss of blood and uterine epithelial slough. It lasts 2-7 days and is usually heaviest at the beginning. Normal loss is 20-80ml, median 28ml.

Climacteric The ovaries fail to develop follicles. Without hormonal feedback from the ovary, gonadotrophin levels rise. Periods cease (menopause), usually at ~50 years of age (p 18).

Postponing menstruation (eg away on holiday) Try norethisterone 5mg/12 PO from 4 days before the period is due until bleeding is acceptable, or take 2 packets of combined contraceptive Pills without a break.

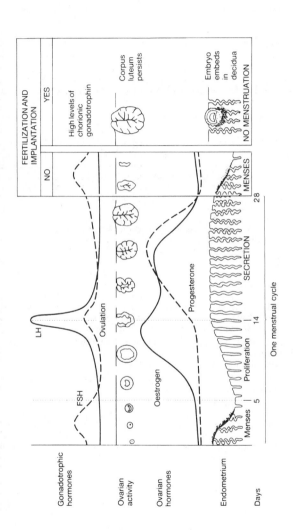

One menstrual cycle

Abnormal menstruation

Primary amenorrhoea (see p 12) This is failure to start menstruating. It needs investigation in a 16-year-old, or in a 14-year-old who has no breast development. For normal menstruation to occur she must be structurally normal with a functioning control mechanism (hypothalamic-pituitary-ovarian axis).

Secondary amenorrhoea (see p 12) This is when periods stop for >6 months, other than due to pregnancy. Hypothalamic-pituitary-ovarian axis disorders are common, ovarian and endometrial causes rare.

Oligomenorrhoea This is infrequent periods. It is common at the extremes of reproductive life when regular ovulation often does not occur. A common cause throughout the reproductive years is polycystic ovary syndrome (p 44).

Menorrhagia (p 14) This is excessive blood loss.

Dysmenorrhoea This is painful periods (± nausea or vomiting). 50% of British women complain of moderate pain, 12% of severe disabling pain.

Primary dysmenorrhoea is pain without organ pathology—often starting with anovulatory cycles after the menarche. It is crampy with ache in the back or groin, worse during the 1st day or two. Excess prostaglandins cause painful uterine contractions, producing ischaemic pain. *Treatment:* Prostaglandin inhibitors, eg mefenamic acid 500mg/8h PO during menstruation reduce contractions and hence pain. Paracetamol is a good alternative to NSAIDs. In pain with ovulatory cycles, ovulation suppression with the combined Pill can help (so dysmenorrhoea may be used as a covert request for contraception). Smooth muscle antispasmodics (eg alverine 60–120mg/8h PO) or hyoscine butylbromide (20mg/6h PO) give unreliable results. Cervical dilatation in childbirth may relieve it but, surgical dilatation may render the cervix incompetent and is no longer used as therapy.

Secondary dysmenorrhoea: Associated pathology: adenomyosis, endometriosis, chronic sepsis (eg chlamydial infection), fibroids—and so it appears later in life. It is more constant through the period, and may be associated with deep dyspareunia. Treatment of the cause is the best plan. Antiemetics may be needed in endometriosis. IUCDs increase dysmenorrhoea.

Intermenstrual bleeding This may follow a mid-cycle fall in oestrogen production. Other causes: cervical polyps; ectropion; carcinoma; cervicitis and vaginitis; hormonal contraception (spotting); IUCD; pregnancy-related.

Post-coital bleeding *Causes:* Cervical trauma, polyps, cervical carcinoma; cervicitis and vaginitis of any cause.

Post-menopausal bleeding This is bleeding occurring later than 1yr after the last period. It must be considered due to endometrial carcinoma until proved otherwise (p 40). Other causes: vaginitis (often atrophic); foreign bodies, eg pessaries; carcinoma of cervix or vulva; endometrial or cervical polyps; oestrogen withdrawal (hormone replacement therapy or ovarian tumour). She may confuse urethral, vaginal and rectal bleeding.

Amenorrhoea

▶ *Always ask yourself 'Could she be pregnant?'*

Primary amenorrhoea (See also p 10.) This may cause great anxiety. In most patients puberty is just late (often familial), and reassurance is all that is needed. Others may have factors which cause secondary amenorrhoea, arising before menarche. A few causes will be structural or genetic so check: ●Has she got normal external secondary sexual characteristics? If so, are the internal genitalia normal (p 6)? ●If she is not developing normally, examination and karyotyping may reveal Turner's syndrome (p 758) or testicular feminization (p 222). The aim of treatment is to help the patient to look normal, to function sexually, and, if possible, to enable her to reproduce if she wishes.

Causes of secondary amenorrhoea ●Hypothalamic-pituitary-ovarian causes are very common as control of the menstrual cycle is easily upset eg by emotions, exams, weight loss, excess prolactin (30% have galactorrhoea), other hormonal imbalances, and severe systemic disease eg renal failure. Tumours and necrosis (Sheehan's syndrome) are rare. ●Ovarian causes: polycystic ovary syndrome (p 44), tumours, ovarian failure (premature menopause), are uncommon. ●Uterine causes: pregnancy-related Asherman's syndrome (uterine adhesions after a D&C). 'Post-pill amenorrhoea' is generally oligomenorrhoea masked by regular withdrawal bleeds.

Tests: Serum LH (↑in polycystic ovary syndrome), FSH (very high in premature menopause), prolactin (↑by stress, prolactinomas and drugs, eg phenothiazines) and TFT are the most useful blood tests. 40% of those with hyperprolactinaemia have a tumour so do skull X-ray ± CT scan (p 58).

Treatment is related to cause. Premature ovarian failure cannot be reversed but hormone replacement (p 18) is necessary to control symptoms of oestrogen deficiency and protect against osteoporosis. Pregnancy can be achieved with oocyte donation and *in vitro* fertilization techniques.

Hypothalamic-pituitary axis malfunction: If mild (eg stress, moderate weight loss): there is sufficient activity to stimulate enough ovarian oestrogen to produce an endometrium (which will be shed after a progesterone challenge eg norethisterone 5mg/8h for 7 days), but the timing is disordered so cycles are not initiated. If the disorder is more severe the axis shuts down (eg in severe weight loss). FSH and LH and hence oestrogen levels are low. Reassurance and advice on diet or stress management, or psychiatric help if appropriate (p 348), and time may solve the problem. She should be advised to use contraception as ovulation may occur at any time. If she wants fertility restored now, or the reassurance of seeing a period, mild dysfunction will respond to clomiphene but a shut-down axis will need stimulation by gonadotrophin releasing hormone (see p 58 for both).

Menorrhagia

This is excessive menstrual blood loss. Pundits favour an objective definition (eg >80ml lost/cycle), but a more holistic approach defines menorrhagia as significant menstrual loss which interferes with the normal life of the woman concerned. 50% of those referred for menorrhagia are depressed/anxious, and it is important to find out which came first, or whether some third factor caused the menorrhagia *and* the depression—is ►anaemia or ►hypothyroidism (look for weight gain, constipation, dislike of cold, bradycardia, dry skin, goitre, odema—see OHCM p 544; do FBC, T4).

Amounts are rarely measured so the diagnosis is, by default, subjective. The woman who passes clots, floods the supermarket floor every month or becomes anaemic clearly has excessive loss, but most complaints are of loss which is in the normal range but is socially inconvenient.

Causes The most likely cause changes with age. In girls, pregnancy and dysfunctional uterine bleeding are likely. With increasing age, think also of IUCD, fibroids, endometriosis and adenomyosis, pelvic infection, polyps. In *perimenopausal women*, consider endometrial carcinoma. Ask about general bleeding problems as she may have a blood dyscrasia. Do a pelvic examination—which may reveal polyps, fibroids, or endometriosis.

Tests None may be needed. FBC, TFT, ultrasound or laparoscopy if pelvic pathology suspected; D&C if premenopausal to exclude endometrial cancer.

Dysfunctional uterine bleeding (DUB) This is heavy and/or irregular bleeding in the absence of recognisable pelvic pathology. It is associated with anovulatory cycles, so is common at the extremes of reproductive life or it may be ovulatory (eg with inadequate luteal phase). If PV is normal and organic pathology is ruled out, this is the diagnosis, by exclusion.

Treatment of menorrhagia Treat any underlying condition. Where the diagnosis is dysfunctional uterine bleeding, treatment depends on age. Reassurance helps. Teenage menorrhagia generally settles without interference as cycles become ovulatory. Unacceptably heavy bleeding may respond to medical therapy, either non-hormonal or hormonal or both. A woman whose family is complete but who cannot wait for the menopause, may prefer a permanent solution—*hysterectomy* (open, vaginal or laparoscopic, p 74) or *endometrial ablation* eg by laser, transcervical resection of the first few mm of myometrium with an electrocautery loop, or coagulation with a rollerball electrode. Ablation is more likely to be effective if the endometrium is thin, and pretreatment with leuprorelin (p 52) may be better than using a progestogen or danazol. ~30% become amenorrhoeic and a further 50% have reduced flow after any of the methods of ablation (there are no large randomized studies showing that one method is superior—but women appear to favour hysterectomy rather than ablation).

Non-hormonal preparations are usually tried first. They are taken during bleeding and reduce loss. Try an anti-prostaglandin eg mefenamic acid 500mg/8h PO pc; CI: peptic ulceration. Antifibrinolytic drugs eg tranexamic acid 1-1.5g/6-8h PO may help. CI: thromboembolic disease.

Hormone therapy: Cyclical progestogens, eg norethisterone 5mg/8h PO, are traditionally used, either in the 2nd ½ of the cycle—eg day 12-26; or if this fails throughout the cycle—eg day 5-26 (she then has a period). No studies have proved them effective. SE: weight gain, breast tenderness, bloating. Combined oral contraceptive pills can be tried but may be contraindicated in older women—who are often the ones to suffer from menorrhagia. Danazol 100mg/6-24h PO is effective, (and expensive), but side effects may be unacceptable: weight gain, acne, muscle pain, amenorrhoea. It may inhibit ovulation but is not reliably contraceptive so advise barrier contraception because it masculinizes the ♀ fetus. ►To stop heavy bleeding from the endometrium use norethisterone 5mg/8h PO.

The premenstrual syndrome

Most women notice that their mood or physical state may be worse pre menstrually. The symptoms may be mild one month and severe the next perhaps depending on external events. Symptoms tend to be worse in the 30s and 40s and improve on the combined Pill. About 3% of women regularly have cyclical symptoms so severe that they cause major disrup tion to their lives: premenstrual syndrome (PMS) or tension (PMT).

Causes There are many hypotheses but little proof. Most suggest some form of imbalance between oestrogen and progesterone. Other theories propose abnormalities in levels of other hormones or electrolytes, nutri tional factors and psychosocial causes. There is evidence that CNS sero tonin systems are blunted premenstrually.

Symptoms Commonest symptom patterns are tension and irritability depression; bloating and breast tenderness; carbohydrate craving and headache; clumsiness. Almost any symptom may feature.

Diagnosis Suggest symptom diary. If she has PMS her symptoms are worst before periods, are relieved by menstruation and there is at least one symptom-free week afterwards. Diaries may also reveal psychiatric dis orders (which may be worse premenstrually) or menstrual disorders.

Treatment[1] Simply to acknowledge her problem, listen, and reassure may be all that is needed to enable her to cope. Are her partner and children understanding? Can she rearrange work schedules to reduce stress premen strually? Some women find self-help groups supportive. Health measures eg improved diet, reducing smoking and drinking, increased exercise and self-relaxation, often help. Herbal remedies are not scientifically tested but some find them helpful, eg sage and fennel for irritability.

Any medical treatment evokes a great placebo effect, improving 90% in some studies. Pyridoxine 100mg/24h PO, for the symptomatic period or continuously, is often particularly helpful with mood changes and headache (note: high doses can cause reversible peripheral neuropathy).

For severe cyclical mastalgia consider: 1 Reduce saturated fats eaten: these increase the affinity of oestrogen-receptors for oestrogen.[2]
2 Gamolenic acid 120mg/12h PO. 1 capsule=40mg in evening primrose oil.
3 Bromocriptine 2.5mg/12h PO days 10–26, even if prolactin normal.
4 Danazol 100–200mg/12h PO for 7 days before menstruation (see below).

Premenstrual progestogens eg progesterone suppositories 200mg/12h PR or PV, or dydrogesterone 10mg/12h PR/PV, may provide general relief. Some benefit from suppression of ovulation with the combined Pill, oestrogen patches or implants with cyclical progesterone, or danazol 200mg/24h (s nausea, weight gain, masculinization of the ♀ fetus, so advise barrie contraception). Avoid diuretics unless fluid retention is severe (when spironolactone 25mg/6h PO days 18–26 of the cycle is the drug of choice)

For premenstrual depression, there is a single small randomized, double blind trial indicating that serotonin-enhancing antidepressants help (flu oxetine 20mg/day PO), with 30% experiencing complete remissions.[3] Othe treatments for PMT include mefenamic acid (250–500mg/8h PO from day 16 until 3 days into period). This may help fatigue, headache, pains, and mood. Goserelin (OHCM p 392) may help severe PMT but symptom return when ovarian activity recommences and after 6 months' use bone thinning can be detected. It is better used to predict the severely affected women who may benefit from hysterectomy with oophorectomy (these women can then have oestrogen replacement).[4]

1 D Sanders 1988 in *Women's Problems in General Practice*, OUP   2 *Drug Ther Bul* 1992 30 1
3 DB Menkes 1992 *BMJ* ii 346   4 P O'Brien 1992 in *Gynaecology*, Churchill Livingstone, 325–39

The menopause and HRT

The climacteric is the time of waning fertility leading up to the last period (which is the menopause).

Problems All are related to falling oestrogen levels:
- Menstrual irregularities are common before periods finally stop, often because cycles are anovulatory.
- Vasomotor disturbances cause hot flushes, sweating and palpitations. Flushes are brief but unpleasant and may occur every few minutes for more than ten years, disrupting life and sleep.
- Atrophy of oestrogen-dependent tissues (genitalia, breasts) and skin. Vaginal dryness can lead to vaginal and urinary infection, dyspareunia, traumatic bleeding, stress incontinence and prolapse.
- Osteoporosis. The menopause accelerates bone loss which predisposes to fracture of femur neck, radius and vertebrae in later life (p 676).
- Arterial disease becomes more common after the menopause.
- Attitudes to the menopause vary widely; psychosocial problems eg, irritability, depression, 'empty nest syndrome', often exacerbate, and are exacerbated by, the menopause.

Management 20% of women seek medical help.
- Is it the menopause? Thyroid and psychiatric problems may present similarly. Measure FSH in younger women (very high in menopause).
- Counselling helps psychosocial problems and makes physical symptoms easier to bear. Are the family understanding?
- Menorrhagia may respond to treatment. A D&C is required if irregular bleeding is abnormal (it may be difficult to decide).
- Contraception should be continued for a year after the last period. PoP, IUCD and barrier methods are suitable.
- Hot flushes may respond to clonidine 50–75μg/12h PO, HRT, or tibolone.
- Vaginal dryness responds to oestrogen.

Hormone replacement therapy (HRT) Oestrogen is not a panacea for all problems, but it is effective for flushes and atrophic vaginitis. It postpones menopausal bone loss and protects against cardiovascular disease (controversial[1,2]▣) and ovarian cancer; it *may* increase rates of breast carcinoma.³ Benefits of HRT are many; most women should at least consider treatment.

Women with a uterus should also receive progestogens, eg norgestrel 150μg/24h, PO for 12 days out of 28 to reduce risk of endometrial carcinoma; this produces withdrawal bleeds—if unacceptable, consider tibolone 2.5mg/day PO, a preparation which aims not to cause bleeds—or continuous combined oestrogen/progestogen combinations, eg Kliofem® (oestradiol 2mg and norethisterone 1mg).⁴,⁵ For both these, either wait for 1yr after the last period, or, if changing from cyclical HRT, wait until after 54yrs of age; bleeding is common, in the first 4 months of usage—reassure; but if after 8 months, do an endometrial biopsy. Kliofem® is said to be useful if cyclical HRT is causing 'premenstrual' symptoms. NB: there may be advantages in using the more natural lipid-friendly progestogen, dydrogesterone (no androgenic, mineralocorticoid, or oestrogenic actions)—eg Femoston 2/10® or (2/20 tabs, i e 2mg 17β oestradiol tabs, each with added 20mg dydrogesterone after day 14)—but see p 66 for possible SE (DVT).

Contraindications to HRT: ● Breast cancer ● Undiagnosed PV bleeding.
● Dubin-Johnson/Rotor syn. ● Breast-feeding ▶BP↑ is *not* a CI. NB: if previous spontaneous DVT/PE check for thrombophilia (OHCM p 597).

Side-effects: Weight↑; 'premenstrual' syndrome; cholestasis; vomiting.

Annual check-up: Breasts; weight; any abnormal bleeding?

Creams, pessaries, and rings are useful for vaginal symptoms, eg Ovestin® (oestriol 0.1%) 1 applicator-full PV daily for 3 weeks, then twice weekly. They are absorbed but, if used intermittently, progestogens are probably unnecessary. If creams are unacceptable, consider an oestrogen-containing

vaginal ring (eg Estring®) replaced every 3 months, for up to 2yrs.
Vagifem® is an oestradiol 25µg vaginal tablet daily for 2 weeks, and then twice weekly, with reassessment every 3 months.
Transdermal patches are less 'medical' but are expensive and women with wombs still need progestogen, eg as a patch (Estracombi®) or tablets. Oestradiol patches supply 25–100µg/24h for 3–4 days. SE: dermatitis.
Oestradiol implants (Surgical) 25mg lasts ~36 weeks, 100mg lasts ~1yr.

When is HRT *particularly* desirable? ▶Make a point of offering HRT to:
Those who have had premenopausal bilateral oophorectomy.
Hysterectomy (ovaries often fail eg 5yrs after a hysterectomy).
Those with ↑risk of osteoporosis (eg smokers or if inactive); see p 676.
Those with ischaemic heart disease or risk factors such as hypercholesterol-
aemia or diabetes mellitus.
Those with rheumatoid arthritis.

1 WF Posthuma 1994 *BMJ* i 1268   2 MH Duncan 1994 *BMJ* ii 191   3 P Belchetz 1989 *BMJ* ii 1467
4 *Drug Ther Bul* 1991 **29** 77   5 M Rees 1994 *Lancet* **343** 250

Termination of pregnancy (ToP)

Worldwide, one-third of pregnancies are terminated. Good contraceptive services reduce but do not abolish the demand. More than 140,000 ToPs are performed annually in Britain (to which this page applies).

Legal constraints The 1967 and 1990 Human Fertilization/Embryology Acts allow termination *before 24 weeks* if this reduces the risk:

1 To the woman's life, or: 2 To her physical or mental health—or:

3 Of adverse physical or mental health to her existing children, or:

4 Of the baby's being seriously handicapped, mentally or physically.

Clause 1 technically covers any early termination as this is safer than birth, but 90% of ToPs are carried out via *Clause 2* ('social'). 2 doctors must sign form HSA 1. Parental consent (and the patient's) is needed for those under 16.

There is no upper time limit if there is: ●Risk to the mother's life.
●Risk of grave, permanent injury to the mother's physical/mental health (allowing for reasonably forseeable circumstances).
●Substantial risk that if the child were born it would suffer such physical or mental abnormalities as to be seriously handicapped.

<1% of ToPs are performed after 20wks, usually after amniocentesis, or when very young or menopausal mothers have concealed, or not recognized, pregnancy. ToPs after 24 weeks may only be carried out in NHS hospitals.

Before ToP ►She has to live with the decision for the rest of her life.
●Counselling to help her reach the decision she will least regret.
●Is she definitely pregnant? Do a VE or scan to confirm dates.
●Is a termination what she really wants? Why? Has she considered the alternatives—discuss their implications. What about her partner? Ideally give her time to consider and bring her decision to a further consultation. If she chooses termination:
●Screen for chlamydia. Untreated, 25% get post-op salpingitis, p 50.
●Discuss contraception (eg start Pill next day if she wishes).
●Arrange follow-up. If RhD-ve she needs anti-D (p 109). The time that she would have delivered may particularly raise emotions.

Methods The commonest method for first trimester abortions is dilatation of the cervix followed by curettage or vacuum aspiration of uterine contents. Mortality is low (1:100,000) as is infection risk.

For 2nd trimester abortions, labour is induced using intravaginal prostaglandin gel, or pessaries eg gemeprost 1mg/3h to a maximum of 5 in 24h. Oxytocin to stimulate contractions and surgical removal of retained placenta may be required. The procedure may be prolonged, painful and distressing. Alternatively the uterus may be evacuated surgically using forceps. It is unpleasant to do and the considerable dilatation of the cervix required may damage it, even after prostaglandin priming. After 14 weeks mortality and morbidity rise steeply with gestation.

4-stage medical ToP with an antigestagen + prostaglandin (Only if gestation <9 weeks, in fully licensed clinics): ●Counselling and ultrasound ●Supervised mifepristone 600mg PO (RU486, an anti-progesterone) disimplants the conceptus (CI: smokers >35yrs old; avoid aspirin and NSAIDs for 12 days) ●A gemeprost 1mg pessary 36–48h later completes abortion (or misoprostol, see below). ◄Only 3% abort before this stage. ●Follow-up and scan at 12 days. Surgery is needed in 5%. Psychiatric morbidity appears to be the same as with surgical ToP. Note that the oral route for the prostaglandin component does not produce such a high expulsion rate—eg compared with vaginal misoprostol 4 tablets (=800µg) high into the vagina. ◄►

◄►► H El-Refaey 1995 *NEJM* 332 983-7

Abortion

Abortion is the loss of a pregnancy before 24 weeks' gestation. 20–40% of pregnancies miscarry, mostly in the first trimester.

Management of early pregnancy bleeding Consider the following:
- ▶Is she shocked? There may be blood loss, or products of conception in the cervical canal (remove them with sponge forceps).
- ▶Could this be an ectopic pregnancy? (See p 24.)
- Has pain and bleeding been worse than a period? Have products of conception been seen? (Clots may be mistaken for products.)
- Is the cervical os open? The external os of a multigravida usually admits the tip of the finger.
- Is uterine size appropriate for dates?
- Is she bleeding from a cervical lesion and not from the uterus?
- What is her blood group? If RhD-ve she will need anti-D (p 109).

If symptoms are mild and the os is closed it is a *threatened abortion*. Rest is advised but probably does not help. 75% will settle. Threatened abortion (especially 2nd trimester[1]) is associated with risk of subsequent preterm rupture of membranes[2] and preterm delivery—so book mother at a hospital with good neonatal facilities.[1]

If symptoms are severe and the os is open it is an *inevitable abortion* or, if most of the products have already been passed, an *incomplete abortion*. If bleeding is profuse, consider ergometrine 0.5mg IM. If there is unacceptable pain or bleeding, or much retained tissue on ultrasound, arrange evacuation of retained products of conception (ERPC)—otherwise an expectant policy is justified for most (>75%) women if gestation is <13 weeks.[3]

Missed abortion: The fetus dies but is retained. There has usually been bleeding and the uterus is small for dates. Confirm with ultrasound. ERPC or prostaglandin evacuation is required.

The diagnosis of early pregnancy bleeding is not always straightforward. An ultrasound scan will help but pregnancy tests remain positive for several days after fetal death.

Mid-trimester abortion This is usually due to mechanical causes eg cervical incompetence (rapid, painless delivery of a live fetus), uterine abnormalities; or chronic maternal disease (eg DM, SLE).

After a miscarriage ▶Miscarriage may be a bereavement. Give the parents space to grieve, and to ask why it happened and if it will happen again.

Most early pregnancy losses are due to abnormal fetal development; 10% to maternal illness eg pyrexia. Second trimester abortion may be due to infection, eg CMV (p 98–100). Bacterial vaginosis has recently been implicated. Most subsequent pregnancies are normal although at increased risk. 3 abortions without a normal pregnancy merit referral for genetic, immunological and anatomical investigation. See p94.

An incompetent cervix may be strengthened with a cervical encirclage suture at ~16 weeks of pregnancy. It is removed prior to labour.
The right time to try again is when the parents wish to do so.

Septic abortion This generally follows a backstreet abortion. It presents like acute salpingitis (p 50) and is treated similarly. Start broad-spectrum antibiotics prior to uterine curettage, eg co-amoxiclav (ampoules are 1.2g of which 1g is amoxycillin and 200mg clavulanic acid; give 1.2g/6h IV) and metronidazole (eg a 1g suppository/8h).

1 P Sipila 1992 *Br J Obstet Gyn* **99** 959   2 J Konje 1992 *J Obstet Gynae* **12** 150   3 S Nielsen 1995 *Lancet* **345** 84 & *E-BM* **1** 13

Ectopic pregnancy

The fertilized ovum implants outside the uterine cavity. The UK incidence is 9:1000 pregnancies and rising; worldwide rates are higher. ~10% of maternal deaths are due to ectopic pregnancy.

Predisposing factors Anything which slows down the ovum's passage to the uterus increases the risk: damage to the tubes by salpingitis or previous surgery; previous ectopic; endometriosis; the presence of an IUCD; the PoP (p 66), GIFT (p 58).

Site of implantation 97% are tubal. Most implant in the ampulla, 25% in the narrow inextensible isthmus (so tend to present early and to rupture). 3% implant on the ovary, cervix or peritoneum.

Natural history The trophoblast invades the tubal wall, weakening it and producing haemorrhage which dislodges the embryo. If the tube does not rupture, the blood and embryo are shed or converted into a tubal mole and absorbed. Rupture can be sudden and catastrophic, or gradual, giving increasing pain and blood loss. Peritoneal pregnancies may survive into the third trimester, and may present with failure to induce labour.

Clinical presentation ►Always think of ectopics in a sexually active woman with abdominal pain or bleeding.

There is generally about 8 weeks' *amenorrhoea* but an ectopic may present before a period is missed. Tubal colic (trying to dislodge the pregnancy) causes *abdominal pain which precedes vaginal bleeding*. Blood loss may be dark ('prune juice', as the decidua is lost from the uterus) or fresh. The ectopic may rupture the tube with sudden severe pain, peritonism and shock. More often there is gradually increased vaginal bleeding, and bleeding into the peritoneum producing shoulder-tip pain (due to diaphragmatic irritation) and pain on defaecation and passing water (due to blood in the pelvis). The patient may be faint, with a tender abdomen (95%), enlarged uterus (30%), cervical excitation (50%), adnexal mass (63%). Examine gently, to reduce risk of rupture.

Tests:[1] No test is 100% reliable. Do a pregnancy test if necessary. Serum progesterone <15mg/ml, ultrasound showing an empty uterus or free fluid in the pelvis, needle aspiration of blood from the rectovaginal pouch (via the vagina—culdocentesis) are strongly suggestive. In 20% ultrasound shows an ectopic pregnancy. The higher the index of suspicion, the quicker the progress to laparoscopy. The pregnancy is removed at laparotomy or laparoscopy, conserving the tube if possible. Is anti-D needed? (p 109).

►►Shock from a ruptured ectopic can be fatal. *Immediate* laparotomy is necessary as only clamping the bleeding artery will relieve it. If you suspect an ectopic, put up an IVI; if already shocked put up 2 (14 or 16G). Give colloid as fast as possible (use pressure bag to ↑flow), followed by blood–O-ve if desperate, but usually better to wait for group compatible. Take immediately to theatre.

Measures to reduce missing ectopic pregnancies[1]
- Always send uterine curettings at ERPC (p 72) for histology.
- If histology does not confirm uterine failed pregnancy, recall the patient (Ensure rapid return of histology results.)
- When ultrasound reports suggest a missed abortion but the fetus has not been seen—think: could this be an ectopic?

1 S Norman 1991 *Br J Obs Gynae* **98** 1267

Trophoblastic disease

Hydatidiform mole and choriocarcinoma are forms of trophoblastic disease. A fertilized ovum forms abnormal trophoblast tissue, but no fetus. The growth may be benign (mole) or malignant (choriocarcinoma).

26

It is genetically paternal but has a 46XX karyotype. Rarely, a triploid, partial mole is found with a fetus (usually abnormal). Partial moles do not seem to develop into choriocarcinoma.

Hydatidiform mole The tumour consists of proliferated chorionic villi which have swollen up and degenerated. Since it derives from chorion, it produces human chorionic gonadotrophin (HCG) in large quantities. This gives rise to exaggerated pregnancy symptoms and a strongly +ve pregnancy test. Incidence is 1.54:1000 births (UK). It is more common at extremes of maternal age, after a previous mole, and in non-Caucasians. A woman who has had a previous mole is at increased risk for future pregnancies; 0.8–2.9% after one mole, 15–28% after 2 moles.

Presentation is usually with a pregnancy which is 'not quite right': she may have severe morning sickness or first-trimester pre-eclampsia (20%); the uterus is large for dates in 50% and feels doughy instead of firm. There is no fetal heart. Ultrasound: 'snowstorm' in the uterus.

50% present with vaginal bleeding which is often heavy and/or prolonged. Molar tissue is likely to be aborted. It looks like a mass of frogspawn.

Abdominal pain may be due to huge theca-lutein cysts in both ovaries. These may rupture or tort. They take ~4 months to resolve after molar evacuation. HCG resembles TSH, and may cause hyperthyroidism. ▶Mention this to the anaesthetist as thyrotoxic storm can occur at evacuation.

Treatment of moles is by removal of molar tissue. The uterus, which is soft and liable to perforation, is evacuated using gentle suction, followed by a repeat evacuation ~10 days later. She should then avoid pregnancy and contraceptive Pills for a year while HCG levels are monitored. Levels should return to normal within 6 months. If they do not, either the mole was invasive (myometrium penetrated) or has given rise to choriocarcinoma (10%). Invasive mole may metastasize eg to lung, vagina, brain, liver and skin. Both conditions respond to chemotherapy.

Choriocarcinoma This highly malignant tumour occurs in 1:40000 deliveries. 50% follow a benign mole, 20% follow abortions and 10% follow normal pregnancy. *Presentation:* May be many years after pregnancy, with general malaise (due to malignancy and to raised HCG); or uterine bleeding; or with signs and symptoms from metastases, which may be very haemorrhagic, eg haematoperitoneum, or cannonball shadows on CXR. Pulmonary artery obstruction via tumour emboli may lead to pulmonary artery hypertension (haemoptysis and dyspnoea).[1]

Treatment: Choriocarcinoma in UK is treated at 3 specialist centres and is extremely responsive to combination chemotherapy based on methotrexate. Outlook is excellent and fertility is usually retained.

▶Persistent vaginal bleeding after a pregnancy requires investigation to exclude choriocarcinoma.

1 M Seckl 1991 *Lancet* ii 1313

The vulva

Pruritus vulvae Vaginal itch is distressing and embarrassing.

Causes: There may be a disorder causing general pruritus (p 578) or skin disease (eg psoriasis, lichen planus). The cause may be local: infection and vaginal discharge (eg candida); infestation (eg scabies, pubic lice, threadworms); or vulval dystrophy (lichen sclerosis, leukoplakia, carcinoma). Symptoms may be psychogenic in origin. Obesity and incontinence exacerbate symptoms.[1]
Postmenopausal atrophy does not cause itch.

The history may suggest the cause. Examine general health and look for widespread skin conditions. Examine the vulva and rest of the genital tract, under magnification if possible, and take a cervical smear. Take vaginal and vulval swabs and test for glycosuria. Biopsy if in doubt about diagnosis.
▶Scratching and self-medication may have changed the appearance.

Treatment: This is often unsatisfactory. Treat the cause if possible. Reassurance can be very important. Advise her to avoid nylon underwear, chemicals and soap (use aqueous cream) and dry with a hair dryer. A short course of topical steroids eg betamethasone valerate cream 0.1% may help. Avoid any topical preparation which may sensitise the skin, so give antipruritics orally if needed, eg promethazine 25–50mg/24h.

Lichen sclerosis This is due to elastic tissue turning to collagen after middle age. The vulva gradually becomes white, flat and shiny. There may be an hourglass shape around the vulva and anus. It is intensely itchy. It does not appear to be pre-malignant.

Treatment: Testosterone cream 3% may reduce symptoms, but is not easily available.[1] Betamethasone valerate cream 0.1% may help; vulval ablation may be needed to relieve itch.

Leukoplakia There are white patches on and around the vulva due to skin thickening and hypertrophy. It is itchy. It should be biopsied as it may be a pre-malignant lesion. *Treatment:* Topical corticosteroids (problems: mucosal thinning, absorption.)

Carcinoma of the vulva 95% are squamous. They are rare and occur mostly in the elderly.
   Vulval malignancy has a pre-invasive phase, vulval intra-epithelial neoplasia (VIN), which may be itchy. ~5% progress to invasive carcinoma. VIN is associated with human papilloma virus (HPV) infection. There may not be visible warts but 5% acetic acid stains affected areas white. If VIN is found on biopsy, examine the cervix and check the breasts (>10% have coexistent neoplasia elsewhere, most commonly cervical). *Treatment* is aimed at symptom control by ablating the lesion chemically (5-fluorouracil cream, p 582), by laser, or surgically. Simple vulvectomy may be overtreatment.[1]
   An indurated ulcer with an everted edge suggests carcinoma. It may not be noticed unless it causes pain and bleeding, so it often presents late (50% already have inguinal lymph node involvement).

Treatment is radical vulvectomy (wide excision of the vulva + removal of inguinal glands). Post-operative healing may require skin grafts. 5-yr survival is 75% for lesions <2cm without node involvement; otherwise <50%.

1 D Luesley 1989 *Update* **39** 908

Vulval lumps and ulcers

Causes of vulval lumps Varicose veins; boils; sebaceous cysts; keratoacanthomata (rare); viral warts (condylomata acuminata); condylomata lata (syphilis); primary chancre; molluscum contagiosum; Bartholin's abscess or cyst; uterine prolapse or polyp; inguinal hernia; varicocele; carcinoma.

Vulval warts (Human papilloma virus. HPV—usually spread by sexual contact.) Incubation: weeks. Her partner may not have obvious penile warts. The vulva, perineum, anus, vagina or cervix may be affected. Warts may be very florid in the pregnant and immunosuppressed. HPV types 16, 18 and 33 can cause vulval and cervical intraepithelial neoplasia, so she needs annual cervical smears and observation of the vulva. Warts may also cause anal carcinoma (OHCM p 154). Treat both partners. Exclude other genital infections. Warts may be destroyed by diathermy, cryocautery or laser. Vulval and anal warts (condylomata acuminata) may be treated weekly in surgeries and GU clinics with 15% podophyllin paint, washed off after 30min (CI: pregnancy). Only treat a few warts at once, to avoid toxicity. Self-application with 0.15% podophyllotoxin cream (Warticon® 5g tubes—enough for 4 treatment courses—is supplied with a mirror): use every 12h for 3 days, repeated up to 4 times at weekly intervals if the area covered is <4cm². Relapse is common. NB: HPV types 6 and 11 may cause laryngeal or respiratory papillomas in the offspring of affected mothers (risk 1:50–1:1500; 50% present at <5yrs old).

Treat both partners. Exclude other genital infections. Warts may be destroyed by diathermy, cryocautery or laser. Vulval and anal warts may be treated with weekly applications of 20% podophyllin paint, washed off after 30mins (CI: pregnancy).

Urethral caruncle This is a small red swelling at the urethral orifice. It is caused by meatal prolapse. It may be tender and give pain on micturition. *Treatment:* Excision or diathermy.

Bartholin's cyst and abscess The Bartholin's glands and their ducts lie under the labia minora. They secrete thin lubricating mucus during sexual excitation. If the duct becomes blocked a painless cyst forms; if this becomes infected the resulting abscess is extremely painful (she can't sit down) and a hugely swollen, hot red labium is seen. *Treatment:* The abscess should be incised, and permanent drainage ensured by marsupialization, ie the inner cyst wall is folded back and stitched to the skin. *Tests:* Exclude gonococcal infection.

Vulvitis Vulval inflammation may be due to infections eg candida (p 48), herpes simplex; chemicals (bubble-baths, detergents). It is often associated with, or may be due to, vaginal discharge.

Causes of vulval ulcers: Always consider syphilis. *Herpes simplex* is common in the young. Others: carcinoma; chancroid; lymphogranuloma venereum; granuloma inguinale; TB; Behçet's syndrome; aphthous ulcers; Crohn's.

Herpes simplex Herpes type II, sexually acquired, classically causes genital infection, but type I transferred from cold sores can be the cause.
The vulva is ulcerated and exquisitely painful. Urinary retention may occur. *Treatment:* Strong analgesia, lignocaine gel 2%, salt baths (and micturating in the bath) help. Exclude coexistent infections. Acyclovir topically and 200mg 5 times daily PO for 5 days shortens symptoms and infectivity. Reassure that subsequent attacks are shorter and less painful. Prescribe acyclovir cream for use when symptoms start. For herpes in pregnancy, see p 100.

The cervix

This is is the part of the uterus below the internal os. The endocervical canal is lined with mucous columnar epithelium, the vaginal cervix with squamous epithelium. The transition zone between them—the squamo-columnar juction—is the area which is predisposed to malignant change.

Cervical ectropion This is often called erosion, an alarming term for a normal phenomenon. There is a red ring around the os because the endocervical epithelium has extended its territory over the paler epithelium of the ectocervix. Ectropions extend temporarily under hormonal influence during puberty, with the combined Pill, and during pregnancy. As columnar epithelium is soft and glandular, ectropion is prone to bleeding, to excess mucus production and to infection. *Treatment:* Cautery will treat these if they are a nuisance; otherwise no treatment is required.

Nabothian cysts These mucus retention cysts found on the cervix are harmless. *Treatment:* Cautery if they are discharging.

Cervical polyps These pedunculated benign tumours usually arise from the endocervical epithelium and so may cause increased mucus discharge or postcoital bleeding. *Treatment:* In young women they may be simply avulsed, but in older women treatment usually includes D&C to exclude intrauterine pathology.

Cervicitis This may be follicular or mucopurulent, presenting with discharge. *Causes:* Chlamydia (up to 50%),[1] gonococci, or herpes (look for vesicles). Chronic cervicitis is usually a mixed infection and may respond to antibacterial cream. Cervicitis may mask neoplasia on a smear.

Cervical screening Cervical cancer has a preinvasive phase: cervical intraepithelial neoplasia (CIN—not to be pronounced 'sin'). Papanicolaou smears collect cervical cells for microscopy for dyskaryosis (abnormalities which reflect CIN). A smear therefore identifies women who need cervical biopsy. The degree of dyskaryosis approximates to the severity of CIN (Table 1, p 34). ~50% of CIN I lesions return to normal but most CIN III lesions progress to invasive carcinoma. This may take ~10yrs, but may happen much faster in young women.

A programme of 2 smears in the first year of sexual activity, then 3-yearly smears should reduce the incidence of invasive carcinoma by 90%. The trouble is, those most at risk are the hardest to trace and persuade to have screening—eg older women, smokers, and those in inner cities. One of the great achievements of UK primary care and its software houses is that these women are now being reached: 83% of the eligible UK population is now screened—a figure which no other country exceeds, and mortality here is starting to fall.

Taking a smear Explain to the woman the nature and purpose of the test, and how results will be conveyed to her. It may be sensible to warn that results are not categoric or unequivocal.

The cervix is visualized with a speculum (p 2). Are there any suspicious areas? If so, carry on with the smear and indicate this on the referral form, but do not wait for its results before arranging further care.

Cells are scraped from the squamo-columnar transformation zone with a special spatula or brush, then transferred to a slide and fixed at once. Good technique is needed (make sure that all 4 quadrants of the cervix are sampled); it is best to learn by instruction from an expert at the bedside.

1 D Taylor-Robinson 1994 *BMJ* i 150

Cervical carcinoma

▶*Aim to detect pre-invasive disease.* ~ 1600 women die each year of cervical carcinoma in the UK. Many have had no smear—but this is changing: see p 32.

Risk factors for neoplasia Smoking, multiple sexual partners (in both sexes) spreading human papilloma virus (HPV 16, 18 and 33). HPV oncogenes (E6) bind to the tumour-suppressor gene *p53* (the guardian of the genome, p 752) so promoting tumorigenesis. Thus cervical carcinoma is a sexually transmitted disease. (Neither partner may have visible warts.)

Management of abnormal smears (Guidelines in Table 2) Either a repeat smear or colposcopy and biopsy are needed, depending on likelihood of the smear reflecting CIN III or microinvasive disease (<3mm).

▶Abnormal smears cause anxiety and guilt. Explain. Give support.

Treatment of preinvasive carcinoma The cervix is examined using a colposcope (× 10 binocular microscope). Abnormal epithelium has characteristic blood vessel patterns and stains white with acetic acid. Punch biopsies are taken for histology. CIN is destroyed by cryotherapy, laser, cold coagulation or electrodiathermy (the latter requires general anaesthesia). These give ~ 90% cure rates with one treatment. She needs annual smears for at least 10 years. If the squamocolumnar junction cannot be seen, or if microinvasive carcinoma is found on histology, the abnormal tissue is removed by cone biopsy, which may be curative. Colposcopy does not detect adenocarcinoma.

Invasive disease Once a disease of the over-50s, this is now seen in under-40s. Most are squamous carcinomas. ~ 5% are adenocarcinomas (from endocervical epithelium, with unknown risk factors). Spread is local and lymphatic. *Stage I* tumours are confined to the cervix. *Stage II* have extended locally. *Stage III* have spread to pelvic wall & lower 1/3 of vagina. *Stage IV* have spread to bladder or rectum. Most present in stages I or II.

Diagnosis ▶Overt carcinoma is rarely detected on a smear. Non-menstrual bleeding is the classic symptom. The early tumour is firm. It grows as a friable mass which bleeds on contact.

Treatment of invasive carcinoma Wertheims's hysterectomy or radiotherapy both treat the tumour and lymph nodes, and achieve good cure rates at stages I (80% 5-yr survival) and II (60%). Radiotherapy causes vaginal stenosis: so encourage intercourse within 2 months of treatment (with use of lubricant if necessary). Follow-up with annual smears. Stage III requires extensive radiotherapy—exenteration is probably *not* indicated (p 73). Treatment at Stage IV is palliative. Terminal problems are pain, fistulae, and obstruction of gut and ureters.

Histology of cervical pre-malignant disease (CIN = cervical intraepithelial neoplasia)

	Papanicolaou class	Action	Histology
I	Normal	Repeat in 3 years (Unless clinical suspicion)	0.1% CIN II–III
II	Inflammatory	Repeat in 6 months (Colposcopy after 3 abnormal)	6% CIN II–III
	Mild atypia	Repeat in 4 months (Colposcopy after 2 abnormal)	20–37% CIN II–III
III	Mild dyskaryosis	Colposcopy	50% CIN II–III
	Moderate dyskaryosis	Colposcopy	50–75% CIN II–III
IV	Severe dyskaryosis 'Positive' 'Malignant cells'	Colposcopy	80–90% CIN II–III 5% Invasion
V	Invasion suspected	Urgent colposcopy	50% Invasion
	Abnormal glandular cells	Urgent colposcopy	?Adenocarcinoma or endometrium

CIN I = mild dysplasia; CIN II = moderate dysplasia; CIN III = Severe dysplasia/carcinoma-*in-situ*

The table above shows comparative terms and the recommended action. The third column, headed histology, shows the percentage of smears in each Papanicolaou (cytological) class which have more serious lesions (CIN II or III) on histology. 6% of inflammatory smears have serious pathology, hence the recommendation for colposcopy if inflammation persists.

In 1986, the term *borderline nuclear abnormality* was introduced to use in cases of genuine doubt as to the neoplastic nature of any change. New guidelines are issued[1] for labs to request recall, eg at 3-6 months, or to advise colposcopy referral, eg after 2 smears with borderline nuclear abnormality, in which the number of affected cells has not reduced over time; colposcopy in these cases should be seen as part of the screening process, and should not necessarily prompt treatment. This goes part of the way to avoid false negatives, as well as to avoid the problems of overtreatment.

In the UK, 5-6 million smears are done each year—and the death rate from cervical carcinoma has fallen to <1900/yr (a fall of ~15% in the last decade). 2.4% of smears are reported as mild dyskaryosis, and 2.2% as borderline nuclear abnormalities. Because most of these will eventually need colposcopy because of further smear results and 48% can be expected to have CIN II or III, and as a significant proportion may be lost to follow-up, it is suggested that *immediate* colposcopy with directed biopsy is indicated.[2] However, this would be very expensive, and anxiety-provoking. The cost in the USA of adopting this policy is ~$1 billion, and the benefits are uncertain as no randomized trials have been done.[3]

Terminology used in reporting smears has changed over the years, and different countries use different nomenclature—eg the Bethesda system.[4]

Neural networks for cervical smears? Until recently, computers in the lab and surgery have acted as no more than glorified filing clerks and secretaries; but as reading smears is labour-intensive, and boring, mistakes are inevitably made, particularly false negatives, and exaggeration of abnormalities (perhaps to avoid blame if time reveals malignancy to have been present). Neural network systems such as PAPNET realize the advantages of a machine (speed—1 slide/min—with objectivity, thoroughness, freedom from boredom, and 24h/day operation) without allowing machines to take over. The program bar-codes slides, focuses them, digitizes images, selects the most interesting/difficult parts of the slide, records their x-y coordinates, and, if needed, signals that this is the area the cytologist should look at, whether they be individual cells, or clusters. Neural networks such as our own brain have ~10^{15} connections, whereas man-made systems have only up to one-billionth of this number, but this is still sufficient for reasonable accuracy, tolerance of conflicting information in a single specimen, and the ability to learn from experience. It is this ability which sets these networks apart from statistical or rule-based systems (if X is present and out if Y is also present). The disadvantage of rule-based systems is the rules have to be explicit—and in cervical cytology this is too tough a prerequisite to be realized.[5,6] For severe dysplasia, ca-*in-situ*, and suspected invasive carcinoma, PAPNET gives accuracy rates (compared with histology) of 38%, 35% and 72% respectively, compared with 40%, 20% and 62% for conventional analysis. PAPNET will also go though your negative slides, and tell you which ones may be false -ves and need reviewing by a cytologist.

The disadvantage of these networks are that they are difficult (and expensive) to train—and they are only as good as the experience they have.

Work. Part. Rep. 1994 *J Clin Path* 34 481-92 2 WP Souter 1994 *BMJ* ii 591 3 M Shafi 1994 *BMJ* ii 590 *JAMA* 1989 262 931-4 5 R Dybowski 1995 *Lancet* 346 1203 6 J Wyatt 1995 *Lancet* 346 1175

The uterus

Endometritis Uterine infection is uncommon unless the barrier t$\cdot$ ascending infection (acid vaginal pH and cervical mucus) is broken, e after abortion and childbirth, IUCD insertion or surgery. Infection ma involve Fallopian tubes and ovaries.

Presentation: Lower abdominal pain and fever; uterine tenderness o. bimanual palpation. Low-grade infection is often due to chlamydia. Test: Do cervical swabs and blood cultures. *Treatment:* Give antibiotics (eg doxy cyline 100mg/24h PO with metronidazole 400mg/8h PO).

Endometrial proliferation Oestrogen stimulates proliferation of th endometrium during the first half of the menstrual cycle; it then come under the influence of progesterone and is shed at menstruation. $\cdot$ particularly exuberant proliferation is associated with heavy menstrua bleeding and polyps.

If high oestrogen levels continue (eg anovulatory cycles) th endometrium becomes hyperplastic ('cystic glandular hyperplasia'— histological diagnosis after D&C). Eventually it will break down, causin irregular bleeding (dysfunctional uterine bleeding). *Treatment:* Cyclic. progestogens (p 14).

In older women proliferation may contain foci of atypical cells whic' may lead to endometrial carcinoma (p 40).

Pyometra This is a uterus distended with pus. Salpingitis may produc pyosalpinx and pyometra, or outflow blockage may lead to inflammatio and secondary infection in the uterus. *Treatment:* It is relieved by drainag and treatment of the cause.

Haematometra This is a uterus filled with blood due to outflow obstruc tion. It is rare. The blockage may be an imperforate hymen in the youn (p 6); carcinoma; or iatrogenic cervical stenosis eg after cone biopsy.

Endometrial tuberculosis Genital tract tuberculosis is rare in Britair except among high-risk groups (Asian immigrants). It is blood-borne an usually affects first the Fallopian tubes, then the endometrium.

It may present with acute salpingitis if disease is very active, or wit infertility, pelvic pain and menstrual disorders (eg amenorrhoe oligomenorrhoea). There may be pyosalpinx.

Exclude pulmonary disease with CXR. *Treatment* is medical with ant tuberculous therapy as for disease elsewhere (OHCM p 198-200). Repe$\cdot$ endometrial histology after one year.

Vaginal carcinoma

These tumours are usually squamous. They are commonest in the uppe third of the vagina. Presentation is usually with bleeding. Clear cell aden carcinoma was thought to be associated with intrauterine exposure t diethylstilboestrol but risk is low (0.1-1:1000). Spread is local and b lymphatics. Treatment is usually radiotherapy.

Fibroids

Fibroids are benign tumours of the smooth muscle of the uterus (leiomyomas). They are often multiple, and may vary in size from seedling size to large tumours occupying a substantial part of the abdomen. They start as lumps in the wall of the uterus but may grow to bulge out of the wall so that they lie under the peritoneum (subserosal, 20%) or under the endometrium (submucosal, 5%), or become pedunculated. Fibroids are common (20% of women have fibroids), increasing in frequency with age, and in non-Caucasians.

Natural history Fibroids are oestrogen-dependent. Consequently they enlarge in pregnancy and on the combined Pill and atrophy after the menopause. They may degenerate gradually or suddenly (red degeneration). Occasionally they calcify ('womb stones'). Rarely, they undergo sarcomatous change—usually causing pain, malaise, bleeding, and increase in size in a postmenopausal woman.

Presentation Many are asymptomatic.
- *Menorrhagia:* Fibroids often produce heavy and prolonged periods. They do not generally cause intermenstrual or post-menopausal bleeding.
- *Fertility problems:* Submucosal fibroids may interfere with implantation ('natural IUCD'). Large or multiple tumours which distort the uterine cavity may cause abortion should pregnancy occur.
- *Pain:* This may be due to torsion of a pedunculated fibroid, producing symptoms similar to that of a torted ovarian cyst. 'Red degeneration' following thrombosis of a fibroid's blood supply usually only happens in pregnancy and produces severe pain, fever and local tenderness until degeneration is complete.
- *Mass:* Large fibroids may be felt abdominally. They may press on the bladder, causing frequency, or on the veins, causing oedematous legs and varicose veins. Pelvic fibroids may obstruct labour or cause retention of urine.

Treatment In many women treatment is not needed.
Menorrhagia due to fibroids tends to respond poorly to anti-prostaglandins, progestogens or danazol. Women who have completed their families may opt for hysterectomy. In younger women a reversible menopausal state may be induced with LHRH analogues eg buserelin nasal spray 100µg/4h (expensive)[1] or goserelin 3.6mg by monthly subcutaneous injection (preferred method—intermittent buserelin use can increase their size rather than shrink them). Bone demineralization can occur with their use, so their place is in shrinking fibroids prior to surgery; those unfit for surgery; or those desiring later pregnancy. Side effects are menopausal symptoms; fertility (and fibroids) return when the drug is stopped. Alternatively the fibroids may be surgically shelled out (myomectomy); torrential bleeding necessitating hysterectomy is a complication.

Red degeneration requires only analgesia until symptoms settle.

Torsion may resemble an acute abdomen, requiring urgent surgery.

1 D Baird 1988 *BMJ* i 1684

Endometrial carcinoma

▶Post-menopausal bleeding must be investigated as the cause may be endometrial carcinoma.

Carcinoma of the uterine body is less common than carcinoma of the cervix. It usually presents after the menopause. Most tumours are adenocarcinomas and are related to excessive exposure to oestrogen unopposed by progesterone. There is marked geographical variation: North American:Chinese ratio = 7:1.

Risk factors ●Obesity ●Nulliparity ●Diabetes ●Unopposed oestrogen therapy ●Late menopause ●Pelvic irradiation ●Functioning ovarian tumour ●Tamoxifen[1]* ●FH of breast, ovary, colon carcinoma

Presentation This is usually as post-menopausal bleeding (PMB). Any woman with a history of PMB has a 10–20% risk of genital cancer. It is initially scanty and occasional, perhaps with a watery discharge. Over time, bleeding becomes heavier and more frequent. Pre-menopausal women may have intermenstrual bleeding, but 30% have only menorrhagia.

Diagnosis Post-menopausal bleeding is an early sign, and generally leads a woman to see her doctor, but examination is usually normal. Endometrial carcinoma can sometimes be seen on a smear. The diagnosis is made by uterine sampling (p 41) or curettage. All parts of the uterine cavity must be samples; send *all* material for histology. Hysteroscopy enables visualization of abnormal endometrium to improve accuracy of sampling. Sceptics claim it may cause spread through the Fallopian tubes to the peritoneum.

Pathology Most tumours start in the fundus of the uterine cavity. The tumour spreads slowly to the uterine muscle. With time, it may reach the cervix or peritoneum and may metastasize to the vagina (5%), ovary (5%), and any of the pelvic lymph nodes (7%).

Staging I: Tumour is confined to the body of the uterus.
II: As for I but with cervical involvement.
III: Tumour has extended beyond the uterus but not beyond the pelvis.
IV: Extension outside the pelvis. Bowel and bladder may be affected.

Treatment Stages I and II may be cured by total hysterectomy with bilateral salpingo-oophorectomy and/or radiotherapy if unfit for surgery (5-yr survival: Stage I 72%, Stage II 56%). Post-operative vault irradiation reduces vault recurrence. In advanced disease consider radiotherapy and/or high-dose progestogens, eg medroxyprogesterone acetate 250mg/24h PO, which shrinks the tumour (SE: fluid retention). Radiotherapy may either be given pre-operatively (caesium or radium rods inserted into the uterus and upper vagina) or post-operatively (external radiation).

Recurrent disease usually presents in the 1st 2–3yrs. Common sites are pelvic (in non-irradiated patients), lung, bone, inguinal and supraclavicular nodes, vagina, liver, peritoneal cavity. Surgical exenteration, radiotherapy, and medroxyprogesterone may all be of use. Cytotoxics may be used for those who fail to respond to the above.

*Note: tamoxifen 20mg daily reduces annual risk of breast cancer recurrence by 27%.[2] Risk of endometrial carcinoma from taking tamoxifen is 1.2 per 1000 person-years.[2] Counsel to report abnormal vaginal bleeding.

1 FE van Leeuwen 1994 *Lancet* 343 448   2 1994 *Current Problems in Pharmacovigilance* (CSM) 20 13

Endometrial sampling

This relatively new bedside technique is replacing D&C (p 72) in the investigation of post-menopausal bleeding, perimenopausal irregular bleeding, and unexpected bleeding patterns in women on hormone replacement therapy because it is cheap, reliable, and gives quick results without the need for anaesthesia.[1] It is less useful in menorrhagia in women with regular cycles, as pathology is less common. It is not indicated if <35 years old.

A sample is obtained using a side-opening plastic cannula in which a vacuum is created by withdrawal of a stopped central plunger mechanism. As the cannula is then withdrawn and rotated within each quadrant of the uterine cavity, endometrial tissue is sucked into its interior, through the hole in the side of the cannula. Successful insertion is possible in 90–99% of women (D&C is possible in 99%). Adequate samples will be obtained in 91% of these, and in 84% of those for whom postmenopausal bleeding (PMB) was the indication (at D&C an adequate sample is only achieved in 58%, and only in 45% if the indication was PMB). 95% of endometrial cancers will be detected by either procedure. Abandon the procedure if it is impossible to enter the uterus successfully, or if it causes too much pain.

Technique

1 Bimanual examination to assess size and position of uterus (p 2).
2 Bend cervical cannula to follow the curve of the uterus.
3 Insert device, watching the centimetre scale on the side; observe resistance as enter the internal os (at 3–4 cm) and then as the tip reaches the fundus (eg at 6cm if postmenopausal or 8cm in an oestrogenised uterus).
4 When the tip is in the fundus, create a vacuum by withdrawing plunger until the stopper prevents further withdrawal. Then move sampler up and down in uterus, rotate and repeat to sample whole cavity.
5 Remove cannula, and expel tissue into formalin. Send for histology.

Management Reassure those in whom the results show normal or atrophic endometrium and those in whom tissue was insufficient for diagnosis. If those with PMB re-bleed refer for hysteroscopy (polyps or a fibroid will be present in 20%). Those with simple hyperplasia on histology can be treated with cyclical progesterones (but refer if >55yrs to search for exogenous oestrogen source). Refer those with polyps or necrotic tissue on histology for hysteroscopy and curettage; and those with atypical hyperplasia or carcinoma for hysterectomy and bilateral salpingo-oophorectomy. Those in whom endometrial sampling was impossible or abandoned should have intravaginal uterine ultrasound to establish endometrial thickness (<5mm normal in the postmenopausal; refer if >5mm).

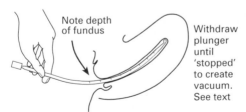

Note depth of fundus

Withdraw plunger until 'stopped' to create vacuum. See text

We thank Genesis Medical Ltd for permission to reproduce the diagram

A Coulter 1993 BMJ i 236

Ovarian tumours

Any of the ovary's many tissue types may become neoplastic.

Benign tumours (94%). They are usually cystic. 24% of all ovarian tumours are functional cysts. Others: endometriotic cysts (5%—p 52), theca-lutein cysts (p 26); epithelial cell tumours (serous and mucinous cystadenomas—40%); mature teratomas (from germ cells—20%); fibromas (solid—5%).

Malignant tumours (6%). 5% are cystadenomas which have become malignant. 0.5% are a group of rare germ cell or sex cord malignancies (p 44). 0.5% are secondaries, eg from the uterus, or the stomach (Krukenberg tumours—in which spread is transcoelomic, ie, in the case of the abdomen, via the peritoneum).

Presentations are varied, depending on size, form and histological type:
Asymptomatic—chance finding (eg on doing a bimanual for a smear test).
Swollen abdomen—with palpable mass arising out of the pelvis which is dull to percussion (and does not disappear if the bladder is catheterized).
Pressure effects (eg on bladder, causing urinary frequency).
Infarction/haemorrhage—this mimics torsion (see below).
Rupture ± local peritonism. Rupture of a large cyst may cause peritonitis and shock. Rupture of a malignant cyst may disseminate malignant cells throughout the abdomen. Rupture of mucinous cystadenomas may disseminate cells which continue to secrete mucin and cause death by binding up the viscera (pseudomyxoma peritonei).
Ascites—shifting dullness suggests malignancy or Meigs' syndrome (p44). If tense, ascites may be hard to distinguish from a mass.
Torsion—to twist, a tumour must be small, and free on a pedicle. Twisting occludes the venous return but the arterial supply continues to engorge the tumour, and cause great pain (with a high WBC). Tumours may twist and untwist, giving a history of intermittent pain. If the pain is not too severe, a firm tender adnexal swelling may be felt.
Endocrine or metastatic effects—Hormone-secreting tumours may cause virilization, menstrual irregularities or post-menopausal bleeding.

Management *Ultrasound* may confirm the presence of a mass (though, as with bimanual examination, it can be difficult to tell an ovarian mass from a uterine one). It may show whether it is cystic or solid.
Laparoscopy may distinguish a cyst from an ectopic pregnancy or appendicitis. Note: laparoscopy is not advised if malignancy is possible, due to seeding along the surgical tract. *Fine needle aspiration* may be used to confirm the impression that a cyst is benign. Urgent *laparotomy* is required when a cyst problem presents as an acute abdomen.

Any cyst not positively identified as non-neoplastic should be removed as seemingly benign tumours may be malignant. In younger women *cystectomy* may be preferable to oöphorectomy. In post-menopausal women if one ovary is pathological both are removed. For management in pregnancy see p 150.

Ovarian tumours: pathology

Functional cysts These are enlarged or persistent follicular or corpus luteum cysts. They are so common that they may be considered normal if they are small (<5cm). They may cause pain by rupture, failing to rupture at ovulation, or bleeding.

Polycystic ovary syndrome (PCOS) The cause of this common, poorly-understood syndrome is unknown. The result appears to be a vicious circle of ovarian, hypothalamic-pituitary, and adrenal dysfunction. Hormonal cycling is disrupted and the ovaries become enlarged by follicles which have failed to rupture.

The patient may be obese, virilized, with acne and irregular or absent menses (Stein-Leventhal syndrome). She may present with infertility. Investigation may show LH↑, testosterone↑. Any feature may be normal. Laparoscopy or ultrasound show characteristic ovaries.

Treatment is symptomatic: clomiphene will induce ovulation (p 58), the combined Pill will control bleeding. Hirsutes may be treated cosmetically, or with an anti-androgen eg cyproterone 2mg/day, as in Dianette® (avoid pregnancy). Weight loss helps.

Serous cystadenomas These develop papillary growths which may be so prolific that the cyst appears solid. They are commonest in women aged between 30 and 40 years. About 30% are bilateral and about 30% are malignant.

Mucinous cystadenomas These are the commonest large ovarian tumours and may become enormous. They are filled with mucinous material and rupture may cause pseudomyxoma peritonei (p 42). They may be multilocular. They are commonest in the 30-50 age group. About 5% will be malignant.

Fibromas These are small, solid benign fibrous tissue tumours. They are associated with Meigs' syndrome (OHCM p 702) and ascites.

Teratomas These arise from primitive germ cells. A benign mature teratoma (dermoid cyst) may contain well differentiated tissue eg hair, teeth. 20% are bilateral. They are most common in young women. Poorly differentiated, malignant teratomas are rare.

Other germ cell tumours (all malignant and all rare): non-gestational choriocarcinomas (secrete HCG); ectodermal sinus tumours (yolk sac tumours—secrete α-fetoprotein); dysgerminomas which are histologically similar to seminomas of testis.

Sex-cord tumours (rare and usually of low-grade malignancy): These arise from cortical mesenchyme. Granulosa-cell and theca-cell tumours produce oestrogen and may present with precocious puberty, menstrual problems or post-menopausal bleeding. Arrhenoblastomas secrete androgens.

Ovarian carcinoma

This is rare, but more women die from it (it is the fourth commonest cause of cancer-related death in Western women) than from carcinoma of the cervix and uterine body combined because in 75% it causes few symptoms until it has metastasized, often to the pelvis with omental and peritoneal seedlings (± lymphatic spread via the para-aortic glands).

Incidence 1 in 2500 women >55yrs; 1 in 3800 of women >25yrs. 1% of women have familial disease. A woman with 2 close relatives affected has a 40% risk of getting the disease (so liaise with a gynaecologist; see below). It is commoner in those with many ovulations (early menarche, nullipara).

Presentation Symptoms are often vague and insidious and include abdominal pain, discomfort and distension.

Diagnosis This is made at laparotomy, although malignancy is likely if ultrasound scan shows an irregular or solid tumour, bilateral disease, or the presence of ascites.

Staging at laparotomy *Stage I:* Disease limited to 1 or both ovaries.
Stage II: Growth extends beyond the ovaries but confined to the pelvis.
Stage III: Growth involving ovary and peritoneal implants outside pelvis (eg superficial liver), or +ve retroperitoneal or inguinal nodes.
Stage IV: Those with distant metastases (including liver parenchyma).

80% present with stage III or IV disease. 5-yr survival: Stage I, 67%; Stage II, 42%; Stage III and IV, 14.4%.

Treatment This depends on the type of tumour. Cystadenocarcinomas (80%) are treated with surgery and chemotherapy which aims for cure. Surgery removes as much tumour as possible: the less left, the more effective is chemotherapy and the better the prognosis. In a young woman with early disease, the uterus and other ovary may be left, for fertility; if the tumour involves both ovaries, uterus and omentum are removed.

Chemotherapy for ~6 months post-op is usual. Cisplatin (eg ≥50mg/m^2) or carboplatin IV are used (± doxorubicin & cyclophosphamide).[2] Carboplatin causes less nausea, vomiting and neuropathy and is not renally toxic, and may not require hospital admission. Leukopenia and platelets↓ are dose limiting side-effects. Relapse is common—and paclitaxel (from Pacific yew trees) *may* be useful here.[3] Radiotherapy is sometimes used.

Further treatment may involve a 'second look' laparotomy, further chemotherapy, or radiotherapy. Colloidal gold may control ascites.

Palliative care involves relief of symptoms which are generally due to extensive peritoneal disease.

Screening and prevention There is no really good screening test. Transvaginal, but not abdominal, ultrasound combined with measurement of tumour bloodflow, can differentiate between benign and malignant neoplasms at an early stage, but is time-consuming. Carcino-embryonic antigen 125 is insufficiently sensitive or specific. A blood test based on a 2-allele polymorphism in intron 3 of gene p53 (explained on page 752) has given a specificity of 98% and a sensitivity of 16%.[4]

Some advocate prophylactic oophorectomy when an older woman is having abdominal surgery. The combined oral contraceptive Pill reduces the risk of ovarian malignancy by up to 40%.

1 C Williams 1992 *BMJ* i 1501   2 M Gore 1994 *Lancet* 343 339   3 HL Long 1994 *Mayo Clin Pro* 69 384 & *Drug Ther Bul* 1994 32 41 4 I Runnebaum 1995 *Lancet* 345 994

Vaginal discharge

A non-offensive discharge may be physiological. Most discharges are smelly and itchy and due to infection. A very foul discharge may be due to a foreign body (eg forgotten tampons, or beads in children).

Note the details of the discharge. Is she at risk of, or afraid of, a sexually transmitted disease (STD)? See OHCM p 216–220. If she has an STD, contact tracing is needed—consult a genito-urinary clinic.

Do a speculum examination and take swabs: endocervical (and special medium) for chlamydia; cervical swabs for gonorrhoea (OHCM p 220).

▶Discharges rarely resemble their classical descriptions.

Physiological discharge Its increase with puberty, sexual activity, pregnancy and the combined Pill may cause much anxiety.

Thrush (*Candida albicans*) Thrush is the commonest cause of discharge which is classically white curds. The vulva and vagina may be red, fissured and sore, especially if there is an allergic component. Her partner may be asymptomatic. Pregnancy, contraceptive and other steroids, immunodeficiencies, antibiotics and diabetes are risk factors—check for glycosuria.
Diagnosis: Microscopy reveals strings of mycelium or typical oval spores. Culture on Sabouraud's medium.
Treatment: A single imidazole pessary eg clotrimazole 500mg, plus cream for the vulva (and partner) is convenient. She may need reassurance that thrush is not necessarily sexually transmitted.

Thrush is often recurrent; consider nystatin vaginal pessaries for 14 nights (messy) or fluconazole 150mg PO as a single dose (avoid if pregnant). Candida elsewhere (eg mouth, natal cleft) of both partners may cause reinfection. Live yoghurt (on a tampon) or acetic acid jelly pH 4 eg Aci-Jel® may prevent or relieve mild attacks. Avoidance of chemicals around the vulva (no bathsalts); wiping the vulva from front to back; cotton underwear, double rinsed; and vinegar baths, may help.

Trichomonas vaginalis (TV) This produces vaginitis and a thin, bubbly, fishy smelling discharge. It is sexually transmitted. Exclude gonorrhoea, which often coexists. The motile flagellate may be seen on a wet film (× 40 magnification), or cultured. *Treatment:* Metronidazole 200mg/8h PO for 7 days or 2g PO stat; treat the partner; if pregnant, use the 7-day régime.

Bacterial vaginosis There is offensive discharge with fishy odour, from cadaverine and putrescine. Vaginal pH is >5.5. The vagina is not inflamed and pruritus is uncommon. Mixed with 10% potassium hydroxide on slide under the nose, a whiff of ammonia may be emitted. Stippled vaginal epithelial 'clue cells' may be seen on wet microscopy. There is altered bacterial flora—overgrowth eg of *Gardnerella vaginalis*, *Mycoplasma hominis*, peptostreptococci, *Mobiluncus* and anaerobes, eg *Bacteroides* species—with too few lactobacillae. There is increased risk of preterm labour and intraamniotic infection in pregnancy. *Diagnosis:* This is by culture.
Treatment: Metronidazole 2g PO once or clindamycin 2% vaginal cream, 1 applicatorful/night PV for 7 doses. If recurrent, treating the partner may be helpful. If pregnant, use metronidazole alone at 200mg/8h PO for 7 days.

Vaginal discharge in children This is often from infection due to atrophic vaginitis. Threadworms may cause pruritus. Always consider sexual abuse. Gentle rectal examination may exclude foreign body. For prolonged discharge examination under anaesthesia may be needed (a paediatric laryngoscope can be used as a speculum).

Pelvic infection

Pelvic infection affects the Fallopian tubes (salpingitis) and may involve the ovaries and parametra. 90% of infections are sexually acquired, commonly chlamydia (60%; many are asymptomatic—but infertility or ectopic pregnancy may be the result[1]) and gonococcus. Organisms cultured from infected tubes are commonly different from those cultured from ectocervix, and are usually multiple. 10% follow childbirth or instrumentation (insertion of IUCD, ToP) and may be streptococcal. Infection can spread from the intestinal tract during appendicitis (Gram –ve and anaerobic organisms) or be blood-borne (tuberculosis).

Salpingitis Patients with *acute salpingitis* may be most unwell, with pain, fever, spasm of lower abdominal muscles (she may be most comfortable lying on her back with legs flexed) and cervicitis with profuse, purulent or bloody vaginal discharge. Heavy menstrual loss suggests endometritis. Nausea and vomiting suggest peritonitis. Look for suprapubic tenderness or peritonism, cervical excitation, and tenderness in the fornices. It is usually bilateral, but may be worse on one side. *Sub-acute infection* can be easily missed, and laparoscopy may be needed to make either diagnosis.

Management ▶Prompt treatment minimizes complications. Take endocervical and urethral swabs. Admit for blood cultures and IV antibiotics if very unwell. She needs bed rest, adequate fluid intake and antibiotics eg doxycycline 100mg/24h PO (substitute erythromycin 500mg/6h PO if pregnant) and metronidazole 400mg/8h PO. This covers gonococcus, but if that is cultured, it is advisable to add procaine penicillin 4.8g IM stat with probenecid 1g PO. Trace contacts and ensure they seek treatment (seek help of genito-urinary clinic).

Complications If response to antibiotics is slow, consider laparoscopy. She may have an abscess (draining via the posterior fornix prevents perforation, peritonitis and septicaemia—but laparotomy may be needed). Inadequate or delayed treatment leads to chronic infection and to long-term tubal blockage (so 12% infertile after 1 episode, 23% after 2, 54% after 3). Advise that barrier contraception protects against infection. Ectopic pregnancy rate is increased 10-fold in those who do conceive.

Chronic salpingitis Unresolved, unrecognized or inadequately treated infection may become chronic. Inflammation leads to fibrosis, so adhesions develop between pelvic organs. The tubes may be distended with pus (pyosalpinx) or fluid (hydrosalpinx).

Pelvic pain, menorrhagia, secondary dysmenorrhoea, discharge and deep dyspareunia are some of the symptoms. She may be depressed. Look for tubal masses, tenderness and fixed retroverted uterus. Laparoscopy differentiates infection from endometriosis.

Treatment is unsatisfactory. Consider long-term broad spectrum antibiotics (eg tetracycline 250mg/6h PO 1 hour before food for 3 months), short-wave diathermy and analgesia for pain, and counselling.[2] The only cures are menopause or surgical removal of infected tissue.

1 D Taylor-Robinson 1994 *BMJ* i 150   2 C Douglas 1989 *Update* **38** 480

Endometriosis

Essence Foci of endometrial glandular tissue, looking like the head of a burnt match, are found beyond the uterine cavity, eg on an ovary ('chocolate cyst'), in the rectovaginal pouch, uterosacral ligaments, on the surface of pelvic peritoneum, and rarely in the umbilicus, lower abdominal scars and distant organs eg lungs. If foci are found in the muscle of the uterine wall, the term adenomyosis is used. It occurs in ~12% of women; not all have symptoms, and in the asymptomatic, it may be physiological.

Cause Possibly retrograde menstruation (Sampson's theory) explaining its association with age—typically 40-44yrs, ie those with most menstrual exposure, its negative association with pregnancy and the Pill,[2] and its distribution in the pelvis—but not its appearance elsewhere. Women with endometriosis have raised autoantibody levels. Endometriotic foci are under hormonal influence so are suppressed during pregnancy, regress with the menopause, and bleed during menstruation. Blood may be confined in cysts (which look brown, hence 'chocolate cysts'). If the blood is free it is intensely irritating, provoking fibrosis and adhesions.

Presentation Endometriosis may be asymptomatic, even though extensive. Pelvic pain is the commonest symptom (classically cyclical, at the time of periods). It may be constant, eg if due to adhesions. Secondary dysmenorrhoea and deep dyspareunia are common. Periods are often heavy and frequent, especially with adenomyosis. Patients may present with infertility. Extra-pelvic endometriosis presents with pain or bleeding at the time of menstruation at the site of the pathology, eg haemothorax.

Diagnosis On vaginal examination a fixed retroverted uterus or nodules in the uterosacral ligaments+general tenderness suggest endometriosis. An enlarged, boggy, tender uterus is typical of adenomysis. Typical cysts or peritoneal deposits at laparoscopy differentiates it from chronic infection.

Treatment If asymptomatic, then do not treat. Otherwise, try surgery or hormones (if not pregnant or lactating). In the UK, consider suggesting joining the Endometriosis Society,[3] as treatment can be long and difficult.

Hormonal treatment aims to suppress ovulation for 6-12 months during which time lesions atrophy. **Danazol** up to 400-800mg/24h PO is effective but SE may be unacceptable at high doses (it is a testosterone derivative with anabolic and androgenic effects so expect weight gain, acne, greasy skin, hirsutism, bloating and fluid retention). Advise barrier contraception as it masculinizes the female fetus. **Gestrinone** 2.5-5mg PO twice weekly is an alternative. LHRH analogues, eg **buserelin** nasal spray 300μg/8h or **goserelin** 3.6mg by monthly SC injection produce a reversible artificial menopause but bone demineralization occurs. SE: menopausal symptoms, headaches, depression, libido↓, breast tenderness, drowsiness, acne, dry skin, ovarian cysts (if so, stop), urticaria, changes in body hair, osteoporosis—so only give for <6 months. An alternative gonadotrophic-releasing hormone agonist with similar side-effects is **leuprorelin acetate**—dose: 3.75mg every 4 weeks IM or SC starting during the 1st 5 days of a cycle, for <6 months. Endometriotic resolution is similar to danazol, but SE are less (flushes, headaches, vaginal dryness, emotional lability). Progestogens, eg **norethisterone** (Primolut N®) 5-10mg/12h PO continuously for 9 months may suppress symptoms temporarily by mimicking pregnancy.[1]

Surgical treatment ranges from local excision or diathermy of endometriotic tissue, to total hysterectomy with bilateral salpingo-oophorectomy, depending on the site of lesions and the woman's wish for future fertility.

Prognosis Endometriosis is chronic or relapsing, being progressive in 50%. There is no evidence that treatment helps fertility.

1 DG Limb 1995 *Update* **50** 158   2 M Vessey 1993 *BMJ* ii 182   3 35 Belgrave Sq, London WIX 8QB

Prolapse

A prolapse occurs when weakness of the supporting structures allows the pelvic organs to sag within the vagina. The weakness may be congenital, but it usually results from stretching during childbirth. Poor perineal repair reduces support (p 146). Weakness is exacerbated by menopausal atrophy and by coughing and straining. They may cause distressing incontinence and be a nuisance but are not a danger to health—except for third degree uterine prolapse with cystocele when ureteric obstruction can occur.

Types of prolapse are named by the structures sagging. Several types may coexist in the same patient.

Cystocele The upper front wall of the vagina, and the bladder which is attached to it, bulge. Residual urine within the cystocele may cause frequency and dysuria.

Urethrocele If the lower anterior vagina wall bulges the urethra is displaced, impairing the sphincter mechanisms (p 70) and leading to stress incontinence. Does she leak when she laughs?

Rectocele The middle posterior wall, which is attached to rectum, may bulge through weak levator ani. It is often symptomless, but she may have to reduce the herniation prior to defecation by putting a finger in the vagina.

Enterocele Bulges of the upper posterior vaginal wall may contain loops of intestine from the pouch of Douglas.

Uterine prolapse With *1st degree prolapse* the cervix stays in the vagina. In *2nd degree prolapse* it protrudes from the introitus when standing or straining. With *3rd degree prolapse* (procidentia) the uterus lies outside the vagina. The vagina becomes keratinized and the cervix may ulcerate.

Symptoms: 'Dragging' or 'something coming down' is worse by day. Cystitis, frequency, stress incontinence and difficulty in defecation may occur depending on the type of prolapse. Examine vaginal walls in left lateral position with a Sims' speculum, and ask the patient to bear down to demonstrate the prolapse.

Prevention: Lower parity; better obstetric practices.

Treatment: Mild cases may improve with reduction in intra-abdominal pressure, so encourage her to lose weight, stop smoking and stop straining. Improve muscle tone with exercises or physiotherapy, and, if postmenopausal, topical oestrogens eg dienoestrol cream 0.01% as often as required (try twice weekly).

Severe symptomatic prolapse is best treated surgically. Incontinence needs to have the cause treated (so arrange urodynamic studies to plan the best type of surgery). Repair operations (p 72) excise redundant tissue and strengthen supports, but reduce vaginal width. Is she sexually active? If so, surgery must compromise between reducing prolapse and maintaining width. Marked uterine prolapse is best treated by hysterectomy.

Ring pessaries may be tried as a temporary measure or for the very frail. Use the smallest diameter which keeps the prolapse reduced. Insert into the posterior fornix and tuck above the pubic bone (easier if the ring has been softened in hot water first). Problems: discomfort, infection, ulceration (change every 6 months).

Infertility: causes and tests[1]

▶ Infertility can be devastating to both partners, and its investigation a tremendous strain. Sympathetic management is crucial.

90% of young couples having regular intercourse conceive within a year. Fertility decreases with age. High fertility in one partner can compensate for low fertility in the other, so in many of the remaining 10% both partners are subfertile. Consider:

Is she producing healthy ova? (Anovulation causes 21%.)
● Is he producing enough, healthy sperm? (Male factors causes 24%.)
● Are the ova and sperm meeting? (Tubal cause 14%, hostile mucus 3%, sexual dysfunction 6%.)
● Is the embryo implanting?

Endometriosis is thought to be the cause in 6%. The cause is 'unexplained' in 27% of couples. With 'unexplained' infertility 60-70% of women will achieve conception within 3 years.

Initial management It takes two to be infertile. See both partners.

Ask her about: Menstrual history, previous pregnancies and contraception, history of pelvic infections or abdominal surgery. *Ask him about:* Puberty, previous fatherhood, previous surgery (hernias, orchidopexy, bladder neck surgery), illnesses (venereal disease and adult mumps), drugs, alcohol intake, job (is he home when ovulation occurs?). *Ask both about:* Technique, frequency, and timing of intercourse (nonconsummation is a rare problem); feelings about infertility and parenthood; previous tests.

Examination Check the woman's general health and sexual development and examine the abdomen and pelvis. If the sperm count is abnormal, examine the man for endocrine abnormalities, penile abnormalities, varicoceles; confirm there are 2 normal testes (size: $3.5-5.5 \times 2.1-3.2$ cm).

Tests for ovulation If cycles are regular ovulation is likely. The only proof of ovulation is pregnancy. It is possible for a follicle to luteinise without rupturing, so tests may be positive in the absence of an ovum. Negative results imply failure to ovulate. *Test examples:* Visualizing follicle development or change to secretory endometrium on ultrasound; finding 'ovulatory' mucus at midcycle (like raw egg white); detecting LH surge (eg Clearplan® kit); detecting a luteal rise in progesterone to >30nmol/l on day 21; detecting a rise in basal body temperature at mid-cycle (temperature charting is difficult to do and may create anxieties).

Blood tests Check rubella status and immunize if non-immune. If you suspect anovulation check: ●Blood prolactin (if high may be due to prolactinoma; do skull X-ray) ●FSH (>10U/l indicates a poor response to ovarian stimulation; it may indicate primary ovarian failure) and LH (for polycystic ovary syndrome) ●TFT.

Semen analysis Unless there has been a satisfactory post-coital test, analyse semen after 3 days' abstinence for: ●Volume (mean 2.75ml now, formerly 3.4ml in the 1940s[1]). ●Sperm count/morphology. ●Antibodies. ●Infection. Normal count >20 million sperm/ml, >40% motile and >60% normal form; mean count = 66 million/ml (113 million/ml in the 1940s; this average is falling at a rate of about 100 sperms/ml every hour—which may be due to an environmental influence[1]—or more frequent ejaculations[2]—or a data artefact[3]). Examine 2 specimens as variation may be considerable. Reduced counts require specialist referral.

1 E Carlsen 1992 *BMJ* ii 609   2 H Menger 1994 *BMJ* i 14409   3 P Bromwich 1994 *BMJ* ii 19

Subfertility options: abbreviations

FEC   Fetal egg child (offspring from an egg taken from the ovary of
        an aborted fetus)
ICSI   Intra-cytoplasmic sperm injection (directly into an egg)
IVF   *In vitro* fertilization
GIFT   Gamete intra-fallopian transfer
MESA   Microepididymal sperm aspiration (from testis, eg post-
        vasectomy)
PESA   Percutaneous epididymal sperm aspiration (rather like MESA,
        but using a 22G butterfly needle inserted into the epi-
        didymis—so scrotal exploration is not required).[1]
POST   Peritoneal oozyte sperm transfer
SUZI   Subzonal sperm injection (directly into an egg)
TET   Tubal embryo transfer
TUFT   Trans-uterine fallopian transfer
ZIFT   Zygote intra-fallopian transfer

1 Y Khalifa 1996 *BMJ* i 5

Infertility: tests and treatment

Tests of tubal patency

1 Laparoscopy and dye test. The pelvic organs are visualized and methylene blue dye is injected through the cervix. If the tubes are blocked proximally they do not fill with dye. With distal block there is no 'spill' into the peritoneal cavity.

2 A hysterosalpingogram (contrast X-ray) demonstrates uterine anatomy and tubal 'fill and spill' It is unpleasant and requires premedication. False positives may occur with tubal spasm. Give antibiotics eg cephradine 500mh/8h PO with metronidazole 1g/12h PR for 24h before and 5 days after procedure to prevent pelvic infection.

The postcoital test Carried out in the immediate pre-ovulatory phase (when cervical mucus is profuse, acellular and exhibits ferning and Spinnbarkeit—the ability for it to be drawn out), performed 6–12h after intercourse, cervical mucus is removed from the cervix and examined under high-power (× 400) microscopy. A satisfactory test (ovulatory mucus, >10 motile sperm/high-power field) shows that his sperm are adequate, coitus is effective and the cervical mucus does not contain antibodies.

Treatment of infertility Treatment is directed at the cause.

Azoospermia is unresponsive to treatment. A low sperm count may be improved by avoiding tobacco and alcohol, by keeping testes cool (avoid hot baths and tight pants). Will they consider donor insemination (AID=artificial insemination) or donor?

Problems of sperm deposition (eg impotence) can be circumvented by artificial insemination using the partner's sperm (AIH).

Hyperprolactinaemia (OHCM p 562) is treated by removing the cause if one is found (pituitary macro-adenoma, drugs); if not, give bromocriptine 1mg/24h PO, increasing the dose until blood prolactin is normal.

Anovulation is managed by stimulating follicle development using clomiphene 50–200mg/24h PO on days 2–6 inclusive. SE: visual disturbance, abdominal pain due to ovarian hyperstimulation; warn about risk of multiple pregnancy. Human chorionic gonadotrophin (HCG) resembles LH and may be needed to rupture the ripe follicle. If clomiphene fails to produce fertility gonadotrophin injections or LHRH analogues may be tried.

Tubal problems may be remedied by surgery but results are poor.

Assisted fertilization The couple require psychological stability (and finance). The 'take-home baby rate' averages 10% and fetal loss, ectopic pregnancy, obstetric problems, multiple births and fetal abnormality are all commoner than normal.

In vitro fertilization (IVF) is used for tubal (and other) problems. The ovaries are stimulated and ova are collected (by transvaginal aspiration under transvaginal ultrasound guidance), fertilized, and embryos are returned to the uterus as an outpatient procedure.

Gamete intrafallopian transfer (GIFT) may be tried where tubes are patent, eg 'unexplained infertility'. Gametes are placed in the tubes by laparoscopic cannulation. Pregnancy rate ~28% per cycle treatment. ZIFT (zygote or early cleaved embryo transfer) can give higher pregnancy rates of up to 48% in women with unexplained infertility.

The possibility of adoption should not be forgotten. Those who remain childless may value counselling or a self-help group.[1]

---

1 In UK, National Association for the Childless, 318 Summer Lane, Birmingham B19 3RL

Male infertility[1]

Normal spermatogenesis takes place in the seminiferous tubules of the testes. Undifferentiated diploid germ cells (spermatogonia) multiply and are then transformed into haploid spermatozoa. The total duration of a cohort of spermatogonia to develop into spermatozoa is 74 days. FSH and LH are both important for initiation of spermatogenesis at puberty. LH stimulates Leydig cells to produce testosterone. Testosterone and FSH stimulate Sertoli cells to produce essential substances for metabolic support of germ cells and spermatogenesis.

Spermatozoa A spermatozoon has a dense oval head (containing the haploid chromosome complement), capped by an acrosome granule (contains enzymes essential for fertilization) and is propelled by the motile tail. Seminal fluid forms 90% of ejaculate volume and is alkaline to buffer vaginal acidity. Only about 200 sperm from any ejaculate reach the middle third of the Fallopian tube, the site of fertilization.

Male infertility Male factors are the cause for infertility in ~24% of infertile couples. In reality, most are subfertile. Only a small number of men have an identifiable cause. Causes include: (% cited in 1 study)

- Idiopathic oligo/azoospermia (16%): Testes are usually small and FSH ↑. No specific treatment is available; many have been tried.
- Asthenozoospermia/teratozoospermia (17%): In asthenozoospermia sperm motility is reduced due to structural problems with the tails. Teratozoospermia indicates an excess of abnormal forms.
- Varicocoele (17%): this is controversial for varicocoele is found in 15% of males, most of whom have normal fertility.
- Genital tract infection (4%): gonococci, chlamydia and Gram −ve enterococci can cause adnexal infection (eg with painful ejaculation, urethral discharge, haematospermia, dysuria, tender epididymes, tender boggy prostate). Confirm by semen culture, urethral swab or finding >1 million peroxidase +ve polymorphs/ml semen. Treat both partners for 4 weeks with erythromycin, doxycycline, or norfloxacin (eg 400mg/12h PO).
- Sperm autoimmunity (1.6%): Risk factors for antibodies: vasectomy, testis injury, genital tract obstruction, family history of autoimmune disease. Most are on sperm membranes or in seminal fluid, but may occur in the woman. Prednisolone 20mg/12h PO on day 1–10 and 5mg on day 11 and 12 of his partner's cycle for 6 months reduces antibody levels.
- Congenital (cryptorchidism, chromosome disorders—2%): Klinefelter's account for 50% chromosome disorders. For optimal fertility undescended testes should be fixed in the scrotum before 2yrs of age.
- Genital tract obstruction (1.8%): Azoospermia, normal sized testes with normal or high FSH suggests this. It may follow infection, vasectomy, or be congenital (as in cystic fibrosis). It may be amenable to surgery eg epididymovasostomy to bypass epididymal obstruction.
- Systemic/iatrogenic (1.3%). ●Coital disorders (1%).
- Gonadotrophin deficiency (0.6%): This is the only cause of testicular failure consistently treatable by hormone replacement.

Examination Look at body form, secondary sexual characteristics, and for gynaecomastia. Normal testicular volume is between 15–35ml (compare with Prader orchidometer®). Rectal examination may reveal prostatitis.

Tests Semen analysis (p 56). Post-coital tests with mid-cycle cervical mucus may indicate antibody presence. Plasma FSH will distinguish primary from secondary testicular failure. Testosterone and LH levels are indicated where androgen deficiency is suspected. Agglutination tests to detect antibodies are available.

Treatment The treatment of treatable causes is given above.

1 F C Wu 1992 *Gynaecology*, Churchill Livingstone, 355–74

Contraception

▶*Any method, even coitus interruptus, is better than none.* 90% of young women having regular unprotected intercourse become pregnant within a year. Properly used, contraception reduces this to <1% (combined Pill, injectables, sterilization); around 2% (PoP, IUCD, cap, sheath); 12% (mucothermic)—success rates may be much higher;[1] 15% (withdrawal). Failure rates may be much higher for user-dependent methods.[2]

The ideal contraceptive would be 100% effective, without side effects, readily reversible and not need medical supervision. The aim is to find the best compromise for each woman depending on her age, health, and needs. Methods available: ●'Natural methods' (no intercourse near time of ovulation)—acceptable to Catholic Church; also free, requiring no 'pollution of the body' with drugs ●Barrier methods (low health risk but need high user motivation) ●Hormonal (complex health risks but highly effective) ●IUCD (convenient and effective where not contraindicated), and sterilization (very effective but effectively irreversible).

Natural (mucothermic) methods[2] involve monitoring physiological phenomena to determine the fertile time (6 days prior to ovulation; the life of a sperm) to 2 days afterwards (the life of the ovum). Cervical mucus becomes clear and sticky at the beginning of the fertile time and dry at ovulation (consistency is altered by semen and vaginal infections). Basal body temperature ↑~0.3°C after ovulation (affected by fevers, drugs, recent food or drink). Additional observations, eg mittelschmertz (p 74), changes in the cervix, improve accuracy. Natural methods require regular cycles, dedication and self-control, but they *do* work. In the UK, it is easy to find a teacher of this method.[3]

Barrier methods ▶The main reason for failure is not using them.
Sheaths reduce transmission of most STDs but not those affecting the perineum. Caps give some protection against gonorrhoea and chlamydia but not syphilis or herpes. Some spermicides inactivate HIV *in vitro*.

●Sheaths are effective when properly used with spermicides. Unroll onto the erect penis with the teat or end (if teatless) pinched to expel air. This prevents bursting at ejaculation.

●Caps come in several forms. Diaphragms stretch from pubic bone to posterior fornix. She should check after insertion that the cervix is covered. Cervical caps fit over the cervix (so need a prominent cervix). They should be inserted <2h before intercourse, be kept in place >6h after. Use with a spermicide. Some find them unaesthetic. Problems: UTIs, rubber sensitivity.

●Cervical sponges These are simple but often fail.

●The female condom (eg Femidom®) Prescription and fitting are not needed. The manufacturer claims excellent barrier properties to HIV, bacteria and sperm. It might be suitable for contraception when a Pill has been missed, or if there is dyspareunia, or the perineum needs protection, eg post-childbirth—or if she is latex-allergic to traditional condoms. One reason for failure is that the penis goes alongside it, rather than in it; another, that it gets pushed up in the vagina[4] or may fall out.[5] They can be noisy.[5]

●Spermicides should not be used without barriers as they provide inadequate contraceptive cover.

1 RE Ryder 1993 *BMJ* ii 723   2 Information from the Family Planning Association, 27-35 Mortimer Street, London, WIN 7RJ   3 UK tel. numbers: 0171 371 1341, 0121 627 2698, 01222 754628   4 *Drug Ther Bul* 1993 **31** 15   5 N Ford 1993 *Br J Fam Plan* **19** 187

The intrauterine contraceptive device (IUCD)

IUCDs (loops, coils) are plastic shapes ~3cm long around which copper wire is wound, and carrying a plastic thread from the tail, eg Novagard®. They inhibit implantation and may impair sperm migration. They need changing every 3–5 years. They have replaced the larger, non-copper bearing 'inert' types (eg Lippes loop®), which caused more complications but did not need changing (and so are sometimes still found *in situ*).

Most of those who choose the IUCD (6%) are older, parous women in stable relationships, in whom the problem rate is low.

Problems with IUCDs 1 They tend to be expelled by a uterus which is nulliparous or distorted (eg by fibroids). 2 Ectopic pregnancy is more likely. 3 They are associated with pelvic infection and infertility, following sexually transmitted disease—or sometimes introduced during insertion. 4 They tend to produce heavy, painful periods. Contraindications aim to exclude women at high risk of these problems, also those who are immunosuppressed (eg HIV+ve), have Wilson's disease (OHCM p 712), or those with cardiac lesions (unless the coil is inserted under antibiotic cover).

Insertion Skilled insertion minimizes complications. Each device has its own technique, so read the instructions carefully and practise beforehand.

An IUCD can be inserted any time, (or as postcoital contraception), as long as the woman is not pregnant, but it is more likely to be expelled during a period or shortly after childbirth. Determine the position of the uterus. Then insert a uterine sound to assess the length of the cavity. Then insert the IUCD, placing it in the fundus. This may cause crampy pains. Once the coil is in place, the threads are cut to leave 3cm visible in the vagina. Teach her to feel the threads after each period.

▶Insertion of IUCDs may provoke 'cervical shock' (from increased vagal tone). Have IV atropine available.

Follow-up The first 3 months are the peak time for expulsion and the coil should be checked after the next period, then 6-monthly. Threads are often easier to feel than to see.

Lost threads The IUCD may have been expelled, so advise extra contraception and exclude pregnancy. Then sound the uterus. If you find the coil you may be able to bring down the threads with a special instrument. If you cannot find the threads, they may have dropped off. If you cannot find the coil, seek it on ultrasound. A coil may perforate the uterus, usually after insertion and often with little pain (remove eg laparoscopically).

Infection This can be treated with the device in place, but if removed do not replace it for 6 months. Rarely, *Actinomyces* is found on smear test. Change the coil, cut threads off old one and send coil for culture. If positive, remove new coil and give penicillin V 250mg/6h PO for 90 days.

Pregnancy >90% are intrauterine. Remove coil, if you can as soon as pregnancy is diagnosed to reduce risk of miscarriage (20% if removed early, 50% if left), and to prevent risk of late septic abortion.

Removal Alternative contraception should be started (if desired) prior to removal, as there may be a fertilized ovum in the tubes. At the menopause, remove after at least 6 months' amenorrhoea.

The future This may lie with IUCDs carrying hormones, eg the Mirena® device (carries levonorgestrel—its local effect is said to make implantation less likely, and periods lighter; perhaps 20% may experience reversible amenorrhoea; reliability is said to match that of sterilization). It lasts 3yrs. There may be less risk of ectopic pregnancy. Warn about spotting ± heavy bleeding (NB: bleeding may become *scanty* or *absent* after a few cycles).

Hormonal contraception ('the Pill')[1]

~3 million women in the UK take the combined oral contraceptive (CoC)—the method of choice for 50% of younger users. CoCs contain oestrogen (usually ethinyloestradiol) with a progestogen, either in fixed ratio or varying through the month (phased). Low dose formulations (≤30μg oestrogen) are the norm. The combined Pill is taken daily for 3 weeks followed by a week's break. This inhibits ovulation and produces a withdrawal bleed in the pill-free week. When prescribing CoCs, pay attention to these areas:

History Why does she want the Pill? Does she know about risks? These must be explained. Has she considered the alternatives? Are there any contraindications in her or her family eg thrombotic disorders? Ask about smoking and drugs, and about contact lenses (Pill may cause dryer eyes). Does she have any anxieties, eg weight-gain?

The Pill, smoking, and thromboembolism Death due to the Pill in 35 year olds is 8 times more common if they smoke (but still as safe as child birth). Can she be helped to stop smoking? If not, stop the Pill at 30.

Absolute contraindications Any disorder predisposing to venous or arterial problems (ie abnormal lipids; prothrombotic disorders such as quite common APC resistance (activated protein C);[1] many cardiovascular problems, except mild non-Pill related hypertension and varicose veins); liver disease; focal migraine—ie *any* focal sensory or motor signs or symptoms migraine attacks for >72h (status migrainosus[2]) or those requiring ergotamine; diseases exacerbated by sex steroids; recent trophoblast disease undiagnosed uterine bleeding; gross obesity; immobility (stop Pill 4 weeks before till 2 weeks after very major surgery—provided alternative contraception is realistic, eg barrier methods or depot progesterone, p 67).

Relative contraindications Prescribe with caution in those with a family history of DVT, myocardial infarction, ↑BP and breast cancer; with severe migraine or migraine requiring sumatriptan;[2] epilepsy; sickle-cell disease diabetes; oligomenorrhoea; illnesses causing diarrhoea (eg Crohn's); in heavy smokers and possibly those with abnormal cervical smears. NB: if a woman gets a 1st attack of migraine while on the combined Pill, stop it and observe closely: restart cautiously only if there are no sequelae.[3]

Drugs interfering with the Pill *Liver enzyme inducers*, eg anti-convulsants griseofulvin, and rifampicin reduce efficacy by ↓circulating oestrogen. In short-term use, take extra precautions (p 66) whilst used and for 7 days after (4 weeks for rifampicin). With long-term use consider using 50μg oestrogen Pills with higher doses of progesterone eg Norinyl-1®. Use another type of contraceptive if on long-term rifampicin. *Antibiotics:* p 66.

Benefits of the Pill Very effective contraception; lighter, less painful 'periods'; reduced premenstrual syndrome. In long-term users, reduced risk of ovarian tumours (carcinoma ↓ 40%) and endometrial carcinoma (↓ 50%); less pelvic infection and endometriosis.

Serious disadvantages The risks of arterial and venous disease are increased eg DVT, myocardial infarction. Risk of death due to the Pill increases sharply over 40yrs: 1:2500 for non-smokers; 1:500 for smokers The relationship with breast cancer is complex but rates in those <35yrs are increased (from ~0.2% to ~0.3%).

For contraception in those who are HIV+ve, see p 155.

1 E MacGregor 1995 *Brit J Fam Plan* **21** 16   2 J Guillebaud 1995 *Brit J Fam Plan* **21** 16 & 1994 **20** 24
3 K Pasi 1995 *Lancet* 345 1437

Hormonal contraception—further details

Do BP 6 monthly. Check weight, and breasts—eg if >35yrs. Do a cervical smear if due (p 32). Adverse effects are dose-related: aim to give the lowest dose Pill that gives good cycle control (ie no breakthrough bleeding).

Pill problems, and progesterones to consider[1,2] If there are problems which may be related to combined pills containing norethisterone or levonorgestrel (break-through bleeding, acne, headaches, weight gain) consider a newer progesterone, if there are no risk-factors for DVT (body mass index >30 (p 472), immobility, marked varicose veins, past DVT, family history of unexplained thrombosis). Examples are Femodene®, Femodene ED®, Marvelon®, Minulet®, Mercilon®, Triadene®, and Triminulet®. They are implicated in an increase in risk of DVT of 1.4–2.7-fold (95% confidence interval)[3] so users, who must be the choosers, need to know about, and accept the risk, which you should document (excess risk is ~15/100 000 users/yr, with ~2% of these thromboses leading to serious consequences).[3]

Switching from PoP: *If to CoC*, start on day 1 of period. Use condom for 7 days. *If to a PoP*, start as the old pack finishes. **Switching from CoC:** *If to a CoC*, start immediately with no pill-free interval. *If to a PoP*, start immediately the CoC packet is finished.

Starting CoCs and PoPs Day 1 of cycle, or on the day of ToP, or 3 weeks post-partum or 2 weeks after full mobilization after major surgery.[1] With day 1–3 CoC starting régime, contraceptive cover is immediate; use other precautions for 7 days with PoP and if CoC started day 4 or 5 of period.

Stopping the Pill Tell to stop at once if she develops: ●Sudden severe chest pain ●Sudden breathlessness (or cough with bloody sputum) ●Severe calf pain in 1 leg ●Severe stomach pain ●Unusual severe prolonged headache. Smokers should stop at 30; low-risk non-smokers may continue into their 40s. 66% of women menstruate by 6 weeks, 98% by 6 months; women who are amenorrhoeic after the Pill were usually so before.

Special cases If >12h late (or diarrhoea), continue Pills but use condoms too for 7 days (+ the days of diarrhoea)—if this includes Pill-free days, start next pack *without a break*. Use extra precautions during and for 7 days after antibiotics. If tablet from the 1st 7 days in pack is forgotten, also consider post-coital contraception. Vomiting <3h post Pill: take another.

Post-coital options ('The sheath split') 1 Placing an IUCD within 5 days 2 *Morning*[+60b] *after pills* (levonorgestrel 250µg + ethinyloestradiol 50µg as Schering PC4®—or Ovran®). Use <72h after 1 episode of unprotected intercourse. Take 2 tabs at once and 2 tabs 12h later (SE: vomiting). Failure rate 2–4% mid-cycle. Advise barrier methods until next period. Arrange future contraception. See again at 5 weeks. Has she had a period?

Progesterone only Pill (PoP, 'mini-pill') Low-dose progestogen renders cervical mucus hostile to sperm. In some women it also inhibits ovulation, improving efficacy but causing erratic bleeding, its worst side effect. The risk of ectopics is increased. It is less effective than the combined Pill except in older, less fertile women. It can be used by most women with medical problems contraindicating the combined Pill, and by breast feeding mothers. Take at same time daily ± 3h, otherwise extra precautions are needed for 7 days. A regular lifestyle or using an alarm watch helps.

Postnatal Start 21 days after birth: eg CoC if not breast feeding; PoP, Depo-Provera® (or ?Norplant®) if breast feeding. IUCD: fit ~4 weeks post-partum.

At the menopause Stop PoP if >50yrs old with >1year's amenorrhoea (2yrs if <50yrs old)—as a rough guide. In this last year, a spermicide and sponge is probably adequate due to declining fertility. As CoC masks the menopause, aim to stop all women at 50yrs, and replace with a non-hormonal method. An FSH >30IU/litre indicates that the ovaries have retired.

1 *Drug Ther Bul* 1992 30 41 2 J Guillebaud 1995 BMJ ii 1111 3 CSM 1995 BMJ ii 1232 & K Mcpherson 1996 i 6

Depot and implant hormonal contraception

Depot progestogen ('the injection') is simple, safe and very effective but has suffered much from adverse publicity. 2 preparations are available: *medroxy progesterone acetate* (Depo-Provera®) 150mg given deep IM 12 weekly, start during the first 5 days of a cycle (or at 5 days post-partum if bottle-feeding, 6 weeks if breast feeding) or *norethisterone enanthate* (Noristerat®) 200mg into gluteus maximus 8 weekly (it is licensed for short-term use only, but can be given immediately post-partum when use of Depo-Provera® can cause very heavy bleeding). Exclude pregnancy biochemically and use condoms for 14 days after late injections (eg if she turns up 1 week late).

Contraindications: Pregnancy; abnormal undiagnosed vaginal bleeding; acute liver disease; severe cardiac disease.

Advantages of injectable progesterone injections:
- No oestrogen content
- Fewer ovarian cysts
- Reduces PMS (p 16)
- No compliance problems
- Reduced endometriosis
- Secret
- Good when GI disease
- Abolishes menorrhagia eventually
- Suppresses ovulation (so protects against ectopics)
- Can be used when breast feeding.
- 5-fold protective effect against endometrial carcinoma.

Problems: There may be irregular vaginal bleeding. This usually settles with more prolonged use and amenorrhoea often then supervenes, so encourage perseverance. With Depo-provera® 33% of women have amenorrhoea after 6 months use; 50% after 12 months, and 60% after 18 months (figures are 14%, 27%, and 33% respectively for Noristerat®). If very heavy bleeding occurs, exclude pregnancy; give injection early (but >4 weeks from previous dose) and give oestrogen if not contraindicated (eg Premarin® 1.25mg/24h PO for 21 days or a combined oral contraceptive pill). There is some concern that long-term use (eg >5yrs) in older women may be associated with bone demineralization so consider checking plasma oestrogens and use of oestrogen replacement if oestrogens are low in these women. Other problems include weight gain and acne.

Special uses: Depot injections may be particularly useful:
- To cover major surgery. If given when stopping combined Pill 4 weeks before surgery it gives contraceptive cover for the next 8 or 12 weeks.
- Sickle-cell disease (reduces incidence of sickle-cell crises).
- Epileptics (inducement of liver enzymes from medication immaterial). If on rifampicin for TB give Depo-provera® injection every 10 weeks.
- After vasectomy whilst awaiting partner's sperm-free ejaculates.
- Where bowel disease may affect oral absorption.

There may be some delay in return of ovulation on stopping injections (median delay 10 months) but long term fertility is not reduced.

Implants Progesterone implants are now available with the advantage that they give 5 years' contraception with one implantation. Norplant® consists of 6 capsules each containing levornogestrel 36mg which are implanted under local anaesthetic into the upper arm. If removed (may be difficult to do), plasma levornogestrel becomes undetectable within 2 days.[1] The main side-effect is irregular menstrual bleeding (during the first year this affects most women). Its action is by changing cervical mucus (ie similar to PoP).☐ Only 10% of users become amenorrhoeic. These therefore provide reversible long-term contraceptive cover with minimal medical intervention—except for those in whom suppression of ovulation leads to symptomatic ovarian cysts (these do not usually require surgery).

1 J Thompson 1993 *Brit J Fam Plan* **19** 195

Sterilization

Sterilization is popular form of contraception—perhaps because it is reliable, without one having to think about it. ~100,000 women and ~90,000 men are sterilized annually in the UK. ~25% of women rely on sterilization for contraception.

Ideally see both partners and consider the following:

●Alternative methods. Do they know about the Cap, the PoP, depot progesterone injections and implants?
●Consent. Is it the wish of both partners? Legally only the consent of the partner to be sterilized is required but the agreement of both is desirable.
●Who should be sterilized? Does she fear loss of her femininity? Does he see it as being neutered? If the woman is requesting sterilization, is she really wanting a hysterectomy or would she benefit from one?
●Irreversibility. Reversal is only 50% successful in either sex and tubal surgery increases the risk of subsequent ectopics, so the couple should see sterilization as an irreversible step. The sterilizations most regretted are those carried out on the young and at times of stress, or immediately after pregnancy (termination or delivery). Are their children healthy? Is the marriage stable?
●If she wishes to be sterilized at Caesarean section, explain that it will only be done if the baby is normal and healthy.
●Failure rate. Currently 2 pregnancies per 1000 sterilizations, but higher immediately after a pregnancy due to better healing.
●Side effects. A women who has been on the Pill for many years may find her periods unacceptably heavy after sterilization.

▶Record in the notes something along the lines of: 'Knows that it is irreversible and that there is a small failure rate'.

Female sterilization The more the tubes are damaged, the lower the failure rate and the more difficult reversal becomes. In the UK, most sterilizations are carried out laparoscopically with general anaesthesia. The tubes may be clipped, ringed or diathermied. Alternatively the tubes may be divided and ligated through a mini-laparotomy incision, or clipped via the posterior fornix ('colpotomy'). An IUCD should be left in place until the next period in case an already fertilized ovum is present.

Vasectomy This is a simpler procedure than female sterilization and can be performed on an outpatient basis. The vas deferens is identified at the top of the scrotum, about 1cm is removed, and the cut ends doubled back before being ligated. Bruising and haematoma are complications.

The major disadvantage of vasectomy is that it takes up to 3 months before sperm stores are used up. 2 ejaculates negative for sperm should be obtained before other methods of contraception are stopped. Reversal is most successful if within 10 years of initial operation.

Urinary malfunction

Control of bladder function Continence in women is maintained in the urethra by the external sphincter and pelvic floor muscles maintaining urethral pressure higher than bladder pressure. Micturition occurs when these muscles relax and the bladder detrusor muscle contracts.

Urge incontinence 'If I've got to go I've got to go', and she goes frequently. The bladder is 'unstable' with high detrusor muscle activity. It occurs in nulliparous and parous women. Usually no organic problem is found. Rarely, the cause is neurogenic.

Stress incontinence Small quantities of urine escape as intra-abdominal pressure rises, eg sneezing. It is much commoner in parous women as childbirth denervates the pelvic floor. Examination may reveal prolapse (p 54) or incontinence (ask her to cough).

Management
- Exclude a UTI and glycosuria.
- History and examination will not tell you whether the problem is stress or urge or both: arrange urodynamic studies.
- Urge improves with 'bladder training' (gradually increasing the time interval between voiding). She will need encouragement. Restricting fluids, and antimuscarinic drugs eg oxybutynin 2.5mg/12h PO for 2 weeks; increasing stepwise to a maximum of 5mg/12h PO in the elderly (5mg/8h non-elderly) may help. (SE: dry mouth, constipation, blurred vision, arrhythmias; CI: glaucoma). Propantheline is an alternative.
- Mild stress incontinence responds well to pelvic floor exercises or physiotherapy eg with weighted vaginal cones—insert the heaviest that can be retained (base up) for 15mins twice daily, graduating to heavier cones (max 100g) as tone improves, to improve muscle tone. A vaginal tampon supports the bladder neck, stopping leaks while playing sport.
- Surgery is required for severe stress symptoms to increase intraurethral pressure (by bringing the upper urethra into the zone of abdominal pressure when previously it was below the pelvic floor) and to reduce prolapse. Operations include: urethroplasty, anterior repair, transabdominal colposuspension.
- In some post-menopausal women urgency, increased frequency and nocturia may be helped by topical oestriol 0.1% cream PV used nightly for 3 weeks then twice weekly long term.
- When there is a mixed picture of stress/urge incontinence treat the detrusor instability first because this can be made worse by operations for stress incontinence.[1]

True incontinence Continuous leakage of urine may be due to congenital anatomical abnormalities, eg ectopic ureters, or to acquired problems eg vesico-vaginal fistula due to trauma (usually post abdominal hysterectomy in developed countries, after prolonged labour with vertex presentation in developing countries), malignancy or radiotherapy. If surgery is not possible, seek the help of the incontinence adviser.

Urethral syndrome Symptoms of cystitis occur with a -ve MSU, often associated with intercourse, and may be due to subinfective numbers of organisms being massaged into the urethra. Is she adequately lubricated? Try using Vaseline® over the urethra (but not if sheaths are used). Micturating before and after intercourse, different coital positions (her on top) and trimethoprim 100mg PO prior to intercourse may help.

1 L Cardozo 1994 *Prescribers' J* 34 134

Voiding difficulties[1]

Symptoms
- Poor flow
- Straining to void
- Chronic retention
- Intermittent stream
- Hesitancy
- Overflow incontinence
- Incomplete emptying
- Acute retention
- UTI from residual urine

▶Remember faecal impaction as a cause of retention with overflow.

Causes
CNS: These may be suprapontine (eg stroke); due to cord lesions (cord injury, multiple sclerosis); peripheral nerve (prolapsed disc, diabetic or other neuropathy); or reflex, due to pain (eg with herpes infections).
Drugs: Especially epidural anaesthesia; also tricyclics, anticholinergics.
Obstructive: Early oedema after bladder neck repair is a common cause. A retroverted gravid uterus, fibroids, ovarian cysts, urethral foreign body, ectopic ureterocele, bladder polyp or carcinoma are other causes.
Bladder overdistension.

Tests
- Do an MSU to exclude infection.
- Uroflowmetry (a rate of <15ml/sec for a volume of >150ml is abnormal).
- Ultrasound for residual urine.
- Cystourethroscopy

Treatment
This depends upon the cause. Acute retention may require catheterization (using a suprapubic catheter if it is expected to be staying in place for several days). For persistent conditions (eg neurological conditions) self-catheterization techniques may need to be learned.[1][2] A silver catheter (not disposable!) may be ideal for this.

Drugs may be used to relax the urethral sphincter or stimulate the detrusor muscle (eg prazosin 0.5mg/12h PO, 1st dose given before bed, increasing to 2mg/12h PO after several days if needed; or indoramin 20mg/12h PO increasing every 2 weeks by 20mg to a maximum of 100mg/12h PO). Operative measures may overcome some of the obstructive causes eg urethrotomy for distal urethral stenosis—although this is not often found in women.

For male voiding difficulty, see *Urinary retention and catheters*, OHCM p 134. See also *Urinary incontinence* and *agents for detrusor instability*, OHCM p 74.

---

1 G Hunt 1993 *The User's Guide to Intermittent Self Catheterisation*, BMA/Family Doctor Publications Ltd (PO Box 118, London WC2N 5BG)   2 *Drug Ther Bul* 1991 **29** 37

Gynaecological operations

Hysterectomy The commonest route is *abdominal hysterectomy*, which is usually total, (uterine body + cervix removed), but may be *Wertheim's* (extended to include local lymph nodes and a cuff of vagina—for malignancy); or, rarely, subtotal (the cervix is left behind—and may become malignant). At *vaginal hysterectomy* the uterus is brought down through the vagina. It is therefore performed for prolapse when the uterus is small. Healthy ovaries are usually conserved, especially in young women, unless the hysterectomy is for an oestrogen-dependent tumour. Complication: residual ovary syndrome (pain, deep dyspareunia, premature ovarian failure). *Laparoscopic hysterectomy* was developed for patients in whom problems with open surgery are anticipated (eg extensive adhesions with endometriosis); but it turns out that most of these patients can have a vaginal hysterectomy—which is much quicker (77min *vs* 131min[1]), with a similar complication rate.

Most hysterectomies are performed for dysmenorrhoea, and rates vary widely. Femininity and sexuality are bound up with the uterus. Reduced ability to produce orgasms after total abdominal hysterectomy is greater than after subtotal hysterectomy (leaving the cervix behind).￼ Subtotal hysterectomy is also a faster and cleaner—but remember to go on smearing these cervixes. ▶Women who are counselled, and make their own decision about surgery, are less likely to have regrets. Operative mortality: ~0.06%. A randomized trial (*N*=204) found hysterectomy does not cause ↑ psychological/sexual morbidity compared with ablation (below) at 1yr post-op.[2]

Manchester repair (Fothergill's operation) Pregnancy is still possible after this operation for uterine prolapse. The cervix is amputated and the uterus is supported by shortening the ligaments.

Dilatation and curettage (D&C) The cervix is dilated sufficiently to admit a curette to scrape out a sample of endometrium for histology. D&C is now largely a procedure for diagnosing abnormal bleeding (but outpatient endometrial sampling may make D&C unnecessary). Evacuation of retained products of conception from the uterus after miscarriage ('ERPC'), or termination of early pregnancy, are carried out by dilatation and suction.

Hysteroscopy As an out-patient alternative to D&C, a hysteroscope can be inserted through the cervix into the uterus to visualize the endometrium. 'Blind' samples may then be taken using a sampler.

Hysteroscopic endometrial ablation by laser or diathermy (under GA) reduces bleeding by achieving a deliberate Asherman's syndrome (p 12); as an alternative to hysterectomy, it has fewer complications.[3] Endometrium may be thinned pre-op by leuprorelin or danazol. By 4 months 10% have menorrhagia again.[4] Complications: haemorrhage; infection (eg late necrotizing granulomatous endometritis); uterine perforation; haematometra, vesicovaginal fistula, fluid overload from irrigation fluid can cause ↑BP, ↓Na$^+$, pulmonary oedema, CNS symptoms, and haemolysis.[5] See p 14.

Laparoscopy Gas (to separate the viscera) is inserted through a small umbilical incision and the laparoscope and instrumentation are inserted suprapubically. This procedure allows visualization of the pelvic organs and is used for diagnosis of pelvic pain and ectopic pregnancy. The patient is spared a full laparotomy unless needed for treatment. A 'lap and dye' demonstrates tubal patency. Sterilization and hysterectomy may be laparoscopically carried out, and ectopic pregnancies sometimes treated.

Colporrhaphy or 'repair' The lack of support from the vaginal wall in cases of prolapse is rectified by excising a piece of the redundant mucosa and plicating levator ani. The operation may be combined with Manchester repair or vaginal hysterectomy. The more mucosa is removed, the tighter the vagina. Enquire before surgery if she is sexually active. Catheterisation circumvents post-operative retention of urine.

Cone biopsy A cone of tissue (point inwards) is cut out around the external cervical os, using knife or laser. This removes neoplastic tissue for histology, and may be curative. Complications: (immediate) bleeding, (long-term) cervical stenosis or incompetence.

Pelvic exenteration Consider this option when initial surgery fails to control neoplasia of the cervix, vulva, or vagina. It involves removal of the pelvic organs—ie ultra-radical surgery, which should only be contemplated if there is a chance of cure. Do your best to establish whether disease has spread to the pelvic sidewall or nodes, eg with MRI or CT scans and intra-operative biopsy with frozen section: if so, exenteration is probably not worthwhile. Only ~20% of possible candidates for surgery meet this criterion: in addition the patient should be quite fit, and ideally have a supportive partner. We know that palliative exenteration in those with unresectable disease is not worthwhile.[5] 5-yr survival: ~50%. Operative mortality: ~5%. Complications: GI obstruction/fistulae; urinary fistulae. Remember to give full pre-operative counselling about colostomies, and sexual function (refashioning of the vagina *may* be possible).

1 E Richardson 1995 *Lancet* 345 36   2 DA Alexander 1996 BMJ **i** 280   3 S Pinion 1994 *BMJ* **ii** 979
3 N Dwyer 1993 *Br J Obs Gynae* **100** 237   4 *Drug Ther Bul* 1994 **32** 70   5 N Saunders 1995 *Lancet* **345** 5

Chronic pelvic pain

This can be a cause of much misery to women of reproductive age. The history is usually of longstanding pelvic pain with secondary dysmenorrhoea and deep dyspareunia. The pain may cause, or be exacerbated by, emotional problems. She may be depressed.

Laparoscopy may reveal a likely cause: chronic pelvic infection, endometriosis, adenomyosis, adhesions or congested pelvic veins. If it does not (or if all gynaecological causes have been surgically removed) the cause may be gastrointestinal: consider irritable bowel syndrome (OHCM p 518).

Pelvic congestion Lax pelvic veins become painfully congested with blood. The pain is worse when she is standing, walking (gravity fills the veins), and premenstrually. It is typically variable in site and intensity and there is unpleasant post-coital ache. She is maximally tender to deep palpation over the ovaries. The vagina and cervix may appear blue due to congestion and there may be associated posterior leg varicosities. The dilated veins may be demonstrated by venography or laparoscopically.

Relief may be difficult, though explanation helps ('pelvic migraine'). Medroxyprogesterone 50mg/24h PO for 3 months reduces pain (SE amenorrhoea, weight gain, bloating); migraine remedies (OHCM p 416) and relaxation may be tried.[1] When symptoms are very severe bilateral ovarian vein ligation may cure,[2] as may hysterectomy with bilateral salpingo-oophorectomy with hormone replacement therapy post-operatively.[3]

Mittelschmerz this is mid cycle menstrual pain which may occur in teenagers and older women around the time of ovulation—from the German 'mittel' (=middle) and 'schmerz' (=pain).

Dyspareunia

This means pain during intercourse. There may be a vicious circle in which anticipation of pain leads to tense muscles and lack of lubrication, eyand so to further pain.

▶The patient may not volunteer the problem so ask about intercourse. Her attitude to pelvic examination may tell you as much as the examination itself. Ask her to show you where the problem is. If the problem is actually vaginismus do not insist on examination and consider counselling and sex therapy (p 380).

Dyspareunia may be superficial (around the introitus). This is often due to infection so look for ulceration and discharge. Is she dry? If so is the problem oestrogen deficiency (p 18) or lack of sexual stimulation? Has she had a recent post-partum perineal repair? A suture or scar can cause well-localized pain which is cured by removing the suture and injection of local anaesthetic. If the introitus has been rendered too narrow, she may need surgery.

Deep dyspareunia is felt internally (deep inside). It is associated with endometriosis and pelvic sepsis; treat the cause if possible. If the ovaries lie in the rectovaginal pouch, (or after hysterectomy), they may be subject to coital thrusts; try other positions or ventrosuspension if a 'cure' can be obtained with trial use of a Hodge pessary.

1 C Farquar 1989 *Br J Obs Gyn* **96** 1153 2 J Hobbs 1990 *Br J H Med* **43** 200 3 R Beard 1991 *Br J Obs Gyn* **98** 988

2. Obstetrics

*The term *pregnancy-induced hypertension with proteinuria* is tending to replace the older term *pre-eclampsia*. We have not followed this trend here because to do so obscures the vital fact about pre-eclampsia: it may lead on to eclampsia. We favour *pre-eclampsia* because the term is short and sends the shadow of a shiver down our spines, being a constant reminder of how dangerous it can be.

The Cochrane Childbirth & Pregnancy Database Obstetrics is as much an art as a science, but when questioning standard practices, it is likely that someone somewhere has researched the issue, and one good starting point is to consult this database disseminated by the Cochrane Centre.[1,2]

1 *Via* Update Software, Oxford, OX44 7QB 2 RJ Lilford 1994 *BMJ* i 1448

The essence of reproductive health

Pregnancy is a risky affair, not only for babies, but also for mothers. The textbook causes of maternal mortality are pulmonary embolism, eclampsia, haemorrhage, and infection—with all the other causes being rare. But if an obstetrician could be granted one wish, it would not be to abolish these; rather it would be to make every pregnancy *planned* and *desired by the mother*. World-wide, a woman dies every minute from the effects of pregnancy, and most of these women never wanted to be pregnant in the first place—but did not have to hand the means for contraception. So the real killers are poverty, ignorance, and the real solutions entail literacy and economic growth. Any obstetric or governmental initiatives in reproductive health which do not recognize these facts are doomed only to operate at the margins, and their good ideas are all too often vitiated by the reality operating in mothers' homes and workplaces.

Gravidity and parity—definitions

Gravidity refers to the number of pregnancies that a woman has had (to any stage). Parity refers to pregnancies that resulted in delivery beyond 28 weeks' gestation. An example of the short-hand way of expressing pregnancies before and after 28 weeks is: para 2+1. This means that she has had 2 pregnancies beyond 28 completed weeks' gestation, and 1 which terminated prior to 28 weeks. If she is not pregnant at the time of describing her she is gravida 3, but if she is pregnant now she is gravida 4. Twins present a problem as there is controversy as to whether they count as 1 for both parity and gravidity or should count as 2 for parity.

It is unclear whether the cut off point in the above definitions should now be 24 weeks, to harmonize with the new definition of stillbirth (p 136). In general, aim to use proper English rather than the short-hand described above—which is open to varying interpretations. For example when presenting a patient try something along these lines: 'Mrs Cottard is a 32-year-old lady who is 15 weeks into her 4th pregnancy; the third ended in a miscarriage at 17 weeks, and the others came to term with normal deliveries of children who are now 2 and 8.' The bald statement 'Para 2+1' is ambiguous, incomprehensible to the patient, and misses the point that the patient is now approaching the time when she lost her last baby.

Relevant pages in other chapters: Neonatology p 230–46; breast feeding p 178; rhesus haemolytic disease p 242; ectopic pregnancy p 24; abortion and termination of pregnancy p 20-2; trophoblastic disease p 26; examination of the neonate p 176; pre-term and light-for-dates babies p 180; varicella (chickenpox) in pregnancy p 216.

The placenta

Th placenta is the organ of respiration, nutrition and excretion for the fetus. It produces hormones for maternal wellbeing and immunologically protects the fetus by preventing rejection and allowing the passage of IgG antibodies from the mother.

Development Placental development is complete by 70 days after ovulation. At term the placenta weighs 1/7th the weight of the baby. It has a blood flow of 600ml/min. The placenta changes throughout pregnancy as calcium is deposited in the villi and fibrin on them. Excess fibrin may be deposited in diabetes and rhesus disease, so decreasing fetal nutrients.

Placental types *Battledore* insertion is where the umbilical cord inserts into the side of the placenta. *Velamentous* insertion (1%) is where the umbilical vessels pass within the membranes before insertion. If these vessels break (as in vasa praevia) it is fetal blood which is lost. *Placenta succenturia:* (5%) There is a separate (succenturiate) lobe away from the main placenta which may fail to separate normally and cause a PPH or puerperal sepsis. *Placenta membranacea* (1/3000) is a thin placenta all around the baby. As some is in the lower segment it predisposes to APH. It may fail to separate in the third stage of labour. *Placenta accreta:* There is abnormal adherence of all or part of the placenta to the uterus—termed *placenta increta* where there is placental infiltration of the myometrium or *placenta percreta* if penetration reaches the serosa. These latter 3 types predispose to PPH and may necessitate hysterectomy.

Placenta praevia The placenta lies in the lower uterine segment.

Associations: Large placenta (eg twins); uterine abnormalities and fibroids; uterine damage (eg multiparity, former surgery including Caesarean section). Ultrasound at <24 weeks' gestation shows a low-lying placenta in 28% but lower segment development later in pregnancy results in only 3% being low-lying at term.

Terminology: Major (old III and IV degrees) with placenta covering the internal os requires Caesarean section for delivery. Minor (old I and II) where the placenta is in the lower segment but not across the internal os: aim for normal delivery. Presentation may be as APH (separation of the placenta as the lower segment stretches causes bleeding) or as failure for the head to engage with high presenting part. Problems are with bleeding and with mode of delivery as the placenta obstructs the os and may shear off during labour, or may be accreta (5%), especially after previous Caesarean section (>24%).[1] Caesarean section should be consultant performed or supervised.[2]

Placental function tests The placenta produces progesterone, human chorionic gonadotrophin, and human placental lactogen. Human chorionic gonadotrophin is used for detection of early pregnancy. Human placental lactogen assay (previously used to try to monitor placental function), and urinary oestrogens assay (to monitor the fetoplacental unit) have fallen out of favour due to lack of sensitivity, and are now being replaced by biophysical tests (p 118).

1 S Clarke 1985 *Obs & Gynecol* **66** 89   2 DoH 1994 *Report on Confidential Enquiries into Maternal Deaths in the United Kingdom 1988-90* HMSO

Physiological changes in pregnancy

Hormonal changes *Progesterone*, synthesized by the corpus luteum until 35 post-conception days and by the placenta mainly thereafter, decreases smooth muscle excitability (uterus, gut, ureters) and raises body temperature. *Oestrogens* (90% oestriol) increase breast and nipple growth, water retention and protein synthesis. *Human placental lactogen* promotes growth hormone release and insulin secretion, but decreases insulin's peripheral effects, liberating maternal fatty acids (so sparing maternal glucose use). It also stimulates mammary growth and maternal casein, lactalbumin and lactoglobulin production. The maternal thyroid often enlarges due to increased colloid production. Thyroxine levels, see p 158. Pituitary secretion of *prolactin* rises throughout pregnancy. Maternal *cortisol* output is increased but unbound levels remain constant.

Genital changes The 100g non-pregnant uterus weighs 1100g by term. Muscle hyperplasia occurs up to 20 weeks, with stretching after that. The cervix may develop ectropion ('erosions'). Late in pregnancy cervical collagen reduces. Vaginal discharge increases due to cervical ectopy, cell desquamation, and ↑ mucus production from a vasocongested vagina.

Haemodynamic changes *Blood:* From 10 weeks the plasma volume rises until 32 weeks when it is 3.8 litres (50% >non-pregnant). Red cell volume rises from 1.4 litres when non-pregnant to 1.64 litres at term if iron supplements not taken (↑18%), or 1.8 litres at term (↑30%) if supplements are taken—hence Hb falls due to dilution ('physiological anaemia'). WCC (mean 10.5×10^9/l), platelets, ESR (up 4-fold), cholesterol, β-globulin and fibrinogen are raised. Albumin and gamma-globulin fall.

Cardiovascular: Cardiac output rises from 5 litres/min to 6.5–7 litres/min in the first 10 weeks by increasing stroke volume (10%) and pulse rate (by ~15 beats/min). Peripheral resistance falls (due to hormonal changes). BP, particularly diastolic, falls during the first second trimesters by 10–20mmHg, then rises to non-pregnant levels by term. With increased venous distensibility, and raised venous pressure (as occurs with any pelvic mass), varicose veins may form. Vasodilatation and hypotension stimulates renin and angiotensin release—an important feature of BP regulation in pregnancy.[1]

Other changes Ventilation increases 40% (tidal volume increases from 500–700ml), the increased depth of breath being a progesterone effect. Oxygen consumption increases only 20%. Breathlessness is common as maternal P_aCO_2 is set lower to allow the fetus to offload its CO_2. Gut motility is reduced resulting in constipation, delayed gastric emptying and with a lax cardiac sphincter, heartburn. Renal size increases by ~1cm in length during pregnancy.

Frequency of micturition emerges early (glomerular filtration rate↑ by 60%), later from bladder pressure by the fetal head. The bladder muscle is lax but residual urine after micturition is not normally present. Skin pigmentation (eg in linea nigra, nipples, or as chloasma—brown patches of pigmentation seen especially on the face), palmar erythema, spider naevi, and striae are common. Hair shedding from the head is reduced in pregnancy but the extra hairs are shed in the puerperium.

Pregnancy tests Positive from the 1st day of the 1st missed period, until ~20 weeks of pregnancy, they remain positive for ~5 days after abortion or fetal death. Otherwise, the false +ve rate is low. They detect the β subunit of human chorionic gonadotrophin in early morning urine, so are positive in trophoblastic disease (p 26).

---

1 P August 1995 *Lancet* 345 896

Plasma biochemistry in pregnancy[1,2]

Centile	Non-pregnant 2.5	97.5	Trimester 1 2.5	97.5	Trimester 2 2.5	97.5	Trimester 3 2.5	97.5
Na$^+$ mmol/l	138	146	135	141	132	140	133	141
Ca^{2+} mmol/l	2	2.6	2.3	2.5	2.2	2.2	2.2	2.5
*corrected	2.3	2.6	2.25	2.57	2.3	2.5	2.3	2.59
Albumin g/l	44	50	39	49	36	44	33	41

Other plasma reference intervals (not analysed by trimester)

	Non-pregnant	Pregnant
Alkaline phosphatase	3–300IU/l	up to 450IU/l **
Bicarbonate mmol/l	24–30	20–25
Creatinine μmol/l	70–150	24–68
Urea mmol/l	2.5–6.7	2–4.2
Uric acid μmol/l	150–390	100–270

*Calcium corrected for plasma albumin (see OHCM p 572).
**Occasionally very much higher in apparently normal pregnancies.[3]

C-reactive protein does not change much in pregnancy.
For thyroid changes, see p 157.

1 B Berg 1984 *Acta Obstet Gynecol Scand* 63 583–6   2 *OTM* 1987, OUP, page 29.2   3 AH Gowenlock 1988 Varley's *Practical Clinical Biochemistry* 1988, page 535, Heinemann, London

Normal labour

From 30 weeks the uterus has Braxton-Hicks contractions. These are non-painful 'practice' contractions to ~15mmHg pressure (in labour pressure is ~60mmHg) and are commonest after 36 weeks.

Normal labour is that occurring after 37 weeks' gestation. It should result in the spontaneous vaginal delivery of the baby within 24h of the onset of regular spontaneous contractions. It is often heralded by a 'show', ie a plug of cervical mucus and a little blood as the membranes strip from the os. The membranes may then rupture.

The first stage of labour is the time from the onset of regular contractions until the cervix is fully dilated (no cervix felt around the head). The cervix initially *effaces* (becomes shorter and softer) before it dilates. A satisfactory rate of dilatation from 3cm dilated is 1cm/h. The first stage generally takes up to 12h in a primip, and 7.5h in a multip. During the first stage check maternal pulse, BP and T° half-hourly; assess the contractions every 15min, their strength (you should not be able to indent the uterus with the fingers during a contraction) and their frequency (ideally 3-4 per 10min, lasting up to 1min). Carry out vaginal examination every 4h to assess the degree of cervical dilatation, the position and the station of the head (measured in cm above the ischial spines) and note the degree of moulding (p 86). Note the state of the liquor (see p 128). Test maternal urine 4-hourly for ketones and protein. If the mother becomes ketotic set up an IVI and give her 10% dextrose. Measure the fetal heart rate (if not being continuously monitored) every 15min. Note the rate before, during and immediately after a contraction.

The second stage is the time from complete cervical dilatation until the baby is born (see Movement of head in labour, p 86).
  The mother will have a desire to push and will use her abdominal muscles with the Valsalva manoeuvre to help move the baby. As the head descends, the perineum stretches and the anus gapes. Normal time for second stage is 45-120min in a primip, 15-45min in a multip. The aim is to prevent a precipitate delivery (and so intracranial bleeding) by pressure over the perineum.

The amount of delay before cutting the cord little affects term babies, but holding the baby 20cm below the introitus and delaying clamping for 30sec results in higher haematocrit levels, so reducing transfusion and oxygen supplement requirements in premature babies.[1]

The third stage is delivery of the placenta. As the uterus contracts to a 20-week size after the baby is born the placenta separates from the uterus through the spongy layer of the decidua basalis. It then buckles and a small amount of retroplacental haemorrhage aids its removal.
Signs of separation: Cord lengthens, rush of blood (retroplacental haemorrhage) from the vagina, uterus rises and contracts in the abdomen (felt with hand), and the uterus becomes more globular. Routine use of syntometrine (ergometrine maleate 500µg IM + oxytocin 5U IM) as the anterior shoulder of the baby is born has decreased third stage time (to ~5min), and has also decreased the incidence of PPH, but may cause problems for undiagnosed twins. Examine the placenta to check it is complete.

1 S Kimmond 1993 *BMJ* i 172

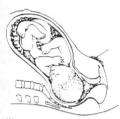

(1)
1st stage of labour. The cervix dilates. After full dilatation the head flexes further and descends further into the pelvis.

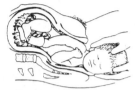

(4)
Birth of the anterior shoulder. The shoulders rotate to lie in the anteroposterior diameter of the pelvic outlet. The head rotates externally, 'restitutes', to its direction at onset of labour. Downward and backward traction of the head by the birth attendant aids delivery of the anterior shoulder.

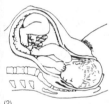

(2)
During the early second stage the head rotates at the level of the ischial spine so the occiput lies in the anterior part of pelvis. In late second stage the head broaches the vulval ring (crowning) and the perineum stretches over the head.

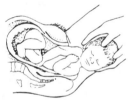

(5)
Birth of the posterior shoulder is aided by lifting the head upwards whilst maintaining traction.

(3)
The head is born. The shoulders still lie transversely in the midpelvis.

Abdominal palpation

The uterus occupies the pelvis and cannot be felt *per abdomen* until 12 weeks' gestation. By 16 weeks it lies about half way between the symphysis pubis and the umbilicus. By 20–22 weeks it has reached the umbilicus. In a primigravida, the fundus lies under the ribs by 36 weeks. At term the uterus tends to lie a little lower than at 36 weeks due to the head descending into the pelvis. Some attendants prefer to measure the symphysis fundal height (SFH) in cm from the symphysis pubis. From 16 weeks the SFH increases ~1cm/week. As a guide, the SFH is the gestation in weeks minus 3. It may be used as a rough guide to find babies small for gestational age (p 104).

On inspecting the abdomen note any scars from previous operations. Caesarean section scars are usually Pfannenstiel ('bikini-line'), although occasionally in very fat women they lie vertically. Laparoscopy scars are just below and parallel to the umbilicus. It is common to see a line of pigmentation, the linea nigra, extending in the midline from pubic hair to umbilicus. This darkens during the first trimester (the first 13 weeks).

Palpating the abdomen Measure the SFH and listen to the *fetal heart*. After 32 weeks palpate laterally to assess the lie, then bimanually palpate over the lower uterine pole for presentation and degree of engagement. Pawlik's grip (examining the lower pole of the uterus between the thumb and index fingers of the right hand) can also be used for assessing the degree of engagement. Watch the patient's face during palpation and stop it if it causes pain. Obesity, polyhydramnios and tense muscles make it difficult to feel the fetus. Midwives are skilled at palpation, so ask them if you need help.

It is important to determine the *number of fetuses* (see p 122), the *lie* (longitudinal, oblique, or transverse), the *presentation* (cephalic or breech), and the *engagement*. Note the amount of liquor present, the apparent size of the fetus, and any contractions or fetal movements seen or felt.

Auscultation The fetal heart may be heard by Doppler ultrasound (Sonicaid) from ~12 weeks and with a Pinard stethoscope from ~24 weeks.

Engagement The level of the head is assessed in 2 ways—engagement, or fifths palpable abdominally. Engagement entails passage of the maximum presenting diameter through the pelvic inlet. Fifths palpable abdominally states what you can feel, and makes no degree of judgement on degree of engagement of the head. In primigravida the head is usually entering the pelvis by 37 weeks, otherwise causes must be excluded (eg placenta praevia or fetal abnormality). In multips the head may not enter the pelvis until the onset of labour.

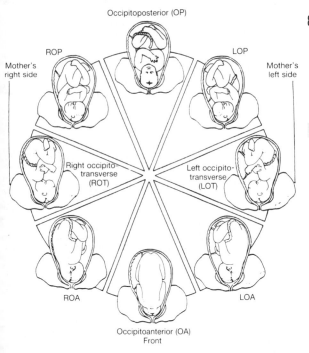

Occipitoposterior (OP)

ROP

LOP

Mother's right side

Mother's left side

Right occipito-transverse (ROT)

Left occipito-transverse (LOT)

ROA

LOA

Occipitoanterior (OA)
Front

Position

Occipitoanterior	*Occipitolateral*	*Occipitoposterior*
Back easily felt	Back can be felt	Back not felt
Limbs not easily felt	Limbs lateral	Limbs anterior
Shoulder lies 2cm from midline on opposite side from back	Midline shoulder	Shoulder 6–8cm lateral, same side as back
Back from midline=2–3cm	6–8cm	≥10cm

►The fetal heart is best heard over its back (left scapula).

Pelvis and head

The ideal pelvis This has a round brim, a shallow cavity, non-prominent ischial spines, a curved sacrum with large sciatic notches and sacrospinous ligaments >3.5cm long. The angle of the brim is 55° to the horizontal, the AP diameter at least 12cm and transverse diameter at least 13.5cm. The subpubic arch should be rounded and the intertuberous distance at least 10cm.

The true pelvis Anteriorly there is the symphysis pubis (3.5cm long) and posteriorly the sacrum (12cm long).

Zone of inlet Boundaries: Anteriorly lies the upper border of the pubis, posteriorly the sacral promontory, laterally the ileopectineal line. Transverse diameter 13.5cm; AP diameter 11.5cm.

Zone of cavity This is the most roomy zone. It is almost round. Transverse diameter 13.5cm; AP diameter 12.5cm.

Zone of mid-pelvis Boundaries: Anteriorly is the apex of the pubic arch, posteriorly the tip of the sacrum, laterally the ischial spines (the desirable distance between the spines is >10.5cm). Ovoid in shape, it is the narrowest part.

Zone of outlet The pubic arch is the anterior border (desirable angle >85°). Laterally lie the sacrotuberous ligaments and ischial tuberosities, posteriorly the coccyx.

The head: terms The *bregma* is the anterior fontanelle. The *brow* lies between the bregma and the root of the nose. The *face* lies below the root of the nose and supraorbital ridges. The *occiput* lies behind the posterior fontanelle. The *vertex* is the area between the two fontanelles and the two parietal eminences.

Moulding The frontal bones can slip under the parietal bones which can slip under the occipital bone so reducing biparietal diameter. The degree of overlap may be assessed vaginally.

Presentation:	Relevant diameter presenting:	
Flexed vertex	suboccipitobregmatic	9.5cm
Partially deflexed vertex	suboccipitofrontal	10.5cm
Deflexed vertex	occipitofrontal	11.5cm
Brow	mentovertical	13cm
Face	submentobregmatic	9.5cm

Movement of the head in labour (normal vertex presentation)
1 Descent with increased flexion as the head enters the cavity. The sagittal suture lies in the transverse diameter of the brim.
2 Internal rotation occurs at the level of the ischial spines due to the grooved gutter of the levator muscles. Head flexion increases. (The head rotates 90° if occipitolateral position, 45° if occipitoanterior, 135° if occipitoposterior.)
3 Disengagement by extension as the head comes out of the vulva.
4 Restitution: as the shoulders are rotated by the levators until the bisacromial diameter is anteroposterior, the head externally rotates the same amount as before but in opposite direction.
5 Delivery of anterior shoulder by lateral flexion of trunk posteriorly.
6 Delivery of posterior shoulder by lateral flexion of trunk anteriorly.
7 Delivery of buttocks and legs.

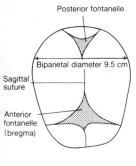

Posterior fontanelle

Biparietal diameter 9.5 cm

Sagittal
suture

Anterior
fontanelle
(bregma)

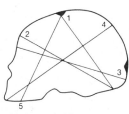

1 Suboccipitobregmatic 9.5 cm
  flexed vertex presentation
2 Suboccipitofrontal 10.5 cm
  partially deflexed vertex
3 Occipitofrontal 11.5 cm deflexed vertex
4 Mentovertical 13 cm brow
5 Submentobregmatic 9.5 cm face

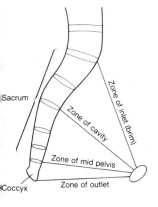

Sacrum

Coccyx

Zone of inlet (brim)

Zone of cavity

Zone of mid pelvis

Zone of outlet

Dystocia

Dystocia is difficulty in labour. There may be problems with *the passenger* (large baby, see *shoulder dystocia*, p 128, abnormal presentation), *the passage* (for ideal pelvis see p 86) or of *propulsion* (the uterine powers). Cephalo-pelvic disproportion results if diameters are unfavourable (p 86).

The pelvis The ideal pelvis has a round brim (ie gynaecoid), but 15% of women have a long oval brim (anthropoid). A very flat brim is less favourable (platypoid); occurring in 5% of women over 152cm (5ft), it occurs in 30% of women <152cm. Spinal scoliosis, kyphosis, sacralization of the L5 vertebra, spondylolisthesis and pelvic fractures may all affect pelvic anatomy. Rickets and polio were formerly important causes of pelvic problems. Suspect pelvic contraction if the head is not engaged by 37 weeks in a Caucasian primip.

The presentation Cephalic presentations are less favourable the less flexed the head. Transverse lie and brow presentations will always need Caesarean section: face and OP (p 126) presentations may deliver vaginally but are more likely to fail to progress. Breech presentation is particularly unfavourable if fetus >3.5kg.

The uterine powers Contractions start in the fundus and propagate downwards. The intensity and duration of contractions are greatest at the fundus, but the contraction reaches its peak in all parts of the uterus simultaneously. Normal contractions occur at a rate of 3 per 10min, they should last up to 75sec, the contraction peak usually measures 30-60mmHg, and the resting uterine tone between them should be 10-15mmHg. Uterine muscle has the property of retraction. The shortening of the muscle fibres encourages cervical dilatation.

Uterine dysfunction Contractions may be hypotonic (low resting tone, low contraction peaks) or they may be normotonic but occur too infrequently. These dysfunctions can be corrected by augmentation with oxytocin (p 112). Whenever oxytocin is used discuss with a senior obstetrician. Pain and fear cause release of catecholamines which can inhibit uterine activity. Thus adequate analgesia is needed (p 114) and may speed the progress of labour.

Cervical dystocia Failure of cervical dilatation may be due to previous trauma, repair, cone biopsy and cauterization. It is difficult to distinguish from failure to dilate due to uterine dysfunction though the latter should respond to syntocinon (note the important difference between primips and multips, p 112). The treatment for cervical dystocia is delivery by Caesarean section.

Consequences of prolonged labour Neonatal mortality rises with prolonged labour as does maternal morbidity (especially infection). With modern management of labour, careful monitoring of progress in labour (p 112) takes place to diagnose delay early, and treat it as necessary, to prevent prolonged labour occurring.

►When there is dystocia, ask 'is safe vaginal delivery possible?'

Minor symptoms of pregnancy

▶Before prescribing any drug, think—Is it necessary. Is it safe?

Symptoms and signs in the first 10 weeks: Early symptoms are amenorrhoea, nausea, vomiting and bladder irritability. Breasts engorge, nipples enlarge (darken at 12 weeks), Montgomery's tubercles (sebaceous glands on nipples) become prominent. Vulval vascularity increases and the cervix softens and looks bluish (4 weeks). At 6-10 weeks the uterine body is more globular. Temperature rises (<37.8°C).

Headaches, palpitations, and fainting are all more common during pregnancy. Sweating and feeling hot are also common, and are due to a more dilated peripheral circulation. *Management:* Increase fluid intake and take plenty of showers.

Urinary frequency is due to pressure of the fetal head on the bladder in later pregnancy. Exclude UTI.

Abdominal pain: See p 135. *Breathlessness* is common. See p 80.

Constipation tends to occur as gut motility decreases. Adequate oral fluids and a high fibre diet help combat it. Avoid stimulant laxatives—they increase uterine activity in some women. Increased venous distensibility and pelvic congestion predispose to *haemorrhoids* (if they prolapse, rest the mother head down, apply ice packs and replace them), and *varicose veins.* Resting with feet up, and properly worn elastic stockings help.

Reflux oesophagitis and heartburn occur as pyloric sphincter relaxation allows irritant bile to reflux into the stomach. Cigarettes and spices should be avoided, small meals taken, and antacids used.

3rd trimester backache: Due to pelvic ligament and muscle relaxation, pain tends to be worse at night. A firm mattress, flat shoes, standing with back straight and pelvic support from physiotherapy all help.

Carpal tunnel syndrome (p 662) in pregnancy is due to fluid retention. Advise wrist splints until delivery cures the problem.

Itch/itchy rashes are common (up to 25%) and may be due to the usual causes (OHCM p 48, check LFTs—see p 158) or to pruritic eruption of pregnancy (PEP = *prurigo of pregnancy*)—an intensely itchy papular/plaque rash on the abdomen and limbs. PEP is most common in 1st pregnancies beyond 35 weeks' gestation. Emollients and weak topical ease it. Delivery cures it. If vesicles are present, think of *pemphigoid gestationis* (PG): a rare (1:50,000) condition which may cause fatal heat loss and cardiac failure; the baby may be briefly affected; refer early (prednisolone may be needed). PG recurs in later pregnancies, eg if there is a different father.

Ankle oedema: This is very common, almost normal manifestation of pregnancy. Measure BP and check urine for protein (pre-eclampsia, p 96). Check legs for DVT. It often responds to rest and leg elevation. Reassure that it is harmless (unless pre-eclampsia).

Leg cramps 33% gets cramp the latter ½ of pregnancy and are severe in 5%, often worse at night. Raising the foot of the bed by 20cm will help.

Nausea affects ~80%. *Vomiting* occurs in ~50%. It may start by 4 weeks and declines over the next weeks. At 20 weeks 20% may still vomit.[1] Most respond to frequent small meals, reassurance, and a stress-free environment. It is associated with good outcome (fewer fetal losses).

1 S Whitehead 1992 *J Obstet Gynae* 12 364

Pre-pregnancy counselling

The aim of pre-pregnancy care is to help parents embark upon pregnancy under conditions most likely to ensure optimal wellbeing for the fetus. At

its simplest level this means ensuring that a woman is rubella immune prior to pregnancy, but areas covered include:

Pre-pregnancy counselling

The aim of pre-pregnancy care is to help parents embark upon pregnancy under conditions most likely to ensure optimal wellbeing for the fetus. At its simplest level this means ensuring that a woman is rubella immune prior to pregnancy, but areas covered include:

- ●Optimal control of chronic disease (eg diabetes) before conception.
- ●Discontinuation of teratogens prior to conception (p 159).
- ●Medication to protect the fetus from abnormality (eg folate supplements for neural tube defects, p 228 and below).
- ●Provision of expert information for those known to be at risk of abnormality so pregnancy or its avoidance is an informed choice and management of pregnancy with respect to any desirable tests (eg chorionic villus sampling p 166) is planned. Regional genetic services give detailed pre-pregnancy counselling for a multitude of genetic conditions (p 210-2).

For all women advice concerning the following may be relevant:

Diet To prevent neural tube defects and cleft lip all should have folate-rich foods and folic acid 0.4mg daily (eg Preconceive®) in the months before conception—until 13 weeks' gestation[1] (5mg/day PO if history of neural tube defect). These foods have >0.1mg of folic acid per serving: brussels sprouts, asparagus, spinach, kale, blackeye beans, fortified breakfast cereals. Avoid liver and vitamin A (vitamin A embryopathy risk).[2] Other vitamins at the time of conception may help prevent malformation.[3]

Cigarette smoking may decrease ovulations, result in abnormal sperm production, reduce sperm penetrating capacity, ↑rates of miscarriage (2-fold), and is associated with preterm labour and lighter-for-dates babies (mean is 3376g in non-smoker; smoker: 3200g). Reduced reading ability in smokers' children up to 11yrs shows that longer term effects may be important. About 17% of smoking mothers stop before or in pregnancy.

Alcohol consumption High levels of consumption are known to cause the fetal alcohol syndrome (p 226). Moderate drinking has not been shown to adversely affect the fetus but alcohol does cross the placenta and may affect the fetal brain. Miscarriage rates are higher amongst drinkers of alcohol. How to cut consumption: see p 454.

Spontaneous abortion (SA) 12% of 1st pregnancies spontaneously abort. Rates after 1 SA are increased to ~24%; after 2 SA to ~36%; after 3 to ~32% and after 4 to ~25%—so chances of a future pregnancy succeeding are about 2 in 3 Pregnancy order of SA/live pregnancies is also relevant: the more recent a live birth the more likely next time will be successful.

Recurrent spontaneous abortion Defined as ≥3 spontaneous recurrent abortions. Causes include maternal anatomical abnormality, maternal infection, genetic abnormality, maternal disease (debilitating disease diabetes and thyroid problems, lupus anticoagulant present), and luteal deficiency (for abortions within the first 8 weeks). A potential cause may be found in ~50% and many of these are treatable.

Pre-pregnancy counselling is best done out by those who will look after the pregnant woman, but those with chronic medical conditions may require liaison between obstetrician and physician. In computerized practices it may be worth looking for likely beneficiaries of pre-pregnancy counselling by carefully perusing lists generated for these conditions:

●Diabetes mellitus	●Epilepsy	●BP↑	*Genetic history eg.*
●Tropical travelers	●Rubella-susceptible	●SLE	●Spinabifida etc
●Frequent abortion	●Pet-owners (toxoplasmosis)	●PKU	●Thalassaemia
			●Duchenne's
			●Cystic fibrosis

1 DoH 1992 Letter from Chief Medical Officer PL/CMO(92) 18 and G Shaw 1995 *Lancet* 346 393
2 DoH 1990 Letter from Chief Medical Officer PL/CMO(90)10   3 A Czeizel 1993 *BMJ* i 645

Antenatal care

The aims of antenatal care are to: ●Detect any disease in the mother ●Ameliorate the discomforts of pregnancy ●Monitor and promote fetal wellbeing ●Prepare mothers for birth ●Monitor trends to prevent or detect early any complications of pregnancy: BP is the most important variable (eclampsia, p 96). Whenever the utility of other variables (weight, fundal height, urine analysis) is scrutinized, they are found to be insensitive, and not to fulfil the criteria for a good screening test (p 430).[1]

Who should give antenatal care? The answer is whoever will get on best with the woman concerned. It is not true that *all* women *must* see a doctor at least once per pregnancy or labour: midwives do very well on their own (with fewer interventions), calling in doctors *only if a specific need arises*.[2]

The first antenatal visit is very comprehensive. ►Find an interpreter if she does not speak your language. *History:*
●Usual cycle length; LMP (a normal period?); expected delivery date ≈ 1yr and 7 days after LMP minus 3 months—unreliable if her last period was a withdrawal bleed; to allow for a cycle which is, say, 2 days *shorter* than 28 days, subtract eg 2; add 2 or whatever the cycle is *long* by, if it is long.
●Contraception; drugs; past history, eg surgery to the abdomen or pelvis.
●Fertility problems; outcome and complications of other pregnancies.
●Is there family history of diabetes, BP↑, fetal abnormality or twins?
●Is she poor (eg gas/electricity supply cut off)? Unmarried? Unsupported?
●Is the diet adequate? Advise on avoiding pâtés and soft cheese (OHCM p 222) and vitamin overdose, eg no vitamin tablets or liver (p 94).

Examination: Check heart, lungs, breasts, nipples, BP, and abdomen. Consider pelvic examination (unless dates uncertain (unless dates will be confirmed later by scan), or if a cervical smear is needed. Any varicose veins?

Tests: Blood: Hb, group (antibodies if Rh-ve, p 242), syphilis and rubella serology, sickle test if black, α-fetoprotein/triple test (p 228 & 167), Hb electrophoresis + HBsAg (p 154 & 100) if relevant. Offer HIV test, with counselling (OHCM p 213). MSU for protein, glucose and bacteria. If she is foreign or is a TB contact or is a hospital worker, consider CXR after 14 weeks.

Ultrasound: in many units an ultrasound scan is routine (and an event for the parents) at 18 weeks, to confirm dates, look for fetal abnormalities and exclude ovarian cysts. If the mother is over 35yrs consider amniocentesis at 16–18 weeks to exclude Down's syndrome (p 166).

Suggest: Parentcraft/relaxation classes; dental visit. Enquire about problems and anxieties. Consider need for *iron and folate* (p 154, 94).

Advise on: Smoking, alcohol, a healthy diet (OHCM p 482), and taking adequate rest. Ensure knowledge of social security benefits. Accustomed exercise and travel are OK (not to malarious areas, p 159) up to 34 weeks on most airlines. Intercourse is fine if there is no vaginal bleeding.

Later visits Check weight, urine for albumin and glucose, BP, fundal height, fetal movements (primips feel them ~19 weeks, earlier if multip), fetal heart—and lie and presentation from 32 weeks. Do Hb, eg at 28 & 36 weeks; look for Rh antibodies. Visits are monthly from 16 to 32 weeks, every 2 weeks until 36 weeks; then weekly. There is evidence that this is *too* often for most tests, but not often enough for BP and albuminuria.[3]

The head is usually engaged (p 84) by 37 weeks in Caucasian primips (if not, consider: large (or malpositioned) head, small pelvis or obstruction, placenta praevia, or wrong estimation of dates).

At 36 weeks some check the pelvis clinically in primips. The pelvis is favourable if the sacral promontory cannot be felt, the ischial spines are not prominent, the suprapubic arch and base of supraspinous ligaments both accept 2 fingers and the intertuberous diameter accepts 4 knuckles.

1 P Steer 1993 *BMJ* ii 697   2 MJ Rowley 1996 *E-BM* i 84   3 DG Daniel 1993 *BMJ* ii 1214

Hypertension in pregnancy

Essential hypertension Commoner in older multips, and present before pregnancy. You may need to modify or start therapy to keep BP <140/90mmHg. Severity and any association with pre-eclampsia (eg BP rise ≥+30/+15mmHg from baseline, progressive hyperuricaemia, proteinuria, or clotting activation) determines fetal outcome. Those with essential hypertension are 5× more likely to develop pre-eclampsia than normotensive woman. ►*When symptoms are episodic, think of phaeochromocytoma.*

Pre-eclampsia Terminology: see p 76. This is pregnancy induced hypertension (PIH) with proteinuria ± oedema. It is a multisystem disorder originating in the placenta. The primary defect is failure of trophoblastic invasion of spiral arteries, so the arterial wall does not distend enough to allow sufficient blood flow to the placenta in late pregnancy, and increasing BP is a mechanism which partially compensates for this.[1] Pre-eclampsia also affects hepatic, renal and coagulation systems. It develops after 20 weeks and usually resolves within 10 days of delivery. Eclampsia (1 in 2000 maternities) is a major cause of maternal death and fetal morbidity/mortality. Pre-eclampsia is asymptomatic, so frequent screening is vital. It may recur in a subsequent pregnancy.

Predisposing factors *Maternal:* Primiparity; previous severe pre-eclampsia, family history of pre-eclampsia or eclampsia; short stature (<155cm) or overweight; age <20 or >35yrs; pre-existing migraine, BP↑ or renal disease increase risk. The incidence is lower in smokers. *Fetal:* Hydatidiform mole (↑BP at 20 weeks); multiple pregnancy; placental hydrops (eg rhesus disease). All have increased placental bulk.

Effects of pre-eclampsia Plasma volume is decreased. There is increased peripheral resistance. There may be placental ischaemia. If the BP is >180/140mmHg microaneurysms develop in arteries. DIC may develop. Oedema may develop suddenly (eg sudden weight gain of 1kg). Proteinuria is a late sign, indicating renal involvement which may be detected earlier by monitoring uric acid levels (levels >0.29μmol/l at 28 weeks, >0.34μmol/l at 32 weeks, >0.39μmol/l at 36 weeks suggest pre-eclampsia). Initially glomerular filtration is normal and only serum uric acid is raised but later urea and creatinine both increase. The liver may be involved (contributing to DIC). The placenta may develop infarcts. Fetal asphyxia, abruption, and small babies (p 104 & p 119) may also occur.

Symptomatic pre-eclampsia It may mimic a viral illness with headache, chest pain, epigastric pain, vomiting and visual disturbance. There may be tachycardia, shaking, hyperreflexia, and irritability. The mother is now in danger of generalized seizures (eclampsia) and urgent admission and treatment must occur. Death may be imminent from stroke (commonest), hepatic, renal or cardiac failure.

Management: early phase of pre-eclampsia Admit mothers if the BP rises by >30/20mmHg over their booking BP; if BP reaches 160/100; if ≥140/90 with proteinuria, or if there is growth retardation. In hospital, measure BP 2-4-hourly, weigh daily, test all urine for protein, monitor fluid balance, check uric acid, renal, liver and placental function, and platelets (beware if falls <110 × 10⁹/l)[2] regularly, and do regular antenatal cardiotocography with ultrasound to check growth, and biophysical profiles. All women with pre-eclampsia should be given H₂-blockers when they go into labour.[4] In asymptomatic pre-eclampsia, raised BP may be treated with drugs, eg methyldopa[3] 250mg-1g/6h under supervision in hospital if BP rises to 170/110mmHg, in order to buy time for fetal maturation, and if all other variables are satisfactory. If signs worsen deliver the baby (liaise with paediatricians). Delivery is the only cure.

Managing impending eclampsia It is almost impossible to foretell eclamptic convulsions—so focus on ensuring that the patient is in a specialist unit, and, if possible, delivered before this stage is reached. If BP needs reducing, hydralazine 10mg IM/IV slowly may be given. (Chewing a 10mg nifedipine capsule may be a fast-acting alternative.) Close and repeated monitoring of BP is essential, and hydralazine can be repeated in boluses of 5mg IV or by continuous IVI (50mg in 500ml 5% dextrose gives a 1mg/10ml solution—give at 1-3mg/h). Labetalol is an alternative to hydralazine. Dose: 50-100mg IV every 20-30min or 10-160mg/h as IV infusion.[3]

Whatever the hour, summon your consultant (not just the registrar, and do not make do with telephone advice).[2] ►*Do not leave the bedside.*

Catheterize and watch urine output. If the mother fits, protect the airway and give magnesium sulphate, as below.

Eclampsia and the use of magnesium sulphate ►The big randomized Collaborative Eclampsia Trial[5] clearly showed magnesium sulphate to be better than diazepam *once a seizure has occurred,* making further seizures much less likely. Maternal mortality was non-significantly lower with magnesium (17 deaths out of 453 *vs* 23 diazepam deaths out of 452). One régime is 4g IVI over 5 mins, then 5g IM into each buttock per 4h (provided respiratory rate is >16/min, urine output >25ml/h, and knee jerks present), continued for 24h.[5] Diazepam may still have a rôle in suppressing convulsions that do not spontaneously remit, or those recurring repeatedly at short intervals. 5-10mg can be given by *slowly* IV, repeated every 15-20mins to keep the patient drowsy but rousable. Once therapeutic levels of magnesium have been achieved, diazepam should not be needed.

In status epilepticus, take to ITU to paralyse and ventilate.

Pitfalls in the management of eclampsia and pre-eclampsia
- Believing that the disease will behave predictably, and that blood pressure is a good marker of disease activity. It is not.
- Ignoring mild proteinuria (eg 1+). Once this is present in every sample the patient may be dead within 24h.
- Believing that antihypertensive treatment stops the progression of pre-eclampsia. It does not. Delivery is the only cure. Diuretics may further deplete the already reduced plasma volume, and are especially contraindicated (except in the rare event of left ventricular failure or laryngeal oedema complicating pre-eclampsia).
- Believing that delivery removes risk. In the UK, half of eclamptic fits occur post-partum of which half are >48h post-partum. Continue vigilance until clinically and biochemically normal. Avoid early discharge.
- Ergometrine should not be used for the third stage (syntocinon may be used). Ergometrine further increases BP so would risk stroke.
- Significant blood loss should be rapidly replaced by blood as the mother is especially susceptible to hypovolaemia.

1 DA Clark 1994 *Lancet* 344 969   2 DoH 1989 *Confidential Enquiry into Maternal Deaths*, p94, HMSO
3 1993 *Drug Ther Bul* 31 53   4 DoH 1994 *Confidential Enquiry into Maternal Deaths*, p95, HMSO
5 Collaborative Eclampsia Trial 1995 *Lancet* 345 1455 & EBM 1996 1(2) 39,44   6 R Fox 1995 *BMJ* ii 1433

Antenatal infection

A variety of maternal infections, mostly viral, may affect the fetus. These were known as the 'TORCH' infections—Toxoplasmosis, Other (eg syphilis), Rubella, Cytomegalovirus (CMV), Herpes (and hepatitis). The first 5 are acquired in the antenatal period; herpes and hepatitis are usually acquired perinatally. The term TORCH is now obsolete as other agents are known to be equally important (HIV; chickenpox, p 216).

Rubella Vaccination aims to prevent rubella-susceptibility.
►Asymptomatic reinfection can occur making serology (as below) essential in all pregnant rubella contacts. Routine antenatal screening detects those requiring puerperal vaccination (avoid pregnancy for 3 months: the vaccine is live). Symptoms (p 214) are absent in 50% of mothers. The fetus is most at risk in the first 16 weeks' gestation. 50-60% of fetuses are affected if maternal primary infection is in the 1st month of gestation: <5% are affected if infection is at 16 weeks. Risk of fetal damage is much lower (<5%) with re-infection.[1] Cataract is associated with infection at 8-9 weeks, deafness at 5-7 weeks (can occur with 2nd trimester infection), cardiac lesions at 5-10 weeks. Other features: purpura, jaundice, hepatosplenomegaly, thrombocytopenia, cerebral palsy, microcephaly, mental retardation, cerebral calcification, microphthalmia, retinitis, growth disorder. Abortion or stillbirth may occur. If suspected in the mother seek expert help. Compare blood antibody levels taken 10 days apart and look for IgM antibody at 4-5 weeks from incubation period.

Cytomegalovirus (CMV) In the UK, CMV is a commoner cause of congenital retardation than rubella. Maternal infection is mild or asymptomatic. The fetus is most vulnerable in early pregnancy. Up to 5/1000 live births are infected, of whom 5% will develop early multiple handicaps, and have cytomegalic inclusion disease (with non-specific features resembling rubella syndrome, plus choroidoretinitis). Another 5% develop later handicap. Reducing women's exposure to toddlers' urine (the source of much infection) during pregnancy can limit spread.

Toxoplasmosis 40% of fetuses are affected if the mother has the illness (2-7/1000 pregnancies); the earlier in pregnancy the more the damage. Symptoms are similar to CMV. Fever, rash, and eosinophilia also occur. If symptomatic, the CNS prognosis is poor. All affected babies (diagnose by serology) should have pyrimethamine, 21 days of 0.25mg/kg/6h PO + sulphadiazine 50mg/kg/12h PO with folic acid supplements to protect against pyrimethamine's folate antagonism. ►Get expert help in pregnancy to consider spiramycin ± sampling of fetal cord blood, eg at 21 weeks for IgM (this may indicate serious fetal infection).

HIV Babies have a 15% chance of acquiring HIV if the mother is +ve (higher risk in Africa; but HIV-2 is less likely than HIV-1 to be passed on). Offer mothers tests for HIV, only after counselling (OHCM p 212). Serology depends on detecting IgG and IgM antibodies (false +ve in up to 1.6%). Those with +ve serology should not be given steroids to accelerate fetal lung maturity. ►Vertical transmission can occur during vaginal delivery (first twins are twice as commonly infected as 2nd), and Caesarean section is protective, possibly cutting neonatal infection by almost half[2]—as does zidovudine 100mg 5 times a day PO antepartum, and, during labour (2mg/kg IVI over 1h, and then 1mg/kg IVI until delivery), and then 2m/kg/6h PO for the baby for 6 weeks. Bottle feeding reduces vertical transmission—possibly up to 50% (but bottle feeding is hazardous in some communities).

Serology is difficult as transferred maternal antibodies persist up to 18 months in uninfected infants—but gene amplification via a polymerase chain reaction can detect neonatal HIV proviral sequences. Culture is also possible—but not all infected children give +ve cultures at birth, and specialized investigations may need to be performed, eg the detection of IgA (which does not cross the placenta) by modified Western blotting.[1]

Intrauterine HIV infection is associated with prematurity and growth retardation (in the Third World, but not in Europe). Clinical illness appears sooner than in adult AIDS (eg at aged 6 months)—with hepatosplenomegaly, failure to thrive, encephalopathy, recurrent fevers, respiratory diseases (interstitial lymphocytic pneumonitis), lymphadenopathy, salmonella septicaemia, pneumocystis and CMV infection. Death is usually from respiratory failure or overwhelming infection. Mortality: 20% at 18 months.[2]
▶For tips on how to prevent transmission in labour, see p 155

Intrauterine syphilis Routine maternal screening occurs; if active infection is found treat the mother with procaine penicillin, eg as half a 1.8g ampoule of Bicillin®, IM daily for 10 days. Neonatal signs: rhinitis, snuffles, rash, hepatosplenomegaly, lymphadenopathy, anaemia, jaundice, ascites, hydrops, nephrosis, meningitis. Nasal discharge: examine for spirochaetes; X-rays: perichondritis; CSF: increased monocytes and protein with +ve serology. Give procaine penicillin 37mg/kg/24h IM for 3 weeks

Listeria This affects 6–15/100,000 pregnancies—more during epidemics. Maternal symptoms: Fever, shivering, myalgia, headache, sore throat, cough, vomiting, diarrhoea, vaginitis. Abortion (can be recurrent), premature labour, and stillbirth may occur. Infection is usually via infected food (eg milk, soft cheeses, pâté). ▶Do blood cultures in any pregnant patient with unexplained fever for ≥48h. Serology, vaginal and rectal swabs do not help (it may be commensal there). See OHCM p 223.
   Perinatal infection usually occurs in the 2nd or 3rd trimester. 20% of affected fetuses are stillborn. Fetal distress in labour is common. An early postnatal feature is respiratory distress from pneumonia. There may be convulsions, hepatosplenomegaly, pustular or petechial rashes, conjunctivitis, fever, leucopenia. Meningitis presents more commonly after perinatal infection. Diagnosis is from culture of blood, CSF, meconium, and placenta. Infant mortality: 30%. Isolate baby (nosocomial spread can occur). Treat for 2 weeks with amoxycillin and gentamicin.

1 GL Gilbert 1991 *Infectious Diseases in Pregnancy and the Newborn Infant* Harwood, Switzerland
2 European Collaborative Study 1994 (ML Newell) *Lancet* **343** 1464

Perinatal infection

Hepatitis B virus (HBV) Although chronic carriers were uncommon in the UK, with the rise of drug addiction and a large immigrant population it is becoming more frequent, and some are now recommend that all mothers be screened. If the mother develops acute infection in the mid- or third trimester there is a high risk of perinatal infection. Most infection is, however, contracted at birth and the infants of known carriers should be given hepatitis immune globulin 0.5ml IM within 12h of birth, and hepatitis B vaccine 0.5ml within 7 days of birth and at 1 and 6 months.

Herpes hominis 80% is due to herpes simplex type II virus. 50% of babies will be infected during birth if the mother has active cervical lesions. Take viral cultures weekly (from 36 weeks) from the cervix of all women with a history of herpes. If virus is present, aim for elective Caesarean section. When membranes rupture spontaneously still perform Caesarean if within 4h of rupture. Neonatal infection usually appears at 5-21 days with vesicular pustular lesions, often at the presenting part or sites of minor trauma (eg scalp electrode). Periocular and conjunctival lesions may occur. With systemic infection encephalitis (focal fits or neurological signs), jaundice, hepatosplenomegaly, collapse and DIC may occur. Infected neonates should be isolated and treated with acyclovir (p 256). Seek expert help.

Chlamydia trachomatis Associations: low birth weight, premature membrane rupture, fetal death. ~⅓ of infected mothers have affected babies. Conjunctivitis develops 5-14 days after birth and may show minimal inflammation or purulent discharge. The cornea is not usually involved.
Complications: Chlamydia pneumonitis, pharyngitis or otitis media.
Tests: Immunofluorescence or culture. Giemsa smear: many inclusions.
Treatment: 1% tetracycline ointment or drops 6 hourly for 3 weeks, + erythromycin 10mg/kg/6h PO for ~3 weeks eliminates lung organisms. Give parents/partners erythromycin. During pregnancy, erythromycin is used, but amoxycillin 500mg/8h PO for 1 wk works OK, and is better tolerated.[2]

Gonococcal conjunctivitis Infection is usually within 4 days of birth—with purulent discharge and lid swelling (± corneal hazing). There is risk of corneal rupture and panophthalmitis. Treatment: Infants born to mothers with known gonorrhoea should have benzylpenicillin 30mg/kg IM stat or a cephalosporin such as cefotaxime, and chloramphenicol 0.5% eye-drops within 1h of birth. For active gonococcal infection give benzylpenicillin 15mg/kg/12h IM and 3-hourly 0.5% chloramphenicol eye-drops for 7 days. Isolate the baby. There is always a very small risk of aplastic anaemia with chloramphenicol drops, so it is worth asking a microbiologist if there is a good alternative in the light of swab results.[1]

Ophthalmia neonatorum This is purulent discharge from the eye of a neonate <21 days old. Originally referring to Neisseria gonorrhoea there are many causes: chlamydiae, herpes virus, staphylococci, streptococci and pneumococci, E coli and other Gram -ve organisms. Tests: In a baby with a sticky eye take swabs for bacterial and viral culture, microscopy (look for intracellular gonococci) and chlamydia (eg immunofluorescence).

Clostridium perfringens Suspect this in any complication of criminal abortion—and whenever intracellular encapsulated Gram +ve rods are seen on genital swabs. It may also complicate death in utero, and it may also infect haematomas, and other anaerobic sites. Clinical features: Endometritis → septicaemia/gangrene → myoglobinuria → renal failure → death.[2]
Management: ●Surgical debridement of all devitalized tissue. ●Hyperbaric O₂. ●High-dose IV benzylpenicillin (erythromycin if serious penicillin allergy). The use of gas gangrene antitoxin is controversial. Seek expert help.

For perinatal listeria and streptococcal infection, see p 260.

1 M Doona 1995 BMJ i 1217 2 M Alary 1994 Lancet 344 1461

Prematurity

▶This is a leading cause of perinatal mortality and morbidity.

Premature infants are those born before 37 weeks' gestation. Prevalence: ~6%, of which ~2% are before 32 weeks—when neonatal problems are greatest. In 25%, delivery is elective (p 110). 10% are due to multiple pregnancy; 25% are due to APH, cervical incompetence, amnionitis, uterine abnormalities, diabetes, polyhydramnios, pyelonephritis or other infections. In 40% the cause is unknown, but abnormal genital tract colonization (bacterial vaginosis) with ureaplasma and *Mycoplasma hominis* is implicated, either as a risk factor[1] or risk marker.[2]

Management of preterm rupture of membranes Admit; take MSU, and HVS—using a sterile bivalve speculum. If liquor is not obvious its presence is suggested if nitrazine sticks (which are pH sensitive) turn black (false +ve with infected vaginal discharge, semen, blood and urine). In 80%, membrane rupture will be followed by onset of labour. The problem with the 20% who do not go into labour is balancing the advantages of remaining *in utero* for increasing maturity and surfactant production against the threat of infection (which accounts for 20% of neonatal deaths after premature rupture of membranes). Intrauterine infection supervenes after membranes have ruptured in 10% within 48h, 26% by 72h, 40% by >72h. Prophylactic antibiotics may delay labour.[3] If infection develops, take blood cultures and give IV antibiotics (eg ampicillin 500mg/6h IV + netilmicin 150mg/12h IV) and expedite labour (p 110). If labour supervenes it should be allowed to progress. In the few cases where liquor ceases to drain for more than 48h the mother can be gradually mobilized.

Management of preterm labour In 50% contractions cease spontaneously. Treating the cause (eg pyelonephritis) may make it cease. Attempts to suppress contractions (tocolysis) are unlikely to succeed if membranes are ruptured or the cervix >4cm dilated. Use between 24–34 weeks. Consider transfer to hospital with SCBU facilities. Call paediatrician to attend to the baby at birth.

Tocolytic drugs Absolute CI: chorioamnionitis, fetal death or lethal abnormality, condition (fetal or maternal) needing immediate delivery. Relative CI: fetal growth retardation or distress, pre-eclampsia, placenta praevia, abruption, cervix >4cm. The ritodrine régime (simpler than salbutamol): give 50µg/min IV via a pump, increase by 50µg every 10min until contractions stop, or SE (maternal pulse >130 beats/min, FHR↑ by >20 beats/min, maternal systolic BP falls by >20mmHg, tremor, nausea, vomiting, flushing, pulmonary oedema from overhydration: diuretics may be needed). Usual dose: 150–350µg/min. Pulmonary oedema and hyperglycaemia can also occur, so monitor fluid intake, and blood glucose, especially if diabetic. Continue infusion for 12–48h after contractions cease. Oral β-agonists probably do *not* add anything.[3] Monitor: maternal pulse, T°, contractions and FHR. Salbutamol is an alternative (SE similar). Nifedipine 30mg PO stat (+20mg PO 90min later if still contracting), then 20mg/8h PO is as effective. SE: ↓BP, flushing, transient tachycardia.

Glucocorticoids Betamethasone or dexamethasone 12mg/12h IM for 48-72h promotes fetal surfactant production, lowering mortality and complications of RDS (p 240) by 40–50%;[4] they also help close patent ductuses. Use before 34 weeks only. Avoid if severe pre-eclampsia, infection, or if HIV+ve.

Delivery For recommendations on cord cutting, see p 82.

1 *Lancet* Ed 1992 **340** 1387   2 PE Hay 1994 *BMJ* i 295   3 B Mercer 1995 *Lancet* **346** 1271   3▶G Macones 1995 *Obstet Gyn* **85** 313 & *E-BM* 1995 **1** 12   4▶ PA Crowley 1996 *E-BM* **1** 92

Intrauterine growth retardation

▶When talking to parents, avoid the term *retardation* as this may imply to them the inevitability of mental handicap—which is not the case.

Distinguish premature babies from those who are small for gestational age (SGA): they are at risk from different problems after birth.

Causes of growth retardation Growth retarded (SGA) neonates are those weighing < the 10th centile for their gestational age (table on p 294).

Predisposing factors: Multiple pregnancy; malformation; infection; maternal smoking, diabetes, hypertension (eg pre-eclampsia), severe anaemia, heart and renal disease. About 10% are to mothers who will only ever produce small babies. Where placental insufficiency has been the cause, head circumference is relatively spared (the baby has been starved). This is then called asymmetrical intrauterine growth retardation (IUGR).

Antenatal diagnosis 50% are not detected before birth, and many babies suspected of IUGR turn out not to have it. Measuring changes in fundal height from the symphysis pubis is a reasonable method of measuring fetal growth, especially if used with centile charts. Oligohydramnios (p 120) and poor fetal movements are other indications of placental insufficiency. If growth retardation is suspected, growth *in utero* can be monitored by serial ultrasound examinations of head circumference and abdominal circumference. If umbilical cord Doppler blood flows are normal the outcome of growth retarded pregnancies is better (fewer premature births and stillbirths). Those with abnormal Dopplers may benefit from maternal low-dose aspirin (p 119) eg 75mg/24h PO. Biophysical profile monitoring (p 118) and antenatal cardiotocography (p 116) are used to try to detect those babies who are becoming hypoxic *in utero* and who would benefit from delivery. Advise the mother to stop smoking, to use a fetal kick chart (p 116) and to take plenty of rest.

Labour and aftercare Growth retarded fetuses are more susceptible to hypoxia, so monitor in labour (p 116). After birth, temperature regulation may be a problem, so ensure a warm welcome; nurse those <2kg in an incubator. After been relatively hypoxic *in utero* the Hb at birth is high, so jaundice is more common. They have little stored glycogen so are prone to hypoglycaemia. Feed within 2h of birth and measure blood glucose before each 3-hourly feed. If hypoglycaemic despite regular feeds, transfer to a special care unit. They are more susceptible to infection. Birth reveals those for whom abnormality was the cause of growth retardation.

Distinguishing growth retardation from prematurity Before 34 weeks' gestation there is no breast bud tissue, from then it develops at 1mm diameter/week. Ear cartilage develops between 35 and 39 weeks so premature babies' ears do not spring back when folded. Testes lie in the inguinal canal at 35 weeks, in the scrotum from 37. Labia minora are exposed in premature girls. Skin creases on the anterior 1/3 of the foot appear by 35 weeks (on anterior 2/3 by 39, and all over from 39). 'Prems' have red, hairy skin. Vernix is made from 28 weeks and is maximal at 36 weeks. Prems do not lie with legs flexed until 32 weeks. All limbs are flexed from 36 weeks.

▶▶Fetal distress

Fetal distress signifies hypoxia. Prolonged or repeated hypoxia causes fetal acidosis. An early sign may be the passage of meconium in labour (p 128). Other signs that the fetus may be hypoxic are a fetal tachycardia persisting above 160 beats per minute (tachycardia may also occur if the mother has a high temperature or is dehydrated). Hypoxia may also be reflected by loss of variability of the baseline in the fetal heart rate trace and slowing and irregularity of the heart rate (especially late decelerations—p 116). ▶▶If the heart rate falls below 100 beats per minute urgent assessment is required. Hypoxia may be confirmed by the use of fetal blood sampling (p 116). When significant hypoxia appears to be present (eg pH <7.24), deliver promptly (eg by the quickest route available, eg Caesarean section or vaginal extraction). In complete anoxia the pH falls by 0.1 unit per minute.

▶▶Obstetric shock

Most obstetric shock is associated with severe haemorrhage. It should be remembered that with abruption actual bleeding may be far in excess of that revealed *per vaginam* (p 108). Other causes of shock may be: ruptured uterus (p 134), inverted uterus (p 140); amniotic fluid embolus (p 143), pulmonary embolism, adrenal haemorrhage and septicaemia.

Septicaemia may lack classical signs (eg pyrexia) and must be considered where profound persisting shock is present,[1] be appropriately investigated (eg blood cultures), and treated, eg amoxycillin/clavulanic acid 1g/6h IV + metronidazole 500mg/8h IVI (+ gentamicin 1.5mg/kg/8h IV given over >3min; do levels, but not needed acutely, see OHCM p 180). Prompt resuscitation is required (see individual pages for management). Renal function and urine output should always be measured after shock has occurred (p 160). A late complication can be Sheehan's syndrome (also called Simmonds' disease) whereby pituitary necrosis leads to lack of thyroid stimulating hormone, adrenocorticotrophic hormone and the gonadotrophic hormones hence leading to hypothyroidism, Addisonian symptoms and genital atrophy.

---

1 DoH 1991 *Report on Confidential Enquiries into Maternal Deaths in the UK 1985-7* HMSO pp 67 & 146

Post-maturity (prolonged pregnancy)

Prolonged pregnancy is defined as that exceeding 42 completed weeks or more from the LMP. Affecting 5–10% of pregnancies it is associated with increased perinatal mortality (5/1000 37–42 weeks, 9.7/1000 >42 weeks). The dangers are placental insufficiency, and problems during labour due to the fetus being larger (25% are >4000g), the fetal skull more ossified and less easily moulded, the passage of meconium more common (25–42% labours), and fetal distress more common. Stillbirth rate is ↑ × 4, neonatal deaths ↑ × 3, and neonatal seizures ↑ × 10 in pregnancies going beyond 42 weeks compared with birth between 37 and 42 weeks.

Antepartum monitoring with use of kick charts (p 116), cardiotocography, and biophysical profiles (p 118) can be used to try to detect fetuses who may be becoming hypoxic. Without evidence to suggest problems, induction should be recommended at 41+ weeks[1] after which time there are higher rates of Caesarean section for those managed 'conservatively'. 'Membrane sweeps' remain a valid physiological prelude to induction; more women go into spontaneous labour if as much membrane as possible is swept from the lower segment at examination between 41–42 weeks than if examined unswept.[2] Monitor the fetus in labour (p 116).

Postnatal signs of post-maturity: Dry, cracked, peeling, loose skin; decreased subcutaneous tissue; meconium staining of nails and cord.

Prelabour rupture of membranes at term[3]

2–4% of term births are complicated by prelabour rupture of membranes without labour spontaneously starting within 24h. It is usual practice when labour does not ensue within hours of membrane rupture to start labour with an oxytocin drip to reduce the possibility of ascending infection. It is helpful to give vaginal prostaglandins when the cervix appears unfavourable. Routine use of prostaglandins does not reduce the rate of Caesarean section (but does reduce time to delivery). Waiting 12h after the membranes rupture before instituting oxytocin is associated with lower rates of Caesarean section (rate ~11%) as opposed to early use of oxytocin (rate ~15–20%). When delay is intended maternal temperature must be monitored and oxytocin started should pyrexia develop.

1 JM Grant 1994 *Br J Obs Gynae* **101** 99   2 H Allott 1993 *Br J Obs Gynae* **100** 898–903 & M El-Torkey 1992 *Ibid* **99** 543 & JM Grant 1993 *Ibid* 100 889–90 & M Griffiths 1995 *BMJ* ii 257
3 S Duncan 1992 *Br J Obs Gynae* **99** 543

▶▶Antepartum haemorrhage (APH)

Traditionally this has been defined as bleeding at >28 weeks' gestation. Any bleeding in pregnancy is associated with increased perinatal mortality. Severe bleeds can cause maternal death.

▶Avoid vaginal examination: placenta praevias may bleed catastrophically.

Dangerous causes Abruption, placenta praevia, vasa praevia (here the baby may bleed to death).
Other uterine sources: Circumvallate placenta, placental sinuses.
Lower genital tract sources: Cervical polyps, erosions and carcinoma, cervicitis, vaginitis, vulval varicosities.

Placental abruption ('Accidental haemorrhage') Part of the placenta becomes detached from the uterus. The outcome depends upon the amount of blood loss and degree of separation. The cause is unknown but it is associated with pre-eclampsia, may recur in a subsequent pregnancy (6%), is commoner in smokers and may complicate external cephalic version. Bleeding may be well localized to one placental area and there may be delay before bleeding is revealed.
Consequences: Placental insufficiency may cause fetal anoxia or death. Compression of uterine muscles by blood causes tenderness, and may prevent good contraction at all stages of labour, so beware a PPH (which occurs in ~25%). Posterior abruptions may present with backache. There may be uterine hypercontractility (>7 contractions per 15min). Thromboplastin release may cause DIC. Concealed loss may account for maternal shock after which beware renal failure and Sheehan's syndrome (p 106).

Placenta praevia (For terminology and complications see p 78). The placenta lies in the lower uterine segment. Bleeding is always revealed.

Distinguishing *abruption*	from *placenta praevia:*
Shock out of keeping with the visible loss	Shock in proportion to the visible loss
Pain constant	No pain
Tender, tense uterus	Uterus not tender
Normal lie and presentation	Both may be abnormal
Fetal heart: absent/distressed	Fetal heart usually normal
Coagulation problems	Coagulation problems rare
Beware pre-eclampsia, DIC, anuria	Small bleeds before large

Note: the risk of PPH is increased in both conditions. The lower segment may not contract well after a placenta praevia.

Management of APH Always admit. ▶▶If bleeding is severe call emergency ambulance, put up IVI, take bloods and raise legs. Give O_2 at 15 litres/min via mask with reservoir. On admission, if shocked give fresh ABO Rh compatible or O Rh-ve blood (eg 6U, 2 IVIs) fast until systolic BP >100mmHg. Catheterize bladder; keep renal output >30ml/h. Call anaesthetist to monitor fluids (CVP lines help). *Summon expert help urgently.* If bleeding is severe, *deliver*—Caesarean section for placenta praevia (sometimes for abruption, or induction). Beware PPH.

For milder bleeding, set up IVI, do Hb, crossmatch, coagulation studies and U&E. Check pulse, BP and loss regularly. Establish diagnosis (ultrasound of placenta, speculum examination). If placenta praevia is the diagnosis, keep on the ward until delivery (usually by Caesarean section at 37–38 weeks). If pain and bleeding from a small abruption settles and the fetus is not compromised the woman may go home (after anti-D, if indicated), but treat as 'high risk' pregnancy thereafter. Arrange follow-up.

Using anti-D immunoglobulin[1]

Dose

250U for gestations <20 weeks, 500U if >20 weeks. Give into deltoid (buttock absorption is too slow) as soon as possible after incident, at latest within 72h. After 20 weeks do Kleihauer test (FBC bottle of maternal blood; fetal RBCs therein are less susceptible to lysis, enabling them to be counted, so measuring the bleed's volume).

Postnatal use

1 500U is the normal dose after 20 weeks' gestation. 37% of Rh-ve women give birth to Rh-ve babies and these women do not need anti-D.

2 Anti-D should be given to all Rh-ve women where the baby's group cannot be determined (eg macerated stillbirths), or if circumstances are such that the baby's group is unknown 72h post delivery.

3 Do a Kleihauer test on all those eligible for anti-D. You cannot get round this requirement by giving everyone a double dose of anti-D, hoping that this will be enough. Such a policy would waste anti-D (volunteer donor-dependent) and would undertreat some patients.[2] 500U anti-D can suppress immunization by up to 4ml of fetal red cells (8ml of fetal blood), but 1% of women have transplacental haemorrhage (TPH) of >4ml, especially after manual removal of placenta, and with Caesarean section. A Kleihauer test is especially important in stillbirth, as massive spontaneous transplacental haemorrhage can be the cause of fetal death. Where >4ml TPH is suggested by the Kleihauer screen, a formal estimation of the TPH volume is required and 500U anti-D given for every 4ml fetal cells transfused (maximum 5000U anti-D at 2 IM sites/24h). Liaise with the Regional Transfusion Service in the UK. Check maternal blood every 48h to determine clearance of cells and need for continuing anti-D.

4 Anti-D should not be given to mothers known prenatally to have antibodies to anti-D.

5 Any mother receiving anti-D prenatally (see below), should also receive it postnatally unless she delivers a rhesus negative baby.

Use of anti-D in abortion in rhesus negative mothers

1 Anti-D should be given to all having surgical terminations of pregnancy unless they are already known to have anti-D antibodies.

2 Anti-D should always be given where spontaneous abortion is followed by instrumentation.

3 Anti-D should be given where spontaneous complete abortion occurs after 12 weeks' gestation.

4 With threatened abortion after 12 weeks give anti-D; if bleeding continues intermittently give anti-D 6-weekly until delivery.

5 With threatened abortion before 12 weeks, determine the mother's group and give anti-D if she is negative, and the pregnancy seems viable. Give anti-D within 96h of bleed.

Use of anti-D in pregnancy in Rh-ve mothers

1 Significant TPH may occur with chorionic villus sampling, external cephalic version, APH, amniocentesis, and abdominal trauma. Use 250U before 20 weeks' gestation, 500U after 20 weeks.

2 Anti-D should be given in cases of ectopic pregnancy.

3 For threatened abortion see above.

1 UK Blood Transfusion Working Party 1991 *Prescribers' J* 31 137   2 D Lee 1993 *BMJ* ii 1145

Induction of labour

5-20% of UK labours are induced artificially, usually because it has been decided that to remain *in utero* is relatively more risky for the fetus than to be born, but in some it is because of risk to the mother. 75% of inductions are for hypertension, pre-eclampsia, prolonged pregnancy or rhesus disease. Other indications are diabetes, previous stillbirth, abruption, fetal death *in utero* and placental insufficiency.

►Inducing mothers at 41+ weeks reduces stillbirth rates.

Contraindications Cephalopelvic disproportion which is absolute, malpresentations other than breech or face presentation, fetal distress, placenta praevia, cord presentation, vasa praevia, pelvic tumour, previous repair to cervix. Cone biopsy requires caution.

Cervical ripeness When an induction is being planned the state of the cervix will be assessed. In 95% of women at term the cervix is ripe. If primips are induced with an unripe cervix (Bishop's score ≤3, see below) the rates of prolonged labour, fetal distress and Caesarean section are increased. This is less marked in multips.

Modified Bishop score	0	1	2
Cervical dilatation (cm)	0	1-2	3-4
Length of cervix (cm)	>2	1-2	<1
Station of head (cm above ischial spines)	- 3	- 2	- 1
Cervical consistency	firm	medium	soft
Position of cervix	posterior	middle	anterior

A score of >5 is 'ripe'. An unripe cervix may be ripened using one prostaglandin (PGE2) vaginal tablet (3mg) the evening before or the morning of induction. If antenatal fetal heart rate monitoring is indicated, this should commence before prostaglandin insertion. If there is failure to ripen (occurs in 12%) PGE2 may be repeated 6-8h later. If the cervix still remains unripe consider Caesarean section. PGE2 may stimulate uterine contractions or precipitate labour.

Once the cervix is ripe, rupture the membranes (amniotomy) and start intrapartum fetal heart rate monitoring using a scalp clip or pulse oximetry (less invasive, see OHCM p 658). Oxytocin is given IV in 5% dextrose using a pump system (eg Ivac®). Infusions start at 2 milliunits (mu) per min, doubling every 20min until effective uterine contractions are produced (usually at a rate of 4-16mu/min: occasionally 32mu/min may be necessary). Beware uterine hyperstimulation and the use of large volumes of IV fluid (if >4 litres, there is risk of water intoxication—ie confusion, convulsions and coma). When the cervix is 5cm dilated the uterus is more sensitive to oxytocin and 8mu/min may be sufficient to maintain contractions. Note: the Dublin régime (p 112) results in most women going into spontaneous labour.

Problems of induction ●Iatrogenic prematurity. ●Infection (use antibiotic cover (p 152) in women with heart lesions as risk of endocarditis). ●Bleeding (vasa praevia). ●Cord prolapse (eg with a high head at amniotomy). ●Some will lead to Caesarean section. ▣

Partograms

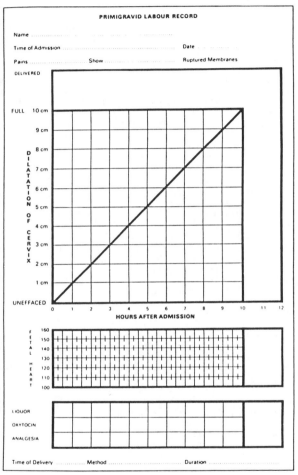

PRIMIGRAVID LABOUR RECORD

Name ....

Time of Admission .... Date ....

Pains .... Show .... Ruptured Membranes

DELIVERED

FULL 10 cm

9 cm

8 cm

7 cm

6 cm

5 cm

4 cm

3 cm

2 cm

1 cm

DILATATION OF CERVIX

UNEFFACED

HOURS AFTER ADMISSION

0 1 2 3 4 5 6 7 8 9 10 11 12

FETAL HEART

160 150 140 130 120 110 100

LIQUOR

OXYTOCIN

ANALGESIA

Time of Delivery .... Method .... Duration ....

The above graph shows the simple Dublin partogram ('simple' because it does not show 'latent' phase of labour). It has a steep x/y gradient of ratio 1:1.

Less steep ratios (eg 2:1) may predispose to premature intervention, as does inclusion of the 'latent' phase on the partogram.[1]

1 R Cartmill 1992 *Lancet* **339** 1520

The active management of labour[1]

Dublin is a centre of excellence for labouring mothers because of its coherent and very successful labour management plan, which leads to a low Caesarean section rate (eg 5% *vs* 14% for other centres). This depends on only ever having women who are actually in labour on labour ward, and regularly assessing them to check that cervical dilatation is progressing well—and taking action if it is not. Primiparous women with single, cephalically presenting babies, who are not achieving cervical dilatation satisfactorily have augmentation of uterine contractions with the use of oxytocin. Women are kept informed of their progress and are given an estimated time of delivery and are assured that delivery should be within 12h of entering the labour ward. Caesarean section is performed on those for whom delivery is not imminent at 12h. By use of this régime it is possible to keep a Caesarean section rate below 5% and a forceps rate of ~ 10%, with the vast majority of women effecting delivery themselves.

Labour This is defined as painful uterine contractions accompanied by a 'show' or cervical dilatation. Women arriving who may not be in labour are accommodated overnight on antenatal wards. Those in labour are admitted to the labour ward and thereby committed to delivery.

Augmentation This recognizes that a primiparous uterus is an inefficient organ of birth that is *not* prone to rupture—but the multiparous uterus is an efficient organ but *is* rupture-prone. Oxytocin may be used to enhance efficiency in the primiparous uterus without danger to the mother. Oxytocin may be dangerous to multips as delay is likely to be from obstructed labour, so a decision to use it must only be made by senior obstetricians.

Examination Women are examined at admission and hourly for the first 3h by a single examiner (may be by rectal examination). Progress is assessed at 1h; artificial rupture of membranes is performed to aid progress and allow liquor viewing. Examination 1h later will result in augmentation of labour by oxytocin in primips if 1cm extra cervical dilatation has not been achieved. Examination is subsequently every 2h. Oxytocin will be used if primip dilatation does not progress at 1cm/h or if there is delay in descent in the involuntary pushing phase of 2nd stage.

Oxytocin A uterus in labour is sensitive to oxytocin. Dublin uses a simple régime. 10U oxytocin in 1 litre of 5% dextrose is started at 10 drops/min, increasing by 10 drops/min every 15min to a maximum of 60 drops/min (with UK standard giving sets 60 drops/min gives 40 mu/min) unless there is fetal distress or uterine hyperstimulation (>7 contractions/15min). Only 1 bag is used ensuring that oxytocin is not used for >6h and water intoxication cannot occur (associated with use of >3 litres 5% dextrose).

Induction by artificial rupture of membranes (ARM). Women return to the antenatal ward to await onset of labour. Those 10% who do not go into labour spontaneously are returned to labour ward after 24h and oxytocin is used exactly as above. 90% of these women also deliver within 12h, thereby keeping Caesarean section rates low.

The partogram This graphs labour's progress (p 111). Dublin uses a simple partogram with time (on *x*-axis) plotted against cervical dilatation, and records fetal heart rate, state of liquor, analgesia, and oxytocin use. Many partograms are much more complicated.

Meconium in liquor At ARM this suggests placental insufficiency. Meconium other than light staining in good liquor volume prompts fetal blood sampling and scalp clip electronic monitoring, or prompt Caesarean section (fetal blood pH low, or very thick meconium).

1 K O'Driscoll 1986 *Active Management of Labour*, Baillière Tindall

The opposite page depicts Dublin practice for primiparous mothers with
term babies of vertex presentation with a normal head. The authors stress
the benefit of their policies both to mothers and also from the financial
management point of view as it allows them to have a personal midwife
for each mother throughout the time that she is in labour, but with unit
costs per delivery of only one-third of costs of UK hospitals.

Midwife monitoring is normally every 15min using a Pinard scope; electronic monitoring is used if there is meconium staining of liquor.
Membranes are ruptured so that the liquor can be regularly inspected.

Many units in the UK often have quite different policies. For example,
some would argue that artificially rupturing membranes may not be
entirely beneficial as the liquor cannot cushion the fetal head and there is
evidence that artificial rupture of membranes is associated with more fetal
heart rate decelerations.[1] ARM may also facilitate spread of HIV from mother
to baby (p 155). A recent review of current knowledge suggests that of early
amniotomy, early use of oxytocin, and the provision of professional support throughout first labours, it is the latter which is the most important
for reducing rates of operative delivery.[2]

1 J Barrett 1992 *Br J Obstet Gynae* **99** 5-9   **2** JG Thornton 1994 *BMJ* **ii** 366 (here figure 2 was incor-
rectly captioned figure 3, and *vice versa*–a vital point in understanding this paper!)

Pain relief in labour

▶Adequate pain relief in labour is the obstetrician's greatest gift to womankind. Not everyone wants natural birth—and do not make people feel guilty about requesting pain relief. A good anaesthetic agent in labour must be harmless to mother and baby, allow maintenance of maternal cooperation, and must not affect uterine contractility.

Education That given by the National Childbirth Trust meets all these criteria. Education about labour reduces fear; breathing exercises and relaxation techniques teach the mother ways to combat pain herself.

In the 1st stage of labour **narcotic injections** give adequate pain relief in 60% of women. The most commonly used is pethidine 50–150mg IM (not if birth expected in <2–3h as neonatal respiratory depression may occur—reversible with naloxone 0.01mg/kg IV—this may be repeated after 3min if needed, or 0.1–0.2mg may be given as a single dose IM). Onset of analgesia is within 20min and should last for 3h. SE: nausea and vomiting, disorientation, delayed gastric emptying, neonatal respiratory depression. CI: mother on MAOI antidepressants (p 340).

Inhalational agents Nitrous oxide can be used throughout labour. Nitrous oxide 50% in oxygen (Entonox®) is self-administered using a demand valve. CI: those with a pneumothorax.

Pudendal block (sacral nerve roots 2, 3 and 4) uses 8–10ml of 1% lignocaine injected 1cm beyond a point just below and medial to the ischial spine each side. It is used with perineal infiltration for instrumental delivery, but analgesia is insufficient for rotational forceps.

Spinal block These may be used for rotational delivery or Caesarean section. For details and contraindications, see p 776.

Epidural anaesthesia See p 776. Pain relief is by anaesthetizing pain fibres carried by T11–S5. Epidurals must only be set up once labour is established (cervix >3cm). Set up IVI first, give 500ml Hartmann's solution to prevent ↓BP. Check pulse, BP, respirations, contractions and fetal heart every 15min after the epidural is set up. Top-ups are required ~2-hourly. Epidurals may be very helpful for the following: OP position (p 126), breech, multiple pregnancy, preterm delivery, pre-eclampsia, forceps delivery, inco-ordinate uterine contractions.

Problems: For those due to technique, see p 778. There may be postural hypotension (IVI, nurse 15° to left side), urinary retention (catheterize regularly), motor paralysis (pelvic floor muscle paralysis reduces rotation and voluntary effort in 2nd stage, so increasing need for forceps). After delivery urinary retention; headache (especially after dural puncture); and, longer term, backache may all be problems.

Combined spinal epidural (CSE) anaesthesia[1] gives quicker pain relief, with little or no motor blockade in most mothers, allowing standing, walking, sitting, and voiding urine. Using the intrathecal route (27G Whiteace needle) apparently does not cause a significant rise in the incidence of headache (only ~0.13%). The patient can control the amount dose, and this leads to a dose-reduction of 35%, and reduces motor blockade.[2] Women should inform their midwife if they notice light-headedness, nausea, or weak legs. Spontaneous delivery rates are not better than with traditional epidural.[3] ▶Skilled anaesthetic help is vital.

Transcutaneous electrical nerve stimulation (TNS) Electrodes placed on the back, give electrical pulses to stimulate the large nerve fibres to the brain, blocking pain impulses. It needs careful instruction but is under the mother's control. It is most effective in the 1st stage.

1 R Collis 1994 *Int J Obs Anesth* 3 75   2 A Shennan 1995 *Lancet* 345 1514   3 P Steer 1995 *BMJ* ii 1209

Fetal monitoring

In high-risk pregnancy, antepartum cardiotocography and biophysical profiles by ultrasound (p 118) are used to monitor fetal activity and responsiveness. The aim is to detect intrauterine hypoxia prenatally.

Kick charts: In the last 10 weeks of pregnancy the mother can count the fetal movements daily and record the time at which the baby has made 10 movements. Although a poor predictor of fetal outcome, less than 10 movements in 12 hours may be a warning sign of problems and the mother should attend her obstetric unit for cardiotocography.

Cardiotocography: Doppler ultrasound detects fetal heart beats and a tocodynamometer over the uterine fundus records any contractions. A continuous trace is printed over ~30min (eg a paper speed of 1 or 2cm/min) with the mother lying semi-recumbent or in left lateral position or in the half-sitting position. A normal trace in an afebrile, mother at term who is not having drugs has a base rate of 110–150 beats/min, with a variability of 5–25 beats/min, and at least 2 accelerations (a common response to movement or noise) of an amplitude ≥15 beats/min over a 15–20min period. (Fetal heart rate falls by ~1 beat/min/week from 28 weeks.) An abnormal fetal heart rate is characterized by a baseline tachycardia of >170 beats/min or bradycardia (<100 beats/min), or a sinusoidal pattern. Loss of variability (<5 beats/min over a 40min period) and recurrent late or atypical decelerations are associated with fetal compromise and are the most important prognostic indicators of the test.[1]

Tests need to be done every 24h antenatally to identify the changing fetal heart rate pattern associated with hypoxia.

Intrapartum monitoring Death and disability due to complications of labour occur in <1:300 labours.[2] Intrapartum fetal heart rate monitoring aims to detect patterns known to be associated with fetal distress. The diagnosis of fetal distress may be supported by the passage of fresh meconium and fetal hypoxia on blood sampling.

The fetal heart rate is traditionally monitored by using a Pinard stethoscope between contractions, but with this method decelerations related to contractions may be missed.

Continuous fetal heart rate monitoring has poor predictive value, overdiagnosing fetal distress even if used with fetal blood sampling. Its value is uncertain even in high risk labours—but it is widely used. Meta-analyses suggest that routine monitoring ↓rates of neonatal seizures, but increases operative deliveries. [3] Monitoring policies vary. Low risk labours may be monitored intermittently. If high risk, monitor throughout labour, ideally with scalp electrode or pulse oximetry. *Indications:* Abnormal admission trace; antenatal problems; use of oxytocin; fresh meconium (p 128). *Disadvantages:* Limited maternal mobility and discomfort.

▶A combination of tachycardia, loss of baseline variability and shallow late decelerations is the best predictor of hypoxia.

Management of a poor trace: 1 Lie the mother on her left side and give O₂. Stop syntocinon. 2 Take fetal blood sample. If you do not have this facility, consider rapid delivery if the trace does not improve.

Fetal blood sampling: Fetal acidosis reflects hypoxia. Scalp blood pH of 7.3–7.4 is normal. If 7.25–7.29 repeat after 45min. If 7.2–7.24 consider Caesarean section. Levels <7.2 require immediate delivery unless in second stage when a level as low as 7.15 may be acceptable.

2. Obstetrics

Fetal heart rate patterns and their clinical significance[4]
The normal pattern is described opposite. Accelerations suggest intact sympathetic activity and are rarely associated with hypoxia.

Loss of baseline variability Baseline variability of >5 beats/min shows response to vagal tone, sympathetic stimuli and catecholamines in a well oxygenated fetal brainstem. Loss of baseline variability may reflect a preterm fetus who is sleeping, drug effects (eg diazepam, morphine, phenothiazine) or hypoxia.

Baseline tachycardia Heart rate >170 beats/min is associated with maternal fever or β-sympathomimetic drug use, chorioamnionitis (loss of variation too), and acute/subacute hypoxia. Persistent rates >200 are associated with fetal cardiac arrythmia.

Baseline bradycardia Heart rates <110 beats/min is rarely associated with fetal hypoxia (except in placental abruption). It may reflect ↑ fetal vagal tone, fetal heart block, or, if spasmodic, cord compression.

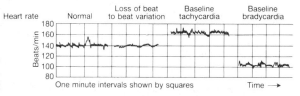

Early decelerations coinciding with uterine contractions reflect increased vagal tone as fetal intracranial pressure rises with the contraction. **Late decelerations**, when the nadir of the deceleration develops some 30sec after the peak of the uterine contraction, reflect fetal hypoxia, the degree and duration reflecting its severity. **Variable decelerations**, both in degree and relation to uterine contractions, may represent umbilical cord compression around the limbs or presenting part.

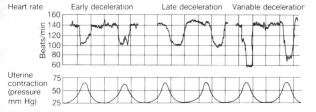

1 N Patel 1989 *Obstetrics* page 373–80, Churchill Livingstone **2** P Steer 1988 *Pregnancy and Risk*, John Wiley **3** SB Thaker 1996 *E-BM* **1** 93 **4** G Rooth 1987 *Int J Gyn Obs* **25** 159–67

Ultrasound

In each ultrasound examination, examine placental site and structure, assess liquor volume and fetal structure and wellbeing (see below).

Early in pregnancy If there is bleeding and pain, ultrasound can confirm intrauterine pregnancy (or ectopic) and the viability of the fetus. Where there is discrepancy between uterine size and dates, gestation can be estimated, a viable fetus ascertained (not missed abortion, p 22), or twins excluded—especially important after the use of fertility drugs or in the presence of hyperemesis gravidarum (also to exclude hydatidiform mole).

Estimation of gestational age Crown–rump length is measured from 6–12 weeks (~10mm at 7 weeks; ~55mm at 12 weeks). From 12 weeks the biparietal diameter can be measured (and femur length from 14 weeks so that there are 2 independent estimations with each procedure). Biparietal diameter measurements to estimate age are most accurate up to 20 weeks (unreliable from 34 weeks). Knowledge of gestational age is important in management of rhesus disease and in diabetic pregnancy (p 156)—and also helps if the date of the LMP is not known, or the cycle is irregular.

Fetal abnormality Many units offer routine scans to find abnormality at ~18 weeks. Is the *routine* use of ultrasound justified? This depends on the skill of the ultrasonographer: in a USA study (N=15,151) no benefits were shown, but only 17% of structural abnormalities were found before 24 weeks; in Europe, benefits are greater as the pick-up rate is 21–84%.[1] Indications for scanning for abnormality include:

- Family history of neural tube defect
- Maternal diabetes or epilepsy
- AFP abnormal
- Oligohydramnios
- Twins
- Polyhydramnios

Biophysical profile scoring Ultrasound of the fetus in the womb over a period of up to 30min aims to assess if the fetus is being affected by acute and chronic asphyxia. With acute asphyxia the fetus loses active physical variables regulated by central nervous system outflow (fetal breathing, gross body movement, fetal flexor tone and heart rate accelerations with fetal body movement). Reduction in amniotic fluid demonstrated by pockets of fluid of less than 1cm depth measured in 2 perpendicular planes is taken to indicate chronic asphyxia. By use of specific criteria for these 5 variables a 'wellbeing' score can be reached. Management protocols according to score then help guide the obstetrician as to when intervention should take place.

Ultrasound is used as an adjunct to the diagnostic procedures of amniocentesis, fetoscopy, cordocentesis and chorionic villus biopsy.

In pregnancies where fetal growth is of concern (p 104) growth can be monitored by regular scans; the abdominal to skull circumference ratio being of interest. Fetal weight can be estimated when planning vaginal delivery of breech presentation (but less accurate for larger fetuses).

In later pregnancy, the lie and presentation of the fetus can be determined. If there is APH, placenta praevia can be excluded and the placenta visualized to look for abruption. With secondary post-partum haemorrhage, retained products of conception may be visualized.

Side-effects Ultrasound is one of the safest tests ever invented; nevertheless, over the years, fears have been expressed about inducing childhood cancers. There is now good evidence that this does not occur.[2] There is one randomized trial indicating that *repeated* ultrasound with Doppler is associated with an increased risk (1.65) of birth weights below the 3rd centile.[2]

1 Oxford Survey of Childhood Ca 1995 *Br J Obs & Gyn* **102** 831 2 JP Newnham 1993 *Lancet* ii 887

Doppler ultrasound and fetal wellbeing[3]

Doppler ultrasound is a technique which can be used to assess circulation on both sides of the placenta. It has not been shown to be of use as a screening tool in routine antenatal care[4] but it has been shown to be of use in high-risk pregnancies. Waveform outlines (velocimetry) have been seen to be abnormal in small babies who ultimately died or were severely ill perinatally, and babies known to be small from real-time ultrasound have been shown not to be at risk if umbilical artery waveforms were normal, but if end-diastolic frequencies are absent the baby is likely to be hypoxic and acidotic. In management of growth retarded babies uncomplicated by other obstetric problems surveillance using Doppler velocimetry is more cost-effective than use of cardiotocography, and may be sufficient as the sole extra means of surveillance (in addition to ultrasound examination of biophysical parameters).[5] ►Consider Caesarean section if umbilical artery Doppler velocimetry shows absent or reversed end diastolic velocities in pregnancies complicated by IUGR and hypertension—to prevent postnatal problems such as cerebral haemorrhage, anaemia, and hypoglycaemia.[6] These changes correlate with placental intervillous space ischaemia and spasm or occlusion of tertiary stem arterioles on the fetal side of the circulation. Placental intervillous ischaemia leads to centralization of fetal circulation so that blood returning from the placenta is shunted to fetal brain, coronary arteries and adrenals. Intervention studies based on knowledge of abnormal umbilical waveform patterns have resulted in babies small for gestational age (SGA) with abnormal waveform patterns being ~500g larger than expected when the mothers were given low dose aspirin rather than placebo from the time of diagnosis. It is also the case that low-dose aspirin delays the onset of hypertension and reduces its severity in primiparous mothers whose babies had abnormal umbilical waveforms. However, aspirin has not been shown to be of benefit in low-risk pregnancies, where it has been shown to increase abruption rates.[7]

1 JP Newnham 1993 *Lancet* ii 887   2 B Ewigman 1993 *NEJM* 239 821   3 Ed 1992 *Lancet* 339 1083
4 J Davies 1992 *Lancet* 340 1299   5 H Alstrom 1992 *Lancet* 340 936   6 VH Karsdorp 1994 *Lancet*
344 1664   7 B Sibai 1993 *NEJM* 329 1213

Amniotic fluid

Amniotic fluid (liquor) Made by amniotic cell secretion and maternal plasma filtrate, volumes are abnormal if the fetus cannot swallow it or contribute to it as urine. By 10 weeks the volume is 30ml, by 20 weeks 300ml, by 30 weeks 600ml. Maximal between 34 and 38 weeks (800–1000ml) it reduces thereafter by 150ml per week.

Oligohydramnios Liquor volume is <200ml. It is rare.
Associations: Prolonged pregnancy, prolonged membrane rupture, placental insufficiency, fetal urethral aplasia or renal agenesis. The (fatal) Potter's syndrome comprises fetal low-set ears, renal agenesis, hypoplastic lungs and amnion nodosum (clumps of fetal squames).

Polyhydramnios (hydramnios) Amniotic fluid exceeds 2–3 litres in 1 in 250 pregnancies. 50% of cases are associated with fetal abnormality, 20% with maternal diabetes. In 30% no cause is found.
Fetal causes: Anencephaly (no swallowing reflex), spina bifida, umbilical hernia, ectopia vesicae, oesophageal or duodenal atresia, hydrops fetalis, hyper-extended attitude.
Maternal causes: Diabetes mellitus, multiple pregnancy. Presenting early in the third trimester, the mother may develop breathlessness and oedema. An abdominal girth of >100cm is suggestive. Ultrasound (p 118) is used to exclude multiple pregnancy and abnormality. Polyhydramnios predisposes to premature labour, malpresentation, cord prolapse, placental abruption when membranes rupture, and PPH (the uterus does not contract well). As indomethacin given to the mother reduces fetal urine production it can be used to treat polyhydramnios (eg 25mg/6h PO). Fetal urine production returns to normal 24h after treatment stops. During labour, check early for cord prolapse. After delivery pass a nasogastric tube in the baby to check the oesophagus is patent.

Amniocentesis Amniocentesis is the sampling of amniotic fluid. It is used for prenatal diagnosis of abnormality; in the management of rhesus disease (p 242); and for estimation of fetal lung maturity (rarely performed now). Lung maturity is assessed from lecithin/sphingomyelin ratio of amniotic fluid. If >2, lungs mature; if <1.5, very immature. Amniocentesis for abnormality diagnosis is carried out at 16 weeks' gestation, when there is enough liquor to give reliable fetal cell culture and time to terminate the pregnancy in the light of results if this is desired. Early amniocentesis at 10–13 weeks can provide satisfactory samples, but has high rates of fetal loss (eg ~5%).[1] Chromosome results take 3–4 weeks to be ready. Real-time ultrasound enables sampling from a good pool of liquor avoiding placental penetration. Using a sterile technique remove 15ml of fluid with a 21G needle. Give anti-D 250U IM to Rh-ve women afterwards. Counsel the woman beforehand.

Indications for amniocentesis: Advanced maternal age (≥35–37yrs for Down's syndrome); a previous child with a neural tube defect (1:20 subsequently affected) or raised maternal serum α-fetoprotein (p 168 & p 228)—though ultrasound is often now used instead to look for neural tube defects; when a parent carries a balanced chromosomal translocation (1 in 4–10 chance fetus affected); risk of recessively inherited metabolic disorder (p 210 & p 211); when the mother carries an X-linked disorder (to determine fetal sex). Chorionic villus sampling (p 166) can provide earlier results for many of the above, allowing earlier termination of pregnancy in light of results. The abortion rate after amniocentesis later, ie 16 week amnocentesis is ~1%. See p 166.

1 A Nicolaides 1994 *Lancet* 344 435

Multiple pregnancy

Incidence Twins: 1/105 pregnancies (33% monozygotic); triplets: 1/10,000.

Predisposing factors Previous twins; family history of twins (dizygotic only); increasing maternal age (<20yrs 6.4/1000, >25yrs 16.8/1000); induced ovulation; race origin (1/150 pregnancies for Japanese, 1/50 for Blacks).

Diagnosis Early in pregnancy the uterus may be felt to be large for dates and there may be hyperemesis. Later there may be polyhydramnios. The signs are that >2 poles may be felt; there is a multiplicity of fetal parts; 2 fetal heart rates may be heard (reliable if beating at rates different by >10 beats/min). Diagnosis may be made or confirmed by ultrasound.

Complications during pregnancy Polyhydramnios (p 120); pre-eclampsia is more common (30% twin pregnancies, 10% singleton); anaemia is more common (iron and folate requirements are increased). There is an increased incidence of APH (6% twins, 4.7% singleton) due to both abruption and placenta praevia (large placenta).

Fetal complications Perinatal mortality for twins is 55.9/1000 (9.7/1000 if single). The main problem is prematurity. Mean gestation for twins is 37 weeks, for triplets 33 weeks. Light-for-dates babies (p 104) are more common (growth rate the same as for singletons up to 24 weeks but may be slower thereafter). Malformation rates are increased 2-4 times. With monozygotic twins, intermingling blood supply may result in disparate twin size and one being born plethoric (hence jaundiced later), the other anaemic. If one fetus dies *in utero* it may become a fetus papyraceous which may be aborted later or delivered prematurely.

Complications of labour PPH is more common (4-6% singletons, 10% twins). Malpresentation is common (cephalic/cephalic 40%, cephalic /breech 40%, breech/breech 10%, cephalic/transverse (Tv) 5%, breech/Tv 4%, Tv/Tv 1%). Rupturing of vasa praevia, increased rates of cord prolapse (0.6% singleton, 2.3% twins), premature separation of the placenta and cord entanglement (especially monozygous) may all present difficulties at labour. Despite modern technology some twins remain undiagnosed, staff are unprepared, and syntometrine may be used inappropriately, so delaying delivery of the 2nd twin. Epidural anaesthesia is helpful for versions.

Management ●Ensure adequate rest (need not entail admission).[1]
●Use ultrasound for diagnosis and monthly checks on fetal growth.
●Give additional iron and folate to the mother during pregnancy.
●More antenatal visits, eg weekly from 30 weeks (risk of eclampsia↑).
●Tell the mother how to identify preterm labour, and what to do.
●Consider induction at 40 weeks. Have IVI running in labour and an anaesthetist available at delivery. Paediatricians (preferably 1 for each baby) should be present at delivery for resuscitation should this be necessary (2nd twins have a higher risk of asphyxia).

Distinguishing monozygous and dizygous twins Monozygous twins are always the same sex and the membrane consists of 2 amnions and only 1 chorion (if in doubt send for histology)

1 G Chamberlain 1991 *BMJ* ii 111

Breech presentations

Breech presentation is the commonest malpresentation. Although 40% of babies are breech at 20 weeks, 25% will be at 32 weeks, and only 3% by term. It is normal in pregnancy for the buttocks to come to lie in the fundus. Conditions predisposing towards breech presentation: contracted pelvis, bicornuate uterus, fibroid uterus, placenta praevia, oligohydramnios, spina bifida or a hydrocephalic fetus. Ultrasound may show the cause and influence the management.

Extended breech presentation is commonest—ie flexed at the hips but extended at the knees. Flexed breeches sit with hips and knees both flexed so that the presenting part is a mixture of buttocks, external genitalia and feet. Footling breeches are the least common. The feet are the presenting part and this type has the greatest risk (5–20%) of cord prolapse.

Diagnosis of breech presentation should be made antenatally. The mother may complain of pain under the ribs. On palpation the lie is longitudinal, no head is felt in the pelvis and in the fundus there is a smooth round mass (the head) which can be ballotted, a sensation akin to quickly sinking an apple in a bowl of water.

External cephalic version (ECV)—turning the breech by manoeuvring it through a forward somersault. Do not turn the baby unless vaginal delivery planned. Traditionally carried out at 34–36 weeks' gestation, it can be used from 37 weeks using tocolysis.[1]

Version contraindications: placenta praevia, multiple pregnancy, APH, small-for-dates babies, in mothers with uterine scars, pre-eclampsia or hypertension (the risk of abruption is increased), who have a bad obstetric history. Give anti-D (500u) to rhesus negative patients.

Mode of delivery Caesarean section is often advocated on the unproven[2] grounds of safety. Aim to discuss all the options with the mother. Consider breech delivery if:[2]

- Experienced obstetrician available
- Estimated fetal wt 2500–3500g by scan
- No evidence of growth retardation
- Good shape/size pelvis on CT pelvimetry
- Labour augmentation/induction avoided
- Good progress in labour stage 1 and 2

- Extended breech presentation (or well flexed if multip)
- No neck hyper-extension
- Normal volume of liquor
- No placenta praevia
- No pre-eclampsia; no distress

If all criteria are met fetal loss may be <1% (possibly much less).[3]

Assisted breech delivery The breech engages in the pelvis with the bitrochanteric diameter (9.5cm) transverse. With further descent through the pelvis, rotation occurs so the bitrochanteric diameter lies anteroposteriorly as it emerges from the birth canal, being born by lateral flexion of the trunk. External rotation then occurs so that this diameter is again transverse. The shoulders enter the pelvis with the bisacromial diameter transverse, rotate through 90° emerging in the AP diameter. The head enters the pelvis with the sagittal suture transverse and rotates 90°. When the body is completely born it is allowed to hang for about 1–2 minutes until the nape of the neck is well seen. The body is then lifted above the vulva by an assistant, the head being delivered with forceps.

▶ Check baby for hip dislocation (↑incidence): also, if vaginal delivery, for Klumpke's paralysis (p 724) and signs of CNS injury.

1 K Mahomed 1991 Br J Obstet Gynae 98 8   2 J Kingdom 1992 BMJ ii 1090   3 JG Thorpe-Beeston 1992 BMJ ii 749 & 1090

Other malpresentations

Occipitoposterior presentation (OP) In 50% of patients the mothers have a long 'anthropoid' pelvis. Diagnosis may be made antenatally by palpation (p 85). On vaginal examination the posterior fontanelle will be found to lie in the posterior quadrant of the pelvis. Labour tends to be prolonged because of the degree of rotation needed, so adequate hydration and analgesia (consider epidural) are important. During labour 65% rotate 130° so that they are occipitoanterior at the time of birth, 20% rotate to the transverse and then arrest ('deep transverse arrest'), 15% rotate so that the occiput lies truly posterior and birth is by flexion of the head from the perineum. Although in 73% delivery will be a spontaneous vaginal delivery, 22% will require forceps and 5% a Caesarean section.

Face presentation Incidence 1:500. 15% are due to congenital abnormality such as anencephaly, tumour of or shortened fetal neck muscles. Most occur by chance as the head extends rather than flexes as it engages. Antenatal diagnosis: the fetal spine feels S-shaped, the uterus is ovoid without fullness in the flanks and there is a deep groove between the occiput and the back. On early vaginal examination, the nose and eyes may be felt but later this will not be possible because of oedema. Most engage in the transverse (mentobregmatic diameter ≈9.5cm). 90% rotate so that the chin lies behind the symphysis (mentoanterior) and the head can be born by flexion. If the chin rotates to the sacrum (mentoposterior), Caesarean section is indicated.

Brow presentation This occurs in 1:2000 deliveries and is often associated with a contracted pelvis or a very large fetus. Antenatal diagnosis: the head does not engage (mentovertical diameter ≈13cm) and a sulcus may be felt between the occiput and the back. On vaginal examination the anterior fontanelle and supraorbital ridges may be felt. Deliver by Caesarean section.

Transverse lie (compound shoulder presentation) This occurs in 1 in 400 deliveries and is usually in multiparous women. Other predisposing factors: multiple pregnancy, polyhydramnios; in primips: arcuate or septate uterus, placenta praevia, contracted pelvis. Antenatal diagnosis: ovoid uterus wider at the sides, the lower pole is empty, the head lies in one flank, the fetal heart is heard in variable positions. On vaginal examination with membranes intact no distinguishing features may be felt, but if ruptured and the cervix dilated, ribs, shoulder or a prolapsed hand may be felt. The risk of cord prolapse is high. External cephalic version (p 124) may be attempted from 32 weeks. If malpresentation persists or recurs Caesarean section will be necessary. Those with persistent instability of lie need hospital admission from 37 weeks (to prevent cord prolapse at home when the membranes rupture) and decision as to elective Caesarean section or induction just after external version (called a stabilizing induction).

▶▶Prolapsed cord

Cord prolapse is an emergency because of the risk of cord compression causing fetal asphyxia. There is an increased incidence at twin deliveries, footling breech delivery and with shoulder presentations. If a cord presentation is noted prior to membrane rupture, carry out Caesarean section. Always be aware when artificially rupturing membranes, when the presenting part is poorly applied, that cord prolapse is a possible risk.

Management The aim is to prevent the presenting part from occluding the cord. ▶▶This may be effected by:
●Displacing the presenting part by putting a hand in the vagina; push it back up (towards mother's head) during contractions.
●Using gravity; either place the woman head down or get her into knee-elbow position (kneeling so rump higher than head).
●Infuse 500ml saline into bladder through size 16 catheter.
●Keep cord in vagina: do not handle it (to prevent spasm).

If the cervix is fully dilated and the presenting part is sufficiently low in pelvis, deliver by forceps (if cephalic) or by breech extraction (by a suitably experienced obstetrician). Otherwise arrange immediate Caesarean section, if the fetus is still alive. There are few other emergencies where speed is so vital.

▶▶Impacted shoulders

Also called shoulder dystocia, this refers to an inability to deliver the shoulders after the head has been delivered.
Associations: Large fetus; post-mature fetus; abnormality; short cord. The danger is death from asphyxia.

Management
●Place the mother in lithotomy with buttocks supported on a pillow over edge of the bed. Lithotomy position increases the pelvic outlet. Give a large episiotomy.
●As an assistant gives firm suprapubic pressure apply firm pressure to the fetal head towards floor. If this fails, check anterior shoulder is under the symphysis (where the diameter of the outlet is widest); if not rotate it to be so and repeat traction. If this fails, rotation by 180° (so posterior shoulder now lies anteriorly) may work. (If the baby dies prior to delivery, cutting through both clavicles—cleidotomy—with strong scissors assists delivery.)

Meconium-stained liquor

In late pregnancy, it is normal for some babies to pass meconium (bowel contents) which stains the amniotic fluid a dull green. This is not significant. During labour, fresh meconium, which is dark green, sticky and lumpy, may be passed. This may be a response to the stress of a normal labour, or a sign of distress, so transfer to a consultant unit (if in a GP unit) and commence continuous fetal heart rate monitoring (p 116). Aspiration of fresh meconium can cause severe pneumonitis. As the head is born, suck out the oropharynx and nose. Have a paediatrician in attendance (p 244) to suck out pharynx and trachea under direct vision using a laryngoscope.

Forceps

Forceps are designed with a cephalic curve, which fits around the fetal head, and a pelvic curve which fits the pelvis. Short-shanked (eg Wrigley's) forceps are used for 'lift out' deliveries, when the head is on the perineum; long-shanked (eg Neville Barnes) for higher deliveries, when the sagittal suture lies in the AP diameter. Kielland's forceps have a reduced pelvic curve, making them suitable for rotation (only in experienced hands).

Conditions of use The head must be engaged; the membranes ruptured; the position of the head known and the presentation suitable, ie vertex or face (mentoanterior); there must not be cephalo-pelvic disproportion (moulding not excessive); the cervix must be fully dilated and the uterus contracting, and analgesia adequate (perineal infiltration for the episiotomy; pudendal blocks may be sufficient for mid-cavity forceps and ventouse deliveries but not for Kielland's). The bladder must be empty.

Indications for use Forceps may be used when there is delay in the second stage: this is frequently due to failure of maternal effort (uterine inertia or just tiredness), epidural analgesia, or malpositions of the fetal head. They may be used when there is fetal distress or a prolapsed cord or eclampsia—all occurring only in the 2nd stage. They are used to prevent undue maternal effort, eg in cardiac disease, respiratory disease, pre-eclampsia. They are used for the after-coming head in breech deliveries.

Technique This should be learnt from demonstration. The following is an *aide-mémoire* for non-rotational forceps. Place the mother in lithotomy position with her bottom just over the edge of the delivery bed. Clean her down, catheterize, check the position of the head. Insert pudendal block and infiltrate the site of the episiotomy (not necessary if she has an epidural). Assemble the blades to check they fit with the pelvic curve pointing upwards. The handle which lies in the left hand is the left blade and is inserted first (to the mother's left side) and then the right: the handles should lock easily. Traction must not be excessive (the end of bed is not for extra leverage!). Synchronize traction with contractions, guiding the head downwards initially. Do a large episiotomy when the head is at the vulva. Change the direction of traction to up and out as the head passes out of the vulva. If the baby needs resuscitation, hand it to the paediatrician. Give vitamin K (p 244).
Complications of forceps: Maternal trauma; fetal facial bruising and VII paralysis (usually resolves).

Ventouse The ventouse, or vacuum extractor, is associated with less maternal trauma than forceps[1] and is preferred worldwide but not in the UK.[2] It may be used in preference to rotational forceps because, as traction is applied, with the cup over the posterior fontanelle, rotation during delivery will occur. It can be used through a partially dilated cervix (primips should be almost fully dilated, multips >6cm), but should not be used if the head is above the ischial spines. A cup is applied with a suction force of 0.8kg/cm². The baby's scalp is sucked up to form a 'chignon' which resolves in 2 days. There is increased rate of fetal cephalhaematoma (p 144) and neonatal jaundice so give vitamin K (p 244).

Follow-up in surviving adults shows no consistent trend to physical or cognitive impairment from low-outlet instrumental deliveries.[3]

1 AH Sultan 1994 *BMJ* i 887   2 J Chalmers 1989 *Br J Obstet Gynae* **96** 505   3 D Seidman 1991 *Lancet* i 1583

Caesarean section

Incidence 5-13% of all labours. Maternal mortality: 0.33/1000. Morbidity is higher—from infection, ileus, thromboembolism, and delay in bonding—unless done under epidural anaesthesia. Factors contributing to this high section rate include over-diagnosis of fetal distress by cardiotocography, fear of litigation should brain damage follow difficult vaginal delivery, and more induced labours (oxytocin may promote fetal distress).

Lower uterine segment incision The almost universal use of this incision makes uterine rupture (p 134) in subsequent pregnancy much less common, allows better puerperal healing, reduces infection as the wound is extraperitoneal, and has lower post-operative complication rates.

Classical Caesarean section (vertical incision) Rarely used. Indications:
●Fetus lies transverse, with ruptured membranes and liquor draining.
●Structural abnormality makes lower segment use impossible.
●Constriction ring present. ●Fibroids (some). ●Some anterior placenta praevia when lower segment abnormally vascular. ●Mother dead and rapid birth desired. ●Very premature fetus, lower segment poorly formed.

Before an emergency section ►Explain to the mother what is to happen.
●Activate the anaesthetist, theatre staff, porters and paediatrician.
●Have the mother breathe 100% O_2 if there is fetal distress.
●Neutralize gastric contents with 20ml of 0.3 molar sodium citrate, and promote gastric emptying with metoclopramide 10mg IV. (NB: there is no time for H_2 agonists to work; ranitidine is kept for elective sections, eg 150mg PO 2h before surgery.) Consider pre-operative emptying of stomach (eg if prolonged labour or opiate given). The stomach should be routinely emptied prior to extubation to minimize risk of postoperative aspiration.[1] See Mendelson's syndrome, p 134.
●Take to theatre (awake); set up IV. Take blood for crossmatch (eg 2u, 6u and 2 IVIs if previous section+anterior placenta praevia—see below).
●Catheterize the bladder. Tilt 15° to her left side on operating table.
●Tell the paediatrician if the mother has had opiates in the last 4h.

Note: halothane should not be used for obstetric procedures because uterine muscle relaxation increases bleeding. Other anaesthetic problems include vomiting on induction (see rapid sequence induction, p 770), and light anaesthesia (out of consideration for the baby) causing paralysed awareness. ►In reducing maternal mortality, the importance of having an experienced anaesthetist cannot be overemphasized.

Indications for *elective* Caesarean sections Known cephalo-pelvic disproportion; placenta praevia; some malpresentations (p 126); after vaginal surgery (suburethral repair; vesico-vaginal fistula repair); with some maternal infections, eg herpes, HIV (p 98). If a repeat section is planned, arrange ultrasound placental localization to exclude placenta praevia as this is more common and more likely to be complicated by placenta accreta (hence risk of massive haemorrhage) in the presence of a uterine scar.

Emergency section may be needed because of antenatal complications, eg severe pre-eclampsia, abruptio placentae (baby still alive). In others, the need becomes apparent during labour: fetal distress; prolapsed cord (if fetus alive); after failed induction; or failure to progress.

Trial of scar in labour: See p 134. Beware oxytocin. *Antibiotic prophylaxis* is controversial. Infections (wounds, endometritis and UTI) is reduced by IV antibiotics (eg 2g cephradine at induction, 1g at 6h and 12h post-op) for both emergency and elective sections.[2] Longer courses do not appear to be superior, nor is use of more expensive 2nd generation cephalosporins.

After Caesarean section In Rh-ve mothers, remove all excess blood from peritoneal cavity. Use a Kleihauer test (p 109) to determine dose of anti-D. Mobilize the mother early. If purported cephalo-pelvic disproportion led to section, consider pre-discharge radiographic pelvimetry.

Complications ●PPH ●Retained placental tissue ●Ureter trauma
●Vesico-uterine fistula ●Fetal injury (eg fracture) ●Colon injury **133**

1 DoH 1994 *Report on Confidential Enquiry into Maternal Deaths in the United Kingdom 1988-90* HMSO
2 M Turner 1992 *J Obs Gynae* **12** (sup 1) 520

▶▶Uterine rupture

Ruptured uterus is rare in the UK (1:1500 deliveries—but 1:100 in parts of Africa). Associated maternal mortality is 5%, and the fetal mortality 30%. ~70% of UK ruptures are due to dehiscence of Caesarean section scars. Lower segment scars are far less likely to rupture (1.4%) than the classical scars (6.4%)—see p 132. Other risk factors: ●Obstructed labour in the multiparous, especially if oxytocin is used. ●Previous cervical surgery. ●High forceps delivery. ●Internal version. ●Breech extraction. Rupture is usually during the 3rd trimester or in labour. *Trial of scar:* a frequent question is: *Is it safer to do a repeat Caesar or to allow a scarred uterus (past Caesar) to labour?* In general, a trial of labour is safer than automatic repeat Caesars—particularly if the thickness of the lower uterine segment is >3.5mm on ultrasound at 36–38 weeks, according to a careful prospective study (N=642).[1]

Signs and symptoms Rupture is usually in labour. In a few (usually a Caesarean scar dehiscence) rupture precedes labour. Pain is variable, some only having slight pain and tenderness over the uterus. In others pain is severe. Vaginal bleeding is variable and may be slight (bleeding is intraperitoneal). Unexplained maternal tachycardia, sudden maternal shock, cessation of contractions, disappearance of the presenting part from the pelvis, and fetal distress are other presentations. Post-partum indicators of rupture: continuous PPH with a well-contracted uterus; if bleeding continues post-partum after cervical repair; and whenever shock is present.

Management If suspected in labour, laparotomy should be carried out, the baby delivered by Caesarean section and the uterus explored at operation. ▶▶●Give O₂ at 15 litres/min via a tight-fitting mask with reservoir. ●Set up IVI. ●Crossmatch 6u of blood and correct shock by fast transfusion. ●Arrange laparotomy. The type of operation performed should be decided by a senior obstetrician; if the rupture is small, repair may be carried out (possibly with tubal ligation); if the cervix or vagina are involved in the tear, hysterectomy may be necessary. Care is needed to identify the ureters and exclude them from sutures. Give post-operative antibiotic cover, eg ampicillin 500mg/6h IV and netilmicin 150mg/12h IV (unless there is renal impairment). 85% of spontaneous ruptures require hysterectomy, but >66% of ruptured scars are repairable.

▶▶Mendelson's syndrome

This is the name given to the cyanosis, bronchospasm, pulmonary oedema, and tachycardia which develops due to inhalation of acid gastric contents during general anaesthesia. Clinically it may be difficult to distinguish from cardiac failure or amniotic fluid embolus. Pre-operative H₂ antagonists, sodium citrate, gastric emtying, cricoid pressure (p 770), the use of cuffed endotracheal tubes during anaesthesia and pre-extubation emptying of stomach aim to prevent it (p 132).

Management ▶▶Tilt the patient head down. Turn her to one side and aspirate the pharynx. Give 100% oxygen. Give aminophylline 5mg/kg by slow IVI and hydrocortisone 1000mg IV stat. The bronchial tree should be sucked out using a bronchoscope under general anaesthesia. Antibiotics eg ampicillin and netilmicin (as above), should be given to prevent secondary pneumonia. Ventilation on intensive care may be needed. Physiotherapy should be given during convalescence

1 P Rozenberg 1996 *Lancet* 347 281

Abdominal pain in pregnancy

▶ With any pain in pregnancy think, could this be the onset of labour?

Abdominal pain in pregnancy may be from ligament stretching or from symphysis pubis strain. Some other causes are given here. In early pregnancy remember abortion (p 22) and ectopic pregnancy (p 24).

Abruption The triad of abdominal pain, uterine rigidity and vaginal bleeding suggests this. It occurs in between 1 in 80 and 1 in 200 pregnancies. Fetal loss is high (up to 60%). A tender uterus is highly suggestive. Ultrasound may be diagnostic (but not necessarily so). A live viable fetus merits rapid delivery as demise can be sudden. Prepare for DIC which complicates 33–50% of severe cases, and beware PPH which is also common. See p 108.

Uterine rupture See opposite, p 134.

Uterine fibroids For torsion and red degeneration, see p 150.

Uterine torsion The uterus rotates axially 30–40° to the right in 80% of normal pregnancies. Rarely, it rotates >90° causing acute uterine torsion in mid or late pregnancy with abdominal pain, shock, a tense uterus and urinary retention (catheterization may reveal a displaced urethra in twisted vagina). Fibroids, adnexal masses, or congenital asymmetrical uterine anomalies are present in 90%. Diagnosis is usually at laparotomy. Delivery is by Caesarean section.

Ovarian tumours For torsion and rupture, see page 150.

Pyelonephritis See p 160.

Appendicitis Incidence: ~1/1000 pregnancies. This is not commoner in pregnancy but mortality is higher (especially from 20 weeks). Perforation is commoner (15–20%). Fetal mortality is ~5–10% for simple appendicitis; ~30% when there is perforation. The appendix migrates upwards, outwards and posteriorly as pregnancy progresses, so pain is less well localized (often paraumbilical or subcostal) and tenderness, rebound and guarding less obvious. Peritonitis can make the uterus tense and woody-hard. Operative delay is dangerous. Laparotomy over site of maximal tenderness with patient tilted 30° to the left should be performed by an experienced obstetric surgeon.

Cholecystitis Incidence 1–6 per 10,000 pregnancies. Pregnancy encourages gallstone formation due to biliary stasis and increased cholesterol in bile. Symptoms are similar to the non-pregnant with subcostal pain, nausea and vomiting. Jaundice is uncommon (5%). Ultrasound confirms the presence of stones. The main differential diagnosis is appendicitis and laparotomy is mandatory if this cannot be excluded. Treatment is similar to the non-pregnant aiming for interval appendicectomy after the puerperium.

Rectus sheath haematoma Very rarely, bleeding into the rectus sheath and haematoma formation can occur with coughing (or spontaneously) in late pregnancy causing swelling and tenderness. Ultrasound is helpful.[1] Laparotomy is indicated if the diagnosis is in doubt.

Abdominal pain may complicate pre-eclampsia by liver congestion. Rarely, in severe pre-eclampsia the liver perforates.

1 G Chamberlain 1991 *BMJ* i 1390

Stillbirth

Stillbirths are those babies born dead after 24 weeks' gestation. Death *in utero* may take place at any stage in pregnancy or in labour. Delivery of these babies presents an emotional strain for both mother and attendant staff as the pain and process of labour may seem so futile and mothers may feel guilty—and that this is in some way a punishment.

Some hours after a fetus has died *in utero* the skin begins to peel. On delivery such fetuses are described as *macerated*, as opposed to *fresh* stillbirths. If left, spontaneous labour usually occurs (80% within 2 weeks, 90% within 3), but it is common practice to induce labour once death is diagnosed, to prevent a long wait for labour for the mother, and minimize risk of coagulopathy. DIC (p 142) is rare unless a retained fetus of more than 20 weeks' gestation has been retained for >4 weeks, but if tests indicate a clotting disorder is present, some authorities recommend heparin (1000U IV/h for up to 48 hours prior to onset of labour[1])—but get expert advice. Stop any heparin prior to labour, or if it supervenes. Ensure excellent supply of freshest possible blood and fresh frozen plasma are available at delivery.

Causes of stillbirth Pre-eclampsia, chronic hypertension or renal disease, diabetes, infection, hyperpyrexia (T° >39.4°C), malformation (11% of macerated stillbirths and 4% of fresh have chromosomal anomalies), haemolytic disease, post-maturity. Abruption and knots in the cord may cause death in labour. In 20% no cause is found.

Diagnosis The mother usually reports absent fetal movements. No heart sounds (using Pinard's stethoscope, Doppler or cardiotocography). There is no fetal movement (eg heart beat) on ultrasound.

Management Labour is induced using prostaglandin vaginally; or extra-amniotically (dose varied with uterine response). Oxytocin IV used concomitantly risks uterine or cervical trauma so use *after* prostaglandin extra-amniotic infusion ceases.[2] Oxytocin infusion alone (p 110) may be used for induction if cervix ripe (Bishop's score ≥4, gestation ≥35 weeks).[2] Amniotomy is traditionally contraindicated as it risks infection. NB: misoprostol is said to be at least as effective as prostaglandin PGE2 for second trimester terminations involving either a dead or living fetus—and is easier to use, and has fewer adverse effects.[3] This use is not licensed.

Ensure good pain relief in labour (if epidural, check clotting tests all normal). Do not leave the mother unattended. When the baby is born it should be wrapped as would any other baby and be offered to the mother to see and to hold—if the mother so desires. A photograph may be taken for her to take home. Unseen babies are difficult to grieve for. Naming the baby and holding a funeral service may also help with grief.

Labour ward procedure (to try to establish cause) Thorough examination of the stillbirth, clinical photographs. Arrange post-mortem examination and placental histology. High vaginal swab for bacteriology. Maternal and fetal blood for TORCH screen (p 98). Maternal blood for Kleihauer acid test, p 109 (detects fetomaternal transfusion—a cause of unexplained stillbirth), and lupus anticoagulant.[2] Examination of fetal blood and skin for chromosomes.

After stillbirth Offer lactation suppression (bromocriptine 2.5mg PO on day 1, then 2.5mg/12h PO for 14 days). Give parents a follow up appointment to discuss causes found by the above tests. Refer for genetic counselling if appropriate.

1 E Letsky 1989 in *Obstetrics* Ed G Chamberlain, page 569, Blackwell   2 V Tindall 1989 in *Progress in Obstetrics and Gynaecology* Vol. 7, Churchill Livingstone   3 JK Jain 1994 *NEJM* 331 290

Helping parents after stillbirth

- A *Certificate of Stillbirth* is required (issued by an obstetrician), which the parents are required to take to the Registrar of Births and Deaths within 42 days of birth. The father's name only appears in the register if the parents are married, or if both parents make the registration.

- The Registrar then issues a Certificate of Burial or Cremation which the parents then give to the undertaker (if they have chosen a private funeral—in which case they bear the cost of the funeral), or to the hospital administrators if they have chosen a hospital funeral—for which the hospital bears the cost. Parents are issued with a Certificate of Registration to keep which has the name of the stillborn baby (if named), the name of the informant who made the registration, and the date of stillbirth.

- Hospitals are directed by the DSS to offer 'hospital' funerals for stillbirths (arranged through an undertaker). If the parents offer to pay for this, the hospital may accept. The hospital should notify the parents of the time of funeral so that they may attend, if they wish. With hospital funerals a coffin is provided and burial is often in a multiple-occupancy grave in a part of the graveyard set aside for babies. The hospital should inform parents of the site of the grave. Graves are unmarked, so should the parents not attend the funeral and wish to visit later it is recommended that they contact the graveyard attendants for the grave to be temporarily marked. Parents may buy a single occupancy grave, if they wish, on which they can later erect a headstone. Hospitals can arrange cremations, but the parents pay for this.

- In addition to arranging a follow-up appointment with the obstetrician to discuss implications for future pregnancy and the cause (if known) of the stillbirth, it is helpful to give parents the address of a local branch of an organization for bereavement counselling, eg SANDS.[1] Grief may take a long time to resolve and parents may find it difficult to contact ordinary medical staff without the 'excuse' provided by asking about baby's ailments.

- Psychological problems are commoner in those who embark upon pregnancy within 6 months of the bereavement. Advise waiting 6 months to 1 year before starting next pregnancy.

Note: in the UK statutory maternity pay and the maternity allowance and social fund maternity payments are payable after stillbirth.

1 SANDS (Stillbirth and Neonatal Death Society), 28 Portland Place, London WIN 4DE, UK (tel. 0171 436 5881)

Post-partum haemorrhage (PPH)

Primary PPH is the loss of greater than 500ml (definitions vary) in the first 24h after delivery. This occurs after ~6% of deliveries. Causes: uterine atony (90%), genital tract trauma (7%), clotting disorders—p 142 (3%). Death rate: 2/yr in the UK.

Factors predisposing to poor uterine contractions •Past history of atony with PPH. •Retained placenta or cotyledon. •Ether or halothane anaesthesia. •A large placental site (twins, severe rhesus disease, large baby), low placenta, overdistended uterus (polyhydramnios, twins). •Extravasated blood in the myometrium (abruption). •Uterine malformation or fibroids. •Prolonged labour. •Poor 2nd stage uterine contractions (eg grand multip). •Trauma to uterus or cervix.

Management •Give ergometrine 0.5mg IV. •Call emergency ambulance unit (p 92)—if not in hospital. Give high-flow O₂ as soon as available. •Set up IVI (2 large-bore cannulae). •Call anaesthetist[1] (a CVP line may help guide fluid replacement, but not if it causes delay). •If shocked give Haemaccel® or fresh blood of the patient's ABO and Rh group (uncross-matched in emergency) *fast* until systolic BP >100mmHg and urine flows at >30ml/h (catheterize the bladder). •Is the placenta delivered? If it is, is it complete? If not, explore the uterus. •If the placenta is complete, put the patient in the lithotomy position with adequate analgesia and good lighting. Check for and repair trauma. •If the placenta has not been delivered but has separated, attempt to deliver it by controlled cord traction after rubbing up a uterine contraction. If this fails, ask an experienced obstetrician to remove it under general anaesthesia. Beware renal shutdown.

If bleeding continues despite all the above, give 10 units of oxytocin in 500ml dextrose saline eg at a rate of 15 drops/min. Bimanual pressure on the uterus may decrease immediate loss. Check that blood is clotting (5ml should clot in a 10ml round-bottomed glass tube in 6min; formal tests: platelets, prothrombin ratio, kaolin-cephalin clotting time, fibrin degradation products). Explore the uterus for possible rupture. If uterine atony is the cause, and the circulation is still compromised, give carboprost 250µg (15-methyl prostaglandin F2α) eg as Hemabate® 1ml deep IM (max 8 doses, each >15min apart). SE: nausea, vomiting, diarrhoea, T°↑; (less commonly—asthma, BP↑, pulmonary oedema).[1] It controls bleeding in ~88%.[2] Rarely, internal iliac artery ligation or hysterectomy is needed to stop bleeding. Ask a haematologist's advice on clotting factor replacement (fresh frozen plasma contains all of them; the cryoprecipitate has more fibrinogen, but lacks antithrombin III).[1]

Secondary PPH This is excessive blood loss from the genital tract after 24h from delivery. It usually occurs between 5 and 12 days and is due to retained placental tissue or clot. Secondary infection is common. Uterine involution may be incomplete. If bleeding is slight and there is no sign of infection it may be managed conservatively—but heavier loss, the suggestion of retained products on ultrasound, or a tender uterus with an open os, requires exploration. Crossmatch 2 units of blood pre-operatively. Give antibiotics (eg ampicillin 500mg/6h IV, metronidazole 1g/12h PR) if there are signs of infection. Carefully curette the uterus (it is easily perforated at this stage). Send currettings for histology (excludes choriocarcinoma).

1 *Drug Ther Bul* 1992 **30** 89   2 A Merrikay 1990 *Am J Obst Gynecol* **162** 205

Retained placenta

Physiological third stage takes 30min. With the use of oxytocic drugs at delivery and controlled cord traction, ie active management, the third stage is complete in 10min in 97% of labours, and a placenta not delivered in 30min will probably not be expelled spontaneously. The danger with retained placenta is haemorrhage.

Management If the placenta does not separate readily, avoid excessive cord traction—the cord may snap or the uterus invert. Check that the placenta is not in the vagina. Palpate the abdomen. If the uterus is well contracted, the placenta is probably separated but trapped by the cervix. Wait for the cervix to relax and release it. If the uterus is bulky, the placenta may have failed to separate. Rub up a contraction, put the baby to the breast (stimulates oxytocin production) or give more Syntometrine®, and empty the bladder (a full bladder causes atony). If the placenta still will not deliver, prepare to remove it manually (delay may precipitate a PPH).

Manual removal Set up IV and crossmatch blood (eg 2u). Call the anaesthetist. The procedure can be done under epidural if *in situ*; halothane assists by relaxing the uterus. Obtain consent. With the mother in lithotomy position, using aseptic technique, place one hand on the abdomen to stabilize the uterus. Insert the other hand through the cervix into the uterus. Following the cord assists finding the placenta. Gently work round the placenta, separating it from the uterus using the ulnar border of the hand. When separated it should be possible to remove it by cord traction. Check that it is complete. Give oxytocic drugs and start antibiotics, eg doxycycline 200mg stat, 100mg/24h and metronidazole 400mg/8h PO. Rarely, the placenta will not separate (placenta accreta) and hysterectomy may be necessary—by a senior obstetrician.

►►Uterine inversion

Inversion of the uterus is rare. It may be due to mismanagement of the third stage, eg with cord traction in an atonic uterus and a fundal insertion of the placenta. It may be completely revealed, or partial when the uterus remains within the vagina. Even without haemorrhage the mother may become profoundly shocked.

Management The ease with which the uterus is replaced depends upon the amount of time elapsed since inversion, as a tight ring forms at the neck of the inversion. With an inversion noted early before shock sets in, replacement by hand may be possible. If shock has ensued, set up a fast IV and infuse colloid or blood. Summon expert help. Under halothane anaesthesia to relax the uterus, hold the uterus in the vagina with one hand. Run two litres of warm 0.9% saline fast into the vagina through cystoscopy tubing (or with a funnel and tube) with an assistant holding the labia encircled tightly around the operator's arm to prevent the fluid running away. The hydrostatic pressure of the water should reduce the uterus. Once the inversion has been corrected, give ergometrine to contract the uterus and prevent recurrence.

►►DIC and coagulation defects

Disseminated intravascular coagulation (DIC) may occur incidentally in pregnancy—or with retention of a dead fetus (of greater than 20 weeks' gestation which has been dead for >1 month); pre-eclampsia; placental abruption; endotoxic shock; amniotic fluid embolism; placenta accreta; hydatidiform mole; prolonged shock from any cause.

Pathogenesis of DIC Thromboplastins are released into the circulation, fibrin and platelets are consumed as intravascular clotting occurs. *Tests:* Kaolin-cephalin clotting time↑ (↓ factors II, V, VII), fibrinogen↓, fibrin degradation products↑. In situations where DIC is a possibility send blood for crossmatch, platelets, partial thromboplastin time or accelerated whole blood clotting time, prothrombin time, fibrinogen estimation and fibrin degradation products. Preliminary results should be available in 30min.

Management Presentation may be as heavy bleeding and shock, and the first measures must be the correction of shock. ►►Give O₂ at 15 litres/min via a tight fitting mask with reservoir. Set up at least 1, preferably 2, wide-gauge IVIs, take bloods as above, and give fresh blood fast (group-compatible blood—available in 5–10min or O Rh–ve blood if desperate). Stored blood is deficient in clotting factors. Give fresh frozen plasma to normalize the kaolin-cephalin clotting time and the prothrombin time. Platelets are indicated with prolonged bleeding and low platelet count. Calcium is sometimes needed to counteract citrate in stored blood (eg 10ml of 10% calcium gluconate IVI, eg after 6U of blood). Seek expert help from a haematologist. The condition is usually self-limiting if the stimulus can be removed. In the case of intrauterine death and abruption (p 108) removal of the uterine contents is the way to correct the the stimulus, and this should be done as promptly as possible.
Mortality: <1% if placental abruption; 50–80% if infection with shock.

Autoimmune thrombocytopenic purpura (AITP) IgG antibodies cause thrombocytopenia (associated with increased bone marrow megakaryocytes) in the mother, and being able to cross the placenta, they may cause thrombocytopenia in the fetus. 52% of babies are affected, with associated morbidity in 12%. 70% are affected if maternal platelets are <100 × 10⁹/l. Exclude systemic lupus erythematosus in the mother (thrombocytopenia may be an early presentation; do DNA binding, OHCM p 672). If maternal platelets fall below 20 × 10⁹/l; 50 × 10⁹/l near delivery,[1] give steroids. Avoid splenectomy during pregnancy (mortality rate 10%). Immunoglobulin IgG 0.4mg/kg IV for 5 days is sometimes used near expected date of delivery, inducing maternal and fetal remission for up to 3 weeks but it is extremely expensive. Aim for non-traumatic delivery for both mother and baby. Neonatal platelet count may fall further in the first days of life, then gradually rise to normal over 4–16 weeks. Treatment is not needed unless surgery is contemplated. Maternal mortality is now negligible due to AITP, but fetal mortality remains (due to intracranial bleeding).

Pregnancy associated thrombocytopenia Mild platelet reduction (100–150 × 10⁹/l) in the 2nd half of pregnancy is not uncommon (7.6% at term in 1 study). Of benign consequence, it resolves post delivery.[1]

1 M Pillai 1993 *Br J Obstet Gynae* **100** 201

Amniotic fluid embolism[1]

This condition, with a mortality up to 80% presents with sudden dyspnoea and hypotension, heralded, in 20%, by seizures. 40% also develop DIC, and of those who survive the initial collapse 70% go on to develop pulmonary oedema (adult respiratory distress syndrome, ARDS, OHCM p 350). An anaphylactic type of response occurs to (possibly abnormal) amniotic fluid in the maternal circulation. *Presentation* is often at the end of the first stage of labour or shortly after delivery but can complicate amniocentesis, or termination of pregnancy, and has even occurred up to 48h post-partum. Previously it was said to be related to uterine hyperstimulation (multiparous short labour and use of oxytocin). This may be anecdotal.

Management ►► The first priority is to prevent death from respiratory failure. Give mask oxygen and call an anaesthetist urgently. Endotracheal intubation and ventilation may be necessary. Set up IVI lest DIC should supervene. Treatment is essentially supportive—important steps are detailed below. Diagnosis may be difficult: exclude other causes of obstetric shock (p 106).

- Cardiopulmonary resuscitation if indicated.
- Give highest available O_2 concentration. If unconscious ventilate and use 100% inspired O_2.
- Monitor for fetal distress.
- If hypotensive, give fluids rapidly IVI to increase preload. If still hypotensive consider inotropes: dobutamine (a better inotrope than dopamine) eg in a dose range of 2.5-10µg/kg/min IVI may help.
- Pulmonary artery catheterization (Swan-Ganz catheter if available) helps guide haemodynamic management.
- After initial hypotension corrected, give only maintenance requirements of fluid to avoid pulmonary oedema from ARDS.
- Treat DIC with fresh whole blood or packed cells and fresh frozen plasma. Use of heparin is controversial,[2] there is insufficient data to warrant routine heparinization.[1]

1 S Clark 1990 *Obstetrical Gynecological Survey* **45** 360   2 DoH 1991 *Report on Confidential Enquiries into Maternal Deaths in the United Kingdom 1985-7* HMSO pages 67 & 146

Birth injuries to the baby

Moulding This is a natural phenomenon rather than an injury. The skull bones are able to override each other (p 86) to reduce the diameter of the head. Moulding is assessed by the degree of overlap of the overriding at the sutures. If moulding is absent, the cranial bones are felt separately. With slight moulding, the bones just touch, then they override but can be reduced; finally they override so much that they cannot be reduced. Excessive moulding during labour indicates cephalo-pelvic disproportion, and can result in intracranial damage.

Cephalhaematoma This is a *subperiostial* swelling on the fetal head, and its boundaries are therefore limited by the individual bone margins (commonest over parietal bones). It is fluctuant. Spontaneous absorption occurs but may take weeks and may cause or contribute to jaundice.

Caput succadaneum This is an oedematous swelling of the scalp, superficial to the cranial periosteum (which does not, therefore, limit its extent) and is the result of venous congestion and exuded serum caused by pressure against the cervix and lower segment during labour. The presenting part of the head therefore has the swelling over it. It gradually disappears in the first days after birth. When ventouse extraction is used in labour a particularly large caput (called a chignon) is formed under the ventouse cup.

Erb's palsy This may result from a difficult assisted delivery, eg shoulder dystocia. The baby's arm is flaccid and the hand is in the 'porter's tip' posture (p 724). Exclude a fractured clavicle and arrange physiotherapy. If it has not resolved by 6 months, the outlook is poor.

Subaponeurotic haematoma Blood lies between the aponeurosis and the periosteum. As haematoma is not confined to the boundaries of one bone, collections of blood may be large enough to result in anaemia or jaundice. They are associated with vacuum extractions.

Skull fractures These are associated with difficult forceps extractions. They are commonest over parietal or frontal bones. If depressed fractures are associated with neurological signs, ask a neurosurgeon if the bone should be elevated.

Intracranial injuries Intracranial haemorrhage is especially associated with difficult or fast labour, instrumental labour, and breech delivery. Premature babies are especially vulnerable. Normally a degree of motility of intracranial contents is buffered by cerebrospinal fluid. Excessive moulding and sudden changes in pressure reduce this effect and are associated with trauma. In all cases of intracranial haemorrhage check babies' platelets. If low, check mother's blood for platelet alloantibodies (PLA1 system). Subsequent babies are at equal risk. IV maternal immunoglobulin treatment is being evaluated.

Anoxia may cause intraventricular haemorrhage (p 232). Asphyxia causes intracerebral haemorrhage (often petechial) and may result in cerebral palsy. Extradural, subdural and subarachnoid haemorrhages can all occur. Babies affected may have convulsions, apnoea, cyanosis, abnormal pallor, low heart rate, alterations in muscle tone, restlessness, somnolence or abnormal movements. Treatment is supportive and expectant. See p 232 & p 234.

Episiotomy and tears

Perineal tears These are classified by the degree of damage caused. Tears are most likely to occur with big babies, precipitant labours, babies with poorly flexed heads, shoulder dystocia, when forceps are used, or if there is a narrow suprapubic arch.

Labial tears These are common and can be very uncomfortable, but heal quickly and suturing is rarely helpful.

1st degree tears These tears are superficial and do not damage muscle. They may not need suturing unless blood loss is marked.

2nd degree tears These lacerations involve perineal muscle and if severe may involve the anal sphincter. They are repaired in a similar fashion to repair of episiotomy (see below).

3rd degree tears Damage extends through the anal mucosa and can extend up into the rectum. Repair is carried out, by an experienced surgeon, under epidural or general anaesthesia. Rectal mucosa is repaired first using catgut inserted from above the tear's apex to the mucocutaneous junction. Muscle is interposed. Vaginal mucosa is then sutured. Severed ends of the anal sphincter are apposed using figure-of-eight stitches. Finally skin is repaired. Avoid constipation post-operatively by using a high-fibre diet and faecal softeners.

Episiotomy This is performed to enlarge the outlet, eg to hasten delivery of a distressed baby, for instrumental or breech delivery, to protect a premature head, and to try to prevent 3° tears (but anal tears are not reduced by more episiotomies in normal deliveries).[1]

The tissues which are incised are vaginal epithelium, perineal skin, bulbocavernous muscle, superficial and deep transverse perineal muscles. With large episiotomies, the external anal sphincter or levator ani may be partially cut, and ischiorectal fat exposed.

Technique: Hold the perineal skin away from the presenting part of the fetus (2 fingers in vagina). Infiltrate area to be cut with local anaesthetic (eg 1% lignocaine). Still keeping the fingers in the introitus, cut mediolaterally towards the ischial tuberosity, starting medially (6 o'clock), so avoiding the Bartholin's glands.

Repair: (See diagrams opposite.) NB: use resorbable sutures, but not glycerol-impregnated cat-gut (risks long-term dyspareunia). In lithotomy, and using good illumination, repair the vaginal mucosa first. Start above the apex using interlocking stitches 1cm apart, 1cm from wound edges. Tie off at mucocutaneous junction of fourchette. Then repair muscles with interrupted stitches to obliterate any dead spaces. Finally close the skin (subcutaneous stitch is more comfortable than interrupted stitches).

Problems with episiotomy: Bleeding (so may increase chance of spread of HIV from mother to baby); infection and breakdown; haematoma formation. For comfort suggest ice packs, salt baths, hair dryer to dry perineum. Superficial dyspareunia (p 74). If labia minora are involved in the skin bridge, the introitus is left too small. If the deep layers are inadequately sutured, the introitus becomes rather rounded exposing the bladder to coital thrusts.

1 T Henriksen 1992 *Br J Obst Gynae* **99** 950

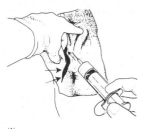

(1)
Swab the vulva towards the perineum. Infiltrate with 1% lignocaine → (*arrows*).

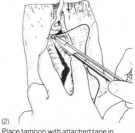

(2)
Place tampon with attached tape in upper vagina. Insert 1st suture above apex of vaginal cut (not too deep as underlying rectal mucosa nearby).

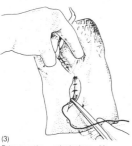

(3)
Bring together vaginal edges with continuous stitches placed 1 cm apart. Knot at introitus under the skin. Appose divided levator ani muscles with 2 or 3 interrupted sutures.

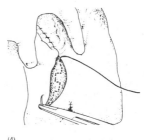

(4)
Close perineal skin (subcuticular continuous stitch is shown here).

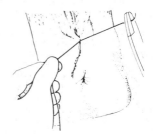

(5)
When stitching is finished, remove tampon and examine vagina (to check for retained swabs). Do a PR to check that apical sutures have not penetrated rectum.

Reproduced from *GP* (27 September 1987) by kind permission.

The puerperium

This is the 6 weeks after delivery. The uterus involutes, from 1kg weight at delivery to 100g. Felt at the umbilicus after delivery; is a pelvic organ at 10 days. Afterpains are felt (especially while suckling) as it contracts. The cervix becomes firm over 3 days. The internal os closes by 3 days, the external os by 3 weeks. Lochia (endometrial slough, red cells and white cells) is passed *per vaginam*. It is red (*lochia rubra*) for the first 3 days, then becomes yellow (*lochia serosa*) then white over the next 10 days (*lochia alba*), until 6 weeks. The breasts produce milky discharge and colostrum during the last trimester. Milk replaces colostrum 3 days after birth. Breasts are swollen, red and tender with physiological engorgement at 3 to 4 days.

The first days If Rh-ve give anti-D (within 72h, see p 109). Check T°, BP, breasts, legs, lochia, fundal height. Teach pelvic floor exercises. Persistent red lochia, failure of uterine involution or secondary PPH (p 138) suggest retained products. Sustained hypertension may need treating (OHCM p 252). ►Check rubella immunity. Give Almevax® 0.5ml SC if non-immune but delay until postnatal exam if anti-D given. Check Hb on postnatal day 1 or ≥day 7 (post-partum physiological haemodilution occurs from day 2-6).

Puerperal pyrexia is a temperature >38°C in the first 14 days after delivery or miscarriage. Examine fully (chest, breasts, legs, lochia and bimanual vaginal examination). Culture MSU, high vaginal swabs, blood and sputum. 90% of infections will be urinary or of the genital tract. Superficial perineal infections occur around the 2nd day. *Endometritis* gives lower abdominal pain, offensive lochia and a tender uterus (on bimanual vaginal exam). Endometritis requires IV antibiotics (see below) and uterine curettage. For breast infection give flucloxacillin 250mg/6h PO early, to prevent abscesses. Suckling or breast expression should continue to prevent milk stagnation. Even if the cause of pyrexia is not apparent, it is wise to treat with amoxycillin 500mg/8h PO or IV and metronidazole 400mg/8h PO.

Superficial thrombophlebitis This affects 1% of women, presenting with a painful tender (usually varicose) vein. Give anti-inflammatories, eg indomethacin 25mg/8h PO pc. Bandage and elevate the leg. Recovery is usual within 4 days. *Deep vein thrombosis:* See p 151.

Puerperal psychosis (1:500 births): This is distinguished from mild depression which often follows birth by a high suicidal drive, severe depression (p 336), mania, and more rarely schizophrenic symptoms (p 358) with delusions that the child is malformed. If an acute organic reaction (p 350) is present, suspect puerperal infection. It may be worthwhile explaining to mothers that the puerperium is not always a time of joy, and they may feel down—and should let the midwife, health visitor or GP know about negative feelings, persistent crying, sleeplessness, and feelings of inadequacy—as early recognition of the problem is the best way to avoid what can be an intensely destructive experience. *Treatment:* See p 408.

The 6-week postnatal examination This gives a chance to: ●See how mother and baby relate. ●Do BP & weight. ●Do FBC if anaemic postnatally. ●Arrange a cervical smear if not done antenatally. ●Check contraceptive plans (p 68) are enacted (pills are best started at about 4 weeks postnatally). ●Ask 'have you resumed intercourse?' Sexual problems are common, and prolonged: ~50% report that intercourse is less satisfactory than pre-pregnancy, with major loss of libido, and dyspareunia the chief complaints.▪
Vaginal examination to check healing is *not* usually needed.

Fibroids in pregnancy

Fibroids 5/1000 Caucasian women have fibroids in pregnancy. They are commoner in negro women. They may cause abortion. They increase in size in pregnancy—especially in the 2nd trimester. If pedunculated they may tort. Red degeneration is when thrombosis of capsular vessels is followed by venous engorgement and inflammation, causing abdominal pain (± vomiting and low grade fever), and localized peritoneal tenderness—usually in the 2nd half of pregnancy or the puerperium.

Treatment is expectant (bed rest, analgesia) with resolution over 4-7 days.

Most fibroids arise from the body of the uterus and do not therefore obstruct labour, as they tend to rise away from the pelvis throughout pregnancy. If large pelvic masses of fibroids are noted prior to labour Caesarean section should be planned. Obstruction of labour also needs Caesarean section.

Ovarian tumours

These are found in ~1/1000 pregnancies. It is easier to distinguish them (lying as they do in the rectovaginal pouch) with an anteverted uterus than with a gravid retroverted uterus. Suspicion of presence of a tumour can be confirmed by ultrasound. Torsion of ovarian cysts is more common in pregnancy and the puerperium. Cyst rupture and haemorrhage into cysts may also occur, but are not more common than at other times. Torsion may present with abdominal pain, nausea, vomiting, shock, local tenderness (usually at 8-16 weeks). 2-5% of tumours are malignant. Suspect malignancy with ruptures (then biopsy other ovary). ~25% of malignant tumours will be dysgerminomas.

Tumours can become necrotic due to pressure on them in labour. Tumours lying in the pelvis can obstruct labour so Caesarean section will be needed unless they are cysts which can be aspirated under ultrasound control before labour.

Asymptomatic simple cysts <5cm can be left until after delivery if watched by ultrasound. Those 5-10cm may be aspirated under ultrasound control (and the aspirate examined cytologically).[1] Other tumours (those which are complex multilocular or with solid portions on ultrasound) should be removed at about 16 weeks' gestation (by which time the pregnancy is not dependent on the corpus luteum and miscarriage is less likely) to exclude carcinoma and prevent complications developing. If the diagnosis is made late in pregnancy and the tumour is not obstructing the pelvis, it is usual to let labour progress normally, and to remove the tumour in the early puerperium because of the risk of torsion then.

1 C Buckley 1989 *Br J Obstet Gynae* **96** 1021

Thromboembolism in pregnancy

Thromboembolism is important because it is one of the 3 leading causes of maternal death (others are eclampsia and bleeding). The risk of thromboembolism is increased 6-fold by pregnancy and it is thought to have an overall incidence of 0.3–1.6% of pregnancies of which 20–50% occur antenatally.[1] It is commoner in older women, those bed-rested, and after Caesarean section. Women with lupus anticoagulant risk arterial and venous thrombosis (may be in atypical veins, eg portal, arm).

151

Pulmonary embolism Small emboli may cause unexplained pyrexia, syncope, cough, chest pain, breathlessness. Pleurisy should be considered due to embolism unless there is high fever or much purulent sputum. Large emboli present as collapse with chest pain, breathlessness and cyanosis. There will be a raised JVP, third heart sound and parasternal heave.

Tests: Chest X-ray and ECG may be normal (apart from showing a tachycardia, the most consistent finding). ECG changes of deep S-wave in lead I and Q-wave and inverted T-wave in lead III can be caused by pregnancy alone. Blood gases may be helpful ($P_aO_2\downarrow$; $P_aCO_2\downarrow$).
Ventilation/perfusion lung scans are safe in pregnancy and diagnostic.

Treatment: Massive emboli may require prolonged cardiac massage and consideration of pulmonary embolectomy. Give heparin 10,000u IV, then 1000–2000u/h IVI by 0.9% saline pump. Monitor APPT (OHCM p 596). After 3–7 days on IV heparin the choice is between maintenance on long term heparin (eg 10,000u/12h SC with careful monitoring) or warfarin (between 13 and 36 weeks' gestation or if >7 days post-partum). Heparin has the disadvantage of osteopenia with prolonged use and can cause thrombocytopenia and alopecia. Warfarin, however, is teratogenic (Conradi-Hünermann syndrome, p 744) and needs changing over to heparin after 36 weeks' gestation. SC heparin is continued through labour with no increased risk of PPH, but epidural anaesthesia is contraindicated. Heparin post-partum is reduced to 7500u/12h SC. Continue for 6 weeks post-partum (or if postnatal thrombosis, for a minimum of 6 weeks).

Deep vein thrombosis (DVT) For signs and symptoms see OHCM p 92. Often symptoms are not clear-cut. Investigation is by contrast or radionuclide venography (ultrasound is too insensitive and ^{125}I fibrinogen must be avoided in pregnancy). Treatment is with heparin used throughout the pregnancy (as above), or warfarin. Meticulous control with regular INR estimations is necessary if warfarin is used.

Prophylaxis Most UK obstetricians use prophylaxis if there has been past-pregnancy thromboembolism (and some will, if thromboembolism on the Pill). Others decide that risks associated with long-term heparin or warfarin (see above) are too great antenatally[1] and only give antenatally if the patient has a history of recurrent thromboembolism, but will give IV dextran during labour, induction or Caesarean section (after taking plenty of blood in case crossmatch is needed) and then use heparin for 6 weeks post-delivery or heparin then warfarin (as above). IV dextran does not preclude use of epidurals. SC heparin may also be given when women are at high risk antenatally because of being bed-rested.

Recurrent thromboembolism Women with a history of recurrent thromboembolism, especially if arterial, atypical sites or associated with poor obstetric history should be screened for lupus anticoagulant and cardiolipin antibody presence. With recurrent thrombosis consider other causes of increased susceptibility to embolism: antithrombin III deficiency, protein C deficiency, protein S deficiency, abnormal fibrinogen forms and homocystinuria.

1 M de Swiet 1989 in *Medical Disorders in Obstetric Practice* pages 166–88, Blackwell

Cardiac disease in pregnancy

During pregnancy cardiac output increases to a maximum of 30-40% above non-pregnant levels. This is effected by ↑ heart rate and stroke volume.

Heart disease affects 0.5-2% of pregnant women. Examine the heart carefully early in all pregnancies (note normal variants associated with pregnancy, below): ask the opinion of a cardiologist if there is doubt: ●Past history (eg congenital heart disease, rheumatic fever). ●Previous Kawasaki disease (now a more common cause of acquired heart disease than rheumatic fever). ●Murmurs (other than those below).
Of maternal deaths due to heart disease, 60% occur after delivery. Cardiac failure can occur at any stage in pregnancy but risk increases as pregnancy advances and is most in the early puerperium.

Eisenmenger's syndrome (p 748), 1° pulmonary hypertension, and inoperable cyanotic heart disease are all associated with increased maternal mortality so advise avoidance of pregnancy. Termination may be medically recommended with these.

Seek specialist advice when managing patients with prosthetic valves requiring anticoagulation. Use of heparin pre-conception until 16 weeks may still risk valve thrombosis, but warfarin use at this time is associated with fetal malformation (p 744) and increased fetal loss.

Antenatal management Regular visits to cardiologist and obstetric combined clinic. Prevent anaemia, obesity and smoking. Ensure sufficient rest. Treat hypertension. Treat infections early. Give antibiotic cover for dental treatment (OHCM p 314). Examine carefully to exclude pulmonary oedema and arrhythmias at all visits. Heart failure requires admission.

Labour If patients have good cardiac reserves prior to labour, risks during labour are low. Have oxygen and drugs to treat cardiac failure to hand. Avoid the lithotomy position because of increased venous return after labour (the best position is semi-sitting). Aim for vaginal delivery at term with a short second stage (lift out forceps or ventouse). Give antibiotic cover eg ampicillin 500mg IV and gentamicin 80mg IV repeated twice at 8-hourly intervals, if valve or septal defect. Pain relief should be adequate. Epidurals are safe if hypotension is avoided. Beware large volumes of IV fluids. Avoid ergometrine. Caesarean section should not be attempted (except during eclampsia) if the patient is in failure. Cardiac failure is most likely within the first 24h after delivery, so especially careful attention should be paid to the patient during this time.

Cardiac failure Symptoms and signs of cardiac failure appearing in pregnancy will require admission, bed rest, and treatment with diuretics, and, in some cases, digoxin. If acute failure develops, give 100% oxygen, nurse semi-recumbent, and give frusemide 40mg IV slowly (<4mg/min), morphine 10mg IV, and aminophylline 5mg/kg by slow IVI (over >15min). If there is no improvement, consider ventilation. (There is no specific information on the use of nitrates in labour.)

Arrhythmias *Atrial fibrillation:* This may occur with mitral stenosis. It requires admission to hospital. Treatment is as for the non-pregnant and may include digitalization or cardioversion. *Supraventricular tachycardia:* This may precipitate cardiac failure. If Valsalva manoeuvre and carotid massage fails, anaesthetize and cardiovert.

Cardiological signs associated with pregnancy which may be normal

- Peripheral pulses are increased in volume.
- Neck veins pulsate more vigorously (but JVP should not be raised).
- Apex beat is more forceful (but <2cm lateral to midclavicular line).
- Oedema is extremely common (dependent).
- 1st heart sound is loud: a 3rd heart sound can be heard in 84%.
- An ejection systolic murmur is heard in 96% of women.
- Systolic or continuous murmurs over R and L 2nd intercostal spaces 2cm from sternal edge, modified by stethoscope pressure, are thought to be due to blood flow in mammary vessels.
- Venous hums may be heard in the neck (modified by posture).
- CXR may show slight cardiomegaly, increased pulmonary vascular markings, distension of pulmonary veins due to ↑ cardiac output.
- Q-waves and T-wave inversion in lead III are not abnormal.

Anaemia in pregnancy

▶Even a small PPH may become life-threatening if the mother is anaemic. Anaemia predisposes to infection, and makes heart failure worse. Worn-out, anaemic mothers may not cope with their offspring. Anaemia is the main cause of perinatal problems associated with malaria; above all, anaemia is a leading mechanism by which poverty exacts its morbid toll in pregnancy.

WHO definition of anaemia of pregnancy Hb < 11g/dl. By this standard 50% of women not on haematinics become anaemic. The fall in Hb is steepest around 20 weeks' gestation, and is physiological (p 80); indeed failure of Hb to fall below 10.5g/dl (but not further than ~9.5g/dl) indicates ↑risk of low birth weight or premature delivery.[1]

Who is prone to anaemia? Those who start pregnancy anaemic eg from menorrhagia, hookworm, malaria, with haemoglobinopathies; those with frequent pregnancies, twin pregnancy, or a poor diet.

Antenatal screening includes Hb estimation at booking, at 28 and 36 weeks. In black patients do sickle-cell tests, in others of foreign descent consider Hb electrophoresis for other haemoglobinopathies. From malarious areas consider malaria, and thick films. See p 159.

Treatment Pregnancy increases iron needs by 700–1400mg (per pregnancy), provided for by a pregnancy-induced 9-fold increase in iron absorption.[2] The routine use of daily oral iron has never been shown to offer any clinical benefit[3]—and harm is possible.[2] On this view, we should only offer iron to those likely to be iron deficient (see above). Parenteral iron may be given (to those with iron deficiency anaemia unable to tolerate oral iron) as iron sorbitol citric acid 1.5mg/kg IM daily (maximum 100mg/injection) as per Data Sheet at least 24h after stopping oral iron to prevent toxic reaction of headache, nausea and vomiting; or as a whole dose slow IVI of iron dextran (5%) under careful observation; the IV dose in ml = [0.0476 × wt in kg × (14.8 – Hb)] + 16. Give a small test dose first. CI: asthma, renal or liver disease, previous pyelonephritis. The rise in haemoglobin takes place over 6 weeks, so for late severe anaemia (Hb < 9g/dl) blood transfusion may be necessary. One unit of blood increases the Hb by ~0.7g/dl.

Thalassaemia These globin chain production disorders are found in Mediterranean, Indian and South-east Asian populations. Although anaemic, never give parenteral iron as iron levels are high. Seek expert advice as to use of oral iron and folate. β-thalassaemia does not affect the fetus but in homozygotes regular transfusions sustain life only until young adulthood. There are α chains in fetal HbF, so in α-thalassaemias the fetus may be anaemic or, if severe, stillborn. Mothers carrying lethally affected hydropic fetuses risk severe pre-eclampsia, and delivery complications due to a large fetus and bulky placenta. Prenatal diagnosis is possible for thalassaemias anticipated by parental blood studies.

Sickling disorders can affect blacks of African origin, Saudi Arabians, Indians and Mediterranean populations. Sickle-cell trait is not usually a problem. Sickle-cell disease predisposes to abortion, preterm labour, stillbirth, crisis. There is a chronic haemolysis (eg Hb 6–9g/dl). Regular 3–4U blood transfusions every 6 weeks (so a problem is development of atypical antibodies) prevents crises. Infection may induce crises, and dehydration exacerbate them; treat with exchange transfusions (OHCM p 588). Sickle-cell haemoglobin C disease is a milder variant of sickle-cell disease. Hb levels usually near normal so women may be unaware they are affected. They are still susceptible to sickling crises in pregnancy and the puerperium, so antenatal diagnosis is essential. Prenatal sickle-cell diagnosis is possible.

▶Aim for diagnosis at birth (cord blood) at the latest so that penicillin pneumococcal prophylaxis may be started (OHCM p 588).

1 P Steer 1995 BMJ i 489   2 JF Barrett 1994 BMJ ii 79   3 US Prev.Task Force 1993 JAMA 270 2846

HIV in labour[1]

In sub-Saharan Africa HIV is common (prevalence ~24% in apparently well antenatal patients—and there are millions of AIDS orphans). In Zimbabwe 120,000 HIV+ve mothers give birth each year. Many babies are not infected *in utero* but become so during parturition. Most mothers at risk of passing on HIV to the next generation (vertical transmission) do not want to know their HIV status, or cannot afford to find out. (If their husband is uninfected, what is the chance of him remaining loyal?) In our section on perinatal HIV (p 98) we comment that zidovudine and Caesarean section can prevent vertical transmission—but this is not much help if the nearest hospital is 3 hours away by wheelbarrow, and has only basic drugs. What is needed is much more cost-effective advice.

- Only give blood transfusions if absolutely necessary.
- Avoid any procedure likely to lead to maternal cells contacting fetal blood—eg external cephalic version, and amniocentesis.
- In instrumental deliveries, try to avoid abrasions of the fetal skin. Vacuum extractors may be preferable to forceps.
- When you clamp the cord, ensure there is no maternal blood on it.
- Artificial rupture of membranes and episiotomies should be left to the last possible moment, or avoided altogether.
- Avoid fetal scalp electrodes, and doing fetal scalp blood samples.
- If the membranes have ruptured, and labour is delayed or slow, consider the use of syntocinon to decrease exposure of the fetus to HIV.
- During Caesarean sections, open the last layer by blunt dissection, to avoid minor cuts from to the baby from the scalpel.
- Rinse the baby after birth. Wipe the face carefully, away from mouth, eyes and nostrils.
- Unless there is apnoea, avoid suction catheters to aspirate mucus from the nostrils. The baby's face is likely to be covered with the mother's blood at this stage, and you do not want to force HIV into nostrils.
- Health programmes are likely to end up encouraging breast feeding because there is no satisfactory alternative. Humanized milk is expensive, and immediately indicates to the mother's neighbours that she was HIV+ve. Using humanized milk might also compound problems by removing the one free method of contraception: lactational amenorrhoea.
- Offer advice on avoiding future pregnancies. This is not an easy area. Encouraging the use of condoms is fine, but many will want the added protection of the Pill. IUCDs promote bleeding, and may increase spread to men. This may also be a problem with the progestagen-only pill, but note that the latter may cause less ectropion than the combined pill, and this might be advantageous. Sterilization is the hardest choice, especially when the mother now has no living children because of HIV.

1 D Verkuyl 1995 *Lancet* 346 293

Diabetes in pregnancy

▶Meticulous control around conception reduces malformation rates.

Diabetes may be pre-existing or appear in pregnancy; glycosuria is common (glomerular filtration ↑ and tubular glucose reabsorption↓). Maternal glucose levels in pregnancy are normally very constant (4–4.5mmol/l) except after meals. Random glucose levels can be used for screening: levels >6.4mmol/l if last meal <2h from test, levels >5.8mmol/l if last meal >2h from test being unusual. Do 2 fasting blood glucoses to confirm diabetes (ie if >7.8mmol/l), and, if equivocal (eg >6–<7.8mmol/l), a 75g oral glucose tolerance test (OHCM p 474) and glycated Hb (HbA1c).

Complications *Maternal:* ↑Rates of pre-eclampsia (14.4%), polyhydramnios (25%—possibly due to fetal polyuria), preterm labour (17%—associated with polyhydramnios). Stillbirth near term was common.
Fetal: Malformation rates increase 3–4-fold. Sacral agenesis, almost exclusive to diabetic offspring, is rare (CNS and CVS malformations are much commoner). Babies may be macrosomic (too large) or sometimes growth retarded and are postnatally prone to hypoglycaemia, hypocalcemia, Mg^{2+}↓ and respiratory distress (surfactant production is delayed). They may be polycythaemic (so more neonatal jaundice).

Antenatal care Liaise with diabetologist. Confirm gestation with early ultrasound. Detailed abnormality scan at 18–20 weeks. Fetal echo at 24 weeks. Educate about benefits of normoglycaemia and home glucose monitoring. Insulin needs ↑ (by 50-100%). One approach to normoglycaemia uses Actrapid® 20min before meals with a Novopen®–p 262) + a single daily dose of Human Ultratard®. Admit unless there is good control at home. Avoid oral hypoglycaemics. Monitor fetal growth and well-being by ultrasound and cardiotocography.

Delivery Timing takes into account the control of diabetes, any pre-eclampsia, maturity and size of the baby, and with attention to fetal well-being. Delivery before 38 weeks may result in neonatal respiratory distress. Deliver the baby where there are good neonatal facilities. Traditionally, delivery was at 36–38 weeks to avoid stillbirth; but with close supervision pregnancies may go nearer to (but not beyond) term.

During delivery Avoid acidosis and monitor the fetus (p 116). Avoid maternal hyperglycaemia (causes fetal hypoglycaemia). Monitor glucose; prevent hyperglycaemia with extra insulin if β-sympathomimetics or glucocorticoids are used in preterm labour. Aim for vaginal delivery with a labour of less than 12h. Beware shoulder dystocia with macrosomic babies. With elective delivery, give normal insulin the evening before induction. During labour give 1 litre 7.5% glucose/8h IVI (ie 50ml 50% dextrose in 1 litre 5% dextrose), with 1–2u insulin/h via a pump. Aim for a blood glucose of 4.5–5.5mmol/l. Check glucose hourly. Insulin needs fall during labour (and immediately post-partum). Do a Caesarean section if labour is prolonged. Return to pre-pregnancy insulin doses 24h post-partum. Liaise with paediatricians in case the baby is affected by the above problems.

Post-natal ●Oral hypoglycaemics are contraindicated if breast feeding. ●Encourage pre-pregnancy counselling *before* next pregnancy (p 94) to transfer to insulin. ●Do a post-partum glucose tolerance test. ~5% of those with gestational diabetes will be insulin dependent in <5yrs (higher risk if aged <30, non-obese, no family history, and for first pregnancies).[1] 50% develop non-insulin dependent diabetes in <10yrs.
▶Advising exercise and avoiding obesity and smoking lowers this risk.

1 A Dornhorst 1994 *Br J Obs Gynae* 101 286

Thyroid disease in pregnancy

Biochemical changes during normal pregnancy
- Thyroid binding globulin and T4 output rise to maintain free T4 levels.
- High levels of HCG mimic thyroid stimulating hormone (TSH).
- There is reduced availability of iodine (in iodine limited localities).
- TSH may fall below normal in the first trimester.
- The best thyroid tests in pregnancy are free T4, free T3 and TSH.

Pre-pregnancy hyperthyroidism Treatment options include antithyroid drugs (but 60% relapse on stopping treatment), radioactive iodine (contraindicated in pregnancy or breast feeding: avoid pregnancy for 4 months after use), or surgery. Fertility is reduced by hyperthyroidism.

Hyperthyroidism during pregnancy This is usually Graves' disease. There is increased risk of prematurity, fetal loss, and possibly, of malformations.[1] Severity of hyperthyroidism often falls during pregnancy. Transient exacerbations may occur in the first trimester and post-partum. In the UK, treatment is usually with carbimazole: keep doses as low as possible. Once control is achieved aim to keep dose at ≤10mg/24h PO, keeping T4 at the top of the normal range. Some advocate stopping antithyroid drugs in the last month of pregnancy.[1] Propylthiouracil is preferred post-partum if breast feeding, (less concentrated in breast milk). When hyperthyroidism cannot be controlled by drugs, partial thyroidectomy can be done in the 2nd trimester. Screen all women with a history of Hashimoto's or Graves' disease for TRAb (thyrotrophin receptor antibodies—thyrotrophin is a synonym for TSH); high levels can cause fetal and neonatal hyperthyroidism: blocking them causes hypothyroidism.

Hypothyroidism Relatively rare, it is associated with relative infertility. In untreated hypothyroidism there are increased rates of miscarriage, stillbirth, premature labour[1] and abnormality.[2] T4 requirements may increase up to 50%. Monitor adequate replacement by T4 and TSH measurements each trimester.[1] Requirements return to pre-pregnancy levels post-partum.

Post-partum thyroiditis Prevalence: ~5%. There is transient hyperthyroidism initially followed by hypothyroidism (manifesting 4-5 months post-partum). 40% of those affected develop permanent thyroid failure and 25% of those with transient disease become hypothyroid 3-4 years later. TRAb antibodies are absent. The hyperthyroid phase does not usually need treatment as it is self-limiting. If treatment is required β blockers are usually sufficient. Antithyroid drugs are ineffective because thyrotoxicosis results from thyroid destruction causing increased thyroxine release, rather than increased synthesis. The hypothyroid phase is monitored for >6 months, and the woman treated if she becomes symptomatic. Withdraw treatment after 6-12 months for 4 weeks to see if long term therapy is required.[1] Hypothyroidism may be associated with post-partum depression, so check thyroid status of women with post-partum depression.

Neonatal thyrotoxicosis This occurs in 1% of offspring of women with a history of Graves' disease, even when controlled and drugs are no longer taken; it is due to TSH-receptor stimulating antibodies crossing the placenta. Fetal tachycardia (rate >160 beats per minute) in late pregnancy and intrauterine growth retardation may indicate fetal thyrotoxicosis. In those born to mothers taking antithyroid drugs it may not be manifest until the baby has metabolized the drug (7-10 days post-partum). Test thyroid function in affected babies frequently. Antithyroid drugs may be needed. It resolves spontaneously at 2-3 months, but perceptual motor difficulties, and hyperactivity can occur later in childhood.[1]

1 1995 *Drug Ther Bull* 33 75   2 R Hall 1993 *Br J Obstet Gynae* 100 512

Jaundice in pregnancy

Jaundice occurs in 1 in 1500 pregnancies. Viral hepatitis and gallstones may cause jaundice in pregnancy and investigation is similar in the non-pregnant. Those with Gilbert's syndrome or Dubin–Johnson syndrome (OHCM p 696, & p 700) have a good outcome with pregnancy, although jaundice may be exacerbated with the latter.

Intrahepatic cholestasis of pregnancy Pruritus eg over limbs and trunk and mild jaundice (bilirubin <100μmol/l) occurs in the second half of pregnancy—and may be associated with mild epigastric pain, anorexia, and malaise, with a bleeding tendency if malabsorption leads to lack of vitamin K. Plasma aspartate aminotransferase is mildly raised (<250u/l). Cholestasis and pruritus may resolve after high-dose IV S-adenosyl-L-methionine, suggesting that the defect is metabolic—but trial results have been rather disappointing.[1] There is risk of pre-term labour, fetal distress, and perinatal death, so monitor fetal wellbeing and consider delivery, at ~37 weeks, eg if any sign of fetal compromise. Monitor prothrombin time and give vitamin K if prolonged. Give vitamin K to the baby at birth. Jaundice clears within 4 weeks of delivery. It is a contraindication to oestrogen-containing contraceptive pills and recurs in 40% pregnancies.

Acute fatty degeneration of the liver This is rare but has a grave prognosis. The mother develops abdominal pain, jaundice, headache, and vomiting. It usually occurs after 30 weeks. There is hepatic steatosis with microdroplets of fat in liver cells. There is deep jaundice, uraemia, severe prolonged hypoglycaemia, and bleeding diathesis. If severe, liver, renal and clotting failure cause coma and death. Give supportive treatment for liver and renal failure and treat hypoglycaemia vigorously through CVP line. Correct clotting disorder with vitamin K and fresh frozen plasma. Expedite delivery. Epidural anaesthesia is contraindicated. Beware PPH and neonatal hypoglycaemia. Mortality can be as low as 18% maternal and 23% fetal.

Other causes of jaundice peculiar to pregnancy
- Associated with severe pre-eclampsia (hepatic rupture can occur).
- Rarely complicating hyperemesis gravidarum (can be fatal).
- Toxic hepatitis may occur if halothane is used for anaesthesia: its use should therefore be avoided.

Hepatitis B Check HBsAg in all women with jaundice. Staff should be careful to avoid contact with blood during delivery of those who are positive, and be especially careful with disposal of 'sharps'. Give 200u specific hepatitis B immune globulin (HBIG) IM to the anterolateral thigh to babies of affected mothers at birth—*plus* hepatitis B vaccine (0.5ml H-B-Vax II® into the other anterolateral thigh) at birth, 1 month and 6 months. Offer vaccination to *all* the family. Do serology at 12–15 months old. If HBsAg –ve and anti-HBs is present, the child has been protected.

Hepatitis E This causes maternal mortality in 20–39%, OHCM p 211.

1 E Fagan 1994 *BMJ* ii 1243

Malaria in pregnancy

In any woman who presents with abnormal behaviour, fever, sweating, DIC, fetal distress, premature labour, seizures, or loss of consciousness, always ask yourself: 'Could this be malaria?'.

Tests Do thick and thin films as appropriate. Confirm (or exclude) pregnancy, and seek expert help (eg from Liverpool, below).

Malaria can be particularly dangerous in pregnancy, with serious complications being more common, particularly in the non-immune. In some places, such as Thailand, malaria is the leading cause of maternal mortality. Cerebral malaria has a 50% mortality in pregnancy. Third stage placental autotransfusion may lead to fatal pulmonary oedema. Hypoglycaemia may be a feature of malaria in pregnancy (irrespective of whether quinine has been given).[1]

Other associations between malaria and pregnancy are anaemia, abortion, stillbirth, and low birthweight and prematurity. PPH is also more common. Hyper-reactive malaria splenomegaly (occurs typically where malaria is holoendemic) may contribute to anaemia via increased haemolysis.[2]

Treatment of malaria is as described in OHCM p 196. Chloroquine and quinine (in full dosage) appear to be safe in pregnancy, but beware hypoglycaemia. For *falciparum* malaria, expect to use quinine as chloroquine resistance is widespread. Mefloquine can be used in the second and third trimesters for the treatment of uncomplicated falciparum malaria.

Transfer earlier rather than later to ITU. Women with a haematocrit <20% should get a slow transfusion of packed cells, if compatible, fresh, pathogen-free blood is available. Remember to include the volume of packed cells in calculations of fluid balance. Exchange transfusions may be needed—if facilities are available.

During labour, anticipate fetal distress, fluid-balance problems, and hypoglycaemia. Monitor appropriately.

Prevention is of central importance. Advise mothers not to travel to malarious areas. If this is unavoidable, prescribe prophylaxis (OHCM p 196). Emphasize the importance of preventive measures such as mosquito nets and insect repellents. Chloroquine and proguanil are used in normal doses, if *P falciparum* strains are sensitive—but with proguanil, concurrent folate supplements should be given.

Maloprim® (dapsone+pyrimethamine) is contraindicated in 1st trimester; if given in 2nd or 3rd trimesters, give folate supplements too. Mefloquine should be avoided as prophylaxis (exclude pregnancy before use, and avoid pregnancy for 3 months after use).

▶When in doubt, it is best to telephone an expert, eg, in the UK, from Liverpool (tel. 0151 708 9393).

1 H Gilles 1992 *Management of Severe and Complicated Malaria*, WHO, Geneva   2 I Bates 1994 *Trans Roy Soc Trop Med Hyg* **88** 277

Renal disease in pregnancy

▶Treat asymptomatic bacteriuria in pregnancy. Check that infection and bacteriuria clear with treatment.

Asymptomatic bacteriuria This is found in 2% of sexually active women, and is more common (up to 7%) during pregnancy. With the dilatation of the calyces and ureters that occurs in pregnancy, 25% will go on to develop pyelonephritis. Pyelonephritis can cause fetal growth retardation, fetal death and premature labour. This is the argument for screening all women for bacteriuria at booking. If present on 2 MSUs treatment is given (eg amoxycillin 250mg/8h PO with a high fluid intake). Test for cure after 1 and 2 weeks. If the organism is not sensitive to amoxycillin, consider nitrofurantoin 50mg/6h PO with food.

Pyelonephritis This may present as malaise with urinary frequency or as a more florid picture with raised temperature, tachycardia, vomiting and loin pain. It is common at around 20 weeks and in the puerperium. Urinary infections should always be carefully excluded in those with hyperemesis gravidarum and in those admitted with premature labour. Treatment is with bed rest and plenty of fluids. After blood and urine culture give IV antibiotics (eg ampicillin 500mg/6h IV, according to sensitivities) if oral drugs cannot be used (eg if vomiting). MSUs should be checked every fortnight for the rest of the pregnancy. 20% of women having pyelonephritis in pregnancy have underlying renal tract abnormalities and an IVU or ultrasound at 12 weeks' post-partum should be considered. In those who suffer repeated infection, nitrofurantoin (100mg/12-24h PO with food) may prevent recurrences. Avoid if the glomerular filtration rate is <50ml/min. SE: vomiting, peripheral neuropathy, pulmonary infiltration and liver damage.

Chronic renal disease With mild renal impairment without hypertension there is little evidence that pregnancy accelerates underlying renal disorders. Patients with marked anaemia, hypertension, retinopathy, or heavy proteinuria should avoid pregnancy as further deterioration in renal function may be expected and fetal loss is considerable (up to 60%). Close collaboration between physicians and obstetricians during pregnancy in those with renal disease is the aim. Induction of labour may become advisable in those with hypertension and proteinuria, or if fetal growth is retarded.

Obstetric causes of acute tubular necrosis Acute tubular necrosis may be a complication of any of the following situations:
● Septicaemia (eg from septic abortion or pyelonephritis).
● Haemolysis (eg sickling crisis, malaria).
● Hypovolaemia eg in pre-eclampsia; haemorrhage (APH eg abruption, PPH, or intrapartum); DIC; abortion—or adrenal failure in those on steroids not receiving booster doses to cover labour.

Whenever these situations occur monitor urine output carefully (catheterize the bladder). Aim for >30ml/h output. Monitor renal function (U&E, creatinine). Dialysis may be needed (OHCM p 390).

Epilepsy in pregnancy[1]

▶If seizures occur in pregnancy, think *could this be eclampsia?*

Epilepsy *de novo* is rare in pregnancy. Epilepsy affects ~0.5% of women of childbearing age so a unit with 3000 deliveries per year has ~15 pregnant epileptic women at any one time. Seizure rate during pregnancy increases in ~25% of epileptic women, reduces in ~22% and is unchanged in the remainder. Where seizure rate increases this is often associated with lack of compliance with medication, or sleep deprivation. It is unusual for seizures to recur in pregnancy when preceded by a long seizure-free period.

Complications[2] *Maternal:* Increased risk of: eclampsia and pre-eclampsia; hyperemesis; anaemia; vaginal bleeding; premature labour.
Fetal: Increased risk of: prematurity; stillbirth; neonatal and perinatal death; haemorrhagic disease of the newborn.

Abnormality Epilepsy in the mother (but not the father) is associated with fetal abnormality, particularly cleft lip and palate. Relative risk for a fetus having clefts compared to the non-epileptic population is 1 if the mother develops epilepsy after the pregnancy; 2.4 if she develops it after conception but is not treated with medication; and 4.7 if the fetus has been exposed to anticonvulsant drugs. Children of epileptic mothers taking medication are twice as likely to have some form of significant malformation, giving a malformation rate of 6%; (a 94% chance of the mother having a normal baby). Malformation is commoner if more than one anticonvulsant is taken. Neural tube defects are commoner with carbamazepine and sodium valproate so screen for these (p 166). Phenytoin and phenobarbitone cause congenital heart disease and cleft lip.[2]

Management If a woman has been seizure-free for 2 or 3yrs it may be best to try to withdraw medication prior to a planned pregnancy.[1] For the MRC (Medical Research Council) withdrawal régime, see OHCM p 451. Where anticonvulsants are needed the dose of the chosen drug should be kept as low as possible. Plasma monitoring is necessary as plasma levels may fall (particularly in the 3rd trimester). Care is needed interpreting plasma phenytoin levels as free levels may remain the same despite a reduced bound level. Folate supplements should be given with phenytoin and phenobarbitone; low-dose supplements (~500µg/day) appear sufficient. Régimes for treating status epilepticus are similar to those for the non-pregnant except that the fetus should be monitored. As phenobarbitone, phenytoin, and sodium valproate increase risk of neonatal bleeding give vitamin K to the baby at birth (p 244).

Babies will clear anticonvulsants from their blood at variable rates after birth. Neonates can suffer barbiturate withdrawal with hyperexcitability, tremulousness, impaired suckling and occasionally fits. Usually babies require only supervision, but if very tremulous, or if they fit, they may be given 3–5mg/kg/day phenobarbitone.

Puerperal readjustment of medication may be needed. Mothers may safely breast feed but phenobarbitone is best avoided in lactating mothers, if possible, as it may cause drowsiness in the baby.

1 A Hopkins 1989 *Medical Disorders in Obstetric Practice*, 2ed, page 734–45, Blackwell Scientific Publications 2 *Drug Ther Bul* 1994 **32** 49

Perinatal mortality

The perinatal mortality rate is the number of stillbirths and deaths in the first week of life per 1000 total births. Stillbirths only include those fetuses of >24 weeks' gestation, but if a fetus of <24 weeks' gestation is born and then shows signs of life it is counted as a neonatal death in the UK and is therefore expressed in the figures (if dying within the first 7 days). Other countries use different classificatory criteria—including stillbirths from 20 weeks and neonatal deaths up to 28 days after birth, so it is not always easy to compare statistics.

Perinatal mortality is affected by many factors. Rates are high for *small* (61% of deaths are in babies <2500g) and *preterm* babies (70% of deaths occur in the 5% who are preterm). See p 102 & p 180. *Regional variation* in the UK is quite marked with rates for the Oxford region being considerably lower than for Merseyside. There is a *social class variation* with rates being less for social classes 1 and 2 than for classes 4 and 5. *Teenage mothers* have higher rates than mothers aged 20-29. From 30yrs rates rise until they are 3-fold higher than the low-risk group (20-29 years) by the age of 40-45. *Second babies* have the lowest mortality rates. Mortality rates are doubled for 4th and 5th children, trebled by 6th and 7th (this effect is not independent of social class as more lower social class women have many children). Rates are lower for *singleton births* than for multiple. Rates are higher for the offspring of mothers of *Bangladeshi* or *West Indian extraction* living in the UK, but are the same for those of Pakistani and UK extraction.

Perinatal mortality rates in the UK have fallen over the years from rates of 62.5/1000 in 1930-5 to 12-14/1000 now. Declining mortality reflects improvement in standards of living, improved maternal health and declining parity, as well as improvements in medical care. The main causes of death are congenital abnormalities (22%), unclassified hypoxia (18%), placental conditions (16%), birth problems including cord problems (11%), maternal conditions and toxaemia (8%).

Examples of how changed medical care may reduce mortality are:
- Antenatal detection and termination of malformed fetuses.
- Reduction of mid-cavity procedures and vaginal breech delivery.
- Detection of placenta praevia antenatally.
- Prevention of rhesus incompatibility.
- Preventing progression of preterm labour.
- Better control of diabetes mellitus in affected mothers.
- Antenatal monitoring of 'at risk' pregnancies.

While the challenge must be to reduce morbidity and mortality still further, this should not blind us to other problems that remain, such as the 'over-medicalization' of birth; the problem of reconciling maternal wishes to be in charge of her own delivery with the immediate needs of the baby; and the problem of explaining risks and benefits in terms that both parents understand, so that they can join in with the decision-making process.

Maternal mortality

Maternal mortality is defined in the UK as the death of a mother occurring during pregnancy or labour, or as a consequence of pregnancy within 1 year of delivery or abortion. The International Federation of Gynaecology and Obstetrics excludes deaths more than 42 days from delivery or abortion. Deaths are subdivided into those from 'direct causes'—those in which the cause of death is directly attributable to pregnancy (for example abortion, eclampsia, haemorrhage)—and associated deaths. Associated deaths are subdivided into 'indirect deaths'—those resulting from previous existing disease or disease developed during pregnancy, and which were not due to direct obstetric causes but were aggravated by pregnancy (eg heart disease). The other group—'fortuitous deaths'—are those occurring during pregnancy or puerperium but which are unrelated to pregnancy.

Since 1952 there have been confidential enquiries into maternal deaths. Reports on these are brought out 3-yearly. Since 1979 late deaths (>42 days from delivery or abortion) have been put into a separate group. Prior to 1979, as a result of the enquiries a certain proportion of deaths were considered to have had avoidable factors (this term was used to denote departures from acceptable standards of care by individuals, including patients) but since 1979 the wider term of 'substandard care' has been used to cover failures in clinical care and other factors, such as shortage of resources and backup facilities.

Maternal mortality has almost halved every decade since reports have been issued (deaths per 100,000 maternities have been 67.1 in 1955-7, 33.3 in 1964-6, 11 in 1973-5, and 7 in 1988-90).[1] It has been estimated that maternal deaths rates/100,000 births are 640 in Africa; 420 in Asia; 270 in Latin America; 30 in all developed countries and <10 in Northern and Middle Europe.[1]

In 1988-90, 277 deaths were reported excluding late deaths. Of these 52% were direct obstetric deaths; 34% were indirect obstetric deaths and 14% were fortuitous. Rates varied considerably by region, direct obstetric mortality rate being 2.4/100,000 births in Northern Ireland in 1985-90 to 9.7/100,000 births in North West Thames Regional Health Authority.

In 1988-90 hypertensive disease was the commonest cause of direct death in the UK (18.6% of deaths), pulmonary embolism the cause in 16.6% of direct deaths. Other direct causes: haemorrhage (15.2%); ectopic pregnancy (10.3%); amniotic fluid embolism (7.6%); abortion (6.2%); sepsis excluding abortion (4.8%); anaesthesia (2.8%); and ruptured uterus (1.4%) of direct deaths.

Death rate from Caesarean section for the 1985-90 period was 0.33 per 1000 operations.

The maternal mortality rate was similar to the previous triennium for direct deaths), but the rate of considered substandard care was lower in this triennium. Care was considered substandard in 49% of cases of direct death showing that room for improved care still exists.

1 DoH 1994 *Report on Confidential Enquiries into Maternal Deaths in the United Kingdom 1988-90*, London, HMSO

① directly attribs to preg
② assoc'd DM HT etc.
③ fortuitous.

1988 - 90.   7:100,000.

① hypertensive   1988 - 90.
② PE.
③ haemorrhage.
④ ectopic pregnancy

RIP rate CS.   0.33:1000
                procedures.

substandard care 49%.

Prenatal diagnosis

Parents at high risk of producing an abnormal baby may be offered prenatal diagnosis if that permits more effective treatment of the expected defect, or (more often) if they wish to terminate any abnormal pregnancy.

High-risk pregnancies: maternal age >35 (chromosome defects); previous abnormal baby or family history of inherited condition; high-risk population (eg sickle cell in Afro-Caribbeans).
For details of problems which can be detected, see p 210–11.

Problems ●Anxiety while false positive screening results are investigated. ●Termination of normal pregnancies, eg when aborting male fetuses of carriers of X-linked conditions. ●Most congenital abnormalities occur in low-risk groups and are missed by selective screening. ●The services available, and the population to whom they are offered, vary widely round the country. ●Termination of female fetuses in cultures where males are more highly valued. ●Devaluation of recent moves to promote a more positive view of handicapped or 'special needs' children.

α-Fetoprotein (AFP) AFP is a glycoprotein synthesized by the fetal liver and GI tract. Fetal levels fall after 13 weeks, but maternal (transplacental) serum AFP continues to rise to 30 weeks. Maternal AFP is measured at 17 weeks. In 10% with a high AFP there is a fetal malformation,[1] eg an open (but not closed) neural tube defect, exomphalos, posterior urethral valves, nephrosis, GI obstruction, teratomas, Turner's syndrome (or normal twins). In ~30% of those with no malformation, there is an adverse outcome, eg placental abruption and 3rd trimester deaths.[2] ►Monitor closely. 1 in 40 with a low AFP have a chromosomal abnormality (eg Down's). AFP is lower in diabetic mothers. NB: as this test is non-specific on its own, it is of use for preliminary screening; those with abnormal values may be offered further tests (see below, and p 168 for the 'triple test').

Amniocentesis See p 120. Amniocentesis is carried out under ultrasound guidance. Fetal loss rate is 0.5–1% at ~16 weeks' gestation, but ~5% for early amniocentesis at 10–13 weeks. Amniotic fluid AFP is measured (a more accurate screen for neural tube defects than maternal serum), and cells in the fluid are cultured for karyotyping and enzyme and gene probe analysis. Cell culture takes 3 weeks so an abnormal pregnancy must be terminated at a late stage.

Chorionic villus biopsy At 8–10 weeks, the developing placenta is sampled using a transcervical catheter or a transabdominal needle under ultrasound guidance. Karyotyping takes 2 days, enzyme and gene probe analysis 3 weeks, so termination of abnormal pregnancies is earlier, safer and less distressing than after amniocentesis. Fetal loss rate is ~4%. The transabdominal approach can be used throughout pregnancy, is easier to learn, allows better access to placental site, and allows more villi to be aspirated. It may result in fewer lost pregnancies.[3] The procedure does not detect neural tube defects—and may itself cause fetal malformations.

High resolution ultrasound (See p 118) At ~18 weeks a skilled operator can detect an increasingly wide range of external and internal structural anomalies.

Fetoscopy This is carried out at ~18 weeks under ultrasound guidance. External malformations may be seen, fetal blood sampled and organs biopsied. The fetal loss rate is ~4%.

---

1 F Cunningham 1991 *NEJM* **325** 55   2 D Waller 1991 *NEJM* **325** 6   3 S Smidt-Jensen 1992 *Lancet* **340** 1327

The triple blood test

Maternal serum AFP, unconjugated oestriol and total human chorionic gonadotrophin (HCG) are assessed in relation to maternal age, weight and gestation (determined by ultrasound) to estimate risk of Down's syndrome (and, in some places, of neural tube defects). If the test (and subsequent tests on those found to be at risk) were available to all mothers, detection of Down's syndrome would increase from 15% to about 50% (48% and 58% in 2 large studies, *N*=38,000).[1] [2]

Problems with the triple blood test—the triple conundrum

1 *It may not be the best test:* Some authorities have pointed out that adding unconjugated oestriol to the test could merely increase uncertainty. Also, there is evidence that *free* HCG is a better marker than total HCG.[3] A quadruple tests measures α and β HCG separately. It has also been pointed out that the wrong statistical test was used to demonstrate the test's benefits. Using Fisher's exact test to compare the 5 additional cases detected over and above what would have been detected by screening based on age alone, the Wald figures would not have been significant (*P*=0.24).[4] As the age at which mothers have their babies increases, any relative benefit of the triple test over age-related screening diminishes.

2 *The cost–benefit sums are hard to calculate*—as benefits depend on a high uptake of amniocentesis and termination. The detection rate is only 39% in those <37yrs. If you assume 100% uptake of amniocentesis and 100% uptake of abortion the cost per detected pregnancy is £36,256 (£28,000 if the detection rate is ~60%). This cost increases to £48,780 with 75% uptake of amniocentesis and 90% uptake of termination.[5] The position is further complicated by ethnic differences in reference intervals that have yet to be quantified. For example, the median value of HCG is too low for non-Caucasians (who will therefore have a higher false positive rate).[6]

3 *The emotional cost to the mother is impossible to calculate:* From the parents' point of view, a telling statistic is that 56 out of every 57 women under 37yrs old who had a +ve test, proved, after amniocentesis, *not* to have an affected fetus. Amniocentesis causes fetal loss, and these losses will almost always be of normal babies.

We have no idea of the best way of counselling parents before the test. If you just hand out a leaflet, few will read it, and then when it comes to amniocentesis and termination, many will refuse—and the screening test wastes money, as well as laying health authorities open to litigation: 'I never understood that I might lose a normal baby . . .' The alternative is to provide full details at the time of the initial blood test—what the 'goodies' have always recommended. The irony is that gaining informed consent is then the most expensive part of the test, and one which itself could cause much distress. Imagine an overjoyed expectant mother arriving in the clinic serenely happy in fulfilling her reproductive potential: the quintessence of health. She leaves only after being handed ethical conundrums of quite staggering proportions, involving death, disease, and human sacrifices, and a timetable for their resolution which would leave even the most fast-moving philosopher breathless and disorientated, and which may leave her forever bereft of one of Nature's most generous gifts: the fundamental belief in one's own wholeness.

1 J Haddow 1992 *NEJM* **327** 588   2 N Wald 1992 *BMJ* ii 391   3 K Spencer 1992 *BMJ* ii 769
4 A Allman 1992 *BMJ* ii 768   5 V Macri 1992 *BMJ* ii 768   6 C Ford 1996 *BMJ* i 1040

3. Paediatrics

Principal sources *Archives of Disease in Childhood Pediatrics* (USA), and Advanced Life Support Group 1993 *Advanced Paediatric Life Support*, BMA

Relevant pages in other chapters: Infectious diseases (OHCM p 170–250); perinatal infection p 100; anorexia nervosa 348, play therapy p 384, autism p 402; squint p 486, retinoblastoma p 506; the painful hip in children p 630; the osteochondritides (Perthes' disease p 630, Scheuermann's and Köhler's diseases p 650); congenital dislocation of the hip p 632; scoliosis p 620; deafness p 540; otitis media and glue ear p 536–8; tonsillitis p 556; stridor p 588; acne p 592, osteomyelitis p 644; burns p 688; accidents p 678–80; unusual eponymous syndromes: p 742–759

Child health, and the central facts dominating paediatric medicine
42% of the world's births are in countries with under-5s mortality >140/1000; 20% of infants in these countries die before their 5th birthday.[1] Most of these deaths are preventable, not by applying the complicated measures described here, but by the provision of clean water, enough food, and simple vaccinations, along with the encouragement of breast feeding. Today's girl is tomorrow's (or this afternoon's) mother. If she is to have healthy, happy, well-nourished children, first and foremost she needs to love them—but love isn't all she needs. She need not be wealthy, but being poor is *so often* fatal. To escape from poverty she needs to be literate, know how to plan her family, find a midwife, a friendly doctor and a reasonable school for her children. Family support helps, as does a faithful husband who can provide a rôle-model which does not lead to alcohol, drug addiction, HIV, and preventable accidents—the chief threats to adolescent health.[2]

Introductory note The newcomer to paediatrics is presented with a double difficulty: not only are children in many important respects quite unlike adults, but they are themselves a very heterogeneous group: the neonate is unlike the toddler—who is unlike the school child or the adolescent. A whole book along the lines of this work could be written on each of these groups of children, and still give a rather sketchy account. But the newcomer should know, lest he becomes down-hearted, that no one can say he has complete mastery of this subject. Rather than aiming for completeness, the aim is to be an effective paediatrician. This entails not endless hours in the library, but time on the wards (and in patients' homes, if the reader is lucky enough to have access to general practice patients)—to enlarge on those innate skills of listening and talking to mothers and children in language that each understands. Doctors who look after children have unrivalled opportunities to practise a holistic art, and it is possible that these doctors have a special skill here. We note with great interest that most patients between the ages of 15 and 20 who have acute leukaemia treated by paediatricians are cured—up to 63%, whereas only 32–42% of this age group survive if treated in adult units.

Time on the wards gives the embryo-paediatrician the opportunity to learn that great skill which no book can fully teach: how to recognize the seriously ill child. With the slow and painful accretion of many days and nights 'on call' the paediatrician's flair and instinctive judgment will develop in this regard. This flair can never be had from books alone, nor can it be had entirely without them, for as William Osler said: to study medicine without books is to sail an uncharted sea; to study medicine from books alone is not to go to sea at all. With this metaphor in mind we offer the following set of charts with the warning that they do not all use the same scale, and that it has not been possible to mark every reef. The blank charts are included to encourage new voyages.

1 ME Wegman 1992 *Paediatrics* **90** 835   2 LD Jacobson 1994 *Br J Gen Prac* **44** 420

Eliciting the history

The aim is not only to reach a diagnosis, but also to establish a good relationship with the parents and child, so that if there is nothing the matter with the child the parents feel able to accept this. Conversely, if there is serious or untreatable illness, the aim is to build up trust, so that the parents are able to accept the best advice, and the child feels that he or she is in safe, friendly hands. So if possible, avoid any hurry or distractions. Introduce yourself; explain your rôle in the ward or consulting room.

Presenting complaints Record the child's and mother's own words.

The present illness When and how did it start? Was the child quite well beforehand? How did it develop? What aggravates or alleviates it? Has there been contact with infections? Has the child been overseas recently?

Especially in infants, enquire about feeding, excretion, alertness, and weight gain. After ascertaining the presenting complaint further questioning is to test the various hypotheses of differential diagnosis.

Past health *In utero:* Toxaemia, rubella, Rh disease.
At birth: Prematurity, duration of labour, type of delivery, birth weight, resuscitation required, birth injury, malformations.
As a neonate: Jaundice, fits, fevers, bleeding, feeding problems. *& Scans (U)?*
Ask about later illnesses, operations, accidents, screening tests, drugs, allergies, immunization, travel and drug or solvent abuse.

Development (p 290) Does mother remember milestones reliably? Consanguinity is common in some cultures and may be relevant to disease.

Family history Stillbirths, TB, diabetes mellitus, renal disease, seizures, jaundice, malformations, others. Are siblings and parents alive and well? Find out if a parent has had a myocardial infarction before 40 years old—if so, do serum lipids (>40% of these children will turn out to have hyperlipidaemia: the sooner it is treated, the better).

Social history It may be vital to know who the father is, but very damaging to ask the mother directly. Be prepared to allow information to surface slowly, after chats with friendly nurses. Ask about play, eating, sleeping (excessively wrapped or liable to cold?) and schooling. Who minds the child if the parents work? What work do they do? Ask about their hopes, fears and expectations about the child's illness and his stay in hospital.

▶ *If the family does not speak your language, find an interpreter.*

Is this child seriously ill?^{APLS}

Recognizing that a child is seriously ill and in need of prompt help is a central skill of paediatrics. It can be uncanny to watch the moment of transformation that this recognition brings about in a normally reflective, relaxed-and-easy doctor who is now galvanized into an efficient, relentless device for delivering urgent care—without the waste of a single word or action. If you are new to paediatrics, take every opportunity you can to observe such events, and, later, closely question the doctor about what made him act in the way he did, using this page to prepare your mind to receive and remember the answers he gives.

Serious signs ●Agitation or coma. ●Unreactive or unequal pupils.
●Laboured breathing (uses alae nasae ± sternomastoids); hyperventilation.
●Stridor (with every breath); wheeze; apnoea; Cheyne–Stokes breathing.
●Ashen, blue or mottled complexion.
●Weak/absent peripheral pulses, poor capillary refill. To elicit this, press on a digit for 5sec. Capillary refill normally takes <2sec. NB: BP *and pulse are unreliable markers in early shock.*
●Decorticate (flexed arms, extended legs) or decerebrate (arms and legs extended) posture; opisthotonus (head extended, back arched).

Age (yrs)	Breathing rate (/min)	Pulse (/min)	Systolic BP mmHg
<1	30–40	110–160	70–90
2–5	20–30	95–140	80–100
5–12	15–20	80–120	90–110
>12	12–16	60–100	100–120

Rapid assessment of the seriously ill child (Should take <1 minute.)
Airway: ●Patency.
Breathing: ●Work of breathing *Look:* Rate; recession; accessory muscle use; cyanosis *Listen:* Stridor; grunting; wheeze.
Circulation: ●Pulse rate & volume; capillary refill; skin colour & T°
Level of response: ●A=Alert ●V=responds to Voice ●P=responds to Pain; ●U=Unresponsive (ie 'AVPU'). Also assess pupils & posture.

Causes Sepsis and trauma are common. Others:
●Intussusception (p 198)
●Inborn error of metabolism, eg: hypoglycaemia; U&E disturbance
●Haemolytic uraemic syndrome (p 280)
●Gastrointestinal: D&V; obstruction (volvulus)
●Arrhythmias
●Cardiomyopathies
●Myocarditis
●Congenital heart disease
●DIC (p 244)
●Reye's syndrome (p 756)
Be alert to rare syndromes, eg toxic shock syndrome (OHCM p 222) or haemorrhagic shock encephalopathy syndrome (HSES; abrupt shock, with bloody diarrhoea, coma, convulsions, hepatomegaly, oliguria, and DIC).

▶**Action**—if very ill ●100% O₂ by tight-fitting mask with reservoir.
●Colloid: 20ml/kg bolus IV; repeat if no better. ≥60ml/kg may be needed.
●Do blood glucose (lab & ward test); U&E; FBC; thick film if tropical travel.
●Crossmatch blood if trauma is possible, or patient looks anaemic.
●Do swabs and blood culture before starting blind treatment with IV antibitics, eg ceftriaxone or cefotaxime (p 258).
●If worse, insert CVP. Get expert help. Do blood gases and clotting screen.
●Ventilate mechanically, eg if bicarbonate is to be given (pH <7.15).
●Consider the need for CXR, MSU, and lumbar puncture.
●If perfusion is still poor (in spite of 60ml/kg colloid), or CVP >10cmH₂O, dopamine ± dobutamine may be needed. Vasodilatation with nitroprusside (p 304) may rarely be needed to promote cardiac output.

Examination of the neonate

The aim is to discover any abnormality or problem, and to see if the mother has any questions or difficulties with her baby. The following is a recommended routine before the baby leaves hospital—or during the first week of life for home deliveries.

Before the examination Find out the weight. Was the birth and pregnancy normal? Is mother Rhesus -ve? Find a quiet, warm, well-lit room. Enlist the mother's help. Explain your aims. Does she look angry or depressed? Listen if she talks. Examine systematically, eg from head to toes.

The head Circumference (30-38cm, p 294), shape (odd shapes from a difficult labour soon resolve), fontanelles (tense if crying or intracranial pressure↑; sunken, eg if dehydrated).

The eyes: Check for corneal opacities and conjunctivitis?

The ears: Shape and position. Are they low set (ie below the eyes)? The nose, when pressed, is a useful indicator of jaundice. Breathing out of the nose (shut the mouth) tests for choanal atresia.

The complexion: cyanosed, pale, jaundiced or ruddy?

Mouth: Look inside. Insert a finger. Is palate intact? Is suck good?

Arms and hands Single palmar creases (normal or Down's). Note the porter's tip sign of Erb's palsy (injury to C5 & 6 trunks).

The thorax Watch the respiratory movements. Note any grunting or intercostal recession (respiratory distress). Palpate the precordium and apex beat. Listen to the heart and lungs. Inspect the vertebral column for neural tube defects.

The abdomen Expect to feel the liver and spleen. Are there any other masses? Next inspect the umbilicus. Is it healthy? Lift the skin to assess skin turgor. Inspect the genitalia and anus. Are the orifices patent? 93% of neonates pass urine within the first 24h. Is the urinary meatus misplaced (hypospadias), and are both testes descended? The neonatal clitoris often looks rather large. Bleeding PV may be a normal variant following maternal oestrogen withdrawal.

The lower limbs Test for congenital dislocation of the hip (p 632). Avoid repeated tests as it hurts—and may induce dislocation. Is there any radiofemoral pulse delay (coarctation)? Note talipes (p 632). Are the toes: too many, too few, or too blue?

Buttocks/sacrum Is there a 'mongolian spot'? This is blue and harmless. Tufts of hair ± dimples: is there spina bifida occulta? Any pilonidal sinus?

CNS by handling the baby. Now intuition can be most helpful in deciding whether the baby is ill or well. Is he jittery (hypoglycaemia, infection, hypocalcaemia)? There should be some degree of control of the head. Do the limbs move normally, and is the tone floppy or spastic? The Moro reflex rarely adds important information (and is uncomfortable for the baby). It is performed by sitting the baby at 45°, supporting the head. On momentarily removing the support the arms will abduct, the hands open and then the arms adduct. Stroke the palm to elicit the grasp reflex. Use the ophthalmoscope (red reflex may be absent in retinoblastoma).

Is the baby post-mature, light-for-dates or premature (p 180)?

▶Discuss any abnormality with the mother and father after liaising with a senior doctor.

Minor neonatal problems

Skin Blemishes are common and worry parents. Most are harmless.

Milia: These tiny cream papules on the nose and sometimes on the palate (Ebstein's pearls) are sebaceous cysts. They disappear.

Erythema toxicum (neonatal urticaria): This is red blotches, often with a central white vesicle. It is harmless. Each spot lasts about 24 hours (try ringing one), in contrast to septic spots which develop obvious pustules. If in doubt take a swab.

Miliaria (heat rash): This itchy red rash fades rapidly when the baby is unwrapped.

Stork mark: These are areas of capillary dilatation on the eyelids, central forehead and back of the neck—where the baby is deemed to have been held in the stork's beak. They blanch on pressure and fade with time.

Harlequin colour change: One side of the face or body suddenly flushes for a few minutes. It is a shortlived vasomotor event.

Peeling skin: Common in postmature babies, it does not denote future skin problems. Olive oil prevents skin folds from cracking.

Petechial haemorrhages, facial cyanosis, subconjunctival haemorrhages: These temporary features generally reflect suffusion of the face during delivery.

Swollen breasts: These occur in both sexes and occasionally lactate (witch's milk). They are due to maternal hormones and gradually subside if left alone, but if infected require antibiotics.

The umbilicus It dries and separates through a moist base at about day 7. Signs of infection: odour, pus, periumbilical red flare, malaise. Isolate the baby, take swabs and blood cultures, give antibiotics. Granuloma: exclude a patent urachus and cauterise with a silver nitrate stick

Sticky eye This is common and usually due to an unopened tear duct (p 482), but swab to exclude ophthalmia neonatorum (p 100).

Feeding anxieties Healthy term babies require very little for the first few days and early poor feeding is not an indication for investigation or bottle top-ups. Occasionally it may be necessary to point out to very worried mothers that well babies are brilliant at surviving: babies buried in earthquakes for 4 days are capable of surviving in good health.

New babies may have difficulty coordinating feeding and breathing and briefly choke, gag or turn blue. Exclude disease, check feeding technique (too much? too fast?) and reassure.

Regurgitation is often due to overfilling a tiny stomach with milk and air. Check feeding technique; if bottle-fed, is the teat too big for the mouth or the hole too small or the amount too great?

Winding during feeds may help but is not essential to health.

Red-stained nappy This is usually due to urinary urates but may be blood from the cord or vagina (oestrogen withdrawal bleed).

Sneezing Neonates sneeze to clear amniotic fluid from the nose.

Breast feeding

Factors which make starting breast feeding harder ●Partner's indifference to breast feeding (10% breast feed *vs* ~70% if he approves).[1]

●If mother and baby are separated at night in hospital.
●Urbanization, or whenever the mother is expected to work, and there is no convenient place for breast feeding.
●Sales girls dressed up as nurses to aid product acceptance.
●Rôle-models; if elder sisters are not breast feeding, younger sisters may feel embarrassed about starting.

The advantages of breast feeding (Some are hard to quantify.)
●A lower infant mortality rate (wherever there is poverty).
●Sucking promotes uterine contractions, so avoiding some PPHs.
●Bonding: eye to eye contact aids recognition, and stimulates further emotional input from the mother.
●Breast feeding is a total experience, stimulating all five senses.
●Breast feeding is cheap, and breast milk is clean.
●IgA, macrophages, lymphocytes (with interferon) and lysozyme protect from infection. Acids in breast milk promote growth of friendly lactobacillus in the baby's bowel. Gastroenteritis may be less severe if the mother makes and transfers antibodies (an 'immune dialogue').
●Breast milk contains less Na^+, K^+ and Cl^- than other milk, so aiding homeostasis. If dehydration occurs, risk of fatal hypernatraemia is low.
●Breast feeding reduces the risk of juvenile-onset diabetes mellitus.[1]
●Breast feeding protects against atopic eczema and other allergies.[2]
●Breast feeding may result in increased intelligence quotient scores (possibly due to long chain fatty acids needed for neurodevelopment).[1]
●Some protection in premenopausal years against maternal breast cancer.

Learning to breast feed Keep on the postnatal ward for quite a few days, and have the mother learn with an experienced, friendly midwife.
●Ensure that all of the nipple and areola are inside the mouth.
●Press the baby's chin against the breast.
●Hold the breast away from his nose—to aid breathing.

Why is feeding on demand to be encouraged?
●It keeps the baby happy, and enhances milk production.
●Fewer breast problems (engorgement, abscesses).
Note: sleep is less disturbed if baby is kept in bed with mother.

Contraindications to breast feeding ●An HBsAg +ve mother ●An HIV +ve mother in developed countries ●Amiodarone. ●Antimetabolites ●Opiates.

Problems These include breast engorgement and breast abscess. Treat by keeping the breast empty, eg by hourly feeds or milk expression—as follows. Wash the hands and the milk receptacle. Massage the breast towards the areola. Press behind the nipple. Next press the thumb and finger a short way into the breast, and squeeze them together. Do not slide the fingers over the areola.

If a breast abscess forms, discard the milk if it is pus-like. Give the mother flucloxacillin 250mg/6h PO (it is safe for baby).

Sore nipples are best treated by exposure to the air, after washing in water and padding dry with a soft towel. Resting the nipple for a few feeds or using a rubber nipple shield may help.

Breast milk is *probably* the best food for preterm infants. Give unheated, via a tube (p 246). Add vitamin D 1000u/day and vitamin K (p 244). Phosphate supplements may also be needed. NB: even term babies may develop rickets and hypocalcaemia (eg with fits, recurrent URTIs, lethargy, or stridor), if exclusively breast-fed, unless vitamin supplements are given.[3]

1 *Clinical Paediatrics* 1994 33 214   2 L Saarinen 1995 *Lancet* 346 1065   3 JJ Train 1995 *BMJ* i 48

Bottle feeding

There are few contraindications to breast feeding but many pressures not to (p 178). 64% of mothers start breast feeding, 52% are breast feeding at 2 weeks and 39% at 6 weeks. Most change to bottle because of lack of knowledge or no encouragement.[1]

Advantages of bottle feeding Fathers can bottle feed. Many mothers are reassured to know how much milk the baby is taking. Babies can be fed in public without embarrassment. Asian mothers may follow the traditional belief that colostrum is harmful and bottle feed until the milk comes in.

Infant formula (Cow's milk which has been 'humanized' by reducing the solute load and modifying fat, protein and vitamin content.) Whey-based formula is considered more suitable for infants than more protein-rich casein-based types, but brands are similar and shopping around for a brand which 'suits' the baby is rarely the answer to feeding problems.

Preparing feeds Hands must be clean, equipment sterilized and boiled water used—infective gastroenteritis causes many deaths in poor countries and considerable morbidity in UK. Powder must be accurately measured. Understrength feeds lead to poor growth and overstrength feeds may cause dangerous hypernatraemia, constipation and obesity.

Feeding Babies require about 150ml/kg/24h (30ml=1oz) divided over 4-6 feeds depending on age and temperament. Feeds are often warmed, but there is no evidence that cold milk is harmful. Milk flow should almost form a stream; check before each feed as teats silt up. The hole can be enlarged with a hot needle. The bottle should be held at such an angle that air is not sucked in with the milk.

Cow's milk allergy This develops in some babies—with diarrhoea (may be bloody), vomiting, peri-oral rash, oedema and failure to thrive. Soya milk formula may be substituted. Cow's milk may be cautiously reintroduced at 1 year (can be risky so do in hospital). NB: allergies to soya are also a problem, so a protein hydrolysate may be needed.[2]

Soya milks These containing corn syrup or sucrose—but no lactose, so are indicated in galactosaemia and primary hypolactasia—as well as in vegan families, as they ban all cow's milk. Wysoy® is *not* the solution for infants with eczema: see p 598. Try a protein hydrolysate instead (eg Nutramigen®). Because of their high aluminium content, avoid soya milks if preterm, or if there is a renal problem. Soya is *not* indicated in regrading after simple gastroenteritis.[2]

Weaning Milk contains sufficient nourishment for the first 3 months. 'Solids' should be introduced at 3-6 months by offering cereal or puréed food on a spoon. Cereals should not be added to bottles. At 6 months 'follow-on' (protein-enriched) formula may be used and lumpy food should be started so that the baby can learn to chew. Doorstep milk may be suitable after the 1st birthday.

1 DoSS 1988 *Present Day Practice in Infant Feeding*, HMSO   2 SG Mitton 1994 *BMJ* i 266

Preterm and light-for-dates babies

A neonate whose calculated gestational age from the last menstrual period is <37 completed weeks is preterm (ie premature). Low-birth-weight babies fall into 3 groups:

1 <2500g but appropriate for gestational age.
2 <2500g term infants who are small for gestational age (ie weight below the 10th centile for gestation—see below).
3 Low-birth-weight premature babies, whose weight is still small for their gestational age. Note: 6% of UK infants are <2500g at birth, and 50% of these are preterm. 5% of pregnancies end in spontaneous preterm delivery,[1] and 70% of all perinatal deaths occur in preterm infants (particularly if also growth retarded).

Weeks' gestation	1st born:	boy	girl	Subsequent births:	boy	girl
		Tenth centile weight (grams)				
32		1220	1260		1470	1340
33		1540	1540		1750	1620
34		1830	1790		2000	1880
35		2080	2020		2230	2100
36	......	2310	2210	......	2430	2310
37		2500	2380		2600	2480
38		2660	2530		2740	2620
39		2780	2640		2860	2730
40		2870	2730		2950	2810

Causes of prematurity These are usually unknown. Smoking tobacco, poverty and malnutrition play a part. Others: previous history of prematurity; genitourinary infection (eg Mycoplasma); pre-eclampsia; polyhydramnios; closely separated pregnancies; multiple pregnancy; uterine malformation; placenta praevia; abruption; premature rupture of the membranes. Labour may be induced early on purpose or accidentally (p 110).

Estimating the gestational age Use the Dubowitz score (p 298).

Infants who are small for gestational age Causes and associations: malformations, multiple pregnancy, placental insufficiency (maternal heart disease, hypertension, smoking, diabetes, sickle-cell disease, pre-eclampsia). The incidence of live-born, very-low-birth-weight infants <1500g is ~0.6%. Neonatal mortality (within 28 days of birth) is ~25%. If the weight is 500–600g only about 10% survive, at best.

Gestational age (based on LMP and ultrasound) is more important for predicting survival than the birth weight alone.

Management Transfer in utero to a special centre, if possible. Once born, protect from the cold. Measure blood glucose. Take to SCBU if glucose <2mmol/l. Measure blood glucose before each 3-hourly feed. Tube-feed if oral feeds are not tolerated. If oral feeding is contraindicated (eg respiratory distress syndrome) IV feeding is needed (p 246).

1 B Hibbard 1987 BMJ i 594

Common happenings in childhood[1]

Crying babies Crying peaks at ~8 weeks old (2–3h/day, especially in the evenings). The cries of hunger and thirst are indistinguishable. Feeding at regular intervals is apt to cause crying in many babies, so demand feeding is preferred.[2] Any thwarting of a baby's wishes may lead crying, with bouts of screaming. No doubt there are good survival reasons for Nature making babies cry, but in the 20th century this provision may be counterproductive: prolonged crying (as well as lack of parental support) causes much child abuse, including murder. The crying simply drives the parent crazy, and it is vital to offer help *before* this stage is reached (eg CRY-SIS self-help group, UK tel. 0171 404 5011). The key skill is not to make parents feel inadequte, and to foster a spirit of practical optimism with one or other parent sleeping and gaining rest whenever the child sleeps.

3-month colic Paroxysmal crying with pulling up of the legs, for >3h on ≥3 days/wk comprise the defining *rule of 3s*.[1] Studies have failed to show a GI cause. Cow's milk whey allergy, and parental discord and disappointment with pregnancy have been implicated[2]—but it may be that such stressed parents may simply be *reporting* more colic.[3] *Treatment:* Nothing helps for long. Try movement (eg a sling). Randomized trials[4] favour dimethicone (simethicone) drops (40mg/ml; 0.5–1ml PO before feeds; avoid if <1 month). Dicyclomine has been withdrawn due to side effects. In breast-fed babies, a cow's-milk-free diet for mother may help, as may a change to soya milk (p 179) in bottle-fed babies. Reassure strongly.

Sleep problems are common, worrying, and often ignored. Explanation, reassurance, and support will often suffice, without the need for drugs. If forced, try trimeprazine syrup (≤2mg/kg). Classification:[5]
Sleeplessness: Hunger/colic (infants); lack of routine (pre-school); worries (eg adolescents). Try behavioural treatments before hypnotics.
Night-time attacks: ●Rhythmic movement disorders (head-banging), rarely a developmental disorder ●Arousal disorders (night terrors/sleep walking), often familial. The child awakens frightened, hallucinated and inaccessible. If they follow a pattern, try awakening the child before they occur ●Nightmares ●Nocturnal seizures (investigate and treat).
Day-time sleepiness: Causes: night sleep↓; sleep apnoea (OHCM p 362).

Vomiting Effortless regurgitation of milk is very common during feeds. Vomiting between feeds is also common; if repeated, find the cause, eg gastroenteritis; pyloric stenosis; hiatus hernia (p 272 mucus ± blood in vomit); pharyngeal pouch or duodenal obstruction (there is bile in the vomit). Observing feeding is helpful in deciding if the vomiting is projectile (eg over the end of the cot), suggesting pyloric stenosis.

Nappy rash or diaper dermatitis 4 types:
1 The common ammonia dermatitis—red, desquamating rash, sparing skin creases. The term is a misnomer, as moisture retention, not ammonia is the cause. It often responds to frequent changes of nappy (which should be well rinsed), careful drying, and an emollient cream. Avoid tight-fitting rubber pants. 'One way' nappies may be helpful at night (eg Snugglers®).

2 Candida dermatitis (thrush): this may be isolated from ~½ of all nappy rashes. Its hallmark is satellite spots beyond the main rash. Mycology: see p 590. Treatment: as above, with nystatin or clotrimazole cream (± 1% hydrocortisone cream, eg as Nystaform HC® cream).

3 Seborrhoeic eczematous dermatitis: a diffuse, red, shiny rash extends into skin folds, often associated with other signs of a seborrhoeic skin, eg on occiput (cradle cap). Treatment: as for 1.

4 Isolated, psoriasis-like scaly plaques (p 586), which can be hard to treat.

Agents to avoid: Boric acid and fluorinated steroids (systemic absorption); oral antifungals (hepatotoxic) and gentian violet (staining is unpopular).

Ethnospecific growth charts[4]

It is clear that some populations are inherently shorter than others, and this poses problems when using growth charts. Consider these facts:

- African and Afro-Caribbean 5–11yr-olds height is ~0.6 standard deviations scores (SDS) greater than white children living in England.
- Gujarati children and those from the Indian sub-continent (except those from Urdu or Punjabi speaking homes) have heights ~0.5 SDS less than white children living in England.
- Gujarati children's weight-for-height is ~0.9 SDS less than expected for Afro-Caribbeans, or white children in England—so Gujarati children's weight is ~1.5 SDS less than for white children living in England.
- Urdu and Punjabi children's weight is ~0.5 SDS less than expected for white children living in England.
- Published charts have centile lines 0.67 SDS apart; for height and weight shift the centile lines up by 1 centile line division for Afro-Caribbeans.
- Relabel Gujarati children's weight charts, so the 0.4th centile becomes the ~15th centile, and the 2nd weight centile becomes the ~30th centile.
- For most other Indian sub-continent groups, both height and weight should be shifted downwards: consider relabelling the 0.4th and 2nd centile lines 1.5 and 6th respectively.
- Body-mass index centiles are said to be appropriate for Afro-Caribbeans,[4] but relabel as above for Indian sub-continent children, except for Gujarati speaking children (0.4th and 2nd centiles → 4th and 14th).

Trends towards increasing height with each generation occur at varying rates in all groups, so 3rd generation immigrants may be taller than expected using 2nd generation data. Intermarriage also confuses the issue.

Coeliac disease

Malabsorption often presents with diarrhoea, anaemia (folate or iron deficiency), and failure to thrive (and theoretically, but rarely clinically, rickets). The abdomen may protrude, with eversion of the umbilicus. *Cause:* Enteropathy induced by gluten (found in wheat and barley). Small bowel biopsy: a swallowed Crosby capsule is guided to the small bowel under X-ray control, and fired by suction.

Diagnosis:[5] Villus atrophy while the patient is eating adequate amounts of gluten, followed by full clinical remission on excluding gluten. It is only necessary to perform a gluten challenge if there are doubts about the effectiveness of gluten exclusion. IgA gliadin, anti-reticulin, and anti-endomysium antibodies found at the time of diagnosis and their disappearance after gluten exclusion support the diagnosis.

Treatment: Gluten-free diet: no wheat, barley, rye, oats, or any food containing them (eg bread, cake, pies). Rice, maize, soya, potatoes, sugar jam, syrup and treacle are allowed. Gluten-free biscuits, flour, bread and pasta are prescribable. Even minor dietary lapses may cause recurrence. After 5yrs a cautious reintroduction of gluten may be tried.

Other causes of malabsorption: Giardia; rotaviruses; bacterial overgrowth; milk sensitivity.

1 M Donaldson 1994 *BMJ* i 596   2 H Marcovitch 1994 *BMJ* i 35   3 *Drug Ther Bul* 1989 **27** 97   4 S Chinn 1996 *Lancet* 347 839   5 JA Walker-Smith 1990 *Arch Dis Chi* **65** 909

Diarrhoea

This kills more than 3×10^6 children/yr.[1] Faeces are sometimes so liquid they are mistaken for urine. It is *normal* for breast-fed babies to have liquid stools. Some cow's milks cause harmless green stools. Diarrhoea may be a 1st sign of *any* septic illness.

Gastroenteritis This is an infection of the small intestine which causes diarrhoea (± vomiting). The main danger is dehydration and U&E imbalance (p 250). The rotavirus is the most common cause, and there is often associated otitis media or upper respiratory infection. Other enteric viruses: astrovirus, Norwalk agent, calicivirus. *Treatment:* ▶If dehydrated, see p 250. Weighing the child (to monitor progress, and to quantify the level of dehydration, if a recent previous weight is available). Stop bottle milk (but if breast feeding, continue) and solids, and give oral rehydration mixture, eg Dioralyte®, = 60mmol Na+/l and osmolarity = 240mmol/l, better in non-cholera diarrhoea than standard WHO solutions (90mmol & 311 mmol/l respectively).<2> Milk is re-introduced after 24h (regrading is not required), or sooner if the child recovers and is hungry. Prolonged starvation delays recovery, and early use of full strength feeds is safe even in small babies.[3] If breast fed and not too ill, allow breast feeding to continue (supplies antibodies and maintains milk production).

Complications: Dehydration; malnutrition; temporary sugar intolerance after D&V with explosive watery acid stools. (Rare; manage with a lactose-free diet.)

Tests: Send stools for pathogens, including ova, cysts and parasites.

Prevention: Hygiene, good water and food, education, fly control.

Secretory diarrhoea *Causes:* Infections: bacteria (Campylobacter, Staphylococcus, *E coli*, and, where sanitation is poor, Salmonella, Shigella and *Vibrio cholerae*); giardiasis; rotavirus; amoebiasis; cryptosporidium. Inflammatory bowel disease.

Self-limiting toddler's diarrhoea "Like peas & carrots"—due to intestinal hurry.

Other causes Allergy (coeliac disease; disaccharide, galactose, glucose or lactose intolerance: to find reducing substances mix 5 drops of stool with 10 drops H_2O + 1 Clinitest® tab); antibiotics; deficiencies (copper, Mg^{2+}, vitamins); kwashiorkor.

Causes of bloody diarrhoea Campylobacter; necrotizing enterocolitis (in neonates), intussusception (often <4 years old); pseudomembranous colitis; inflammatory bowel disease (rare, even in older children).

Malnutrition

This is not common in the UK, but, being a major cause of infant mortality and morbidity world-wide, it has a global importance to us all.

Kwashiorkor is due to dietary lack of protein and essential aminoacids. It is characterized by growth retardation, diarrhoea, apathy, anorexia, oedema, skin/hair depigmentation, anaemia and abdominal distension. Serum glucose, K+, Mg^{2+}, cholesterol and albumin are low. *Treatment* depends on re-education (of child, family and politicians) as well as on offering a gradually increasing, high-protein diet, with added vitamins.

Marasmus The deficiency of calories—with marked discrepancy between height and weight. *Presentation:* Abdominal distension, diarrhoea, constipation, infection. *Tests:* Serum albumin↓. In young children a mid-arm circumference <9.9cm (any age) predicts severe malnutrition better than traditional signs (eg being <60% of median weight for age, 85% of median height for age and 70% of median weight for height). *Treatment:* Parenteral feeding may be needed to restore hydration and renal function. Next offer a balanced diet with vitamins. In spite of this, stature and head circumference may remain reduced. NB Kwashiorkor and marasmus may occur together.

1 C Burn 1992 *Bul of WHO* **70** 705 <2> ORS group *Lancet* 345 282 3 F Chew 1993 *Lancet* 341 194

Childhood urinary tract infection (UTI)

Presentation ▶Often the child may be *non-specifically* ill. Infants may present with collapse and septicaemia, and toddlers as 'gastroenteritis', failure to thrive, colic, or PUO. Many children with dysuria and frequency have *no* identifiable UTI, and will often have vulvitis. Of those with UTI, ~35% have vesico-ureteric reflux, ~14% have renal scars, ~10% have stones, ~3% will develop hypertension. <1% develop renal failure. ▶Always check the BP.

Definitions *Bacteriuria:* Any bacteria in urine uncontaminated by urethral flora. It may be asymptomatic. It may lead to *chronic pyelonephritis* and renal failure. UTI denotes symptomatic bacteriuria, which may involve different sites. Loin pain ± fever suggests *pyelonephritis* (look for white cell casts, do blood cultures). *Chronic pyelonephritis* is a histological/radiological diagnosis. Juxtaposition of a cortex scar and a dilated calyx is the key to its diagnosis. New scarring is rare after the age of 4. It is a major cause of renal failure. During micturition, urine may flow up the ureters. This is called *reflux*, and it is identified by a micturating cystogram (requires catheterization). Grading: I Incomplete filling of upper GU tract, without dilatation. II Complete filling ± slight dilatation. III Ballooned calyces.

Epidemiology Annual incidence: boys: 0.17–0.23%; girls: 0.31–1%. (Sex ratios are reversed in neonates.) Recurrence: 35% if >2yrs old. Prevalence of asymptomatic bacteriuria in schoolgirls: 1–2%. Prevalence of radiological abnormalities associated with UTI: 40% (½ have reflux; others: malpositions, duplications, megaureter, hydronephrosis).

Tests[1][2] *Urine collection:* In infants, wash genitals gently in water and tap repeatedly (in cycles of 1min) with 2 fingers just above the pubis, 1h after a feed, and wait for a *clean catch*, avoiding the first part of the stream. Do prompt microscopy and culture. >10^8 organisms/l of a pure growth signifies UTI. Any organisms found on *suprapubic aspiration* are significant (the rigorous way to exclude UTI). Ward ultrasound helps identify a full bladder. Method: clean skin over the bladder; insert a 21–23G needle in the midline 2.5cm above symphysis pubis. Aspirate as you advance the needle. Other devices: *dipslides* (go out of date rapidly), *culture pads* (Microstix®). *Urine bags* applied to the perineum are associated with many false +ves (7–64%, depending on type: Hollister U® bags have valves and are recommended).

Ultrasound (US): As this is non-invasive, it is worthwhile even in 1st UTIs (specificity 99%; sensitivity 43%, but reflux and scarring may be missed). Almost all the significant lesions missed by US either occur in infants <2yrs, or occur with fever and vomiting, so if these are present and US is normal, proceed not to IVU, which is radiation-rich and unreliable, but to: [99]Technetium renography—static for scarring (^{99m}Tc DMSA scan, OHCM p 790), dynamic for obstructive uropathy—and *isotope cystography*.[3] In neonatal males *micturating cystourethrography* is still the best way of excluding reflux.

Treatment *UTI:* ●Trimethoprim 50mg/5ml, 50mg/12h PO for 5 d (halve if <6mths; double if >6yrs), nalidixic acid, or co-amoxiclav. ●Encourage fluids. ●Avoid constipation. ●Encourage full voiding. ●Do repeat MSU.

Treatment of reflux: Ureteric reimplantation can reduce reflux, but scarring remains. Keep on antibiotic prophylaxis.

Prevention Just one episode of reflux of infected urine may initiate chronic pyelonephritis—so *screening* for bacteriuria is useless—the damage happens too quickly. But once a UTI is suggested (eg by urine microscopy) in an infant, treat it *at once* before culture sensitivities are known, as renal damage may be about to happen. Consider trimethoprim prophylaxis while awaiting surgery or investigations. Screen siblings for reflux.

1 M Taylor 1986 *BMJ* i 990 2 UTI Group 1991 *J Roy Col Phys* 25 36 3 A Rickwood 1992 *BMJ* i 663

Childhood leukaemia (ALL)

Acute lymphoblastic leukaemia (ALL) is the commonest childhood leukaemia. For other forms, see OHCM p 536–40. Hodgkin's disease: OHCM p 606.

Causes unknown—but preconception paternal exposure to sawdust, benzene (eg in petrol) and radiation has been implicated.[1][2]

Classification *Common ALL* (75%): Blasts have characteristics of neither T (rosette) nor B lymphocytes, but have a characteristic surface polypeptide. Median age: 4yrs. Prognosis: good (86% 5yr survival).
B-cell ALL: These blasts look like those of Burkitt's lymphoma. They show surface immunoglobulins. The prognosis is poorer but improving (up to 70% disease-free survival).
T-cell ALL: Median age: 10yrs. Prognosis: 40–50% disease-free survival.
Null-cell ALL: Blasts show none of the above features.
Worse prognosis in:[3] Blacks, males, WCC >100 × 10^9/l; Philadelphia translocation [t(9;22)(q34:q11)] seen in 12% (0–15% disease-free at 5yrs).

The patient Presentation may be with pancytopenia (pallor, infection, bleeding), fatigue, anorexia, fever and bone pain (as blasts invade the marrow). The period before diagnosis is often brief (2–4 weeks).

Laboratory findings WCC↑, ↓ or normal. Normochromic, normocytic anaemia ± thrombocytopenia. Marrow: 50–98% of nucleated cells will be blasts. CSF: pleocytosis (with blast forms), protein ↑, glucose ↓. Biochemistry: uric acid ↑, lactate dehydrogenase (LDH) ↑.

Treatment[4] This entails remission induction, CNS prophylaxis, and maintenance therapy. Pneumocystosis is prevented by co-trimoxazole 10mg/kg/12h PO. While the WCC is <1.0 × 10^9/l use a neutropenic régime (OHCM p 530). Drugs used: high-dose cyclophosphamide, cytarabine, methotrexate (intrathecal), mercaptopurine, asparaginase, daunorubicin, etoposide, thioguanine, vincristine, prednisolone. Reinforce remission by use of non-cross-resistant drug pairs, to reduce the leukaemic clone, and lower the frequency of drug-resistant mutants and, therefore, relapse. With the use of increasingly intense régimes, tailored to the patient's risk group, preventive meningeal irradiation has either been omitted or reduced in dose, partly because of fears of inducing delayed neoplasia elsewhere (risk of CNS tumours or secondary acute myeloid leukaemia is ~3–4%). Bone marrow transplantation is not routinely used in ALL in children.

Prognosis Many centres are reporting 75% 5-year event-free survival.

Pitfalls ●Serious infection without fever or tachycardia—eg: zoster, measles, CMV, candidiasis, pneumocystosis. Manage sepsis according to the régime on p 248.
●Hyperuricaemia following massive cell destruction. Prevent with a high fluid intake and allopurinol 10–20mg/kg/24h.
●Ignoring the patient's quality of life. Think of ingenious ways of entertaining the child while reverse barrier nursed.
●Omitting to examine the testes (a common site for recurrence).
●Inappropriate transfusion (leucostasis if WCC >100 × 10^9/l).
●Failing to treat side effects of chemotherapy optimally—eg 5-HT$_3$ receptor antagonists such as (ondansetron) are better than other antiemetics.[5]
●Failing to give support and encourage compliance. (Paediatricians may be better at avoiding this pitfall than adults' physicians, see p 171.)

1 M Gardner 1987 *BMJ* ii 822   2 P McKinney 1991 *BMJ* i 681   3 D Hoelzer 1993 *NEJM* 329 1343   4 GK Rivera 1993 *NEJM* 329 1289   5 GS Dick 1995 *Arch Dis Chi* 73 243-5

Cystic fibrosis

This is the commonest autosomal recessive disease (~1/2000; ~1/22 of Caucasians are carriers); it reflects mutations in the cystic fibrosis transmembrane conductance regulator gene (CSTR) on chromosome 7, which codes for a cyclic AMP-regulated chloride channel. There is a broad range of severity of exocrine gland function, leading to meconium ileus (neonates), lung disease akin to bronchiectasis, pancreatic exocrine insufficiency and a raised Na^+ sweat level—depending in part on the type of mutation (often ΔF_{508}; but other mutations, eg in intron 19 of CFTR, cause lung disease but no increased sweat Na^+).[1] Death may be from pneumonia or cor pulmonale. Many survive to adulthood (median survival is now 30yrs).

Antenatal (p 210) carrier-status testing is possible, as is pre-implantation analysis after *in vitro* fertilization: at the 8 cell stage, 1 cell is removed from the embryo, and its DNA analysed; only embryos without the cystic fibrosis gene are offered for implantation—this may be a more acceptable than terminating fully-formed fetuses after chorionic villous sampling.

Diagnosis ~10% present with meconium ileus as neonates. Most present later with recurrent pneumonia (± clubbing), steatorrhoea (>7g/day/100g of ingested fat), or slow growth. *Sweat test*: Sweat Na^+ >60mmol/l (inaccurate if <100mg of sweat is collected or Na^+:Cl^- discrepancy >20mmol/l)—this test is difficult, so find experienced worker. *Other tests*: CXR: shadowing suggestive of bronchiectasis, especially in upper lobes. Malabsorption screen. Glucose tolerance test. Lung function tests. Sputum culture. Mycobacterial colonization affects up to 20%—consider if rapid deterioration.

Treatment ▶*Offer genetic counselling to every family* (p 212). Long survival depends on antibiotics and good nutrition. *Respiratory problems*: Physiotherapy (× 3/day) must start at diagnosis. Teach parents percussion + postural drainage. Older children learn the forced expiration technique. Organisms are usually *Staph aureus, H influenzae* (rarer), and *Streptococcus pneumoniae* in younger children. *Pseudomonas aeruginosa* acquisition is often late, but eventually >90% are chronically infected.[2] *Berkholderia cepacia* (*Ps cepacea*) is associated with rapid progression of lung disease (prompt diagnosis using polymerase chain reaction on DNA samples may be available: isolate patient). Treat acute infection after sputum culture using higher doses, and for longer than normal. If very ill, ticarcillin (50–60 mg/kg/6h IV) + netilmicin (p 238), or ceftazidime (10–30mg/kg/8h IV) alone may be needed before sputum results are known. Admissions may be reduced by using nebulized ticarcillin and tobramycin at home.[1] If reversible airways obstruction is a problem, give inhaled salbutamol (look for *Aspergillus* in sputum). Lung transplant (heart + lung, or double lung) works for some, with good results limited by donor availability (avoid raising hopes). Ensure pertussis and MMR vaccination.

Gastrointestinal problems and nutrition: Energy requirements are increased by 120–150% because of chronic lung inflammation. Most have steatorrhoea from pancreatic malabsorption and will need enzymes: Pancrex V® powder mixed with food for infants—and Pancrex V Forte® for older children, ≤10 tabs/meal—to give regular, formed, non-greasy bowel actions. If the enzymes are given in microspheres (eg Pancrease®, Creon 25000®) fewer tablets will be needed. Cimetidine 10mg/kg PO ½h before each meal may also be needed. If this controls steatorrhoea, a low-fat diet is not needed, but vitamin supplements are still required (A and D, eg as Abidec® 0.6ml/24h PO for infants or as multivitamin capsules 2/24h PO for older children). The diet should be high calorie/high protein. Fine-bore nasogastric feeding is needed only if weight cannot otherwise be maintained.

If GI obstruction occurs (eg if Pancrex® is omitted), urgent admission to a specialist hospital is needed for medical treatment (laparotomy is contraindicated unless perforation is imminent[1]).

Dibetes mellitus: Insulin may be needed. Adjust the dose to the (optimized) diet, not *vice-versa*. Only try oral hypoglycaemics if nutrition is satisfactory.

Psychological help: Parents and children need expert counselling. The Cystic

Fibrosis Research Trust and regional centres may help here.[3]

Complications Haemoptysis (rarely bad), nose polyps, pneumothorax (in 20%), diabetes, cirrhosis, cholesterol gallstones (10%), pulmonary osteoarthropathy, ♂ infertility (stenotic or absent vas deferens).

Meconium ileus The baby usually presents with vomiting in the first 2 days of life. Distended loops of bowel may be seen through the abdominal wall, and if a firm mass is felt in one of the loops, this indicates that the cause is a hard plug of meconium. In most other causes of GI obstruction, lateral decubitus films will show fluid levels. This is not the case in meconium ileus. Tiny bubbles may be seen in the meconium. *Treatment:* Place a nasogastric tube. Seek expert advice. Surgical excision of the distal ileum (or whatever contains the most meconium) may be needed.

The future This may lie in agents to *reduce viscosity* of sputum with synthetic human DNA-ase which fragments viscous DNA from dead neutrophils in sputum (an example, with modest clinical effect, is Dornase alfa: it has been tried in a once-daily dose via an air compressor with a jet nebulizer.

Lung transplantation is getting safer; considered in those who are deteriorating (FEV$_1$ <30% of expected) despite maximum therapy, provided nutrition is good, and there is no active mycobacterial or aspergilos infection.

Gene therapy aims to deliver normal copies of the cystic fibrosis gene into patients, so allowing them to make CFTR protein. Viral vectors and liposomes have been used to get the gene into cells.

In a sense, the furure is already here: its just too expensive.

1 W Zhou 1994 *NEJM* **331** 974   2 S Fiel 1993 *Lancet* **341** 1070   3 5 Blythe Rd, Bromley, Kent, UK tel. 0181 4647211

Rheumatic fever

This is a systemic febrile illness caused by a cross-sensitivity reaction to Group A β-haemolytic streptococcus, which, in the 2% of the population which is susceptible, may result in permanent damage to heart valves. It is common in the Third World, but is now rare in the West (incidence: ~5–10 per million children), although pockets of resurgence have been noted in the USA in areas of overcrowding[1]—which favours streptococcal spread. Some specific Group A streptococcal serotypes are known to be particularly rheumatogenic—eg type 5 of the M-protein serotypes. Other ubiquitous serotypes appear to be non-rheumatogenic (eg type 12).

Diagnosis The presence of 2 of Jones' major criteria or 1 major and 2 minor + evidence of preceding streptococcal infection: scarlet fever, a throat swab growing β-haemolytic streptococci or a serum ASO titre >333U/l (reference intervals vary) is strong evidence that rheumatic fever is present.

Major criteria (revised[2]):
Carditis (ie 1 of: changed murmur; cardiomegaly; CCF; friction rub)
Polyarthritis (often migratory)
Erythema marginatum (p 580)
Subcutaneous nodules
Sydenham's chorea (p 758)

Minor criteria (revised[2]):
Fever
ESR >20mm/h or C-reactive protein↑
Arthralgia, ie pain but no swelling
ECG: PR interval >0.2 sec.
Previous rheumatic fever or rheumatic heart disease

NB: do not count arthralgia if polyarthritis is being used as a major criterion—likewise for a long P-R interval if carditis is being used. Affected joints may be exquisitely tender. Those most commonly affected are the knees, ankles, elbows and wrists. There are no permanent sequelae. If the child seems to have rheumatic fever but little in the way of carditis, Doppler ultrasonography may reveal inaudible mitral regurgitation.[1] The classical lesion is the endocardial MacCullam patch seen at the base of the posterior mitral leaflet. Aortic, pulmonary and tricuspid valves are affected in descending order of frequency.

Treatment of rheumatic fever
●Bed rest and immobilization (to rest the heart and the joints).
●Relieve joint pain (aspirin, 80mg/kg/day, as in Still's disease, p 758).
●In severe carditis, enlist expert help. Prednisolone (eg 2mg/kg/24h PO) *may* reduce symptoms—but do not expect much impact on sequelae.[2]
●Penicillin for pharyngitis (eg 125mg/6h PO) preceded by one dose of benzylpenicillin (25mg/kg IM or IV).

Prophylaxis Rheumatic fever may be recurrent. Aim to prevent this by giving phenoxymethylpenicillin 250mg PO daily or sulphadimidine 250mg PO daily. Cover operations/dentistry with antibiotics (see below).

Infective endocarditis (IE)

Presenting signs: Fever, splenomegaly, anaemia, rashes, heart failure, microscopic haematuria and murmurs. Further features: p 258.
Treating endocarditis before the organism is known: (Do 3 blood cultures first.) Benzylpenicillin 25mg/kg/4h IV + netilmicin 2mg/kg/8h IV (p 238).
Preventing IE in those with heart lesions: Ampicillin 30mg/kg IM/IV + gentamicin 2mg/kg IM at induction,[3] followed by ampicillin 30mg/kg IM/IV 6h later. Metronidazole 7.5mg/kg IV may also be added. This is recommended for any dentistry, colonic or GU procedure. ▶See OHCM p 314.

1 L Veasy 1987 *NEJM* 316 421 2 G Stollerman 1965 *Circulation* 32 664–8 & *Lancet* 1995 346 391 3 Brit Soc for Antimicrob Chemoth 1992 *Lancet* i 1292 See D Albert 1995 *Medicine* 74 1–12 [meta-analysis]

Some metabolic diseases

Hypothyroidism *Congenital causes:* Genetically determined lack of thyroid stimulating hormone (TSH); congenital goitrous hypothyroidism (eg defective enzymes for thyroxine synthesis or iodine lack); maternal antithyroid drugs (propylthiouracil). Acquired: Hashimoto's thyroiditis; hypopituitarism; X-rays.

The Patient: There may be no signs at birth: the first sign is often prolonged neonatal jaundice (p 236). Inactivity, excessive sleeping, slow feeding, little crying and constipation may occur. Look for coarse dry hair, a flat nasal bridge, a protruding tongue, hypotonia, umbilical hernia, slowly relaxing reflexes, slow pulse and delayed growth and mental development. Other later signs: IQ↓, delayed puberty (occasionally precocious), short stature, delayed dentition.

Tests: T4↓, TSH↑ (but undetectable in secondary hypothyroidism), ^{131}I uptake↓, Hb↓. Bone age is less than the chronological age. As it is undesirable to X-ray the whole skeleton, the wrist and hand are most commonly used. There are a large number of ossification centres. Each passes through a number of morphological stages, and using comparisons with key diagrams from 'normal' populations, a rough bone age may be determined. There is no hard-and-fast answer to the question of how much discrepancy (eg 2 years) between skeletal and chronological years is significant.[1]

Treatment with thyroxine: Infants need ~10µg/kg/day (≤50µg/day); this rises to ~100µg/day by aged 5yrs and to adult doses by 12yrs. Adjust according to growth rate and clinical response. Avoid high TSH levels.

Universal neonatal screening: Filter paper blood spots (from a heel prick) allow early diagnosis, and the prevention of serious sequelae.

Hyperthyroidism The typical child is a girl at puberty. Clinical and laboratory features are described in OHCM p 484. Treatment is with carbimazole eg 15mg/24h PO for 18 months, with the dose adjusted according to response—or propylthiouracil 200mg/m^2/day PO. Relapses may occur.

Thyroid disease in pregnancy and neonates See p157.

Inborn errors of metabolism
These are very numerous, but some helpful generalizations about their acute presentations may be made. Signs include diarrhoea, lethargy, respiratory distress, metabolic acidosis (± odd body smells), jaundice, hypoglycaemia, U&E imbalance, fits and coma. Look for:

Physical sign:	Possible significance:
Hepatosplenomegaly	Eg amino acid and organic acid disorders, lysosomal storage diseases.
Coarse facies	Mucopolysaccharidoses, eg Hurler's syndrome, p 750, gangliosidoses, mannosidoses.
WCC↓, platelets↓	Organic acidurias.
Mental retardation	See p 204.
Failure to thrive	Aminoacidurias, organic aciduria, cystinuria, lactic acidosis, storage diseases.

A metabolic urine screen is helpful, but differs from laboratory to laboratory. It usually includes amino acids, organic acids, carbohydrates and mucopolysaccharides. Request expert laboratory help.

The glycogenoses result from defects in the synthesis and degradation of glycogen. Abnormal stores may be deposited in liver, muscle, heart or kidney. Most types (there are at least 7) are inherited as an autosomal recessive. Types include: von Gierke disease (type I, p 758), Pompe's disease (type II, p 756), Cori disease (type III), Anderson disease (type IV), McArdle disease (type V), Hers disease (type VI) and Tauri disease (type VII).

Acute surgical problems

Congenital hypertrophic pyloric stenosis

This does not (contrary to its name) present at birth, but develops during the first month of life ($\sigma/\varphi \approx 4$). *Presentation:* Vomiting which occurs after feeds, and becomes projectile (eg vomiting over far end of cot). Congenital pyloric stenosis is distinguished from other causes of vomiting (eg gastroenteritis) by 3 important negatives:
- The vomit does not contain bile, as the obstruction is so high.
- No diarrhoea: constipation is likely (occasionally 'starvation stools').
- Even though the patient is ill, he is rarely obtunded: he is alert and anxious—as well as being malnourished, dehydrated, and always hungry.

Diagnosis: *Observe* left-to-right LUQ peristaltic waves during a feed; *palpate* the olive-sized pyloric mass: stand on the baby's left side and palpate with the left hand at the lateral border of the right rectus in the RUQ, during a feed—preferably from the left breast. The baby may be severely depleted of water and sodium chloride and this makes urinary output and plasma Cl^- (also K^+ and HCO_3^-) vital tests, which will guide the paediatrician in his resuscitative measures and will determine when it is safe to perform surgery (the chloride should be >90mmol/l).

Ultrasound detects early, hard-to-feel pyloric tumours[1] (~14% of cases).

Management: Before surgery (Ramstedt's pyloromyotomy) pass a wide-bore nasogastric tube.

Intussusception in children The small bowel telescopes, as if it were swallowing itself. The patient may be any age (usually 5–12 months), and presents with intermittent inconsolable crying, with drawing the legs up (colic). He may (but need not) pass blood PR (like red-currant jelly or merely flecks). A sausage-shaped abdominal mass may be felt. The patient may quickly become shocked, and moribund. Barium enema examination makes the diagnosis (the barium fails to flow proximal to the intussusception, and at this point a knuckle of bowel is outlined). During the examination early intussusceptions (those of <24h duration) may be reduced by the radiologist. Most require surgery. Success depends upon early diagnosis, and an experienced paediatric team and surgeon, skilled anaesthesia, and close monitoring both pre- and post-operatively. At laparotomy any necrotic bowel should be resected.

Management: Resuscitate, crossmatch blood, pass nasogastric tube.

School-children have a different presentation compared with infants: rectal bleeding is less common, and they are more likely to have a long history (>3 weeks) and some sort of contributing pathology (Henoch–Schönlein purpura, Peutz–Jeghers' syndrome; cystic fibrosis, ascariasis, nephrotic syndrome or tumours such as lymphomas—in the latter obstructive symptoms caused by intussusception are the most frequent mode of presentation).[2]

Phimosis In this condition the foreskin is too tight (eg due to circumferential scarring) so that retraction over the glans is impossible. The flow of urine may be impeded, with ballooning of the foreskin. There may be recurrent balanitis. Time and trials of gentle retraction usually obviate the need for circumcision. Note: it is normal to have a simple non-retractile foreskin up to the age of 4yrs.

Post-operative pain relief Morphine IV: child's loading dose: 100μg/kg in 30min, then 10–30μg/kg/h (neonates 5–7μg/kg/h, but same loading dose)—or diclofenac 0.5–1.5mg/kg/12h if over 1yr (eg a 12.5mg suppository every 12h if ~1yr old). Ibuprofen dose: 5mg/kg/6h PO (syrup is 100mg/5ml).[3]

Other surgical problems: appendicitis, hernias, volvulus, torsion of the testis, acute abdomen: see p 272, p 218, OHCM p 132 & p 112.

1 M Rollins 1993 *BMJ* i 1065 2 S Miller 1988 *BMJ* i 518 3 *Drug Ther Bul* 1994 21

199

Precocious puberty

Sometimes, it may be quite normal for puberty to start as early as 8yrs old. The onset of puberty before this time warrants investigation.

Biology Each of the physical signs of puberty may be thought of as a bioassay for a separate endocrine event. Enlargement of the testis is the first sign of puberty in boys, and is due to pulses of pituitary gonadotrophin. Breast enlargement in girls and penis enlargement in boys is due to gonadal sex steroid secretion. Pubic hair is a manifestation of adrenal androgen production. The growth of boys accelerates when the volume of the testis reaches 10-12ml (as measured by comparison with the orchidometer's beads, p 752). Girls start to grow more quickly once their breasts have started to develop. Stage 4 breast development is a prerequisite for menarche (in most girls). This consonance of puberty may go awry. In Cushing's syndrome, pubic hair is 'too much' for the testicular volume; in hypothyroidism, the testes are large (FSH↑ because TSH↑↑) in the presence of a low growth velocity.

Precocious puberty is manifested in boys by a rapid growth of penis and testes, increasing frequency of erections, masturbation, the appearance of pubic hair, changing body odour, and acne. There will be corresponding changes in the secondary sexual characteristics of girls. The most important long-term consequence is short stature caused by early fusion of epiphyses. Enquire about symptoms of more general hypothalamic dysfunction: polyuria, polydipsia, obesity, sleep and temperature regulation disturbances. There may also be manifestations of raised intracranial pressure and visual disturbance.

Precocious puberty is at least 4 times as common in girls, compared with boys. In girls, often no cause is found, but in 80-90% of boys a cause is found. If onset is before 2yrs old the cause may well be a hamartoma in the hypothalamus. These may be revealed as non-enhancing, circumscribed lesions on computerized tomography. Other causes (mostly rare):

- CNS tumours and hydrocephalus
- Post encephalitis or post meningitis
- McCune-Albright syndrome (p 752)
- Craniopharyngioma
- Tuberous sclerosis
- Hepatoblastoma
- Choriocarcinoma
- Hypothyroidism

Investigations Skull X-ray, bone age by skeletal X-ray, CT scan of head, urinary 17-ketosteroids, pelvic ultrasound (girls), T₄.

Management: a physiological approach ▶Refer to a specialist. Initiation of puberty depends on release from inhibition of neurones in the medial basal hypothalamus which secrete gonadotrophin releasing hormone (GnRH), and on decreasing hypothalamic-pituitary sensitivity to -ve feedback from gonadal steroids. These changes are accompanied by a marked increase in frequency and magnitude of 'pulses' of luteinizing hormone (LH), and, to a lesser extent, follicle stimulating hormone (FSH). It is the ability to secrete GnRH pulses at a fast rate which leads to normal gonadal function. Continuous high levels of GnRH paradoxically suppress the secretion of pituitary gonadotrophins, and this forms the basis for treatment of precocious puberty with synthetic analogues of GnRH. After subcutaneous or nasal insufflation there is a reversal of gonadal maturation and all the clinical correlates of puberty (not for pubic hair, as there is no change in the secretion of androgens by the adrenal cortex). There is deceleration in skeletal maturation. Treatment is continued until the average age of puberty or of the menarche (eg ~11yrs old). Families need reassurance that the child will develop normally.

Primary antibody deficiency

Essence Synonyms for this condition include *hypogammaglobulinaemia*, ie not secondary to conditions such as protein-losing enteropathy, chronic lymphatic leukamia, or myeloma. The leading symptom is recurrent infections, needing repeated antibiotics, with the development of chronic infective states such as bronchiectasis and sinusitis. Other features:

- Failure to thrive
- Enteropathy
- Nodular lymphoid hyperplasia (gut)
- Hepatosplenomegaly
- Absent tonsils
- Anaemia.

Most diagnoses are in those >6 years old (adults are also affected).

Prevalence ~4/100,000.[1]

Classification *Common variable immune deficiency:* IgG↓, IgA↓, IgM variable. Typical age at referral: 6–10yrs old (with another peak at age 26–30yrs).[1]
X-linked antibody deficiency: Often ♂ >2yrs old and +ve family history. B-cell development may be abnormal.
IgG subclass deficiencies.
Specific antibody deficiency: Typically IgM, IgG, and IgA normal, but fails to respond to immunizations.
Selective IgA deficiency: Prevalence: 1:700. May be symptom-free.

Suspect antibody deficiency if
- Failure to thrive
- Severe infections needing surgery
- Arthropathy
- Recurrent infections
- Hepatosplenomegaly

When these signs are unexplained, refer to an immunologist, to assess antibody responses to protein and carbohydrate antigens, measure IgG subclasses, and count lymphocytes involved in antibody production (CD4, CD8, CD19, CD23 positive lymphocytes). Immunoglobulin levels are interpreted by age. There is a rôle for watching responses to test vaccinations.[1]

Management Aim to include the patient and the family in the process. Ensure prompt treatment of intercurrent infections. This may include postural physiotherapy, and bronchodilators as well as antibiotics.

Immunoglobulin replacement obviates most complications and is best delivered by a specialist immunologist, after detailed immunological assessment. Many patients can join a self-infusion programme. Before any infusion, active infection should be excluded (to minimize risk of adverse reactions), and a baseline check of transaminase enzymes, creatinine, and anti-IgA antibody titres should be done.[1]

The dose of IV immunoglobulin is determined by the severity and frequency of infections, and the plasma level of IgG. Most receive ~400mg/kg/month, usually as 2 doses, 2 weeks apart.[1] Hydrocortisone and an antihistamine should be to hand during an infusion. SE: headaches, abdominal pain, anaphylaxis.

Intramuscular immunoglobulins are not favoured, but the subcutaneous route is being investigated.[1]

Complications *Chest:* Bronchiectasis, granulomas, lymphomas.
Gut: Malabsorption, giardia, cholangitis, atrophic gastritis, colitis.
Liver: Acquired hepatitis, chronic active hepatitis, biliary cirrhosis.
Blood: Autoimmune haemolysis, ITP, anaemia of chronic disease, aplasia.
Eyes/CNS: Keratoconjunctivitis, uveitis, granulomas, encephalitis.
Others: Septic arthropathy, arthralgia, splenomegaly.

1 HM Chapel 1994 *BMJ* **i** 581 and the Royal College of Pathologists 1995 *Consensus Document for the Diagnosis and Management of Patients with Primary Antibody Deficiency* ISBN 0 9518574 5 2

Delay in walking

Babies usually learn to walk at ~1yr old. If this has not occurred by 18 months, ask yourself 2 questions: Is the child physically normal? Is development delayed in other areas too?

▶Consider Duchenne muscular dystrophy early (genetic counselling *before* next pregnancy). Congenital hip dislocation is another important cause.

Cerebral palsy is a chronic disorder of posture and movement caused by a non-progressive CNS lesion sustained before 2yrs old, resulting in poor motor development ± mental retardation (35%).[1] *Prevalence: 2/1000. Typical survival:* 16yrs whether or not severely retarded (IQ <50).[1] Imaging studies finding periventricular leucomalacia help in prognosis.

Prenatal factors:	Perinatal factors:	Postnatal factors:
APH (with hypoxia)	Birth trauma	Trauma; Intravent-
X-rays; alcohol	Fetal distress	ricular haemorrhage
CMV; rubella; HIV	Hypoglycaemia	Meningoencephalitis
Toxoplasmosis	Hyperbilirubinaemia	Cerebral vein thrombi
Rhesus disease	Failed resuscitation	(from dehydration)

Clinical picture:	Epidemiology:
●Paralysis	⅓ are of low birth wt
●Weakness, ataxia	⅓ have visual defects
●Delayed milestones	⅓ mentally retarded
●Convulsions	⅓ improve with time
●Hearing and speech disorders	⅙ live a normal life

Spasticity suggests a pyramidal lesion; unco-ordinated, involuntary movements and postures (dystonias) suggest basal ganglia involvement; ataxia suggests cerebellar involvement. Most children have either a hemiparesis or a spastic diplegia—eg both legs affected worse than the arms, so that the child looks normal until he is picked up, when the legs 'scissor' (hip flexion, adduction and internal rotation; with knee extension and feet plantar-flexed). Gait is broad based.

Type 1 Ataxic palsies (pure ataxia):	Type 2 Ataxic diplegia:
●Hypotonia ('floppy baby')	●Hypertonia
●Other defects rare	●Other defects common
●Flexor plantars	●Extensor plantars
●Associations: deafness; strabismus	●Associations: trauma
mental retardation (fits rare)	hydrocephalus
●Normal developmental milestones	spina bifida; viruses

Dyskinetic cerebral palsy: Unwanted actions, breaks in movement flow, agonist/antagonist imbalance, poor posture control, hypotonia, hearing loss, dysarthria, poor visual fixation. Fits and IQ↓ are rare.

Assessment and management (p 290) Can he roll over (both ways)? Ability to come to the sitting position independently. Grasp. Transfer of objects from hand to hand. Effective head righting. Ability to shift weight (when prone) with forearm support. IQ. Treat epilepsy (p 268). Use calipers to prevent deformity (eg equinovarus, equinovalgus, hip dislocation due to excessive flexion and adduction). Attempts to show the benefits of neurodevelopmental physiotherapy (directed towards improving equilibrium and righting) have failed to show any benefit over simple stimulation of motor activity.[2] Some parents prefer the Hungarian (Petö) approach[3] where one person (the 'conductor') devotes herself to the child and uses interaction with the child's peers to reinforce successes in all areas: manipulation, art, writing, fine movement, social interaction.

1 A Nicholson 1992 *Arch Dis Chi* 67 1050 2 F Palmer 1988 *NEJM* 318 803 3 R Robinson 1989 *BMJ* ii 1145

Delay in talking

▶When in doubt, test the hearing. *Ensure the result is reliable.*

Speech development After the first few months of life, language differentiation is already occurring. Chinese babbling babies can be distinguished
from European ones.

Before the first year babbling gives way to jargon—plausible 'words' which
have no meaning.
At about the first year, a few words may be used meaningfully.
At 1½yrs 2-word utterances appear ('Daddy come').
At 2, subject–verb–object sentences appear ('I want a pudding').
At 3½, the child has mastered thought, language, abstraction and the
elements of reason, and has a vocabulary of a thousand words at his disposal. This enables him to produce sentences such as 'I give her cake 'cos
she's hungry'.

The rest of life holds nothing to match the intellectual and linguistic
activities of these years. The child's remaining linguistic development is
devoted to conceptually minor tasks, such as learning the subjunctive,
expanding his vocabulary, and entertaining counter-factual hypotheses 'If
I hadn't thrown the cup on the ground, I might have been given a
pudding'.

There is much variation in speech timing: what is clearly abnormal?
Vocabulary size: If the child reaches 3yrs with <50 words, suspect: ●Deafness
●Expressive dysphasia.
●Speech dyspraxia, particularly if there is a telegraphic quality to speech,
  poor clarity, and deterioration in behaviour (frustration).
●Audio-premotor syndrome (APM). The child cannot reflect the sounds
  correctly heard into motor control of larynx and respiration. Instead of
  babbling, the child is quiet, unable to hum or sing.
●Respiro-laryngeal (RL) dysfunction (dysphonia from incorrect vocal fold
  vibration/air flow regulation). The voice is loud and rough.
●Congenital aphonia (rare): the voice is thin, weak and effortful.

Speech clarity: By 2½yrs the mother should understand the child's speech
most of the time. If not, suspect deafness—or:
●Articulatory dyspraxia (the easy consonants are *b* and *m* with the lips,
  and *d* with the tongue—the phonetic components of babbling. This is the
  commonest form of speech problem. ♂/♀=3. Tongue-tie is a possible
  cause, and they will have trouble producing sounds which need elevation
  of the tongue (*d* and *s*)—surgery to the fraenum may be needed as well as
  speech therapy.
●APM or RL dysfunction as above.

Understanding: By 2½yrs he should understand 'Get your shoes', if not suspect: ●Deafness—if the hearing is impaired (eg 25-40 dB loss) secretory
otitis media is likely to be the cause. Worse hearing loss is probably sensorineural. ●Cognitive impairment. ●Deprivation.

Some other causes of speech disorder There are many, ▨eg:

Acquired:	Congenital:
Post meningoencephalitis	Klinefelter's syndrome
Post head injury	Galactosaemia, histidinaemia
Landau Kleffner syndrome	Auditory agnosia
(progressive loss of language + epilepsy)	Floating harbour syndrome

Speech therapy Refer early—and well before school starts.

Handicap

Mental handicap ► *The mother often makes the first diagnosis.*
IQ <35 constitutes a severe learning disability (mental handicap, p 314).

Causes: This is a very large subject, many are very rare. Be prepared to refer to an expert, who may himself need to rely on an electronic database such as the *London Neurogenetics Database.*[1]

Congenital disorders are legion: chromosomal (eg Down's; fragile X, p 752); metabolic (eg PKU p 205). *Acquired:* Perinatal infection p 98–100, birth injury and cerebral palsy, and subsequent trauma, meningitis.

Lead exposure: This is a leading preventable cause of mildly impaired intellect. For example in 2-yr-olds for each 0.48µmol/l plasma increment there is an associated 5–8 point fall in IQ as measured on the Wechsler Intelligence Scale for children (revised[2]). This defect is long-lasting.
Biochemical defects associated with mental retardation—eg:

Homocystinuria: Paraplegia, fits, friable hair, emboli, cataracts; homocystine is found in the urine. Treat with a low methionine, cystine supplemented diet, with large doses of pyridoxine.

Maple syrup urine disease: Hypoglycaemia, acidosis, fits, death. Urine smells of maple syrup, due to defective metabolism of branched chain keto acids. Dietary treatment.

Tryptophanuria: Rough, pigmented skin. Treat with nicotinic acid.

Management: Refer to an expert, so that no treatable cause is missed. Would the family like to be in touch with specialist self-help groups, such as MENCAP?[3] Other members of the family may need special support (eg normal siblings, who now feel neglected). If the IQ is >35, life in the community is the aim.

Physical handicap *Sensory:* Deafness, see p 540.
Blindness: congenital defects are described on p 516. Principal acquired causes of blindness are: vitamin A deficiency, onchocerciasis (p 512), eye injuries, cataract (eg Down's syndrome).

Neurological and musculoskeletal handicap: These can be congenital or acquired. Causes include accidents, cerebral palsy (p 202), spina bifida (p 228), following meningitis, polio, congenital infections (as above), congenital dislocation of the hip, tumours, rare syndromes (p 742–58).

Wheelchairs[4]: Is it for indoors or outdoors? Patient-operated, motorized or pushed? What sort of restraints are needed to prevent the child falling out? If collapsable, how small must it be to get into the family car? Are the sides removable to aid transfer from chair to bed? Can the child control the brakes? Are there adjustable elevated leg rests? Liaise with the physiotherapist and the occupational therapist.

Calipers will allow some patients to stand and walk. Long-leg calipers are required for those with complete paralysis of the legs. The top should be constructed so that it does not produce pressure sores. A knee lock supports the knee in the standing position. An internal coil spring prevents foot drop.

1 M Baraitster 1993 *London Neurogenetics Database* OUP   2 DC Bellinger 1992 *Paediatrics* **90** 855
3 123 Golden Lane, London (tel. 0171 253 9433)   4 Disabled living foundation, 346
Kensington High St, London W14 (tel. 0171 602 2491)

Check-list to guide the management of handicap

Whether based in hospital or in the community the doctor should address each of these points:

- Screening and its documentation.
- Communication with parents.
- Refer to and liaise with district handicap teams, and community paediatrician.
- Ensuring access to specialist services, including dietitian.
- Assessing special needs for schooling and housing.
- Co-ordinating neuropsychological/neurodevelopmental assessments.
- Co-ordinating measures of severity (eg electrophysiology ± CT/MRI).
- Prescribing of special food products, where relevant.
- Promotion of long-term compliance with treatment or education programs.
- Education about the consequences of the illness.
- Offering family planning.
- Pre-conception counselling (with specialist in molecular genetics).
- Co-ordinating prenatal diagnostic tests and fetal assessment.

Phenylketonuria

Cause Absent or reduced activity of phenylalanine hydroxylase—inherited as an autosomal recessive (the gene—on chromosome 12—often mutates, and there is a full range of clinical manifestation from nearly symptom-free to severe mental impairment).

Clinical features Fair hair, fits, eczema, musty urine. The most important manifestations are impaired IQ and cognition.

Tests Hyperphenylalaninaemia (reference interval: 50–120µmol/l). 'Benign' phenylketonuria may be indicated by levels <1000µmol/l.

Treatment See check-list above. Phenylalanine-low diet, to keep phenylalanine to <600µmol/l—by prescribing artificial food-substitutes (amino acid drinks) to give <300mg–8g natural protein/day (depending on age and severity of phenylalanine hydroxylase deficiency).[1]

There is much evidence to suggest that CNS damage (hypomyelination) is proportionate to the degree of phenylketonaemia, but some studies fail to show that stricter diets are associated with higher IQs.[2]

Prevention of manifestations Screen blood at 1 week (using a heel-prick and filter paper impregnation—the Guthrie test).

Maternal phenylketonuria However good the mother's control, subtle neurological impairment is likely.[1] The implication is that control during pregnancy should be even tighter than in childhood. Hence the need for pre-conception counselling.

1 MRC working party 1993 *BMJ* i 115   2 U Michel 1990 *Eur J Pediat* **149** (suppl 1) 34–8

Behavioural questions

▶Only enter battles you can win. If the child can win, be more subtle (eg consistent rewards, not inconsistent punishments). Get health visitor's advice—and ensure that everyone is saying the same thing to the child.

Entrances *Food refusal* and *food fads* are common. Reducing pressure on the child, discouraging parental over-reaction, and gradual enlarging of tiny portions of attractive foods usually prevents malnutrition. Check Hb.
Overeating: Eating comforts, and if the child is short on comfort, or if mother feels inadequate, the scene is set for overeating and life-long patterns are begun. Diets may fail until the child is hospitalized (p 348).
Pica is eating of things which are not food. There are likely to be other signs of disturbed behaviour, or ↓IQ.

Exits *Soiling* is the escape of stool into the underclothing.
Encopresis is the repeated passage of solid faeces in the wrong place in those >4yrs old. Faecal retention is the central event. *Constipation* is delay in defecation; it may lead to distress, abdominal pain, rectal bleeding, abdominal masses, overflow soiling ± 'lavatory blocking' enormous stools (megarectum), and anorexia. Causes: usually poor food, fluid or fibre intake. Rarely Hirschprung's disease (think of this in infants if there is an explosive gush of faeces on withdrawing the finger after rectal exam, or if there is alternating diarrhoea with constipation and abdominal distension, and failure to thrive). *Action:* ●Find out about pot refusal ●Does defecation hurt? ●Is there parental coercion? Try to break the vicious cycle of: large faeces → pain ± fissure → fear of the pot → rectum overstretched → call-to-stool sensation dulled → soiling → parental exasperation → coercion). Exonerate the child to boost confidence for the main task of obeying the call-to-stool to keep the rectum empty (eg by *ducosate sodium,* then a dose of *sodium picosulphate* elixir or polyethylene glycol, followed by maintenance *methyl cellulose tablets* ± prolonged alternate-day *senna*). Psychological help (± family therapy, p 382) is very important. Give the family time to air feelings which encopresis engenders (anger, shame, ridicule).[1]

Enuresis: Bedwetting occurs on most nights in 15% of 5-yr-olds, and is still a problem in 1% of 15-yr-olds—usually from delayed maturation of bladder control; but tests for diabetes, UTI and GU abnormality (p 188) can yield surprises. *Treatment:* Fluid restriction before bed, and the passing of time may be sufficient. A system of rewards for dry nights as documented on a star chart above the child's bed may be effective. Alarms triggered by urine in the bed can allow 65% to be dry after 3 weeks;[2] the high relapse rate (30%) may be reduced by continuing training after dryness. Alarms may be loaned from Community Health/Child Guidance Services or Mini Drinite® (tel. UK 0181 441 9641). Imipramine 25–50mg/24h PO should not be used for >3 months, has side effects, and high relapse rates on discontinuation. Antidiuretic desmopressin 20–40µg (2–4 puffs) at night by nasal spray helps only in the short term, eg to show that dryness is possible—or to allow unembarrassed holidaying. CI: cystic fibrosis, uraemia, BP↑.

Hyperactivity◆ This is a level of activity > parents are happy with. Parents say the child is never still, being 'on the go, as if driven by a motor'. If excessive inattention or poor concentration with impulsivity exists, the wider term *attention deficit disorder* or *minimal brain dysfunction* is used by some—provided the cause is not psychosis or an affective disorder. *Treatment:* Explanation (offer a booklet[4]). Cognitive therapy may increase parental tolerance. Hypoallergenic diets (eg no tartrazine) remain controversial and may cause obsessions in the parents. There is evidence that lead (eg from petrol, see p 205) may contribute to hyperactivity. Stimulant (not depressant) drugs are a last resort (confer with an expert): methylphenidate 0.08–0.25mg/kg/8h PO is said to offer the best benefit-to-hazard ratio,[5] but it is a controlled drug, only available on a named patient basis (SE: sleep reduced, slowing of growth; monitor growth, BP, pulse). Other drugs: pemoline 0.5–2mg/kg/24h PO (effects are delayed for 6 weeks); haloperidol 0.01–0.04mg/kg/12h PO; imipramine 0.2–0.5mg/kg/8h, especially if co-existing anxiety or depression; clonidine (≤5mg/day PO in divided doses, particularly if there is a co-existing tic disorder).[5]

School phobia *Setting:* Emotional overprotection; high social class; neurotic parents; school work of high standard. In truancy, the reverse is true. *Treatment:* Confer with headmistress, parents, and a child psychologist. Escort by an education welfare officer aids the prompt return to school.

Wakefulness A common problem is the child who wakes at 3am, ready to play—or wanting entry to the parent's bed. For those not appreciating these visitations from the pure of heart, the easiest thing is to refuse to play and buy ear plugs to lessen the impact of screaming—or to let the child into the bed. Or try extinguishing the behaviour by attending the child ever more distantly: cuddle in bed → cuddle on bed → sitting on child's bed → voice from doorway → distant voice. Try to avoid hypnotics (such as promethazine, chloral hydrate).

1 T Swanwick 1991 *Br J Gen Prac* 41 514 2 *Drug Ther Bul* 1986 24 6 3 G Clayden 1992 *Arch Dis Chi* 67 340 4 E Taylor 1985 *The Hyperactive Child: a Parents' Guide*, M Dunitz 5 *Drug Ther Bul* 1995 33 57

Screening and child health promotion[1,2]

The main aims: ●Encouraging breast feeding (p 178). ●Monitoring development. ●Immunization. ●Overseeing growth (p 184). ●Parental support. ●Education and reassurance about normal childhood events. ●Talking to the child, and building up a good relationship to be used in later illnesses.

Monitoring The most cost effective times to screen are unknown. A 'best buy' might be checks after birth (p 176), at ~7 weeks, and then:

6–9 months: Hips, testes descent, CVS examination.

18–24 months: Educate on diet, dental care, accidents, walking (look for waddling), social and linguistic milestones; Hb if iron deficiency likely—may well be. Any parental depression?

4 years: Testes descent, CVS examination. Nutrition, dental care.

►At each visit: ●Encourage parent to air queries. ●Ask about (and test for) squint, vision & deafness. Do centile charting.
See Denver developmental test, p 288. Note the milestones below. There is much individual variation. ►Beware reading too much into a single test.

1 month: Lifts head when prone; responds to bell; regards face.

2 months: Holds head at 45° when prone; vocalizes; smiles.

4 months: Uses arm support when prone; holds head steady when sitting reaches out; rolls over; spontaneous smiling.

6 months: Bears some weight on legs; on pulling to sitting, there is no head lag; resists your trying to take a toy from his grasp.

1 year: Just stands; walks using a table's support; clashes cubes; pincer grip; can say 'Mummy' ± 'Daddy'. Plays 'pat a cake'.

18 months: Can walk backwards; scribbles; 2-cube tower. 2–4 words. NB: drooling or throwing things on the floor is abnormal by now.

2 years: Kicks a ball; overarm 'bowling'; gets undressed.

3 years: Jumps; can stand on one foot; copies; can build an 8-cube tower; knows his first and last name; dressing needs help.

4 years: Stands on 1 foot for >4sec; picks the longer of 2 lines.

Immunization Offer the schedule given opposite, noting the items below.

MMR vaccine: Those between 18 months and 5yrs who have not had the vaccine (even if they have had single measles vaccine) may have MMR with the pre-school booster of Diphtheria/Tetanus/Polio—but use a different site. There is no upper age limit to this immunization.

SE: Rash ± fever from day 5–10 for ~2 days (so advise on how to control temperature); occasional non-infectious parotid swelling (from week 3).

Contraindications: fever, pregnancy (advise against this for 1 month), a previous live vaccine within 3 weeks or an injection of immunoglobulin within three months, primary immunodeficiency syndromes (not including HIV or AIDS), those receiving steroids (equivalent to prednisolone ≥2mg/kg/day for >1 week in the last 3 months), leukaemia, lymphoma, recent radiotherapy, anaphylaxis induced by egg, neomycin or kanamycin (vaccine preservative). Non-anaphylactoid allergies are not contraindications, and nor is a past history of seizures or febrile convulsions.

Pertussis immunization SE: Pain and fever. Serious brain damage is so rare, that if it occurs at all, it is very hard to put a figure to. It is equivocally implicated in 1 in 2×10^6 injections.[3]

CI: Past severe reaction to pertussis vaccine—ie indurated redness most of the way around the arm, or over most of the anterolateral thigh (depending on the site of the injection); or generalized reactions: fever >39.5°C within 48h of vaccination, anaphylaxis, prolonged inconsolable screaming, or other CNS signs (eg fits) occurring within 72h.

1 D Hall 1995 *Health for all Children*, 3ed. 2 UK DoH Schedule   3 S Sepkowitz 1994 *BMJ* i 655

Special considerations: most children with idiopathic epilepsy, or a family history of epilepsy (sibling or parent) should be vaccinated. If in doubt, seek expert advice rather than withholding the vaccine. Those with stable CNS conditions (eg cerebral palsy, spina bifida) are especially recommended for vaccination. There is no upper age limit for vaccination. A single-antigen acellular vaccine is available in the UK on a named patient basis from Farillon (tel. 01708 379000).

Notes ►An acute febrile illness is a contraindication to any vaccine. Give live vaccines either together, or separated by ≥3 weeks. Do not give live vaccines if there is a primary immunodeficiency disorder, or if the child is taking steroids (equivalent to >2mg/kg/day of prednisolone).

Children with HIV infection and AIDS: Give these children all the usual immunizations (including live vaccines) except for BCG.

Immunizations The new suggested DoH schedule[1] (L=live vaccine)

3 days: BCG[L] (if TB in family in last 6 months). *OHCM* p 192.

2 months: 'Triple' (pertussis, tetanus, diphtheria) 0.5ml SC; HiB; oral polio[L].
Note: If premature still give at 2 postnatal months.

3 months: Repeat 'triple', HiB and polio[L].

4 months: Repeat 'triple', HiB and polio[L].

12–18 months: Measles/Mumps/Rubella[L] (=MMRII vaccine) for *both* sexes. Use a single dose (0.5ml SC).

4–5yrs: Tetanus, diphtheria & polio[L] booster. MMRII (at a different site) if not already given at 12–18 months.

10–14yrs: BCG[L]. (Also Rubella[L] for girls who have missed MMRII.)

15–18yrs: Polio[L] &Tetanus+low-dose Diphtheria 'Td' boosters. ▣

Adult boosters: Tetanus (p226), polio (p234), rubella (p 214).

Hepatitis B vaccine: Give at birth*, 1 and 6 months, if mother is HBsAg +ve (H-B-Vax II®) 0.5ml IM—into the anterolateral thigh (the adult dose is 1ml into deltoid). *(With 200U hepatitis B immunoglobulin at a different IM site). Other risk groups: see *OHCM* p 512.

1 DoH 1992 *Immunization Against Infectious Disease*, HMSO

Inherited diseases and prenatal diagnosis[1]

▶It is more important to be able to love the handicapped and to respect their carers than it is to prevent the handicap—for in doing the first we become more human. In the second, we risk our own inappropriate deification.

Gene probes These use recombinant DNA technology to link genetic diseases of unknown cause to DNA markers scattered throughout the human genome. Using fetal DNA from amniotic fluid cells (amniocentesis, p 166) in the 2nd trimester, or from chorionic villus sampling (ChVS, p 166) in 1st, it is possible to screen for many genetic diseases (opposite)—eg Huntington's chorea; muscular dystrophy; polycystic kidneys; cystic fibrosis; thalassaemias.

Enzyme defects Many of the inborn errors of metabolism can be diagnosed by incubation of fetal tissue with a specific substrate.

Chromosomal studies can be undertaken on cultured cells or on direct villus preparations. The most important abnormalities are aneuploidies (abnormalities in chromosome number)—eg trisomy 21, 18 and 13.

Screening for chromosomal abnormalities (eg the fragile X syndrome, p 752) may be performed on at-risk mothers who may be carriers.

Nondisjunction After meiosis one gamete contains two number 21 (say) chromosomes and the other gamete has no 21 chromosome. After union of the first gamete with a normal gamete the conceptus will have trisomy 21, and will develop Down's syndrome (but ~50% of such conceptuses suffer early miscarriage). This is the cause in about 95% of Down's babies, and is related to maternal age. Risk at 20yrs: 1 in 2000; 30yrs: 1 in 900; 35yrs: 1 in 365; 40yrs: 1 in 110; 45yrs: 1 in 30; 47yrs: 1 in 20.

Swapping chromosomal fragments Such a cell may end up with the correct amount of genetic material (a balanced translocation); but a gamete from such a cell may have one intact chromosome (say number 21), and another mixed chromosome (say number 14), which, because of the swap, contains material from chromosome 21. After union with a normal gamete, cells will have 3 parts of chromosome 21. This translocation trisomy 21 is the cause in ~2% of babies with Down's syndrome, and is not related to maternal age. If the father carries the translocation, the risk of Down's is 10%; if it is the mother, the risk is 50%. 0.3% of all mothers have this translocation.

Mosaicism A trisomy may develop during the early divisions of a normal conceptus, so that some cells are normal and others have trisomy 21. This accounts for 1–2% of Down's babies. They show variable clinical severity.

Other chromosomal abnormalities Turner's (p 758), Klinefelter's (p 750), Edward's (p 746), and Patau's (p 754) syndromes. In the *cri-du-chat* syndrome there is deletion of the short arm of chromosome 5, causing a high-pitched cry, cvs abnormalities and a 'moon' face.

Down's syndrome[1] *Causes:* See above. *Recognition at birth:* Flat facial profile, abundant neck skin, dysplastic ears, muscle hypotonia and X-ray evidence of a dysplastic pelvis are the most constant features. Other features: a round head, protruding tongue, peripheral silver iris spots (Brushfield's), blunt inner eye angle, short, broad hands (eg with a single palmar crease), and an incurving 5th digit. Widely spaced first and second toes and a high-arched palate are more obvious later. If you are uncertain about the diagnosis, it is best to ask an expert's help, rather than baffle the mother by taking blood tests 'just in case it's Down's'. Associated problems: Duodenal atresia; VSD; patent ductus; ASD (foramen primum defects, p 224); and, later, a low IQ and a small stature. Helping the mother accept her (often very lovable) child may be aided by introducing her to a cheerful mother of a Down's child.[a]

A=amniocentesis; AFP α-fetoprotein; c#=chromosome breakage; CA=chromosome analysis; ChVS=chorionic villous sampling; CP=chromosome puffs; def=deficiency; del=deletion; EA=enzyme assay; EC=echocardiography; F=factor; FBS=fetal blood sample; FSkinB=fetal skin biopsy (fetoscopy); HA=hormone assay; ISCE=increased sister chromatid exchange; IUS=increased ultraviolet sensitivity; LCFA=long chain fatty acid assay; LTT=lymphocyte transformation test; Mass-Sp=Mass spectroscopy; RE=restriction enzymes; RFLP=restriction fragment length polymorphism; SSP=site specific probe; US=high resolution ultrasound; URS=uptake of radiolabelled sulphur-35.

1 MA Cleary 1991 *Arch Dis Chi* **66** 816 2 RW Newton *Management of Down's Syndrome* in *Recent Advances in Paediatrics* Ed TJ David (vol 10), Churchill Livingstone

Examples of prenatal diagnosis

Inborn errors of metabolism
Mucopolysaccharidoses[EA]
Aminoacidopathies: Phenylketonuria[EA]
  Ornithine carbamoyl transferase def[EA]
  Citrullinuria[EA]
  Arginosuccinicaciduria[EA]
  Homocystinuria[EA]
  Maple syrup urine disease[EA]
Carbohydrate disorders:
  Galactosaemia[EA/Mass-Sp]
  Galactokinase def[EA/Mass-Sp]
  Galactose 4-epimerase def[EA/Mass-sp]
  G6PD def[EA], Glycogenoses[EA]
Purine & pyramidine disorders:
  Lesch–Nyhan syndrome[EA/RFLP]
  Adenine phosphoribosyl
    transferase def[EA]
Organic acidurias: 3-Methylcrotonyl
    coenzyme A carboxylase def[EA]
  Methylmalonic acidurias[EA/Mass-Sp]
  Proprionic acidaemia[EA/Mass-Sp]
Lipid disorders: Sialidosis[EA]
  Anderson–Fabry disease[EA/RFLP]
  Farber's disease[EA]
  Fucosidosis[EA], Gangliosidoses[EA]
  Gaucher's disease[EA]
  I cell disease[EA]
  Krabbe's disease[EA]
  Mannosidosis[EA], Wolman's disease[EA]
  Metachromatic leucodystrophy[EA]
  Mucolipidosis III[EA]
  Niemann–Pick disease[EA]
Others: Acid phosphatasia[EA]
  Adrenal hyperplasia (21-hydroxylase
    def)[HA/RFLP] (HLA DNA linkage may
    also be used.)
  Cerebrohepatorenal syndrome[LCFA/EA]
  Cystinosis[URS], Hypophosphatasia[EA]
  Placental steroid sulphatase def
    (X linked ichthyosis)[HA/EA]
Single gene defects
  Cystic fibrosis[EA/RFLP]
  Becker muscular dystrophy[RFLP]
  Huntington's chorea[RFLP]
  Adult polycystic kidney disease[RFLP]
  Generalised neurofibromatosis[RFLP]
  Tuberous sclerosis[RFLP]
  Myotonic dystrophy[RFLP]
  Ataxia telangiectasia[C#]
  Bloom's syndrome[ISCE& IUS]
  Fanconi's anaemia[C#]
  Xeroderma pigmentosum[DNA repair defect]
  Epidermolysis bullosa syndromes[FSkinB]
  Hypohidrotic ectodermal dysplasia[RFLP]
  Multiple endocrine neoplasias[RFLP]

Chromosomal diseases
  Down's syndrome[Amnio/ChVS]
  Trisomies eg 13 & 18[Amnio/ChVS]
  Klinefelter's syndrome[Amnio/ChVS]
  Other sex chromosome
    aneuploidy in males[Amnio/ChVS]
  Turner's syndrome[Amnio/ChVS]
  xxx syndrome[Amnio/ChVS]
  Other sex chromosome
    aneuploidy in Q[A/ChVS]
  Chromosome deletion/regrouping
    -balanced or aneuploid[Amnio/ChVS]
  Fragile (X) syndrome[FBS/CA]
Congenital malformation
  Neural tube defect[Mass-sp/AFP/US]
  Severe micro/hydrocephaly[AFP/US]
  Cardiac defect[ECHO]
  Renal agenesis[US]
  Renal cysts[US]
  Hydronephrosis, Prune belly[US]
  Extrophy of the bladder[US]
  Trachea-oesophageal stenosis
    and fistula[US]
  Exomphalos, Gastroschisis[US]
  Diaphragmatic hernia[US]
  Duodenal atresia[US]
  Skeletal dysplasia[US]
  Cleft lip or palate[US]
  Cystic hygroma[US]
  Teratomas[US], Fetal hydrops[US]
  Many multiple anomalies[US]
Gene defects (known product)
  Duchenne muscular dystrophy
  Sickle-cell disease[SSP/RE/DNAdel]
  Thalassaemias[SSP/RE/DNA del]
Coagulation defects
  Haemophilia A&B[FVII/FIXa/RFLP]
  Factor x deficiency[RFLP]
Immune disorders
  Severe combined immuno-
    deficiency[EA]
  T cell immunodeficiency[RFLP]
Collagen disorders
  Ehlers-Danlos syndrome types
    II & V (some cases)[DNA del]
  Osteogenesis imperfecta types
    I & II (some cases)[DNA del]
Viruses Rubella[Antibody/RSDP]
  CMV[Antibody/IIT]
Others
  α-1-Antitrypsin def[RFLP/SSP]
  Acute intermittent porphyria[RFLP]
  Robert's sy[CP]

Genetic counselling

Goal To provide patients with accurate, up to date information on their condition and its consequences for them and their relatives in such a manner as to allow them to make informed decisions.

►Genetic counselling is a specialism best conducted in regional centres to which you should refer affected families.[1] *Do not lightly do blood tests which might have long-term consequences on young children unless some form of treatment is available. The child may never forgive you for labelling him or her.*

In order to receive most benefit from referral:
- The affected person (proband) should come with family (spouse, parents, children, siblings); individuals can of course be seen alone as well.
- The family should be informed that a detailed pedigree (family tree) will be made, and medical details of distant relatives may be asked for.
- Irrational emotions (guilt, blame, anger) are common. Deal with these sensitively, and do not ignore. Remember: you do not choose your ancestors, and you cannot control what you pass on to your descendants.
- Warn patients that most tests give no absolute 'yes' or 'no' but merely 'likely' or 'unlikely'. In gene tracking, where a molecular fragment near the gene is followed through successive family members, the degree of certainty of the answer will depend on the distance between the marker and the gene (as crossing-over in meiosis may separate them).
- Accept that some people will not want testing, eg the offspring of a Huntington's sufferer—or a mother of a boy who might have fragile X syndrome, but who understandably does not want her offspring labelled (employment, insurance and social reasons). Offer a genetic referral to ensure that her decision is fully informed (but remember: 'being fully informed' may itself be deleterious to health and wellbeing, p 167).

Chromosomal disorders include Down's (trisomy 21, p 210), Turner's (45, X0, p 758) and Klinefelter's (47, XXY, p 750) syndromes. Many genes are involved when the defect is large enough to be seen microscopically.

Autosomal dominants Polycystic kidneys (16p), Huntington's (4p). A single copy of the defective gene is sufficient to cause damage, and so symptomless carriers do not usually occur.

Autosomal recessives Cystic fibrosis (7q), β-thalassaemia, sickle-cell (11p). In general, both genes must be defective before damage is seen, so carriers are common—and both parents must be carriers for offspring to be affected, so consanguinity (marrying relatives) increases risk.

X-linked Duchenne muscular dystrophy, Haemophilia A & B, fragile X (p 752). In female (XX) carriers a normal gene on the second X chromosome prevents the deleterious effects manifesting. The male (XY) has no such protection and so is affected by the disease.

Chromosome nomenclature Autosomes are numbered 1 to 22 roughly in order of size, 1 being the largest. The arms on each side of the centromere are designated 'p' (petite) for the short arm, and 'q' for the long arm. Thus 'the long arm of chromosome 6' is written '6q'.

NB Being pregnant and unwilling to consider termination does *not* exclude one from undergoing useful genetic counselling.

Couple screening' A big problem with counselling is the unnecessary alarm caused by false +ve tests. In cystic fibrosis screening (analysis of cells in mouthwash samples) this is reducible by 97% (0.08% *vs* 3.2%) by screening mother and father together—who need only get alarmed if they *both* turn out to be screen-positive. The trouble with this is false reassurance. Many forget that they will need future tests if they have a different partner, and those who do not are left with some lingering anxieties.[2]

1 In UK, phone 0171 794 0500—to find nearest centre.   2 M Hall 1995 *BMJ* i 353

Measles, rubella, mumps & parvoviruses

Measles *Cause:* RNA paramyxovirus. *Spread:* droplets. *Incubation:* 7–21 days. Infective from first prodromal symptom (catarrh, wretchedness, conjunctivitis, fever) until 5 days after the rash starts. Conjunctivae look glassy; then the semilunar fold swells (Meyer's sign). *Koplik spots* are pathognomonic, grain-of-salt-like spots on the buccal mucosa. They are often fading as the macular rash appears (starts behind ears, eg on days 3–5, then spreads down the body, becoming confluent). *Treatment:* Isolate. Treat any secondary bacterial infection. *Complications:* Febrile fits, meningitis, D&V, keratoconjunctivitis; immunosuppression; in the immunosuppressed there may be fatal giant cell pneumonitis. Intrauterine infection may cause malformations. The most serious complication is the rare encephalitis (headache, lassitude, fits, coma)—≤15% may die; 25% develop fits, deafness or mental retardation. *Prognosis:* Complete recovery in rich countries—but it is a serious, mortal disease in many poor countries. *Prevention:* p 208.

Rubella (known as german measles, german being *germane* ie closely akin to). *Cause:* RNA virus. *Incubation:* 14–21 days. The patient is infective 5 days before and 5 days after the day the rash starts. *Signs:* Usually mild; macular rash; suboccipital lymphadenopathy. *Treatment:* None is usually needed. *Immunization:* (live virus, p 208). Give at least 1000 TCID50 (tissue culture infective dose, 50) eg as 0.5ml Almevax® SC. CI: pregnancy. The best time to immunize girls who have not had the MMR vaccine is from 11–13yrs old. *Complications:* Small joint arthritis. Malformations *in utero* (p 98). Infection during the first 4 weeks: eye abnormalities in 70%; during weeks 4–8: cardiac abnormalities in 40%; weeks 8–12: deafness in 30%.

Mumps *Cause:* RNA paramyxovirus. *Spread:* Droplets and saliva. *Incubation:* 14–21 days. *Immunity:* Lifelong, once infected. The patient is infective 7 days before and 9 days after parotid swelling starts. *Presentation:* Prodromal symptoms (malaise, fever); painful swelling of the parotids, unilaterally at first, becoming bilateral in 70%. *Complications:* Orchitis (± infertility), arthritis, meningitis, pancreatitis, myocarditis. Complete recovery is the rule. *Treatment:* Rest; paracetamol syrup (120mg/5ml) for *high* fever or *severe* discomfort. Dose (per 6h PO): <3 months old: 5–10mg/kg; 3 months–1yr: 60–120mg; 1–5yrs: 120–250mg; 6–12yrs 250–500mg. 60mg suppositories are available. Tepid (not cold) sponging augments paracetamol.[1] Also unwrap and rehydrate (this avoids vasoconstriction). Usually antipyretics are *not* required, and there is evidence of increased mortality in severe sepsis.[2]

Human parvovirus Type B19 is the cause of the 'slapped cheek' syndrome, also known as 'fifth disease' and erythema infectiosum. This is usually a mild, acute infection of children, characterized by malar erythema (a raised, fiery flush on the cheeks) and a rash mainly on the limbs. Constitutional upset is mild. Arthralgia is commoner in adults—who may present with a glandular fever-like syndrome and a false +ve Paul-Bunnell test.[3] Spread is rapid in closed communities. It is also the cause of aplastic crises in sickle-cell disease. It is uncommon in pregnancy but risks fetal death in ~10% and is a cause of hydrops fetalis (monitor AFP for several weeks; if it rises abnormally, do ultrasound). Fetal hydrops is treatable by intrauterine transfusion. Fetal malformation is generally not caused.

Hand, foot and mouth disease is caused by Coxsackie A16 virus. The child is mildly unwell and develops vesicles on the palms and soles and in the mouth. They may cause discomfort until they heal, without crusting. *Incubation:* 5–7 days. *Treatment* is symptomatic.

Note Measles, rubella and mumps are notifiable diseases in the UK. MMR vaccine: see p 208–9. **Herpes infections** See OHCM p 202–3.

Other causes of rashes in children See also skin diseases section (p 574).

- A transient maculo-papular rash is a feature of many trivial childhood viral infections.
- Purpuric rashes: meningococcaemia (p 258); Henoch–Schönlein purpura (p 750); idiopathic thrombocytopenic purpura (check FBC).
- Drug eruptions: maculo-papular rashes in response to penicillins and to phenytoin are particularly common.
- Scabies (p 600); insect bites.
- Eczema (p 588); urticaria (p 578); psoriasis—guttate psoriasis may follow a respiratory tract infection in children (p 586); pityriasis rosea (p 594).
- Still's disease—there is a transient maculopapular rash, fever and polyarthritis (p 758).

1 A Kinmoth 1992 *BMJ* ii 1134   2 F Shann 1995 *Lancet* **345** 338   3 JV Pether 1994 *BMJ* i 595 ◻

Varicella (herpes) zoster

Chickenpox is a primary infection with varicella-zoster virus. Shingles is a reactivation of dormant virus in the posterior root ganglion.

Chickenpox *Presentation:* Crops of vesicles on the skin of different ages, typically starting on back. *Incubation:* 11–21 days. *Infectivity:* 4 days before the rash, until all lesions have scabbed (1 week). Spread is by droplets. It can be caught from someone with shingles. It is one of the most infectious diseases known. 95% of adults will have been infected; immunity is life-long.
Tests: Fluorescent antibody tests and Tzanck smears[■] are rarely needed.
Differential: Hand, foot & mouth dis; insect bites; scabies; rickettsia.
Course and dangers: Starts as fever. Rash appears ~2 days later, often starting on the back: macule → papule → vesicle with a red-surround → ulcers (eg oral, vaginal) → crusting. 2–4 crops of lesions occur during the illness. Lesions cluster round areas of pressure or hyperaemia. Chickenpox is dangerous if immunosuppressed, or in cystic fibrosis, and in neonates.
Treatment: Keeping the child apparently reduces the number of lesions. Trim the nails to lessen damage from scratching. Daily antiseptic for spots may be tried (eg chlorhexidine). Flucloxacillin 125–250mg/6h PO if bacterial superinfection. Antivaricella-zoster immunoglobulin (12.5U/kg IM max 625U) + acyclovir if immunosuppressed or on steroids and within 10 days of exposure (give as soon as possible, see *BNF*). Acyclovir is licensed as a 7-day course in chickenpox—beginning within 24h of the rash.
Acyclovir child dose: (20mg/kg/6h PO), eg <2yrs: 200mg/6h PO; 2–5yrs: 400mg/6h PO. ≥6yrs: 800mg/6h. Adults: 800mg/4h PO, but no night-time dose. Tablets are 200, 400 or 800mg; suspension: 200mg/5ml or 400mg/5ml. In renal failure, ↓ dose. There is *no* clear evidence that acyclovir reduces complications in the immunocompetent—but it may relieve severe symptoms, eg in older patients, or 2nd or 3rd family contacts.[■]
Rare complications:[1] Pneumonia, meningitis, DIC, hepatitis, Guillain–Barré, Henoch–Schönlein, nephritis, pancreatitis, myositis, myocarditis, orchitis, transverse myelitis, CNS thrombosis, purpura fulminans.

Shingles *Treatment:* Oral analgesia. Ophthalmic shingles: p 484. Acyclovir may reduce progression of zoster in the immunocompromised (may be rampant, with pneumonitis, hepatitis, and meningoencephalitis).[1] Acyclovir IVI dose: 10mg/kg/8h (over 1h), with concentration <10mg/ml.

Varicella in pregnancy In the UK chickenpox affects 3 per 1000 pregnancies. Complications (pneumonitis affecting 1 in 400, and encephalitis 1 per 1000 cases) are not commoner in pregnancy, despite pregnancy being an immunocompromised state.[2] Infection within the first 20 weeks (especially 13-20 weeks) may result in congenital varicella (cerebral cortical atrophy, cerebellar hypoplasia, manifested by microcephaly, convulsions and mental retardation; limb hypoplasia, rudimentary digits and pigmented scars). It is rare (~2%), can be mild, and does not occur with maternal shingles. If mother affected from 1 week before to 4 weeks after birth, babies may suffer severe chickenpox. Give the baby zoster immune globulin 250mg IM at birth; if affected, isolate from other babies.

Infection may be prevented by vaccination of susceptible women prior to pregnancy with live varicella vaccine (where available). Use of varicella zoster globulin can prevent infection in 50% when given to susceptible contacts. Give zoster immune globulin 1000mg IM to the pregnant woman within 3 days of exposure to infection. Evidence suggests that infected pregnant women should have acyclovir (above) to ameliorate the course of the disease. The fetus does not appear to be adversely affected. Chickenpox at birth is a problem. Barrier nursing mother causes distress and is of unproven value—but we know that infant's mortality is 30%.[2]

1 A Leung 1994 *Update* 49 277 2 P Venkatesan 1996 *The Practitioner* 240 256

Some gastrointestinal malformations

Oesophageal atresia (± tracheo-oesophageal fistula—TOF) At least 85% of these babies will have a tracheo-oesophageal fistula. 30% will have another abnormality. Signs: cough, airway obstruction (excessive secretions—'blowing bubbles'), abdominal distension, cyanosis and recurrent pneumonias. *Diagnosis:* Inability to pass a catheter into the stomach. X-rays will show it coiled in the oesophagus. Avoid contrast radiology. Use endoscopy instead. *Treatment:* Surgical correction of fistulae, using a cervical incision.

Diaphragmatic hernia There is respiratory distress and bowel sounds in one hemithorax (usually left—so the heart is best heard on the right). Cyanosis augers badly. It can present with difficult resuscitation at birth. It is associated with other malformations (in 50%, eg neural tube defects), trisomy 18, deletion of the small arm of chromosome 27, Pierre Robin (p 226) and Beckwith-Wiedmann (p 743) syndromes. Incidence 1:2200. *Diagnosis:* Prenatally by ultrasound; postnatally by CXR. *Treatment:*
- *Prenatal:* Refer to a fetal surgery centre (tracheal obstruction may be tried: it encourages lung growth, so pushing out other viscera).[1]
- *Postnatal:* Insert a large bore nasogastric tube as soon as you suspect the diagnosis. Face masks are contraindicated (so intubate and ventilate, if needed, with low pressures at a rate of 100/min and minimum positive end-expiratory pressure). Get urgent surgery in an appropriate centre.

Hirschsprung's disease Congenital absence of ganglia in a segment of the colon leads to the passage of infrequent, narrow stools, GI obstruction and megacolon. Faeces may be felt per abdomen, and rectal examination may disclose only a few pellet-like faeces. It is more frequent in males (3:1). *Complications:* GI perforation, bleeding, ulceration. The diagnosis may be made by barium enema or by passing a sigmoidoscope into the aganglionic section, and taking a biopsy (for acetylcholinesterase positive nerve excess). Surgical excision of the aganglionic segment is usually needed. This does not always lead to immediate recovery.

Inguinal hernias These are due to a patent processus vaginalis (the passage which ushers the descending testicle into the scrotum). They present as a bulge lateral to the pubic tubercle. They may be intermittent, appearing during crying. In 25% there are bilateral hernias. The aim is to repair these as soon as possible, before obstruction occurs. Note: there is often an associated hydrocele, and this may be difficult to distinguish from an incarcerated hernia—exploration is required if there is doubt.

Imperforate anus Most girls will have a fistula to the posterior fourchette; most boys will have a fistula to the posterior urethra (and may pass meconium in the urine). Absence of a perineal fistula in boys indicates communication with the urethra, and indicates that a colostomy may be required. Do an IVU to reaveal commonly associated GU abnormalities.

Meckel's diverticulum See p 754.

Mid-gut malrotations Absent attachment of the mesentery of the small intestine may result in a volvulus of the midgut or obstruction of the third part of the duodenum by fibrotic bands. The child may not present for some years, until unexplained vomiting occurs. The passage of blood per rectum heralds midgut necrosis—and is an indication for emergency surgical decompression. See OHCM p 126 for acute gastric volvulus.

1 F Luks 1995 *BMJ* ii 1449

Genitourinary diseases

Undescended testis (2-3% of neonates, 15-30% of prems; bilateral in 25% of these). On cold days retractile testes may hide in the inguinal pouch, eluding all but the most careful examination (eg while squatting, or with legs crossed, or in a warm bath). If it is truly undescended it will lie along the path of descent from the abdominal cavity. Early (eg at 1 year) fixing within the scrotum (orchidopexy) prevents infertility and neoplastia.[1] Intranasal gonadotrophin-releasing hormone gives unreliable results.

Urethral valves present with uraemia (p 280) and a palpable bladder after voiding. The stream is intermittent. Micturating cystogram studies show dilatation of the posterior urethra.

Hypospadias The meatus is on the ventral side of the penis. It may be stenosed. ►Avoid circumcision: use foreskin for repair before school age.

Epispadias The meatus appears on the dorsal side of the penis.

Some congenital and genetic disorders *Horseshoe kidney:* The kidneys are fused in a midline hoop. Symptoms may be absent; or there may be obstructive uropathy and renal infections. IVU diagnosis: kidneys displaced medially; rotated collecting system.

Infantile polycystic kidneys: These may cause obstructed labour and respiratory difficulties, with later uraemia and hypertension. Radiology: collecting tubule cysts (<5mm across). The histology of the liver is always abnormal. Survivors risk UTIs and portal hypertension with haematemesis. Inheritance is recessive. The inheritance of adult polycystic disease (OHCM p 396) is dominant.

Ectopic kidney: This may reveal itself as a pelvic mass, or be seen on IVU. Associations: anorectal abnormalities, UTIs.
Renal agenesis causes oligohydramnios and Potter's facies (p 120) with early death if bilateral. Associations: unicornuate uterus.

Patent urachus: Urine leaks from the umbilicus.

Extrophy of the bladder: The bladder mucosa is exteriorized.

Double ureters: Associations—ureterocele, UTIs, pyelonephritis.

Renal tubular defects: There may be disturbance of one or more tubular functions, eg renal glycosuria, cystinuria, or diabetes insipidus. In renal tubular acidosis conservation of fixed base is impaired, so that there is a metabolic acidosis with alkaline urine. *Symptoms:* failure to thrive, polyuria and polydipsia.

Wilms' nephroblastoma This is the commonest intra-abdominal tumour of childhood (20% of all childhood malignancies). It is an undifferentiated mesodermal tumour of the intermediate cell mass. It may be sporadic, or familial, when Wilms' may be associated with *a*niridia, GU malformations, and *r*etardation (WAGR). The Wilms' tumour gene (WT1 on chromosome 11) encodes a protein which is a transcriptional repressor downregulating IGF-II, an insulin-like growth factor.[2]

Median age at presentation: 3.5yrs. 95% are unilateral. Staging:

I Tumour confined to the kidney	IV Distant metastases
II Extrarenal spread, but resectable	V Bilateral disease
III Extensive abdominal disease.	

The Patient: Features include fever, flank pain, and an abdominal mass. Haematuria is not common. IVU: renal pelvis distortion; hydronephrosis.

Management: Avoid renal biopsy. Nephrectomy, chemotherapy (eg vincristine and doxorubicin) and radiotherapy can be curative.

---

1 UK testic. grp. 1994 *BMJ* i 1393 & D Forman 1994 *BMJ* ii 666 2 G Skuse 1995 *Lancet* 345 902

Ambiguous genitalia

▶This is a rare phenomenon: refer promptly for expert help.

Gender determinants Distinguish between genetic, gonadal, phenotypic (eg affected by sex hormone secretion), and psychological sex.

History and examination Any exposure to progesterone, testosterone, phenytoin, aminoglutethamide? Previous neonatal deaths (the adrenogenital syndrome has recessive inheritance)? Note phallic size, and the position of the urethral orifice. Are the labia fused? Have the gonads descended?

Tests Buccal smear (Barr body present if ♀); white cell mustard stains to make Y chromosome fluoresce. These tests take <24h. Chromosome analysis takes 5–9 days. If there is a phallus and the buccal smear is 'female' the diagnosis is either true hermaphroditism, adrenogenital syndrome or maternal androgens (drugs, tumours). If there is a phallus and the buccal smear is 'male', tell mother that the baby is a boy.

Do not rely on appearances whenever babies have: bilateral cryptorchidism (at term), even if a phallus is present; unilateral cryptorchidism with hypospadias; penoscrotal or perineoscrotal hypospadias. These patients need examination by a paediatric endocrinologist to exclude androgen resistance (eg testicular feminization in which feminization will occur in adolescence in spite of androgens, making it imperative that these children are brought up as girls[1]). NB: if the stretched phallus is <25mm long at birth it is unlikely that it can ever be used for procreation, in spite of the best plastic surgery.

If there is uncertainty because the penis is so short, a paediatric endocrinologist may recommend 3 days' treatment with human chorionic gonadotrophin. If the baby is a boy the penis will grow (possibly even to normal length) after 48h.[1]

True hermaphroditism (10% are XY, 10% mosaic, and 80% XX). It is often best to raise these as girls. Two-thirds will menstruate.

The adrenogenital syndrome (Congenital adrenal hyperplasia—due to the excessive secretion of androgenic hormones because of deficiencies of 21-hydroxylase, 11-hydroxylase, or 3-β-hydroxysteroid dehydrogenase). Cortisol cannot be adequately produced, and the consequent rise in adrenocorticotrophic hormone leads to adrenal· hyperplasia, and over-production of cortisol precursors (which are androgenic).

Clinical and biochemical features: Vomiting, dehydration, and ambiguous genitalia. Boys may show no obvious abnormality at birth, but will exhibit precocious puberty, or may have ambiguous genitalia (reduced androgens in 17-hydroxylase deficiency), or incomplete masculinization (hypospadias with cryptorchidism from 3-β-hydroxysteroid dehydrogenase deficit). Hyponatraemia, (with a paradoxically high urine Na^+) and hyperkalaemia are common. ↑ plasma 17-hydroxyprogesterone in 90%; urinary 17-ketosteroids ↑ (not in 17-hydroxylase deficit).

Emergency treatment: Babies may present with an adrenocortical crisis (circulatory collapse) in early life. ▶▶Urgent treatment is required with 0.9% saline IVI (3–5g Na^+/day), glucose, and hydrocortisone 100mg/2–6h IV (all ages). Give fludrocortisone 0.1mg/day PO.

1 JO Ahlquist 1994 *BMJ* i 1041

Congenital heart disease (CHD)

Incidence 8/1000 births. 2 key questions: *Is the defect compensated?* (unobtrusive); if not, *Is there cyanosis?* Uncompensated defects cause poor feeding, dyspnoea, hepatomegaly, engorged neck veins. *Physiological categories:*[1] ●Left to right shunt (eg ASD; VSD) ●↓Systemic perfusion—weak pulse ± acidosis (eg hypoplastic L heart; aortic stenosis) ●Pulmonary venous congestion (eg mitral stenosis) ●Transposition streaming (of caval return → aorta) ●↓Pulmonary blood flow (eg Fallot's) ●Intracardiac mixing of 'blue' and 'red' blood (eg truncus arteriosus).

Acyanotic causes of uncompensated defects (The shunt is left to right.) Atrioseptal defects (ASD), VSDs, aortopulmonary window, patent ductus arteriosus (PDA). Later pulmonary hypertension may lead to a shunt reversal and subsequent cyanosis (Eisenmenger's syndrome, p 746).

Cyanotic causes (R to L shunt through VSD or ASD.) Fallot's tetralogy (p 746), transposition of the great arteries (TGA), other multiple defects; tricuspid or pulmonary stenosis/atresia associated with shunts.

Tests FBC, CXR, P_aO_2, ECG, 3D echocardiogram, cardiac catheter.

VSD: (25% of cases) Symptoms: as above. Signs: harsh, loud, pansystolic 'blowing' murmur ± thrill; ± a diastolic apical inflow murmur. ECG: left (or combined) ventricular hypertrophy. CXR: pulmonary engorgement. Course: 20% close spontaneously by 9 months (*maladie de Roger*).

ASD: (7% of cases) Symptoms: above. Signs: widely split, fixed S_2 + midsystolic murmur maximum in the 2nd intercostal space, at left sternal edge. CXR: cardiomegaly, globular heart (primum defects). ECG: RVH ± incomplete R bundle branch block.

Patent ductus: Presents with failure to thrive, pneumonias, heart failure, SBE. Signs: collapsing pulse, thrill, loud P_2, systolic murmur in pulmonary area. CXR: vascular markings↑, big aorta. ECG: LVH. Dexamethasone in preterm labour helps close PDAs.

Coarctation of the aorta: Aortic constriction leads to radiofemoral pulse delay, increased BP in the arms (may cause epistaxis), and reduced in the legs, absent foot pulses, ± systolic murmur at the left back, heart failure. CXR: rib notching. ECG: LVH.

Transposition of the great arteries (TGA): Cyanosis, CCF, ± a systolic murmur. CXR: egg-shaped ventricles. ECG: RVH. Balloon atrial septostomy allows oxygenated blood to reach the aorta via the right heart.

Pulmonary stenosis: Pulmonary thrill with systolic murmur. See OHCM p 310.

Treatment ●Treat heart failure in babies with nasogastric feeds, sitting upright, O_2, frusemide eg 1mg/kg/24h IV slowly and digoxin 10μg/kg/24h PO. ●Neonatal cardiologists may advise keeping ductus open in duct-dependent cyanotic conditions with prostaglandin E₁ (0.05–0.1μg/kg/min IV); intubate and ventilate before transfer to specialist unit. During transfer be alert to hypothermia, hypoglycaemia, hypocalcaemia, hypovolaemia. ●Prevent endocarditis (p 194). Open-heart surgery using hypothermia and circulatory arrest is possible at any age, eg for: Fallot's tetralagy, VSD, TGA, total anomalous pulmonary venous drainage (in which pulmonary veins drain eg into the portal system, causing severe CCF). Balloon valvuloplasty is reducing the need for open heart surgery in pulmonary stenosis (there is more of a problem with restenosis and residual incompetence with aortic valvuloplasty). It is also employed in aortic coarctation. Examples of palliative surgery: pulmonary artery banding to restrict blood flow; systemic to pulmonary shunts to enlarge an underdeveloped pulmonary arterial tree, before inserting a valve-bearing conduit.

1 SG Howarth 1993 *Arch Dis Chi* **68** 707

Murmurs and heart sounds in children

Benign flow murmurs are often heard in early systole at the L sternal edge. Lack of other features distinguishes them from malformations.

Questions to ask yourself while listening to the 2nd heart sound (S_2)
- Is it a double sound in inspiration, and single in expiration? (Normal)
- Is S_2 split all the time? (ASD—atrial septal defects)
- Is S_2 never split, ie single? Fallot's; pulmonary atresia; severe pulmonary stenosis; common arterial trunk; transposition of the great arteries (the anterior aorta masks sounds from the posterior pulmonary trunk.)
- Is the pulmonary component (2nd part) too loud? (Pulmonary stenosis)

NB: the 2nd heart sound is more useful diagnostically than the 1st.

Questions to ask yourself while listening to a murmur

Timing: • Ejection systolic crescendo–decrescendo (eg innocent murmur, semilunar valve or peripheral arterial stenosis).
- Pansystolic with no crescendo–decrescendo (VSD, mitral incompetence).
- Late systolic, no crescendo–decrescendo (mitral prolapse, *OHCM* p 306).
- Early diastolic decrescendo (aortic or pulmonary incompetence).
- Mid-diastolic crescendo–decrescendo (↑atrio-ventricular valve flow, eg VSD, ASD; or tricuspid or mitral valve stenosis). An opening snap (*OHCM* p 306) and presystolic accentuation suggest the latter.
- Continuous murmurs—usually crescendo–decrescendo (usually patent ductus, venous hum, or arterio-venous fistula).

Loudness: The 6 grades for systolic murmurs: ►Thrills mean pathology.
1 Just audible in a quiet child in a quiet room. 2 Quiet, but easily audible.
3 Loud, but no thrill.   4 Loud with thrill.
5 Audible even if the stethoscope only makes partial contact with skin.
6 Audible without a stethoscope.

Place:

AS ⟩ sternum ⟨ PS
                  ASD Still's
                  VSD Mitral incompetence

AS=aortic stenosis; PS=pulmonary stenosis

Accentuating or diminishing manoeuvres:
- Inspiration augments systemic venous return (the 'negative' pressure draws blood from the abdomen into the thorax), and therefore the murmurs of pulmonary stenosis and tricuspid regurgitation.
- Expiration augments pulmonary venous return and decreases systemic return, and therefore VSD, mitral incompetence, and aortic stenosis too. In (mild) pulmonary stenosis, the ejection click is augmented by expiration.
- Valsalva manoeuvre ↓ systemic venous return and benign flow murmurs, but ↑ murmurs from mitral incompetence and subaortic obstruction.
- Sitting or standing (compared with lying) ↓ innocent flow murmurs, but ↑ murmurs from subaortic obstruction or from a venous hum (a hum is best heard at the right base or below the left clavicle, or in the neck—it is abolished by gently pressing the ipsilateral jugular; patent ductus murmurs are similar, but do not change with posture).

Cardiac catheter findings

Pulmonary stenosis	RV pressure↑; pulmonary artery pressure↓
+ foramen ovale	as above with R atrial pressure↑ and P_aO_2↓
VSD	RV pressure↑; RV O_2 >R atrial O_2
ASD	right atrial pressure and oxygenation↑ compared to IVC
Patent ductus	RV pressure↑; pulmonary artery O_2 >RV O_2
Fallot's tetrad	see p 748. Pulmonary artery pressure >RV; RV O_2 & P_aO_2↓

Still's murmur A common innocent murmur (peak incidence: 4–5yrs), and may manifest during fever or on routine examination. It is musical, low-pitched, vibratory and mid-systolic, localized between the left sternal edge and the apex. It may radiate to the 2nd intercostal space. Differential diagnosis: VSD (but it is crescendo–decrescendo).

Cleft lip and palate

This is the most common facial malformation. It results from failure of fusion of the maxillary and premaxillary processes (during week 5). The defect runs from the lip to the nostril. It may be bilateral—in which case there is often a cleft in the palate as well, with the premaxillary process displaced anteriorly. Palate clefts may be large or small (eg of the uvula alone). *Incidence:* 0.8–1.7/1000. *Causes:* Genes, drugs (benzodiazepines, antiepileptics, p 161), rubella. Other malformations are present in 50% (eg trisomy 18, 13–15, or Pierre Robin syndrome—in which the mandible is too small, causing cyanotic attacks). *Prevention:* Possibly folic acid ± multivitamins (p94).

Treatment: Ask an orthodontist's opinion. Feeding with special teats may be necessary before plastic surgery (repair of the lip at 3 months, and of the palate at 1yr). Repair of unilateral complete or incomplete lesions often gives good cosmetic results. Try to refer to expert centres.[1] There is always some residual deformity with bilateral lesions. Complications: otitis media, aspiration pneumonia, post-operative palatal fistulae, speech defects (enlist a speech therapist's help). Avoid taking to SCBU (this can hinder bonding).

Other malformations in the head and neck

Eyes Anophthalmos: there are no eyes; rare; part of trisomy 13–15.
Ectopia lentis: presents as glaucoma with poor vision. The lens margin is visible; seen in arachnodactyly, Ehlers–Danlos (p 746), homocystinuria; incidence: <1/5000; autosomal dominant (a-Dom) or recessive (a-R).
Cataract: rubella, Down's, others-recessive or sex-linked.
Coloboma: notched iris with a displaced pupil; incidence: 2/10,000; (a-R).
Microphthalmos: small eyes; 1/1000; due to rubella—or genetic (a-Dom).

Ears Accessory auricles: seen in front of the ear; incidence: 15/1000.
Deformed ears: Treacher–Collins' syndrome (p 758).
Low-set ears: associations—Down's syndrome; congenital heart disease.

Nose and throat Choanal atresia: this presents as postnatal cyanotic attacks. A nasal catheter does not pass into the pharynx because of malformation in the nose; incidence: <1/5000.
Congenital laryngeal stridor: this may be due to laryngeal webs, or laryngomalacia (the larynx is unable to stay open); incidence: <1/5000.
Laryngeal atresia: this presents as the first breaths fail to expand the lungs. Immediate tracheostomy is needed.
Branchial fistula: there is an opening at the front of sternomastoid (a remnant of the 2nd or 3rd branchial pouch); incidence: <1/5000.
Branchial and thyroglossal cysts: see OHCM p 56.

Skull and vertebrae Brachycephaly: short, broad skull from early closure (craniostenosis) of the coronal suture; incidence: <1/1000; a-Dom. Cleidocranial dysostosis: no clavicles, enabling the shoulders to meet. Slow skull ossification, no sinuses, high-arched palate; incidence: <1/5000; a-Dom.
Craniofacial dysostosis: tower skull, beaked nose, exophthalmos.
Klippel–Feil syndrome (p 752): fused cervical vertebra (so the neck is short).

CNS Hydrocephalus: incidence 0.3–2/1000. Neonatal injury, infection, or genes (sex-linked) may cause aqueduct stenosis. Dandy–Walker syn. (p 744) Arnold–Chiari malformation (OHCM p 632).
Microcephaly: causes—genetic, intrauterine infection (eg rubella), hypoxia, irradiation. Incidence: 1/1000. Risk of recurrence: 1/50.

Spina bifida and anencephaly See p 228.

Fetal alcohol syndrome
The severity depends on how much alcohol the mother has drunk during pregnancy. Features: microcephaly, short palpebral fissure, hypoplastic upper lip, absent filtrum, small eyes, IQ↓, cardiac malformations.

3. Paediatrics

Some words about the head and neck
Acrocephaly: This term may be applied to tower headed conditions.
Arhinencephaly: Congenital absence of the rhinencephalon.
Arthrogryposis: this term implies contracture of a joint.
Brachycephaly: the head is too short.
Cephalocele: intracranial contents protrude. Kélé means hernia in Greek.
Craniostenosis=craniosynostosis=premature closure of skull sutures.
Dolicephalic: the head is too long.
Dystopia canthorum: intercanthal distance is increased, but not the inter-pupillary or (bony) interorbital distances.
Holoprosencephaly: hypotelorism with cleft palate.
Lissencephalic: bat-like brain with no convolutions (agyria).
Micrognathia: the mandible is too small.
Metopic suture=frontal suture.
Neurocranium: that part of the skull holding the brain.
Obelion: the point on the saggital suture crossed by a line joining the parietal foramina.
Oxycephalic=turricephaly=acrocephaly=the top of the head is pointed.
Plagiocephaly: An asymmetrical, twisted head—eg associated with irregular closure of the cranial sutures.
Rachischisis: congenital fissure of the spinal column.
Sinciput: the anterior, upper part of the head.
Viscerocranium: facial skeleton.
Wormian bones: supernumerary bones in the sutures of the skull.

1 T Markus 1995 *BMJ* ii 765

Neural tube defects

Spina bifida means having an incomplete vertebral arch.
Spina bifida occulta: the defect is covered by skin and fascia.
Meningocele: dura & arachnoid mater bulge through the defect.
Myelocele: segments of the cord are exposed with no covering.
Myelomeningocele: the cord's central canal is exposed.
Anencephaly: absent skull vault and cerebral cortex.
Encephalocele: part of the brain protrudes through the skull.

Prevalence (% births) USA 0.1%; UK 0.4%; Wales 0.7%; N Ireland 0.9%; it increases with lower social class. Recurrence risks rise 10-fold if one pregnancy has been affected, 20-fold if 2, 40-fold if 3 pregnancies affected; and 30-fold if a parent is affected.

The neurological deficit is very variable, depending on the level of the lesion, and the degree to which the lower cord functions independently from the upper cord. The defect may progress after birth[1] (and after subsequent operations); subsequent hydrocephalus gradually worsens mental performance. A child who learns to walk during his second year may outgrow his ability to support himself during the next years (weight increases as the cube of surface area, power only as its square). Those with lumbosacral myelomeningoceles usually learn to walk with calipers by the age of 3, but few with higher lesions ever walk. An unstable condition exists when there is paralysis below L3, as unopposed hip flexors and adductors are likely to dislocate the hips. Only between 5 and 13% retain their ability to walk.

Surgery Firm guidelines on whom to treat often prove too simple to apply to the individual infant. The final outcome of early closure of the defect depends on the state of the kidneys after multiple UTIs, and the extent of delayed hydrocephalus (requiring ventriculoperitoneal CSF shunts). Early post-operative mortality may account for ~25% of deaths. Many operations may be needed for spinal deformity (often severe and very hard to treat).[1]

Hurdles for the developing child ●Urinary and faecal incontinence. Penile appliances, urinary diversions or intermittent self catheterization (for girls) save laundry and bed sores.
●Social and sexual isolation, if a special school is needed.
●The mother who 'does it all' can prevent maturity developing.
●Immobility. Mobility allowances are small and of little help.

Intrauterine diagnosis A maternal serum α-fetoprotein >90U/ml at 18 weeks' gestation[2] detects about 80% of open spina bifidas, and 90% of anencephalics, but also 3% of normal singleton fetuses, twins, and some with exomphalos, congenital nephrosis, urethral valves, Turner's syndrome, trisomy 13 and oligohydramnios. Amniocentesis and skilled ultrasound increase pick up further.

Prevention In mothers who have already had one affected baby, there is good evidence that folic acid (5mg/day is recommended by DoH)[1] given from before conception (as the neural tube is formed by 28 days, before pregnancy may even be recognized) reduces the risk of recurrence of neural tube defects by 72%.[2] If no previous neural tube defects, 0.4mg of folic acid is now recommended in the months before conception and for 13 weeks after. See p 94.

1 DoH 1991 PL/CMO (91) 11   2 MRC Vitamin Study Research Group 1991 *Lancet* ii 131

►►Resuscitation after delivery^{APLS}

Most new-born babies are perfectly healthy, and the best plan is to return these babies to the mother without interference, so that breast feeding and bonding may start. Those who do not breathe need prompt action to save their lives. Remember mouth to mouth-and-nose breathing if there is no equipment. A paediatrician should attend the following births: Caesarean sections, breeches, multiple pregnancies, forceps, intrapartum bleeding, prematurity, erythroblastosis, eclampsia.

Before birth Familiarise yourself with the equipment available. Heat the resuscitation crib. Ask for a warm blanket. Wash your hands. Explain to the mother that you are here to welcome the baby. It helps if she has 100% O_2 if there is fetal distress.

At birth Determine the Virginia Apgar score (see opposite). ►If baby is in difficulty, do not delay for scoring, but proceed as follows:
●Set a clock in motion ●Cover the baby in the crib ●Extract mucus from the oropharynx with suction. Use a laryngoscope to help with this if the baby may have inhaled meconium ●Assess colour, heart rate, respiratory effort.

If baby is pink, pulse >100, and good respiratory effort give to mother.
Primary apnoea implies pulse <60 + cyanosis. Give O_2 by funnel and wait 1min. *Terminal apnoea* implies a pulse <60, pallor and floppiness. Give gentle suction; bag and mask ventilation or intubation. *Fresh stillbirth*—asystolic terminal apnoea—start full CPR.

►►Neonatal CPR Intubate and ventilate with 100% O_2 (ensure system has a reservoir bag, and a blow-off valve set to 30–40cmH_2O) at 30–40/min.

Start cardiac massage (120 compressions/min, using index and middle fingers 1 finger breadth below internipple line to depress the sternum ~1.5cm). Give 3 chest compressions for each ventilation. If the baby fails to improve give: ●*Adrenaline* 10µg/kg IV or 20µg/kg via ET ●*Bicarbonate* 1mmol/kg IV (4.2%) ●*Adrenaline* 10–30µg/kg IV every 3–5min of arrest.

If there is a poor response think of: ●Hypothermia—dry baby and keep under heat lamp ●Pneumothorax, if ventilated ●Hypoglycaemia—give dextrose 25%, 2ml/kg IV ●Opiate toxicity—give naloxone 10µg/kg IV if mother received opiate; the dose may need repeating every 2–3 min ●Anaemia (heavy fetal blood loss?)—give 20ml/kg 4.5% albumen, then blood ●Does the baby have lung disease or congenital cyanotic heart disease (p 224)? ►Transfer to SCBU for monitoring.

Endotracheal intubation is the key skill. Use a 3.5mm uncuffed, unshouldered tube on term infants; 3.0 if 2500-1250g and 2.5 for smaller infants. Learn from an expert. Always have the size larger and smaller available. Practice on models saves lives.

Prognosis Apgar 0: of the survivors, most survive intact (>80%), and most of those who are neurologically abnormal after birth recover fully. If the Apgar is <4 at 1min 17% die (48% if of low birth weight). If the Apgar is <4 at 5min 44% die.

Apgar score

Score	Pulse	Respiratory effort	Muscle tone	Colour	On suction
2	>100	strong cry	active movement	pink	coughs well
1	<100	slow, irregular	limb flexion	blue limbs	depressed
0	0	nil	absent	all blue	nil

Problems: It is often the resuscitator who does the scoring—and he has more important jobs such as intubating (the Apgar score presumes the baby is unintubated).[1]

1 N Marlow 1992 *Arch Dis Chi* **67** 765

Neonatal intensive care

Cost: ~£600/day

Neonatal intensive care is a technological development of the basic creed of first aid—ABC; A for *a*irway, B for *b*reathing, and C for *c*irculation. Success depends on mastering this trinity, and designing the best *milieu intérieur* to encourage healing and growth without developing mental retardation or spasticity. As more physiological functions are taken over by machines, monitoring the state of the *milieu intérieur* becomes ever more vital. Monitor temperature, pulse, blood pressure (intra-arterial if critical), respirations (continuous read-out device), blood gases—transcutaneous oximetry (SpO_2, OHCM p 658) or intra-arterial electrode, U&E, bilirubin, FBC, daily weight, weekly head circumference.

The patient is usually a premature baby whose mortal enemies are cold, hypoglycaemia (p 238), respiratory distress syndrome (p 240), infection (p 238), intraventricular haemorrhage, apnoea and necrotizing enterocolitis (p 244).

Cold With their small volume and relatively large surface area, this is a major problem for the premature and light-for-dates baby. The problem is circumvented by using incubators which allow temperature (as well as humidity and F_IO_2) to be controlled, and also afford some protection against infection. F_IO_2 is the partial pressure of O_2 in inspired air.

Neonatal apnoeic attacks *Causes:* prematurity; milk aspiration; heart failure; infection; hypoxia; hypoglycaemia; hypocalcaemia.

If stimulating the baby does not restore breathing, suck out the pharynx and use bag and mask ventilation. Avoid high concentrations of O_2 to prevent retinopathy of prematurity, which may follow repeated resuscitations.

Tests in apnoea: CXR; U&E; infection screen; glucose; Ca^{2+}; Mg^{2+}.

Prevention and treatment of apnoea: If aspiration is the problem, give small feeds every 2h, or tube feed. Monitor SpO_2 continuously; if hypoxia develops in spite of an ambient O_2 of 40%, CPAP or IPPV will be necessary. Aminophylline (0.15mg/kg/h IV) may also be useful, but levels must be monitored. Reduce the dose for persistent tachycardia (>165/min). Abrupt withdrawal may cause further apnoea.

Retinopathy of prematurity (RoP)[1] Immature retinal vessels are sensitive to high P_aO_2, as occurs during repeated resuscitations. This may lead to retinal fibrosis, detachment and visual loss. Postnatal age of onset is *later* for the more immature baby. *Prevalence (lower limits):* Babies <1000g: 53%; ≤1250g: 43%; ≤1500g: 35%.

Classification: There are 5 types, depending on site involved, the degree of retinal detatchment, and extent (measured as clock hours in each eye).

Treatment: Cryotherapy has been shown to be beneficial.[2]

Screening: Any baby <1500g or ≤31 weeks' gestation. If ≤25 weeks, screen ophthalmoscopically under 0.5% cyclopentolate pupil dilatation at 6–7 weeks old, and every 2 weeks until 36 postmenstrual weeks. Often one examination is enough for those of 26–31 weeks' gestation. It may be best done by an ophthalmologist.

Intraventricular haemorrhage (IVH) This occurs in >40% of infants with a birth weight <1500g. It may be subependymal, subarachnoid, or directly into a ventricle or brain tissue. Possible causes: birth trauma, hypoxia, bleeding disorder (p 244). Suspect IVH in any neonate who deteriorates in an unexplained way. Opisthotonus, cerebral irritability, shock, deteriorating feeding skills, a bulging fontanelle with a rapidly expanding head, an exaggerated (or absent) Moro reflex, fits and somnolence are telling signs, but a high proportion (~50%) may be 'silent'. *Tests:* Ultrasound and CT confirm the diagnosis. Complications: mental retardation, spasticity. Many survive unscathed. It is uncertain whether late learning and behavioural disorders occur. *Treatment:* Rest, head elevation, and control of fits (p 238).

1 AR Fielder 1992 *Arch Dis Chi* **67** 860   2 Cryotherapy for retinopathy group 1990 *Arch Ophth* **108** 1408

Ventilation of neonates

This is a practical skill which must be learned at the cot side. Nurses are often expert, and may be best placed to teach. Needs of apparently very similar babies will vary, so that what follows is only a rough guide to prepare your mind before receiving teaching. Continuous refinement in the light of transcutaneous and blood gas analysis is needed.

Pressure-limited, time-cycled continuous flow ventilation A continuous flow of gas that is heated and humidified is passed to the infant via an endotracheal tube. The nasotracheal route is preferred (fewer tube displacements), but insertion needs more experience. *Variables:* Air/O_2 mix; maximum inspiratory pressure (P_I); positive end-expiratory pressure (PEEP); inspiratory (T_I) and expiratory (T_E) times. The infant is able to make respiratory efforts between ventilator breaths (intermittent mandatory ventilation, IMV). Aim for a prolonged inspiratory time (1–1.5sec) and a T_I/T_E ratio of ≥1 for bad respiratory distress syndrome. If less severe, a T_I/T_E ratio of <1 may be tried (pneumothorax risk ↓).
Initial settings: Choose to give good chest excursion and air entry on auscultation and adequate transcutaneous O_2 readings. Typical settings might be T_I 1sec, 20–25 cycles/min, inspiratory pressure 14–16cmH_2O, and PEEP 5cmH_2O. Adjust in the light of blood gas analysis.

PEEP (Positive end expiratory pressure) A loaded valve is fitted to the expiratory limb of the ventilator, so airways pressure stays ≥atmosphere. Levels >10cmH_2O are rarely used (venous return to the thorax ↓).

CPAP (Continuous positive airways pressure) Pressure is raised throughout the respiratory cycle, so assisting spontaneous inspiration. With skill, this method has few complications, and is useful as a first stage in ventilating a baby before it is known whether he will need IMV.

PTV (patient-triggered ventilation) With traditional ventilation, the baby may fight the ventilator, trying to expire during inflations: hence the need for paralysis: but this makes the baby entirely dependent on the ventilator, fluid retention is common, and contractures may develop in limbs. One answer is PTV in which inspiratory and end-expiratory pressure is set by the operator, but the rate set by the baby.[1]

Muscle paralysis Pancuronium (0.02–0.03mg/kg IV; then 0.03–0.09mg/kg every 1.5–4h to maintain paralysis) prevents pneumothorax in infants 'fighting the ventilator' (eg needing an unexpectedly high P_I).

Pain relief Consider 5% dextrose IVI+morphine 100–200μg/kg over 30min, then 5–15μg/kg/h.[1][2] This is thought to be safe, and reduces catecholamine concentrations—a possible objective correlate of pain and stress.

Weaning from the ventilator Decrease the rate of IMV and lower the P_I by eg 2cm H_2O decrements. Try extubation when blood gases are adequate with a PEEP of ~2cmH_2O with spontaneous breathing.
Air leak: Air ruptures alveoli and tracks along vessels and bronchioles (pulmonary interstitial emphysema), and may extend intrapleurally (pneumothorax with lung collapse), or into the mediastinum or peritoneum.
Signs: Tachypnoea, cyanosis, chest asymmetry. The lateral decubitus CXR is often diagnostic. Prompt 'blind' needle aspiration of a pneumothorax may be needed. Aspirate through the second intercostal space in the midclavicular line with a 25G 'butterfly' needle and a 50ml syringe on a 3-way tap. If the leak is continuous, use underwater seal drainage.

Retinopathy of prematurity p 232 Bronchopulmonary dysplasia p 235

---

1 A Greenough 1992 *Arch Dis Chi* **67** 69   **2** *Drug Ther Bul* 1994 21

Bronchopulmonary dysplasia

This was first described in 1967, as a syndrome of chronic lung disease in premature babies who had had mechanical ventilation for respiratory distress syndrome. The patient has persistent hypoxaemia, and is difficult to wean from the ventilator.

Tests: CXR: hyperinflation, rounded, radiolucent areas, alternating with thinner strands of radiodensity.

Histology: Necrotizing bronchiolitis and alveolar fibrosis.

Mortality: About 40%.[1]

Early sequelae: Decreased cognition, cerebral palsy, feeding problems. O_2 desaturation during feeds is not uncommon.[2]

Late sequelae: By adolescence/early adulthood the main changes remaining are airways obstruction, airways hyper-reactivity, and hyperinflation.[1]

1 W Northway 1992 *NEJM* **333** 1793   2 L Singer 1992 **90** 380

Neonatal jaundice

To detect jaundice in non-white babies, press the nose.

Hyperbilirubinaemia (<200μmol/l) after the 1st day of life is common and may be 'physiological' (eg in binding free radicals[1]). Mechanisms:

1 Hepatic immaturity in bilirubin uptake and conjugation.
2 Excessive removal and destruction of fetal red cells.
3 A low plasma albumin (so unconjugated bilirubin is unbound).
4 Absence of gut flora impeding bile pigment elimination.
5 Poor fluid intake, or breast feeding (inhibiting factors—and is not a reason to stop breast feeding).

Jaundice within 24h of birth is always pathological. Causes:
- Rhesus haemolytic disease: +ve direct Coombs' test (DCT, p 242).
- ABO incompatibility: (mother O; baby A or B) DCT +ve or –ve. Send maternal blood for haemolysins (also for other rare blood group incompatibilities, eg anti-C, E, c, e, Kell, Duffy).
- Red cell anomalies: congenital spherocytosis (do blood film); glucose-6-phosphate dehydrogenase deficiency (do enzyme test).

Tests in all patients: FBC; film; blood groups; Coombs' test; urine for reducing agents; infection screen, p 238 (an important cause).

Prolonged jaundice (not fading after 7–10 days) Causes:
Hypothyroidism (vital not to miss; do T₄ and TSH); biliary atresia—
▶Exclude this in *all* who continue to be jaundiced after 14 days.
Galactosaemia: urine tests for reducing agents (eg Clinitest®) are +ve, but specific tests for glycosuria are –ve.

Kernicterus May occur if bilirubin is >360μmol/l (lower in prems).
Stage I: Sleepy; reduced suck; lethargic feeding.
Stage II: T°↑; restless; abnormal mouth movement; lid retraction (the sign of the setting sun).
Stage III: Latent phase.
Stage IV: Subsequent cerebral palsy; deafness; IQ↓ (later).

Lesser levels of bilirubin (170–323μmol/l) are unlikely to give rise to permanent problems, and have little effect on later IQ unless the infant is preterm or light-for-dates (eg 0.009–0.018 IQ points/μmol/l).[1]

Treatment Methods of reducing plasma bilirubin Phototherapy lamps may be used if the jaundice is mild. It is not a substitute for exchange transfusion. SE: cold; fluid loss (give 30ml/kg/day extra water).
Start phototherapy at the following plasma bilirubin levels: Term baby at age 48h: 230μmol/l. (At 72h 250μmol/l; at 4 days 275μmol/l at 5 days 300μmol/l.) In prems or low-birth-weight babies reduce these thresholds by at least 25μmol/l. Stop phototherapy when levels fall (eg by >25μmol/l) below these thresholds.

Exchange transfusion Uses warmed blood (37°C), crossmatched against maternal and infant serum, given via an umbilical vein IVI. Aim to exchange 160ml/kg over ~2h (twice the blood volume). IV Ca²⁺ is not needed if citrate phosphate dextrose blood is used. Monitor ECG, U&E, Ca²⁺, bilirubin and blood glucose. Further exchanges may be needed if the bilirubin level continues to rise.
Stop the exchange transfusion if the heart rate fluctuates by >20 beats/min.
▶Ensure that the volumes exchanged always balance.
If anaemic, consider a simple fresh blood transfusion (20ml/kg).
Plasma bilirubin levels at which exchange transfusion is indicated: Term baby at birth, 50μmol/l; at 12h, 125μmol/l; at 24h, 200μmol/l; at 36h, 250μmol/l; at 48h, 325μmol/l; at 72h, 350μmol/l; at 4 days, 375μmol/l; at 5 days, 400μmol/l. If the baby is premature or weighs <2.5 kg lower these thresholds (eg by 50μmol/l, after age 48h).

1 *Lancet* Ed 1991 ii 1242

239

Respiratory distress syndrome (RDS)

Insufficient surfactant is made so that the lungs are unable to stay expanded; re-inflation between breaths exhausts the baby, and respiratory failure follows. It is the major cause of death from prematurity.

Infants at risk: 100% if 24–28 weeks' gestation; 50% if 32 weeks. Also: maternal diabetes, male sex, second twin, Caesarian birth.

Prenatal prevention Amniotic fluid phospholipid analysis showing a lecithin/sphingomyelin ratio <2 and a saturated phosphatidylcholine <500µg/dl may occasionally be used to indicate fetal lung maturity and to guide use of dexamethasone (8mg/8h IM to the mother for at least 24h if labour can be delayed; CI: maternal infection; need for immediate delivery)—so promoting lung maturity.

Signs Worsening tachypnoea (>60/min) in the hours after birth. Increased inspiratory effort, with grunting, flaring of the nasal alae, intercostal recession and cyanosis. CXR: diffuse granular patterns, with air bronchograms. Mild signs subside after ~36h.

Differential diagnosis *Transient tachypnoea of the newborn* is due to excess lung fluid. It usually resolves after 24h. *Meconium aspiration* (p 244); congenital pneumonia (β-haemolytic streptococci); tracheo-oesophageal fistula (suspect if respiratory problems after feeds).

Treatment Learn this skill at the cot side. If gestation <~28 weeks, intubate at birth and, if >700g, consider surfactant by ET tube, eg synthetic pumactant (100mg, repeat at 1h and 24h if needed: monitor P_aO_2 as needs may suddenly↓ OR pork 'poractant' (Curosurf® 100–200mg/kg, ± 2nd dose in 12h). Rock gently to promote spread in the bronchial tree.
- Wrap up warmly and take to SCBU incubator.
- Monitor blood gases (eg transcutaneously). Aim for a P_aO_2 of 7–12 kPa, enhancing the ambient O_2 with a perspex head box.
- If blood gases deteriorate intubate and ventilate (p 234), before the infant becomes exhausted. Start with CPAP, eg 5cmH$_2$O (p 234). A rising P_aCO_2 may indicate either that CPAP is too high, or that paralysis and further ventilation is needed.
- Ventilator setting guide: inspiratory pressure 20cmH$_2$O with 60% O_2; positive end-expiratory pressure 3–5cmH$_2$O; 20–25 breaths/min, with an inspiratory duration of 1sec.
- When the endotracheal tube is connected, check that chest movement is adequate and symmetrical. Listen for breath sounds. Enlist experienced help in making further adjustments.
- Increase the P_aO_2 by increasing pressures (not too high).
- Decrease P_aCO_2 by lengthening the expiratory time, or ↑breath frequency by lessening inspiratory time. ▶If any deterioration, consider: blocked tube, infection, faulty ventilator, pneumothorax.
- Fluids: avoid oral feeding. Give 10% dextrose IVI (p 246).

Signs of a poor prognosis Persistent pulmonary hypertension, large right to left shunt via the ductus.[1]

If, inspite of everything, hypoxia worsens, the baby is dying. Explain this to the mother, emphasizing that the baby will feel no pain. Encourage her to christen the baby. Give 115.5µg/kg/IM of papavaretum 7.7mg/ml (=0.15ml/kg of one 7.7mg ampoule diluted with 9ml water). Disconnect the tubes, so allowing the mother to hold the baby, and, in so doing, to aid her grief.

1 FJ Walther 1992 *Paediatrics* **90** 899

Rhesus haemolytic disease

Physiology When a RhD–ve mother delivers a RhD+ve baby a leak of fetal red cells into her circulation may stimulate her to produce anti-D IgG antibodies (isoimmunization). In later pregnancies this can cross the placenta, causing worsening rhesus haemolytic disease (*erythroblastosis fetalis*) with each successive Rh+ve pregnancy. 1st pregnancies may be affected due to leaks, eg: ●Threatened miscarriage, ●APH, ●Mild trauma, ●Amniocentesis, ●Chorionic villous sampling, ●External cephalic version.

An affected oedematous fetus (with stiff, oedematous lungs) is called a *hydrops fetalis*. Oedema occurs as the liver is devoted to producing new RBCs (so albumin ↓, leading to ↓oncotic pressure).

Clinical Rh disease ►*Test for D antibodies in all Rh–ve mothers, at booking, 28 and 34 weeks' gestation.* Anti-D titres <4U/ml (<1:16) are very unlikely to produce serious disease; it is wise to check maternal blood every 2 weeks. If >10U/ml, get the advice of a referral centre: fetal blood sampling ± intraperitoneal (or, with fetoscopy, intravascular via the cord) transfusion may be needed. Expect fetal Hb to be <7g/dl in 10% of those with titres of 10–100U/ml (75% if titres >100U/ml).[1]

Do regular ultrasound and amniocentesis and if anti-D titre >4U/ml. Timing is vital. Do it 10 weeks before a Rh-related event in the last pregnancy (eg if last baby needed delivery at 36 weeks, expect do amniocentesis at ~26 weeks). Fetuses tolerating high bilirubins may be saved risky transfusions (fatality 2–30%) if monitored by serial measurements of fetal Hb (by fetoscopy) and skilled daily ultrasound, to detect oedema, cardiomegaly, pericardial effusion, hepatosplenomegaly or ascites.

Anti-D is the most common antibody. Others: Rh C, D, E, c, e, Kell, Kidd, Duffy (all are IgG). Relatively low concentrations sometimes produce severe disease.

Signs ●Jaundice on day one (or later)
●Heart failure (oedema, ascites)
●Progressive anaemia; bleeding

●Yellow vernix
●Hepatosplenomegaly
●CNS signs

Tests Hb ↓ ●Direct Coombs' +ve
●Anti-Rh agglutinins present
●Mother Rh–ve, baby Rh+ve
●Anti-Rh titre ↑ in mother

●Bilirubin ↑
●Reticulocytes ↑
●Hypoglycaemia

Exchange transfusion *Indications and technique:* p 236.
If Hb <7g/dl, give the first volume of the exchange transfusion (80ml/kg) as packed cells, and subsequent exact exchanges according to response. ►Keep the baby warm.

Ultraviolet photodegradation of bilirubin (with phototherapy lamp) may be all that is needed for mild cases. Give extra water (30ml/kg/24h PO). Avoid heat loss. Protect the eyes.

Kernicterus See p 236.

Giving Rh–ve mothers anti-D Ig (If the baby is RhD+ve) ►See p 109.

1 MJ Whittle 1992 **67** 65

Necrotizing enterocolitis (NEC)

This is death of the bowel mucosa. Typically it occurs at the end of the first week of life in a premature infant on SCBU (it affects 1–2% of admissions to SCBU). If mild, a little blood and mucus may be passed per rectum. At worst, there is sudden abdominal distension, tenderness (± perforation), shock, DIC, and sloughing of the rectal mucosa. It may be sporadic or epidemic. *Treatment:* If mild, barrier nurse the baby; culture faeces; stop oral feeding; do plain erect and supine abdominal X-rays, to look for oedematous loops of bowel with intramural gas. If more severely affected, crossmatch blood; give metronidazole 7mg/kg/12h IV with penicillin and netilmicin (p 238). Do repeated radiology and measurements of girth. Liaise early with a surgeon. Indications for laparotomy: progressive distension, perforation.

Meconium aspiration

Only thick meconium is important (aspiration pneumonitis). Using a laryngoscope, the pharynx and trachea should be sucked out at birth. Intubate the trachea with the largest comfortable tube. 2–3 intubations with sucking withdrawals are repeated to clear as much meconium as possible. The obstetrician may try compressing the thorax as he passes the baby to the paediatrician, so minimizing inspiratory effort until the airway is clear (but this does not stop diaphragm effort). Ventilation with CPAP or IMV may be needed (p 234). Try a PEEP of 6cm H_2O and an inspiratory time of 0.5–0.75sec. Give penicillin + netilmicin (p 238).

The bleeding neonate

Haemorrhagic disease of the newborn (=Vitamin K deficiency bleeding) occurs in the immediate postnatal period (day 2–7), and is due to lack of vitamin K (no enteric bacteria, or mother taking anticonvulsants). The baby is usually well, but develops unexplained bruising and bleeding. Prothrombin and partial thromboplastin times (PT & PTT) are prolonged; platelets normal. *Prevention:* Vitamin K 500µg PO to *all* babies within 24h of birth. After this time régimes vary. Question local guidelines to discover if the régime uses the readily absorbed mixed micelle preparation, not the cremophor preparation (there is evidence that the latter may not prevent bleeding).[1][2][3] (The parenteral route may be inadvisable because of an uncertain ↑risk of subsequent neoplasia.[1]) Formula milk already contains vitamin K. *Treatment:* Give plasma (10ml/kg IV) and vitamin K for active bleeding.

Disseminated intravascular coagulation (DIC) The baby is usually ill (and often infected), and may have petechiae, oozing from venepuncture sites and GI haemorrhage. Platelets ↓, fibrin degradation products ↑, INR & partial thromboplastin time ↑, schistocytes (fragmented RBCs).
Treatment of DIC: ●Treat cause (eg infection, NEC); vitamin K 1mg IV.
●Platelet transfusion, to keep platelets >30 × 10^9/l.
●Replace clotting factors with fresh plasma (eg 10ml/kg by IVI).
●If bleeding continues, consider an exchange transfusion.

Immune thrombocytopenia Partial thromboplastin time & INR ↔ platelets↓.

1 R von Kries 1994 *Lancet* 343 352   2 Brit Paed Ass 1993 *Drug Ther Bul* 31 80   3 G Draper 1994 *BMJ* i 867

Enteral and parenteral feeding

Enteral feeding *Indications:* Any sick infant who is too ill or too young to feed normally (eg respiratory distress syndrome). It is best achieved by continuous infusion of expressed breast milk (from the baby's own mother) via a silastic nasojejunal tube (the least risk of aspiration). After entering the stomach, the tube enters the jejunum by peristalsis. Confirm its position by X-ray. Other milks: cow with added carbohydrate (eg Cow & Gate Plus®, New Ostermilk Two®); skimmed cow's milk with fats and minerals (eg SMA®); skimmed cow's milk with less whey (eg SMA Gold Cap®), in which the casein to whey ratio is similar to breast milk's. Volume to infuse (for neonates): ~150ml/kg/day. Breast milk is still probably the best food for premature infants—but add vitamin D 1000u/day and vitamin K 2–3μg/day—and extra phosphate if plasma phosphate is <1.5mmol/l.[1]

Indications for IV feeding ●After surgery, trauma, or burns.
●When oral nutrition is poor; eg in ill, low-birth-weight babies.
●Necrotizing enterocolitis (when the gut must be 'rested').

Parenteral nutrition (PN) Day-by-day guide. ►All values are per kg/day.

Type of baby	Day of PN	Age days	PROTEIN Vamin1® (ml)	CARBOHYDRATE ml dextrose 5%	20%[2]	FAT Intralipid 10%(ml)	IONS mmol Na	K	FLUID PN Vol(ml)/ total need[3]
Neonates	1	3	7	–	36	10	2.7	2.4	53/90
and	1	4&5	7	60	22	10	2.7	2.4	99/120
Low-birth-	1	≥6	7	120	6	10	2.7	2.4	143/150
weight	2	4&5	10	20	40	10	2.5	2.3	80/150
babies	2	≥6	10	80	25	10	2.5	2.3	125/150
	3	5	14	–	43	20	2.3	2.2	77/150
	3	≥6	14	60	28	20	2.3	2.2	108/150
	4	≥6	21	20	44	20	2.0	2.1	105/150
	5	>6	28	–	56	30	1.6	1.9	114/150
	M	>6	35	–	52–62	35	1.25	1.8	≥122/150
Infants	1		7	120	6	10	2.7	2.4	143/*
>1 month	2		14	80	23	10	2.3	2.2	127/*
& <10kg	3		21	20	44	20	2.0	2.1	105/*
	M		35	–	52	30	1.25	1.8	117/*
10–30kg	1&2		14	–	15	15	2.0	2.0	44/*
	M		28	–	21–26	20–30	1.5	1.5	69–84/*
>30kg	1&2		14	12	–	10	2.0	2.0	36/*
	M		21		4–30	20–25	1.5	1.5	45–76/*

Note: M=maintenance. *See p 250 for 24h fluid requirement.
1 Vamin® with 10% glucose. 2 If 10% dextrose is used, double the volume. Both dextrose solutions are needed. 3 This is the total volume of fluid required (60 & 75ml/kg for days 1 & 2). Ca^{2+} & other elements are given to infants as Ped-el® 4ml/kg/day if renal function good. Vitamins Solvito N®: add 0.5ml/kg/day (≤5ml/day) to Vamin® (protect from light). Give vitamins A, D, & K as Vitlipid N Infant®, 1ml/kg/day (≤4ml/day), added to Intralipid®. ►These values are a guide only. Individual needs vary greatly. Get the advice of an expert.

Sterility is vital. Prepare using laminal flow units. Give into a central vein.

Regular checks *Daily:* Check weight; fluid balance; U&E; blood glucose; Ca^{2+}. Turbid plasma means too much Intralipid® has been given. Urine electrolytes and osmolality (every 48h from week 1). Test for glycosuria. Change IVI sets and filters. Culture: filters, Vamin® and Intralipid® samples.

Weekly: Length and head circumference; skin fold thickness. LFTs; Mg^{2+}; PO_4^{3-}; alk phos; ammonia; lipids; FBC. Infection screen (p 238).

Complications of IV feeding Infection; acidosis; metabolic imbalances. If plasma $PO_4^{3-}\downarrow$, consider giving PO_4^{3-} (0.25–0.5mmol/kg/day) as the potassium salt. Mix with dextrose, but not Vamin® or trace element mixtures.

Stopping IV nutrition Do in stages to prevent hypoglycaemia.

1 B Wharton 1987 *Nutrition and feeding of Preterm Infants,* Blackwell

▶▶The ill and feverish child

There are 2 distinct questions to ask: *How severe are the symptoms* and *How appropriate is the child's response to the illness?* The symptom may be severe (eg 'terrible diarrhoea all over the cot') without being biologically serious (if the baby is alert, drinking, wetting many nappies, and behaving as usual); but ANY apparently mild symptom should set your internal alarm bell ringing if:[1] [2]

●<Half the usual amount of feed has been taken in the last day.
●There is breathing difficulty, or high-pitched continuous moans.
●There is a history of being pale, mottled, cyanosed, and hot.
●Dull expression (abnormally quiet); drowsiness; dehydration.
●Fewer than 4 wet nappies in the last 24h. ●Blood in the diarrhoea.

For recognizing severe illness, and treating the moribund ▶see p 175.

The above carry particular weight in the immunocompromized (eg neonates; post measles; AIDS; cystic fibrosis; leukaemia; chemotherapy for malignancy; absent spleen; B/T cell dysfunction).

Fever The most common 'emergency' presentation in paediatrics. In most, the cause is a viral infection, otitis media or pharyngitis/tonsillitis (p 556)—and the prognosis is good. Here fever may have some *beneficial* effects (enhanced neutrophil migration and secretion of antibacterials, increased production and activity of interferon, increased T-cell proliferation).[1] The challenge is to identify those feverish children with serious bacterial infections who need prompt treatment.

Bacteraemia This occurs in ~4% of febrile children, sometimes without an obvious focus of infection; this is made more likely if the child is between 3 and 24 months of age with fevers of $\geq40°C$, and a WCC >15 × $10^9/l$ (but using these criteria alone, 50% of those with bacteraemia will be missed). An ESR >30mm/h and a raised C-reactive protein may also indicate bacterial infection. Examples of organisms: *S pneumoniae, H influenzae, N meningitidis.* If blind antibiotics are started (eg if there is shock), they should cover these organisms (p 258 & p 260). Cefotaxime 50–60mg/kg/8h IV over 20min or ceftriaxone (80mg/kg/day IVI over 30min to 4g daily) are examples.

Finding a focus of infection entails a full history and examination, a CXR, and an LP if the child is ill (as described above). If the child is ill and no focus of infection is found, it may be wise to withold antibiotics, and to review the child yourself after 1h. Watching and waiting is not the easiest rôle, and depends critically on the presence of a skilled nurse whose opinions you trust. However, dividends may be great. For example, if a rash or diarrhoea develop, you know you are observing the illness as it unfolds, and not an antibiotic side effect.

Encephalitis (Rarer than septicaemia) This is characterized by clouding of consciousness, odd behaviour and (sometimes) seizures. An infective cause is suggested by fever and meningitis. Consider herpes simplex (for acyclovir treatment, see p 256), mumps, rubella, influenza, measles, mycoplasma, rickettsia and toxoplasmosis (OHCM p 214). Note: non-infective causes (encephalopathy) include kernicterus (p 236), hepatic failure (eg Reye's syndrome), lead poisoning.

▶Urinary infections often present as fever: do an MSU (p 188).

▶Consider malaria whenever the child has been to a malarious area (even for a 'stop-over'). Do not wait to see how the illness unfolds: do a thick blood film at once, and enlist expert help.

In prolonged fever, consider: endocarditis; Still's (p 758); malignancy.

1 MJ Kluger 1992 *Paediatrics* 90 846   2 P Hewson 1990 *Arch Dis Chi* 65 750

Fluid régimes^{APLS}

If tolerated, always use *oral* rehydration. Rehidrat® comes in sachets, and contains glucose, Na^+ and K^+. Show mother how to make it up (water is the vital ingredient!). If breast feeding, continue.

²⁵⁰

Daily IV water, Na & K (mmol/kg/day) MAINTENANCE needs[1]

Age:	Weight:	Water (ml/kg/day):	Na^+:	K^+:
<0.5yr	<5kg	150	3	3
0.5–1yr	5–10kg	120	2.5	2.5
1–3yr	10–15kg	100	2.5	2.5
3–5yr	15–20kg	80	2	2
>5yr	>20kg	45–75	1.5–2	1.5–2

Use dextrose saline for these requirements (0.18% sodium chloride + 4% dextrose; few calories, but prevents ketosis). Pre-existing deficits and continuing loss must also be made good. Reliable input-output fluid balance charts are essential.

Calculating pre-existing deficit ≈ % dehydration × weight × 10—usually as 0.45% saline over 24h (eg 750ml for a 10kg child who is 7.5% dehydrated). Add in K^+ (20mmol/500ml) once the child has passed urine.

Estimating dehydration *Mild:* ●Decreased urine output.
5% dehydration: ●Dry mucous membranes ●Decreased urine output.
10% dehydration: The above + ●Sunken fontanelle (but crying ↑pressure) ●Eyeball pressure ↓ (difficult to assess if young) ●Pulse ↑ ●Hoarse cry.
>10%: ●↑ severity of the above, with: ●Shock ●Drowsiness ●Hypotension. If a recent weight is known, this is useful in quantifying dehydration.

IV fluids for the first 24h in MILD dehydration Give maintenance needs (above), and start oral fluids when possible.

IV fluids for the first 24h in 5–10% dehydration●Give maintenance water requirement (above) + the deficit. Keep the rate <25ml/kg/h.
●Measure and replace ongoing losses (eg from the bowel).
●Monitor U&E on admission, and at 2, 12 and 24h. Also do PCV.

▶▶IV fluid replacement in the first 24h in >10% dehydration
●If IV/intravenous access fails, do a cut-down: find a surgeon urgently.
●0.9% saline (or plasma, if desperate) at 25ml/kg/h IVI, while calculations are performed. Continuously monitor pulse, BP, ECG.
●Continue until BP rises, pulses are felt, and urine flows (catheterize).
●Then give the daily requirement and the fluid deficit as described above, making good continuing loss with 0.18% or 0.45% saline depending on type of dehydration.
●Measure plasma and urine creatinine and osmolality (p 280), and plasma bicarbonate. Metabolic acidosis usually corrects itself.
Guidelines for success: 1 Stay at the bedside; use clinical state + biochemistry to adapt IVI. 2 Be simple 3 Beware hidden loss (oedema, ascites, GI pools). 4 Measure U&E and urine electrolytes often.

Hypernatraemic dehydration:[2] (ie greater water loss than salt, eg from wrongly made feeds). It causes intracellular dehydration (± fits, CNS thrombosis). Rehydrate slowly with 0.45% saline (replace deficit over 48h, lowering Na^+ by <12mmol/l/24h, and giving only 60% of maintenance volume) to avoid brain oedema (p 256). The chief danger is too fast rehydration. Hyperglycaemia is common, but self-correcting.

¹ APLS group 1993 *Advanced Paediatric Life Support* BMA   ² R Jones 1989 *Prescribers' J* 29 183

Poisoning: general management

- Find the name and quantity of the poison ingested. The number of tablets dispensed is often recorded on the bottle. The dispensing pharmacist may also have a record, and he may be able to identify unlabelled tablets.
- When was the poison ingested? Has he or she vomited since? If the child is *comatose*, enlist expert help. Procede thus:
- Place in the *semi-prone recovery position*. Note the pupil size.
- Have *suction, O_2, a laryngoscope* and *endotracheal tube* to hand.
- Do ward test for *blood glucose*. Treat hypoglycaemia (p 264).
- If there is respiratory failure (do blood gases), summon expert help. Intubate and ventilate. Treat shock (with plasma). Note: the mean systolic BP in mmHg (± 1 sd) is 85 (± 15) at birth, 95 (± 15) at 1yr, 100 (± 10) at 4yr and 110 (± 8) at 10yr.
- Record the *level of consciousness*. Glasgow Coma Scale: p 680. Coma Scale if <4 yrs: p 257. Record the time.
- If narcotic poisoning is possible, give *naloxone* 10µg/kg IV, then 100µg/kg every few mins if no response (SC or IM all right, but slower action).
- Send blood and urine for a *drug screen*—and blood alcohol if indicated (give 10% dextrose IVI: the danger is hypoglycaemia).
- Do a gastric lavage (after intubating the trachea) if ingestion has been recent, or if salicylates suspected. Have the child held with his head lower than his trunk. The length of orogastric tube required is estimated by the length from the mouth to the ear lobe and thence to the xiphisternum. Pass the tube and keep the initial aspirate for analysis. Use gravity to drain 50ml of 1.4% sodium bicarbonate solution into the stomach. Aspirate after a few mins and repeat eg 10 times. Do not use lavage if petroleum products or corrosives have been ingested.
- Monitor TPR, BP, urine output and blood glucose frequently.

The alert child It is common to induce vomiting by ipecacuanha 10-15ml mixed with juice (200ml)—but this is of uncertain value. However, activated charcoal (p 254) does reduce absorption.

Some specific poisons For aspirin and paracetamol, see p 254.

- Atropine poisoning (eg from deadly nightshade) causes dilated pupils, dry skin and mouth, tachycardia, excitement. With Lomotil® (atropine and diphenoxylate) the pupils may be constricted.
- Iron poisoning is mild if <50mg/kg elemental iron is ingested (200mg ferrous sulphate tablet ≈ 60mg iron). After emesis, give 60ml milk. If >50mg/kg ingested, measure iron levels in clotted blood urgently. If >90µmol/l, give desferrioxamine 15mg/kg/h IVI up to a total of 80mg/kg/24h. Leave 5g in 50ml water in the stomach after gastric lavage. Beware hypotension.
- Tricyclic poisoning is often fatal with seizures, coma, cyanosis and arrhythmias if >10mg/kg (eg of amitriptyline) is taken.
- Paraffin or petrol (kerosine, gasoline): ►Avoid gastric lavage and emetics. Inhalation may cause lung collapse and consolidation. Do CXR. Lung complications may be delayed 12h.
- ►Contact a National Poisons Information Service.[1]

---

1 eg in UK, Edinburgh, tel. 0131 229 2477

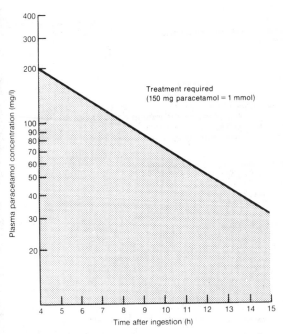

Graph for use in deciding who should receive *N*-acetylcysteine.[2]
►1 tablet of paracetamol=500mg. 150mg=1mmol paracetamol.

▶▶Rising intracranial pressure^{APLS}

Causes Meningoencephalitis; head injury; subdural or extra-dural haemorrhage; hypoxia (eg near-drowning) may all cause sudden and life-threatening rises in ICP. The rise with brain tumours is slower.

Presentation Listlessness; irritability; headache; diplopia; drowsiness; vomiting; tense fontanelle; reduced level of responsiveness—assess by Children's Coma Scale if <4yrs (see opposite), or by Glasgow Coma Scale if >4yrs (p 680); pupil changes (ipsilateral dilatation); rising BP and falling pulse (Cushing reflex); intermittent respiration. Later ie chronic: papilloedema and hydrocephalus.

Management The aim is to prevent secondary damage. Facilitate venous drainage by keeping head in the midline, elevated at ~20°. Give O₂. Set up IVI. Fan and sponge (tepid water) if T°>40°C. Check for hypoglycaemia.
- Intubate and hyperventilate the child (to obtain P_aCO_2 of ~28mmHg if Coma Scale is <8.
- Give mannitol 20% (check it is crystal-free), 7ml/kg IVI over 30min if the child is deteriorating—but discuss this with the neurosurgeon first.
- Control any seizures (p 268).
- Dexamethasone 0.1mg/kg stat, then 0.05mg/kg/6h IV.
- Fluid restriction and diuresis, but avoid hypovolaemia (keep Na⁺ 145–150mmol/l, osmolarity to 300–310, and CVP to 2–5cmH₂O).
- Measure pulse and BP continuously (check that cuff fits well).

Herpes simplex encephalitis This is the one form of encephalitis discussed in detail as it is treatable. ▶*Think of the diagnosis in any febrile child with focal seizures or focal neurological signs associated with deteriorating consciousness.*[1] Patients often present with subtle or non-specific signs. Nasolabial herpetic lesions are often absent. Neurological deficits may be mild or gross (eg hemiparesis). Focal or generalized seizures occur. Tests such as CT and CSF examination are often non-specific (unless polymerase chain reaction is available). MRI is better.[2] EEG may show focal changes and periodic complexes. Herpes simplex antibodies are produced too late to guide management. Brain biopsy is diagnostic but seems rather drastic. It is therefore probably wise to start treatment with acyclovir as soon as the diagnosis is suspected. Give 5–10mg/kg/8h by IVI over 1h (250–500mg/m²). Monitor U&E and urine output, adjusting dose in the light of these (see Data Sheet).

Brain tumours ⅔ are in the posterior fossa. Key features in their diagnosis are a raised ICP with focal signs (below), ± false localizing signs (eg VI nerve palsy, due to its long intracranial course[□]).

Medulloblastoma: This mid-line cerebellar tumour arises from the inferior vermis. It causes truncal ataxia and sudden falls. 75% are boys. It seeds along the CSF pathways. It is radiosensitive.

Brain-stem astrocytoma: Cranial nerve palsies; pyramidal tract signs (eg hemiparesis); cerebellar ataxia; signs of ICP↑ are rare.

Midbrain and third ventricle tumours may be astrocytomas, pinealomas or colloid cysts (cause posture-dependent drowsiness). Signs: behaviour change (early); pyramidal tract and cerebellar signs; upward gaze defect.

Suprasellar gliomas: Visual field defects; optic atrophy; pituitary disorders (growth arrest, hypothyroidism, delayed puberty); diabetes insipidus (DI). Cranial DI is caused by ADH↓, so that there is polyuria and low urine osmolality (always <800mosmol/l) in spite of dehydration.

Cerebral hemispheres: Usually gliomas. Meningiomas are rare. Fits may be the presenting sign. Signs depend on the lobe involved (OHCM p 400).

Tests: EEG, skull X-ray, CT scans ± magnetic resonance imaging.

Treatment options: Excision, CSF shunting, radio- and chemotherapy.

Other space-occupying lesions Aneurysms; haematomas; granulomas; tuberculomas; cysts (cysticercosis); ►abscess—suspect whenever signs of ICP ↑, fever and a leucocytosis occur together; arrange urgent referral.

Other causes of headache Viruses; meningitis; sinusitis (frontal sinus not developed until >10yrs); hypertension (always measure BP); migraine (OHCM p 416; admit if hemiplegic). Signs requiring further assessment:[1]
Headaches of increasing frequency or severity. ●Age <6yrs.
●Headache unrelieved by paracetamol ●Irritable; loss of interest/skills
●Slowing of physical or cognitive development
●Head circumference >97th centile, or greatly out of line

1 JM Hockaday 1990 *Arch Dis Chi* 65 1174   2 RP Murphy 1994 *Prescrib J* 34 12

Children's Coma Scale (<4yrs)[APLS]

Best motor response This has 6 grades:
6 Carrying out request ('obeying command'): The child moves spontaneously, or to your request.

5 Localizing response to pain: Put gentle pressure on the patient's finger nail bed with a pencil then try supraorbital and sternal pressure: purposeful movements towards changing painful stimuli is a 'localizing' response.

4 Withdraws to pain: Pulls limb away from painful stimulus.

3 Flexor response to pain: Pressure on the nail bed causes abnormal flexion of limbs—decorticate posture.

2 Extensor posturing to pain: The stimulus causes limb extension (adduction, internal rotation of shoulder, pronation of forearm)—decerebrate posture.

1 No response to pain.
NB: record the best response of any limb.

Best verbal response This has 5 grades.
5 Orientated: The patient smiles, is orientated to sounds, fixes and follows objects.

	Crying	*Interacts*
4	Consolable	Inappropriate
3	Inconsistently consolable	Moaning
2	Inconsolable	Irritable
1	No response	No response

Score the highest out of either two columns.

Eye opening This has 4 grades.
4 Spontaneous eye opening.

3 Eye opening in response to speech: Any speech, or shout, not necessarily request to open eyes.

2 Eye opening to response to pain: Pain to limbs as above.

1 No eye opening.

An overall score is made by summing the score in the 3 areas assessed. Eg: no response to pain + no verbalization + no eye opening = 3.

►If >4 yrs old, see the Glasgow Coma Scale, p 680.

1 Advanced Life Support Group 1993 *Advanced Paediatric Life Support*, BMA

Suspect this in any ill baby or child. The only signs may be unusual crying, poor feeding, or vomiting. Pay great attention to the fontanelle: in GI vomiting it becomes sunken; with meningitis it may be tense. ▶*Get expert help from your senior and a microbiologist*. If there is any hint of meningococcal disease, give penicillin at once (dose below). On lumbar puncture, if pus issues from the LP needle, start a broad-spectrum IV agent (eg ceftriaxone below) at once, while awaiting microscopy and further advice.

Meningeal signs: Stiff neck (unable to kiss his knees); photophobia; Kernig's sign (resistance to extending the knee, with the hip fully flexed); Brudzinski's sign (hips flex on bending head forward); opisthotonus.

ICP↑: Irritable; drowsy; vomiting; fontanelle tense; pre-hernia, below.

Septic signs: Fever; arthritis; odd behaviour; rashes—any, petechiae suggest meningococcus; shock; cyanosis; DIC; pulse↑; tachypnoea; WCC↑.

Other causes of meningism: tonsillitis, pneumonia, subarachnoid bleed.

Lumbar puncture: Do this at once unless: focal signs; DIC; purpura (top priority is penicillin 50mg/kg max 2g for meningococcus); or if *brain herniation* is near (abnormal postures or breathing; coma scale < 13, p 257; dilated pupils, doll's eye reflexes, BP↑, pulse↓, papilloedema).
NB: preliminary CT is incapable of showing that LP would be safe.[1]

Technique: There is no substitute for learning from an expert physician. What follows are simply some reminders. ●Explain everything to the mother. ●Ask an experienced nurse to help, by positioning the child fully flexed (knees to chin) on the side of a bed, with his back exactly at right angles with it. ●Mark a point between the spinous processes between the iliac crests. ●Drape and sterilize the area; put on gloves. ●Infiltrate 1ml of 1% lignocaine. ●Insert the LP needle aiming towards the umbilicus. ●Catch 4 CSF drops in each of 3 bottles for an urgent Gram stain, culture, virology, and glucose (do blood glucose too). ●Request frequent neurological observations. ●Report to mother.

Tests: (after LP): FBC, U&E, culture blood, urine, nose swabs (and stool, for virology). C-reactive protein: is it >20mg/l? CSF lactate: is it >3mg/l?[1] CXR. Fluid balance, TPR and BP hourly.

Treating pyogenic meningitis before the organism is known ●Protect airway and give high flow O_2 ●Set up IVI: colloid in 10ml/kg boluses.
●Dexamethasone (p 260) for 2 days, starting before 1st antibiotic dose.
●If meningococcaemia likely use antibiotics as detailed on p 260.
●Where available, ceftriaxone 80mg/kg/day IVI over 30min (max 4g daily) is becoming standard.[2,3] Alternative: cefotaxime 50–60mg/kg/8h IV over 15min. Benzylpenicillin may also be used—dose is 30mg/kg/4h slowly IV. Netilmicin may be used in neonates (2mg/kg/8h slowly IV—monitor plasma levels, U&E and creatinine, lowering dose as needed; see *Data sheet*). If >2 months old you can swap chloramphenicol 12–25mg/kg/6h IVI (do levels) for the ceftriaxone (resistance is a problem; SE: aplastic anaemia; high doses risk 'grey baby syndrome'—abdominal swelling, cyanosis, collapse).
●If pre-hernia signs, treat for ICP↑ (p 256), eg mannitol 0.5g/kg IVI.
●Once the organism is known and the right drug started, find its minimum inhibitory concentration (MIC) to the organism *in vitro*.

Complications Disseminated sepsis, subdural effusion, ataxia, paralysis, deafness, mental retardation, epilepsy, brain abscess.

CSF in meningitis	*Pyogenic:*	*Tubercular:*	*Aseptic:*
Appearance	often turbid ±	fibrin web	usually clear
Predominant cell	polymorphs	mononuclear	mononuclear
	eg 1000/mm³	10–350/mm³	50–1500/mm³
Glucose level	<⅔ of blood	<⅔ of blood	>⅔ of blood
Protein	1–5g/dl	1–5g/dl	<1.5g/dl

1 G Rennick 1993 *BMJ* ii 935 **2** UB Schaad 1993 *Lancet* **342** 457 **3** H Peltola 1989 *Lancet* i 1281

Meningitis: the organisms

The meningococcus, the pneumococcus, and, in the unvaccinated, *Haemophilus influenzae* are the great killers, but of these three it is the meningococcus that is the swiftest and the most deadly. The interval between seeming wellbeing and coma may be counted in hours and minutes. If you suspect meningitis, benzylpenicillin 600mg IM (any age) given immediately, before sending to hospital, may save life.

Neisseria meningitidis (the meningococcus) Abrupt onset ± rash (purpuric or not, eg starting as pink macules over the legs). Septicaemia may occur first without meningitis (Waterhouse–Friedrichsen, OHCM p 654), so that an early LP may be normal, and give false reassurance. Arthritis, conjunctivitis, myocarditis and DIC may coexist. Typical age: any. Film: Gram -ve cocci in pairs (long axes parallel), often within polymorphs. Drug of choice: benzylpenicillin 50mg/kg/4h IV. If the penicillin allergic, give cefotaxime (p 258) or chloramphenicol (below). Treat shock with colloid. Prevention: rifampicin 10mg/kg/12h PO (if <1yr, 5mg/kg/12h) for 48h for close contacts or ciprofloxacin, see OHCM p 444, or ceftriaxone 125mg IM (if under 12yrs, 250mg IM stat if over 12yrs). Keep under observation.

Haemophilus influenzae Predisposing events: ENT infections, from which the organism is often recovered (send swabs). Typical age: <4yrs, if unvaccinated. Gram stain: Gram -ve rods in pairs or chains. The lower the CSF glucose, the worse the infection. Drugs: ceftrixone[1] (dose on p 258) or, where resistance is not a problem, chloramphenicol 12-25mg/kg/6h IVI (not in neonates) with ampicillin 30mg/kg/4h IV. Rifampicin (see below) may also be needed. If chloramphenicol is used, try to monitor peak chloramphenicol levels. Aim for 20-25µg/ml. Do not be surprised if you far exceed usual doses to achieve this.[2] (Trough level: <15µg/ml.) As soon as you can, switch to PO (more effective).

Dexamethasone (0.15mg/kg/6h IV for 4 days) may reduce post-meningitis deafness (eg from 15% to 3%[3][4]), as well as promoting early recovery (controversial[5]). This is known to suppress interleukin-1 and tumour necrosis factor which are believed to be the fountainhead of the 'cytokine cascade' which leads to vascular congestion, clotting, prostaglandin synthesis, T-cell proliferation, and endotoxic shock.

Streptococcus pneumoniae Risk factors: upper respiratory infections, pneumonia, skull fracture, meningocele. Typical age: any. Film: Gram +ve cocci in pairs. Drug of choice: benzylpenicillin 25-50mg/kg/4h slow IV–or, if resistance likely (eg parts of Europe and USA), vancomycin (dose in children older than 1 month: 10mg/kg/6h IVI over 1h) with a third-generation cephalosporin (see ceftriaxone, dose on p 256) is advised.[1] Monitor U&E.

E coli This is a cause of meningitis in the neonate (in whom the signs may consist of feeding difficulties, apnoea, fits and shock). Drug of choice: netilmicin 2mg/kg/8h slowly IV (p 238).

Group B haemolytic streptococci may be passed on via the mother's birth canal (so swab mothers whose infants suddenly fall ill–eg at 24h-old). Infection may be delayed for a month. Drug of choice: benzylpenicillin 25-50mg/kg/4h slowly IV.

Perinatal listeriosis (*Listeria monocytogenes*) This presents soon after birth with meningitis and septicaemia (± pneumonia). Microabscesses form in many organs (granulomatosis infantiseptica).

Diagnosis: culture blood, placenta, amniotic fluid, CSF. Treatment: IV ampicillin with gentamicin. Further details and prevention: OHCM p 222.

Mycobacterium tuberculosis can cause CNS infarcts, demyelination (hence cranial nerve lesions) and tuberculomas—also meningitis (with a long pro-drome—lethargy, malaise, anorexia). Photophobia/neck stiffness are likely to be absent. The first few LPs may be normal, or the CSF may show visible fibrin webs, and widely fluctuating cell counts (p 258). *Treatment:* Seek expert help. Dose example: isoniazid 10–20mg/kg/24h IV + rifampicin 10–20mg/kg/24h PO (≤600mg/day) + pyrazinamide 40mg/kg/day + ethion-amide 20mg/kg as 1 daily dose (1h before any food).[2]

Other bacteria: Leptospiral species (canicola); Brucella; Salmonella.

Causes of 'aseptic' meningitis Viruses (eg mumps, echo, herpes, polio), partially treated bacterial infections, cryptococcus (use ink stains).

1 A Tunkel 1995 *Lancet* **346** 1675 & I Eltringham 1996 *Lancet* **347** 537   2 P R Donald 1992 *Paediatrics* **89** 247   3 M Lebel 1988 *NEJM* **319** 964   4 UB Schaad 1993 *Lancet* **342** 457   5 R Finch 1991 *BMJ* i 607

Diabetes mellitus—some questions

'What should the family know before the child leaves the ward?'

●Insulin: the dose to use. Can the parent draw it up accurately, and can the child self-inject? Practise on oranges.

●Diet: What is it? Why is it important? Why must meals be regular? What does mother do if the child is hungry?

●Can blood sugar be monitored accurately? Watch her technique.

●What should she do if the blood sugar is not well controlled?

●Does she know what 'well controlled' means?

●What happens if he has too much insulin? What will happen as hypoglycaemia evolves? What action should be taken? Practise inducing hypoglycaemia on the ward, by omitting breakfast after morning insulin. Explain symptoms as they happen.

●What should happen if the child misses a meal, or is sick afterwards? What happens to insulin requirements during flu?

●Who does mother contact in emergency? Give written advice.

●Is the GP informed of discharge and follow-up arrangements?

●Encourage membership of the nearest Diabetic Association.[1]

'What are the aims of routine follow-up in the diabetic clinic?'

●To achieve normoglycaemia through education and motivation.

●To prevent complications. So check growth and fundi (dilate the pupil). Blood: glucose, glycosylated Hb. Note: it is generally the case that retinopathy develops only after 10yrs of disease.

●If normoglycaemia cannot be achieved, choose the best compromise with the child's way of life and strict glucose control. Methods are available such as the Novopen® which allow great flexibility in the timing and dose of insulin. This device delivers a variable dose (2U/push) of insulin at the push of a plunger, and obviates the need for carrying syringes and drawing up insulin at inconvenient times (eg during a party).

Epidemiology 60% of all cases of type I (insulin dependent) diabetes start between the ages of 4 and 12 years. The mean duration of symptoms before diagnosis is 30 days. Islet cell antibodies are found in HLA-B8 (but not HLA-B15) individuals. If one child in the family has diabetes, the risk to siblings is 5–10%. Diagnostic criteria: see OHCM p 474.

Management of the new diabetic The management of ketoacidosis is described on p 264. If there is no ketosis IV fluids are seldom needed. A suitable initial dose is insulin 0.2U/kg SC followed by 0.35–0.5U/kg/12h SC, given 30min before breakfast and tea. About 70% of the dose should be as a long-acting insulin. With help from the pancreas a 'honeymoon' period may follow in which the insulin requirement may fall to 0.1U/kg/day. Diet: 1500kcal/m² or 1000kcal plus 100–200kcal for each year of age (depending on energy expenditure). Aim for 30% of this with each major meal, and 10% as a bedtime snack. Give 20% of the calories as protein, 50% as unrefined carbohydrate, and 30% as fat. Enlist the help of a paediatric dietitian.

1 In UK: 10 Queen Anne Square, London, WIM OBD (UK tel. 0171 323 1531)

Hypoglycaemic coma There is no firm lower limit to the normal blood glucose, but levels <2mmol/l may lead to coma.

Clinical features: Weakness, hunger, irritability, faintness, sweating, abdominal pain, vomiting, fits and coma.

Causes: As well as diabetes over-treatment, hypoglycaemia may also be spontaneous, or be caused by hyperinsulin states (islet cell hyperplasia or tumour, Beckwith-Wiedmann syndrome, p 743) or metabolic disorders (glycogen storage diseases, glycogen synthetase↓, fructose intolerance, maple syrup urine disease–p 204).

Treatment: Give glucose 0.5 g/kg as a 25% IVI (or via rectal tube if no IV access, with glucagon 0.5–1mg IM or slowly IV). Then give 10% dextrose, 5ml/kg/h. Expect a quick return to consciousness. If this does not occur, give IV dexamethasone (p 256).

Ketoacidotic coma ►Nurse semi-prone, to prevent aspiration.

Clinical features: Listlessness; vomiting; drowsiness; polyuria; polydipsia; weight loss. Examination: dehydration (p 250); deep, rapid respirations; ketotic ('fruity'-smelling) breath; shock, coma.

Bedside tests: Urine: ketones +++; high ward test for blood glucose (may be falsely low if the patient is vasoconstricted).

The 1st 4h: Airway Breathing Circulation. 100% O_2, ECG monitor.

● Weigh; FBC, U&E, glucose, Ca^{2+}, PO_4^{3-}, blood gas. Set up 0.9% saline IVI.

● Give 0.5 ml/kg/min while calculations are performed (+ plasma 10 ml/kg IV if shocked).

● Calculate the fluid requirement. This is the maintenance fluid (eg 100ml/kg/day, p 250) + the water deficit. Assume that the patient is 10% dehydrated, so the water deficit is 100ml/kg.

● Give ⅓ of this volume over the first 4h as a 0.9% saline IVI.

● Determine plasma osmolality [2 ($Na^+ + K^+$)] + glucose + urea. If >340 mosmol/l (hyperosmolar coma) give fluid as 0.45% saline.

● Insulin (eg Actrapid®) If >5yrs: 0.1U/kg stat IV, then 0.1U/kg/h IVI by pump in 0.9% saline. If no pump, give 0.25U/kg IM stat, then 0.1U/kg/h IM. Halve the dose to 0.05U/kg/h if glucose falls faster than 4mmol/h). If <5yrs old, give less insulin (eg 0.05U/kg IVI).

● When urine flows, start IV K^+, as guided by plasma K^+. If <5 mmol/l, give 6mmol/kg/day (keep concentration <30mmol/l). The danger is hypokalaemia. Aim for plasma K^+ of 4–5mmol/l.

● Even in severe acidosis bicarbonate's rôle is controversial (it causes an unfavourable shift in the Hb O_2-dissociation curve). Dose example: if pH <7.0, give bicarbonate 2.5ml/kg of 8.4% IVI over 2h. Recheck pH after 1h, and stop IVI if pH >7.15

● Repeat U&E, glucose. Do CXR, FBC, and do infection screen.

● Monitor TPR; continuous ECG; BP. Pass a nasogastric tube.

● Measure urine output and specific gravity (oliguria p 280).

Subsequent management: Measure glucose hourly. When <15mmol/l change IVI to dextrose saline (4.3% dextrose+0.18% saline).

● Give next ⅓ of the 24h water requirement over the next 8h, and the last ⅓ over the subsequent 12h (oral fluids may then be possible, eg 15–30ml/h PO). NB: some régimes rehydrate more slowly (eg equal hourly rates over 36h) to avoid risking cerebral oedema (suspect if there is a set-back after 2–24h; mannitol is the only treatment which helps[1]).

● Repeat U&E, PCV & blood gases (for acid-base balance) every 2–4h, until consciousness returns, and oral fluids are tolerated.

1 An elevated WCC (even a leukaemoid reaction) may be seen in the absence of infection.

2 Infection: often patients are apyrexial. Perform an MSU, blood cultures and CXR. Start broad spectrum antibiotics early if infection is suspected.

3 Creatinine: some assays for creatinine cross-react with ketone bodies, so plasma creatinine may not reflect true renal function.

4 Hyponatraemia may occur due to the osmotic effect of glucose. If <120mmol/l, search for other causes, eg hypertriglyceridaemia. Hyper­natraemia (>150mmol) may be treated with of 0.45% saline to start with—then 0.9% saline thereafter.

5 Ketonuria does not equate with ketoacidosis. Normal individuals may have ketonuria after an overnight fast. Not all ketones are due to diabetes—consider alcohol, if glucose normal. Test plasma with Ketostix® or Acetest® to demonstrate ketonaemia.

6 Acidosis without gross elevation of glucose may occur, but consider overdose (eg aspirin).

7 Serum amylase is often raised (up to 10-fold), and non-specific abdominal pain is common even in the absence of pancreatitis.

8 Cerebral oedema is a constant threat: prevent this by *slow* return to normal glucose and hydration. Monitor neurological status hourly.

Epilepsy and febrile convulsions

EPILEPSY is a tendency to intermittent abnormal brain activity. It is classified according to whether signs are referable to one part of a hemisphere (partial epilepsy) or not (generalized).

Generalized epilepsy *Tonic/clonic (grand mal):* Limbs stiffen (the tonic phase) and then jerk forcefully (clonic phase), with loss of consciousness.
Absences: Brief (eg 10sec) pauses ('he stops in midsentence, and carries on where he left off'); eyes may roll up; he/she is *unaware of the attack.*
Infantile spasms: Eg at 5 months. Jerks forwards with arms flexed, and hands extended ('Salaam attack'). Repeated every 2–3sec, up to 50 times. Here the EEG is characteristic. Treat with ACTH + nitrazepam or valproate.
Myoclonic fits: 1–7-year-olds; eg 'thrown' suddenly to the ground.

Partial epilepsy Signs are referable to part of one hemisphere.
Elementary phenomena: consciousness is not impaired.
Complex phenomena: (temporal lobe fits) consciousness ↓; automatisms (lip smacking, rubbing face, running); fits of pure pleasure.

Causes of epilepsy (Often none is found) Infection (eg meningitis); imbalance of U&E, Ca^{2+}, Mg^{2+}; hypoxia; hypoglycaemia; toxins, phenylketonuria, flickering lights (eg television); CNS disease—tuberous sclerosis, tumours (<2%), malformations.

Differential diagnosis Arrhythmias, breath-holding, febrile convulsions, migraine, narcolepsy, night terrors, faints, tics, Munchausen's syndrome.

Investigating epilepsy Always have an EEG done by an expert. Do CT only if there are infantile spasms, or unusual features, or CNS signs, or the epilepsy is partial or intractible.[1] (MRI may be more sensitive, but is not so available and may require anaesthesia or sedation—but there is no radiation.)

A FEBRILE CONVULSION is diagnosed if the following occur together:
- A tonic/clonic, symmetrical generalized seizure with no focal features
- Occurring as the temperature rises rapidly in a febrile illness
- In a normally-developing child between 6 months and 5yrs of age
- With no signs of CNS infection or previous history of epilepsy
- When there are less than 3 seizures, each lasting <5mins.

Differential diagnosis: Meningoencephalitis, brain lesions, trauma, hypoglycaemia, hypocalcaemia and hypomagnesaemia.
Lifetime prevalence: ~3% of children have at least one febrile convulsion.
Examination: A complete search to find any possible infection.
Tests: FBC, MSU, CXR, ENT swabs, LP (always if <1.5yrs old). Get help in deciding if an LP is needed. If it is not, review this decision in ~2h. If LP *is* indicated, be sure to exclude signs of ICP↑ (see p 258).
Management: Lie prone; diazepam (p 268);. Cool by tepid sponging when hot; 12mg/kg/4–6h of paracetamol syrup, 120mg/5ml.
Parental education:
- Allay fear (eg a child is *not* dying during a fit).
- Febrile convulsions do not mean epilepsy, or, usually, risk of epilepsy.
- For the 30% who have recurrences (50% if 1st degree relative also affected), try teaching the mother to use rectal diazepam, with one 5mg tube (Stesolid®) if <3yrs, or one 10mg tube if older (0.5mg/kg). This dose may also be used at home *during* seizures.
Further prevention: Prophylactic phenobarbitone is rarely indicated.[2]
Follow-up: Explain that fever with future pertussis and MMR vaccination should prompt oral paracetamol (p 214)—with rectal diazepam to hand.
Prognosis: No complex features: 1% develop epilepsy. <1% of those with febrile convulsions have ≥3 complex features; their risk approaches 50%.

The treatment and prevention of epilepsy See p 268 & p 269.

1 RE Appleton 1995 *Prescribers' J* 35 182   2 H Valman 1993 *BMJ* i 1743

Epilepsy: management^{APLS}

Stepwise treatment of status epilepticus Supportive therapy:
● Secure airway. Set a clock in motion. ● Set up 5% dextrose IVI.
● Measure BP, pulse, blood glucose, blood gases, Ca^{2+} (± Mg^{2+}).

▶ If hypoglycaemic, give glucose 0.5g/kg IV as a 25% solution, then 5–10mg/kg/min as 10% dextrose IVI.

Seizure control:[1] Proceed to the next step only if fits continue.

0min	Estimate the weight of the child ($2 \times$ [age in yrs + 4]).
	Diazepam 0.25–0.4mg/kg slow IV as Diazemuls® (less toxic to veins). If IV access fails, diazepam PR 0.4mg/kg (not as Diazemuls®, absorption is slow)—if using preloaded tubes (Stesolid®) give 5mg if <3yrs, or 10mg if >3yrs.
5min	Diazepam repeated as above.
10min	Paraldehyde 0.4mg/kg PR in arachis oil, or IM (max 5ml/buttock to reduce risk of sterile abscess). Rule of thumb: 1ml/year to max 10ml.
20min	Phenytoin 18mg/kg IVI; max rate 1mg/kg/min.
50min	Diazepam 0.1–0.4mg/kg/h IVI. Seek expert help.
65min	Paralyse and ventilate: use thiopentone infusion.

▶ These times refer to elapsed time on the clock from the 1st drug, *not* gaps between each drug. Some authorities recommend starting ventilation earlier, and always be ready to do this to protect the airway.

Tests Pulse oximetry, cardiac monitor, glucose, blood gases, U&E, Ca^{2+}, magnesium, FBC, platelets, ECG. Consider anticonvulsant levels, toxicology screen, blood ammonia, lumbar puncture, culture blood and urine, EEG, CT, carbon monoxide level, lead level, amino acid levels.[2]

Once the crisis is over, start prophylxis, eg with sodium valproate or carbamazepine, guided by régime on p 269. ▶ Aim to use one drug only. Increase the dose until fits stop, or toxic levels are reached.

Carbamazepine SE: rash (± exfoliation); thrombocytopenia, agranulocytosis, aplastic anaemia (all rare). It induces its own enzymes, so increasing doses may be needed.

Sodium valproate The sugar-free liquid is 200mg/5ml. SE: vomiting, appetite increase, drowsiness, thrombocytopenia (so measure platelets before surgery). Rare hepatotoxicity can be fatal. Routine monitoring of LFTs is needed, usually only in the first 6 months of treatment.

Ethosuximide The syrup is 250mg/5ml. SE: D&V, rashes, erythema multiforme, lupus syndromes, agitation, headache. Good in absence fits.

Phenytoin Not a first choice because: 1 The dose is difficult: a small increase may lead to big increases in plasma concentration. 2 Monitoring of plasma levels is essential. Aim for a range of 40–80μmol/l (10–20mg/l). 3 Toxic effects: nystagmus, diplopia, dysarthria, tremor and ataxia. 4 SE: intellectual deterioration, depression, impaired drive, behavioural disorders, polyneuropathy, acne, coarse facial features, hirsutism, gum hypertrophy, blood dyscrasias, plasma Ca^{2+} and folate↓.

Vigabatrin This GABA analogue blocks GABA transaminase (so GABA ↑ at GABA-ergic synapses, OHCM p 406). Consider adding it to régimes if partial seizures are uncontrolled. Vigabatrin starting dose: 40mg/kg/day PO (sachets and tabs are 500mg). Max: 100mg/kg/day. Blood levels do not help (but monitor concurrent phenytoin: it may fall by ~20%[3]). SE: drowsiness, depression, psychosis, amnesia, diplopia, dizziness.

Education If the fits are well suppressed, educate teachers about lifting misguided bans on swimming, cycling, and other sports.

Drug	Starting dose mg/kg/24h	Target dose for initial assessment of effect mg/kg/24h	Dose increment Size mg/kg/24h	Dose increment Interval in days	Usually effective dose mg/kg/24h	Doses per day	Target trough drug level in plasma mg/l	Target trough drug level in plasma µmol/l
Carbamazepine	5	12.5	2.5	7	10-25	2-3	4-10	17-42
Valproate	10	20	10.0	10	15-40	1-2	Not helpful	
Phenytoin	5	7	1.0	10	5-9	1-2	10-20	40-80
Phenobarbitone	4	6	2.0	10	4-9	1	15-40	60-180
Ethosuximide	10	15	5.0	5	15-40	1	40-99	280-700
Clonazepam	0.025	0.05	0.025	7	0.025-0.1	2-3	Not helpful	

When to use which drug
Tonic/clonic fits: first try sodium valproate or carbamazepine.
Absences: first choice, ethosuximide; second, sodium valproate.
Myoclonic or akinetic fits: sodium valproate or benzodiazepines.
Infantile spasms: prednisolone or nitrazepam (0.2–0.6mg/day PO).
Partial fits: first try carbamazepine; then sodium valproate.

Stopping anticonvulsants See OHCM p 451. The risk of seizure recurrence during the tapering down process is no greater if the tapering period is 6 weeks compared with 9 months. ◻

1 Advanced Life Support Group *Advanced Paediatric Life Support* 1993 BMA   2 M Tunic 1992 *Paed Clin N America* (Oct, page 1014)   3 *Drug Ther Bul* 1990 **28** 95

►►Childhood asthma^{APLS}

Asthma affects >7% of children (>20% wheeze at some time); prevalence↑ if: male, low birth weight, atopy, past bronchopulmonary dysplasia, or mother smoked in pregnancy or there is passive smoking.

Diagnosis There is reversible airways obstruction (peak flow varies by >20%, p 271), with wheeze, dyspnoea, or cough (eg nocturnal, and the only sign). *Differential:* Foreign body, croup, epiglottitis, pneumonia. *Signs of severe asthma:* (Patients may not be distressed). Too breathless to speak or feed; respirations ≥50 breaths/min; pulse ≥140; peak flow ≤33% predicted. *Life-threatening if:* Peak flow <50% of predicted ●Cyanosis ●Silent chest ●Fatigue or exhaustion ●Agitation/reduced level of consciousness.

Treatment: BTS guidelines[1] ●Try to avoid provoking factors. ●Check inhaler technique. ●Address any fears. ●Have self-management plan. ●Check compliance. ●Prescribe peak flow meter (eg if >5yrs old). ●Start at the step most appropriate to severity; move up as needed (or down, if control good for >6 months). ●Prednisolone may be needed at any time (≤40mg/day).

Step 1: Try occasional β-agonists. If needed >daily, add step 2.

Step 2: Add cromoglycate, eg 20mg/8h as inhaled powder.§

Step 3: Swap cromoglycate for budesonide 50–200µg/12h by large volume spacer.◁²▷ Consider 5 days of prednisolone 1–2mg/kg/day PO.

Step 4: ↑Budesonide, up to 400µg/12h, by large volume spacer, ± course of prednisolone ± long-acting β-agonist (salmeterol, 2 puffs/12h).

Step 5: Add slow release xanthine ± nebulized β-agonist ± alternate day prednisolone 5–10mg PO ± ipratropium or β-agonist SC infusion.

Dose examples: β-agonists: Sugar-free salbutamol mixture 2mg/5ml. 5ml/8h (0.1mg/kg/8h if <2yrs). Inhalers are preferable: terbutaline 250µg/4–6h (one puff)—eg with spacer and mask until 8yrs old.

Xanthines: Theophylline, eg 5mg/kg/8h PO (syrup = 60mg/5ml). Try to monitor levels. Bioavailability varies between brands. Slow release theophylline (as Slo-Phyllin®) dose: <12mg/kg/24h PO. Capsules are 60, 125, or 250mg. Their enclosed granules may be spread on soft food.

NB: beware growth retardation from steroids (including from inhalers).[3]

Treatment of severe asthma ►A calm atmosphere is beneficial.

1 Sit the child up; high-flow 100% O_2

2 Terbutaline 4–5mg nebulized with O_2 (2mg if <3yrs)

3 Prednisolone 1–2mg/kg PO (max 40mg)

4*Aminophylline 5mg/kg IV over 20min (omit if already on a xanthine); then IVI aminophylline 1mg/kg/h. Do levels

5*Hydrocortisone 100mg IV/6h; add ipratropium 0.125–0.25mg to nebulizer

6 Oximetry + CXR if SpO_2 ≤92%

7 Treat any pneumonia

8 Chart peak flow before & after each nebulizer treatment

9 Repeat nebulizers as needed, eg every 30min–6h.

10 Take to ITU if falling peak flow, exhaustion, confusion, feeble respirations or coma

*Give these treatments if life-threatening signs are present (above) or the patient is not improving 15–30min after treatment has started.

Before discharge ensure: ●Good inhaler technique. ●Has had discharge régime for >24h. ●Peak flow is >75% of predicted and diurnal variation is <25%. ●Taking inhaled steroids *and* oral soluble prednisolone. ●Written management plan. ●Follow-up by GP in 1 week and in clinic in <4 weeks.[1]

Pitfalls[1] ●Faulty inhaler technique. Watch the patient operate his device.
●Reluctance to diagnose until a serious attack occurs. Formerly, 50% of diagnoses were made after >15 visits to the GP. Things are now improving.
●Inadequate perception of, and planning for, the severe attack.
●Unnoticed, marked diurnal variation in airways obstruction. Always ask about nocturnal waking: it is a sign of dangerous asthma.
●Being satisfied with less than total symptom control.
●Forgetting to start prophylaxis—and not using oral prednisolone early.
●No direct access to hospital. ●Failure to discourage smoking.

Peak flow* (litres/min) in normal boys and girls (5-18 years).

Height (cm)	Mean	Third centile	Height (cm)	Mean	Third centile
100	110		150	360	300
113	160	100	155	400	320
120	210	140	160	420	350
125	240	160	165	450	370
130	260	190	170	470	400
135	290	220	175	500	430
140	310	240	180	520	450
145	350	270	185	550	470
			190	570	500

Source: S Godfrey 1970 *Br J Dis Chest* **64** 15 *To 2 significant figures

Guidelines for continuous IVI of aminophylline (mg/kg/h)—after a loading dose[†] of 6-7.5mg/kg IV. *(Multiplying by 0.85 gives dose of theophylline)*

	Aminophylline (mg/kg/h)
Neonates	0.15
Infants <6 months	0.47
Infants 6-11 months	0.82
Children 1-9 years	0.94
Children over 9 years	0.70

†Omit the loading dose if previous theophyllines have been taken. If the plasma theophylline concentration is known and is subtherapeutic, an additional loading dose may be given. Increasing the dose by 1mg/kg causes an increase in serum theophylline of ~2μg/ml.

Erythromycin *increases* the half-life of aminophylline, as do cimetidine, ciprofloxacin, propranolol, and contraceptive steroids. Drugs which *decrease* the half-life: phenytoin, carbamazepine, barbiturates, and rifampicin. Adjust the dose according to plasma concentrations.

▶Aim for a plasma level of 10-20μg/ml (55-110μmol/l). Serious toxicity (hypotension, arrhythmias, cardiac arrest) can occur at levels ≥25μg/ml.

1 Brit Thor Soc Guidelines *BMJ* 1993 i 779 & *Thorax* 1993 **48** sp. 1-24 §Most UK paediatricians use inhaled steroids before cromoglycate (which they mistrust, A Robins 1995 *BMJ* ii 508 ⌐) ◁2▷ KJ Chau 1995 *Arch Paed Adol Med* **149** 201   3 O Walthers 1993 *Arch Dis Chi* **68** 673 See also UM MacFadyen 1995 *Prescrib J* **35** 174

Abdominal pain

Acute abdominal pain Children often have difficulty in localizing pain, and other factors in the history may be more important. Pointers may be:
●Hard faeces suggest that constipation is the cause (p 206).

●In negroes, suspect sickle-cell disease (do sickling tests).
●In Asian families, suspect TB (do a tuberculin test, p 208).
●In children with pica (p 206), do a blood lead level.
●Abdominal migraine is suggested by periodic abdominal pain with vomiting especially if there is a positive family history.

Common physical causes Gastroenteritis, UTI, viral illnesses (eg tonsillitis associated with mesenteric adenitis) and appendicitis.

Some rarer causes Mumps pancreatitis; diabetes mellitus; volvulus; intussusception; Meckel's diverticulum; peptic ulcer; Crohn's/ulcerative colitis; Hirschsprung's disease; Henoch–Schönlein purpura and hydronephrosis. Consider menstruation or salpingitis in older girls.
►In boys always check for a torted testis.

Investigations ►Always microscope and culture the urine.
Others: consider plain abdominal X-ray; FBC; ESR; IVU, barium studies.

Appendicitis (OHCM p 114) is rare if <5yrs, but perforation rates are high in this group (nearing 90%[1]). Think: how can I tell this from other causes of abdominal pain? ●Clues in the history: increasing pain in right lower quadrant, no previous episodes, anorexia, slight vomiting, absence of cough and polyuria. ●Examination hint: if the child appears well and can sit forward unsupported, and hop, appendicitis is unlikely. ●Tests are unlikely to help when the clinical picture is uncertain.

Gastro-oesophageal reflux oesophagitis May present with regurgitation, apnoea, pneumonia, failure to thrive and anaemia. *Tests:* Evaluation with an oesophageal pH probe is more reliable than barium studies.
Treatment: Drugs may be needed, eg an antacid+sodium/magnesium alginate–eg Infant Gaviscon® dual dose sachets, 1 dose mixed with 15ml boiled (cooled) water. This paste is given by spoon after each breast feed. If bottle-fed, give the dose dissolved in the feed. Children >4.5kg may have a whole dual dose sachet. Carobel® thickens the feeds. If this fails, some experts use cisapride (↑ tone of the gastro-oesophageal sphincter). Most resolve by 6-9 months.

Abdominal distension

Causes: *Air:*	*Ascites:*	*Solid masses:*	*Cysts:*
Faecal impaction	Nephrosis	Neuroblastoma	Polycystic kidney
Air swallowing	Hypoproteinaemia	Wilms' tumour	Hepatic; dermoid
Malabsorption	Cirrhosis; CCF	Adrenal tumour	Pancreatic

Hepatomegaly Infections: many, eg infectious mononucleosis, CMV.
Malignancy: leukaemia, lymphoma, neuroblastoma (see below).
Metabolic: Gaucher's and Hurler's diseases, cystinosis; galactosaemia.
Others: sickle-cell disease, and other haemolytic anaemias, porphyria.

Splenomegaly All the above causes of hepatomegaly (not neuroblastoma).

Neuroblastoma A highly malignant tumour derived from sympathetic neuroblasts. Prevalence: 1:6000–1:10,000. A common mode of presentation is with abdominal swelling. It may develop at any time in childhood, but the prognosis is better, with some spontaneous remissions in those <1yr (25% of patients) and in those with stage I and II disease. Metastatic sites:

lymph nodes, scalp, bone (causing pancytopenia and osteolytic lesions). In 92%, urinary excretion of catecholamines (vanillylmandelic and homovanillic acids) will be raised. *Treatment:* Excision (if possible) and chemotherapy (eg cyclophosphamide and doxorubicin).

Recurrent abdominal pain At least 10% of children >5yrs suffer recurrent abdominal pains interfering with normal activities. The question is: is there organic disease? No cause will be found in most, if looked for, but this should neither encourage complacancy (you may delay a diagnosis of *Crohn's*[2] (OHCM p 516) or *peptic ulcer*) or lead to an over-zealous diagnosis of underlying psychological problems (now thought to be less important—but do consider it: who is present when the pain starts; what, or who, makes the pain better?).[3]

Consider these causes: gastro-oesophageal reflux (above), small bowel dys-motility, gastritis, duodenitis, carbohydrate malabsorption (eg lactose, sorbitol). The rôle of these diseases is unclear in children.

Who to investigate: There are no hard rules. Be suspicious if the pain is unusual in terms of localization, character, frequency, or severity.

1 L Kapila 1991 *Arch Dis Chi* **66** 1270   2 IW Booth 1991 *Arch Dis Chi* **66** 742   3 MS Murphy 1993 *Arch Dis Chi* **69** 409

Anaemia

The clinical problem You have the results of a full blood count, showing a low Hb (eg <11g/dl, p 292). How should you proceed?

1 Take a history (include travel and diet); examine the child.
2 If the MCV is <70fl suspect iron deficiency or thalassaemia. Features suggesting iron deficiency anaemia (IDA): poor diet, low socio-economic class, bleeding, stomatitis, koilonychia. Features suggesting thalassaemia: of Mediterranian or Thai extraction, short stature, muddy complexion, icteric sclerae, distended abdomen (from hepatosplenomegaly), bossed skull, prominent maxillae (from marrow hyperplasia).
3 If the MCV is >100fl suspect folate deficiency (eg taking phenytoin, malabsorption, p 184), or B₁₂ deficiency (breast milk from a vegetarian mother, absent intrinsic factor, malabsorption, worms), or haemolysis.
4 If the MCV is 81–97fl (normocytic), suspect haemolysis, or marrow failure. This may be transient, after infections, or chronic (eg thyroid, kidney or liver failure). Marrow aplasia may follow toxins (eg chloramphenicol) or be inherited (Diamond–Blackfan and Fanconi anaemia, p 744 & 748).
5 Are there any abnormalities in the white cell count or differential? For example, if the eosinophils are >400 × 10⁹/l, the cause of the anaemia may be blood loss from hookworms.
6 Next look at the ESR. This may indicate some chronic disease.
7 Next look at the film, including reticulocyte count. Do thick films for malaria if indicated. Note red cell morphology (OHCM, p 518). A hypochromic microcytosis suggests IDA; target cells suggest liver disease or thalassaemia.
8 Next do further tests as indicated, eg transferrin for IDA; thick films for malaria; sickling tests and Hb electrophoresis for sickle-cell anaemia (do in Blacks); B₁₂; red cell folate.

Iron deficiency anaemia This is common (eg 12% of white children and 28% of Asian children admitted to hospital). The deficiency is mostly mild, but as IQ and CNS performance may be reduced, it is important to treat it. The most marked behavioural effect is that iron deficient babies are less happy than non-anaemic ones. Dietary causes are the most common. In recurrent IDA, suspect bleeding (eg Meckel's diverticulum or oesophagitis). Treatment with ferrous sulphate syrup (60mg/5ml): <1 year 12mg/8h; 1–5 years 24mg/8h; 6–12 years 40mg/8h.

Haemolysis Request expert help, and try to provide the expert with sufficient information to answer these 4 questions:
● Is there evidence of an increased rate of red cell production? (Polychromasia, reticulocytosis, marrow hyperplasia.)
● Is there decreased RBC survival? (Bilirubin ↑, haptoglobins ↓.)
● Is there intravascular haemolysis? (Haemoglobinuria.)
● Is there an inborn error of metabolism (eg G6PD deficiency), spherocytosis or is the defect acquired (usually with +ve Coombs' test)?

Sickle-cell disease in children OHCM p 588. Pain relief:
First try warmth, hydration, and oral analgesia (ibuprofen 5mg/kg/6h PO, codeine phosphate 1mg/kg/4–8h PO up to 3mg/kg/day). If this fails, see on the ward and offer prompt morphine (eg IVI—eg 0.1mg/kg. Consider patient-controlled analgesia (PCA). Start PCA with morphine 1mg/kg in 50ml 5% dextrose, and try a rate of 1ml/h, allowing the patient to deliver extra boluses of 1ml when needed. Do respiration and sedation score every ¼h + pulse oximetry if chest/abdominal pain.[1]

1 R Grundy 1993 *Arch Dis Chi* **69** 256

Purpura in children

If a child is ill with purpura he or she should be presumed to have meningococcaemia (and be given penicillin, p 258), leukaemia or disseminated intravascular coagulation (investigate initially with a visual blood film and WCC). If the child is well, and there is no history of trauma, the cause is likely to be Henoch-Schönlein purpura or idiopathic thrombocytopenic purpura, readily told apart by the normal platelet count in the former. (Aplastic anaemia is very rare.) **275**

Idiopathic thrombocytopenic purpura *Presentation:* Acute bruising, purpura, and petechiae. Less commonly there is bleeding from gums, nose or rectum. If this is present to any large extent, or there is lymphadenopathy, hepatosplenomegaly, or pancytopenia, another diagnosis is likely.

Tests: The peripheral blood film usually shows only simple thrombocytopenia. ITP may be a response to a virus, and it is worth looking for congenital CMV if the patient is <1yr old. Also look for Epstein-Barr virus. There may also be a lymphocytosis. It is not necessary to do a marrow biopsy, unless:
- Unusual features are present. ● Platelet count is <30 × 10⁹/l.
- Platelet count not rising after 2-3 weeks.
- Treatment is contemplated with steroids.

Platelet antibody tests are unhelpful. If there are features of SLE (may be associated) do DNA binding.
CT scans to exclude intracerebral haemorrhage may be needed if there is headache or CNS signs.

Natural history: The course may be gradual resolution over ~3 months, or the course may be more chronic (>6 months in 10-20%). The chronic form is compatible with normal longevity, and normal activities of daily living, provided contact sports are avoided.

Management: 1 Consider admitting to hospital if:
- Unusual features, including excessive bleeding.
- Social circumstances.
- There are rowdy siblings at home who might engage in physical badinage (the risk of intracranial haemorrhage is <1%).

2 Steroids may be tried for 2-3 weeks (eg prednisolone 0.25mg/kg/day PO) there only appears to be a slightly faster return of platelet numbers—and are rarely justified in those whose counts are >30 × 10⁹/l. The mechanism of action by be inhibition of phagocytosis of sensitized platelets. Pulses of high-dose methyl prednisolone may help the chronic form, when this needs treatment to cover an emergency.

3 Immunoglobulin appears to offer no clinical advantage over steroids.

4 Splenectomy may benefit two-thirds of those with chronic ITP who are over 6yrs old, and have had ITP for >1yr.

5 Platelet transfusion is only indicated in those rare instances of severe haemorrhage (± emergency splenectomy).

Henoch-Schönlein purpura (HSP) This presents with purpura (ie purple spots/nodules which do not disappear on pressure—signifying intradermal bleeding) often over buttocks and extensor surfaces. There may be associated urticaria. The typical patient is a young boy. There may be a nephritis (with crescents, in ⅓ of patients—an IgA nephropathy—we could think of HSP as being a systemic version of Berger's syndrome, OHCM 694), joint involvement, abdominal pain (± intusussception), which may be severe enough to mimic an 'acute abdomen'. The fault lies in the vasculature; the platelets are normal. It often follows respiratory infection, and it usually follows a benign course over months. Complications (worse in adults): massive GI haemorrhage, ileus, haemoptysis (rare) and renal failure (rare).

Upper respiratory infections

Stridor and epiglottitis ►►Acute stridor may be a terrifying experience for children and this fear may lead to hyperventilation, which makes symptoms worse. Causes: p 558. The leading causes to be distinguished are epiglottitis and viral croup: see the table below.

Investigations: A lateral neck X-ray may show an enlarged epiglottis (but do not insist on X-rays at the expense of upsetting the child).

Management of suspected epiglottitis: Accompany the child to hospital (complete airways obstruction may occur suddenly).

►*Avoid examining the throat.* This may precipitate obstruction. Once in hospital, summon the most experienced anaesthetist. Ask him to make the diagnosis by laryngoscopy. If the appearances are those of epiglottitis (a cherry-red, swollen epiglottis), consider elective intubation, rather than waiting until obstruction occurs. (A smaller diameter endotracheal tube than normal for that age may be needed—so don't precut all your tubes!) Then do blood cultures. The cause will often be *Haemophilus influenzae* type b, and strains resistant to ampicillin (and less so chloramphenicol) are prevalent, so the only safe initial treatment is a 3rd generation cephalosporin (eg cefotaxime 1–2g/8h IV adults, 25–50mg/kg/8h IV children). Hydrocortisone (1–2mg/kg/6h IV) is often given, but is not of proven value. Expect to extubate after about 24h (longer if the diagnosis turns out to be staphylococcal laryngotracheobronchitis, when flucloxacillin 50mg/kg/6h IV should also be given).

Viral croup	Epiglottitis (supraglottitis)
Onset over a few days	Sudden onset
Stridor only when upset	Continuous stridor
Stridor sounds harsh	Stridor softer, snoring
Voice hoarse	Voice muffled/whispering
Barking cough	Cough not prominent
Likely to be apyrexial	Toxic and feverish (eg T° >39°C)
Can swallow oral secretions	Drooling of secretions

Note: the distinction may not be clear cut. If in doubt, admit to hospital.

Croup Epidemics occur in autumn and spring. Causes: parainfluenza virus (types 1, 2 and 3), respiratory syncytial virus, measles. Symptoms are caused by subglottic oedema, inflammation and exudate.

Management: Most are managed at home. Anecdotal evidence suggests that a warm, humid environment is helpful, but mist tents have lost favour as they frighten children, and subsequent hyperventilation increases distress. Aim for minimal interference and careful observation by an experienced nurse. Watch for restlessness, rising pulse and respiration rate, increasing indrawing of the chest wall, fatigue and drowsiness. This will prompt intubation under general anaesthesia. Nebulized adrenaline 1:1000 (5ml) helps buy time. Nebulized budesonide also helps—acute dose: 3 mths–12yrs: 0.5–1mg/12h (eg 2 250mg Pulmicort Respules®; halve dose if chronic).[1,2]

Diphtheria is caused by the toxin of *Corynebacterium diphtheriae*. It usually starts with tonsillitis ± a false membrane over the fauces. The toxin may cause polyneuritis, often starting with cranial nerves. Shock may occur from myocarditis, toxaemia, or conducting system involvement. Other signs: dysphagia; muffled voice; bronchopneumonia and airway obstruction preceded by a brassy cough (laryngotracheal diphtheria); nasal discharge with an excoriated upper lip (nasal diphtheria). If there is tachycardia out of proportion to fever, suspect a toxin-induced myocarditis. Monitor with frequent ECGs. Motor palatal paralysis also occurs.

Diagnosis is by swab culture of material below the pseudomembrane.

Treatment: Diphtheria antitoxin—10,000–30,000U IM (any age) and erythromycin. Give contacts erythromycin 10mg/kg/6h PO *before swab results are known* (adults 250mg/6h PO). Vaccination has been effective in the UK.

Prevention: Isolate until 3 –ve cultures separated by 48h. There is a resurgence of diphtheria in Moscow and St Petersburg. Those born before 1942 **277** visiting Russia or the Ukraine who have *close local contacts* need a primary *low dose* course. Give a booster (0.5ml of Diftavax®; contains tetanus toxoid too) to diphtheria contacts whose primary immunization was >10yrs ago. Explain that it may not work. Schick tests are no longer available.

Sore throats See p 556. Coryza See *OHCM* p 206.

1 I Doull 1995 *BMJ* **ii** 1244 **2** A Edmunds 1989 *Prescrib J* **29** 192

Lower respiratory infections

Each year ~3 million infants die from acute respiratory infections (WHO).
▶If severely ill, think of TB and HIV (pneumocystosis, OHCM p 334).

Chronic cough Think of pertussis, TB, foreign body, asthma, smoking.

Acute bronchiolitis is the commonest lower respiratory tract infection of infancy. Coryza precedes cough, low fever, tachypnoea, wheeze, apnoea, intercostal recession, and cyanosis if severe. In winter, an epidemic of respiratory syncytial virus is the usual cause, and is identified by immunofluorescence of nasopharyngeal aspirates. Other causes: mycoplasma, parainfluenza, adenoviruses. Those <6 months old are most at risk. Signs that admission is needed are feeding difficulties and tachypnoea >50 breaths/min. *Tests:* CXR (eg hyperinflation); blood gases/oximetry; FBC. *Treatment:* If severe, nurse in 40% O_2, and give IV ampicillin and flucloxacillin to prevent bacterial infection (both 25mg/kg/6h). ~5% need ventilating (mortality ≈1%, rising to ~33% in those with symptomatic congenital heart disease). Nebulized ribavirin may help those who are greatest risk (congenital heart disease and cystic fibrosis).

Pneumonia This may occur at any age, and presents with fever, malaise, feeding difficulties, tachypnoea and cyanosis. Auscultation: crepitations and bronchial breathing (easy to miss). CXR: consolidation. Cavitation suggests TB or a staphylococcal abscess. Take samples (including blood and urine culture) before starting 'blind' antibiotics, eg penicillin G 25mg/kg/4h IV + erythromycin ethylsuccinate eg as Erythroped® syrup (250mg/5ml)—5ml/12h PO for babies ≤2yrs, 10ml/12h from 2–8yrs; if severely affected, give this dose 6-hourly. High-flow O_2 may be needed.
　　Some causes: pneumococcus, haemophilus, staphylococcus, mycoplasma (hence the choice of erythromycin), TB, viruses.

Pulmonary tuberculosis ▶Suspect; if from overseas, and their contacts, or HIV+ve. *Clinical features:* Anorexia, low fever, failure to thrive, malaise. Cough is common, but may be absent. *Diagnosis:* tuberculin tests (p 208); culture and Ziehl-Neelsen stain of sputa (× 3) and gastric aspirate. CXR: consolidation, cavities. Miliary spread (fine white dots on CXR) is rare but grave. *Treatment:* Seek expert help. Example of a 6-month régime: isoniazid 15mg/kg PO 3 times a week + rifampicin 15mg/kg/PO ac 3 times a week + pyrazinamide (first 2 months only) 50mg/kg PO 3 times a week. Monitor renal and liver function before and during treatment. Stop rifampicin if the bilirubin rises (hepatitis). Isoniazid may cause neuropathy (so give concurrent pyridoxine). ▶Explain the need for prolonged treatment. Multiple drug resistance: OHCM p 199.

Whooping cough (*Bordetella pertussis*) is often undiagnosed. Bouts of coughing which end with vomiting which are worse at night, and after feeding suggest the diagnosis—particularly if associated with cyanosis. The whoop is caused by forced inspiration against a closed glottis. Differentiation from pneumonia, asthma, bronchitis and bronchiolitis is aided by the (usual) absence of a fever >38.4 °C and wheeze. Peak age: 3yrs (a smaller peak at 30yrs, ie the reservoir in parents). In the UK, the illness is usually mild, with ~1% needing hospital admission (eg with a secondary pneumonia)—in one careful GP survey of 500 consecutive patients.￼ *Diagnosis:* Culture is unsatisfactory (organisms often die on the way to the laboratory). Direct fluorescent antibody testing of nasopharyngeal aspirates is specific but insensitive. An absolute lymphocytosis is common. Incubation period: 10 days to 2 weeks.

Complications: Prolonged illness may occur (>3 months). Coughing bouts may cause petechiae (eg on the cheek), conjunctival haemorrhage or CNS damage. Deaths may occur (particularly in infants), as may late bronchiectasis.

Treatment and prevention: Erythromycin, salbutamol and steroids have not proved very effective. Erythromycin (as above) is often used in those likely to expose infants to the disease (benefit is unproven). Admit if <6 months old. The risk is apnoea. *Immunization:* p 208.

Renal failure

Acute renal failure Essence: a rapidly rising K^+ & urea often with anuria (<100ml/day) or oliguria (<200ml/m² /day) and ↑BP.

Causes: GU obstruction	Acute tubular necrosis *from:*
Toxins (eg sulphonamides)	●Crush injury ●Burns
Glomerulonephritis (GN) OHCM p 380	●Dehydration ●Shock
Haemoglobinuria; myoglobinuria	●Septicaemia ●Malaria

Plasma biochemistry: K^+↑, urea↑, PO_4^{3-}↑; Ca^{2+}↓, Na^+↓, Cl^-↓.

MSU: Are there red cell casts (=GN)? If no RBCs seen but Labstix +ve for RBCs, consider haemo/myoglobinuria (OHCM p 394).

Other tests: ECG, serum and urine osmolality, creatinine, acid–base state, PCV, platelets, clotting studies (DIC), C₃, ASO titre, ANA.

Radiology: ▶Arrange prompt abdominal ultrasound. Are the ureters dilated (eg stones)? If so, urgent surgery may be required.

Treatment: ●Treat shock and dehydration promptly (p 250).
●If the urine/plasma (U/P) osmolality ratio is >5 the kidneys concentrate well; the oliguria should respond to rehydration.
●If the U/P ratio is low, try for a diuresis: frusemide 1.5mg/kg IV slowly, maximum 20mg/day.
●If BP↑↑: nitroprusside (p 304).
●24h fluid requirement: Avoid overhydration. Replace losses + insensible loss (12–15ml/kg). Aim for weight loss (0.5%/day).
●Give no K^+. Monitor ECG. Tall T-waves and QRS slurring prompt urgent lowering of K^+, with:
  1 Glucose (4g/kg) with soluble insulin (1U/4g of glucose) IVI over 2h.
  2 Resonium A® 0.5g/kg PO.
  3 Calcium gluconate (10%, 0.5ml/kg IV over 5min; monitor ECG: stop IVI if heart rate↓) to counteract electrophysiological effect of hyperkalaemia.
●High energy, protein limited diet to slow catabolism.
●Dialysis indications (a guide, only): K^+ >7mmol/l; urea >40mmol/l; HCO_3^- <13mmol/l; bad hypertension; overhydration.
●Monitor BP. ●Improvement is ushered in with a diuretic phase.

Chronic renal failure *Causes:* UTI, pyelonephritis, glomerulonephritis.

Clinical features: Weakness, tiredness, vomiting, headache, restlessness, twitches, hypertensive retinopathy, fits and coma.

Treatment: ●Correct fluid and electrolyte imbalance.
●If diastolic BP >100mmHg consider propranolol (10–20mg/8–12h PO).
●Vitamin D (50 nanograms/kg of alfacalcidol each day PO) is helpful in reducing phosphate excretion, and promoting Ca^{2+} absorption, thereby preventing 'renal rickets'.
●Confer with experts (haemodialysis, transplants, growth hormone treatment).

Haemolytic uraemic syndrome *Essence:* Microangiopathic haemolytic anaemia, thrombocytopenia, renal failure, and endothelial damage to glomerular capillaries. *Typical age:* 3 months to 3yrs. Epidemics may occur from *E coli* Verocytotoxin. Diarrhoea is not always present.

Other causes: Shigella, HIV, SLE, drugs, tumour, scleroderma, BP↑.

Clinical features: Colitis → haemoglobinuria → oliguria ± CNS signs → encephalopathy → coma. LDH↑. WCC↑. Coombs' –ve. PCV↓. Fragmented RBCs.

Mortality: 5–30%. A few have many relapses—to form a condition rather like thrombotic thrombocytopenic purpura (TTP).

Treatment: Seek expert advice. Treat the renal failure (above). There is no evidence that fibrinolytic agents or anticoagulation help. Relapses in TTP may be preventable by steroids, splenectomy, or vincristine.

Acute nephritis and nephrosis

Acute nephritis Essence: haematuria and oliguria (±hypertension and uraemia) produced by an immune mechanism in the kidney.

Causes: Often β-haemolytic streptococcus, which may cause a sore throat 2 weeks before an acute glomerulonephritis. Others:

- Henoch-Schönlein purpura
- Toxins, heavy metals
- Berger's disease (OHCM p 694)
- Malignancies
- Infections: viruses; SBE; syphilis
- Renal vein thrombosis

Uncomplicated presentation: Age peak: 7 years. Haematuria; oliguria; 50% have hypertension; periorbital oedema; fever; GI disturbance; loin pain.

Complications: 1 hypertensive encephalopathy (restless, drowsy, bad headaches, fits, visual disturbances, vomiting, coma). 2 Cardiac involvement: gallop rhythm, cardiac failure and enlargement, pulmonary oedema. 3 Uraemia: acidosis, twitching, stupor, coma.

Tests: ●Urine microscopy (count red and white cells, hyaline, granular cellular casts; red cell casts mean glomerular bleeding). Phase-contrast detects abnormally shaped red cells, signifying glomerular bleeding. This change may not be present at first. 24h urine for protein and creatinine clearance. Check urine culture, and specific gravity (normal range: neonate 1.012; infants 1.002-1.006; child/adult 1.001-1.035).

●Blood: urea↑ in ²/₃; ESR↑; acidosis. Complement (C3) often ↓ 2-8 weeks after onset (not in Henoch-Schönlein purpura). Find the cause: do ASO titre, antinuclear factor, syphilis serology, blood cultures, virology.

●Other tests: renal ultrasound, renal biopsy—check platelets, clotting and IVU (are there two kidneys?); parents' consent.

Treatment: Restrict protein when in oliguric phase. Give penicillin 10mg/kg/4h IV for the first few days, then PO for 3 months to prevent further streptococcal infection. Measure BP often. Treat severe hypertension (p 304). If encephalopathy, give nitroprusside 0.5-8μg/kg/min IVI (p 304).

Nephrotic syndrome (nephrosis) Essence: oedema, proteinuria (eg 4g/24h), hypoproteinaemia ± hypercholesterolaemia. In 90% the cause is unknown, but any of the causes of nephritis (above) may also cause nephrosis. Histology: usually minimal change.

Clinical features: Anorexia, GI disturbance, infections, irritability; then oedema (periorbital, genital), ascites, oliguria.

Urine: Frothy; albuminous; ± casts; Na⁺↓ (2ry hyperaldosteronism).

Blood: Albumin↓ (so total Ca²⁺↓); urea and creatinine usually normal.

Renal biopsy: Reserve this for older children, haematuria, BP↑, urea↑, if protein loss is unselective (ie large molecular weights as well as small), and treatment 'failures'.

Complications: Pneumococcal peritonitis, so consider Pneumovax II if >2yrs.

Treatment: Limit oedema with high protein (3g/kg/24h) low Na⁺ diet (<50mmol/24h). Consider frusemide (1-2mg/kg/24h slow IV or PO) and spironolactone (1.2mg/kg/12h PO).

For 'minimal change' give prednisolone, eg 1mg/kg/24h PO (not if incubating chickenpox!) for 8 to 12 weeks, or for 1 week after induction of a remission. Half the induction dose is then given eg for 1 month, with subsequent tapering. 90% respond within 8 weeks. ~30% of responders have no relapse. 10-20% of these are cured after up to 4 courses of steroids. The rest may become dependent on steroids: if >0.5mg/kg/day of prednisolone is needed, growth will be slowed, and an agent such as cyclophosphamide is needed (2-2.5mg/kg/day PO, SE: haemorrhagic cystitis, leukopenia, hair loss, infertility). Alternatives are chlorambucil 0.2mg/kg/day and cyclosporin up to 150mg/m²/day (but nephrotoxic, see OHCM p 620).[1]

▶Monitor BP. Meticulous control minimizes progression to renal failure.[1]

1 PD Mason 1994 BMJ ii 1557

Non-accidental injury

Systematic injury has, until recently, been an overtly endemic feature of our culture (eg floggings of 24 lashes of 'the cat' for 'misconduct' in 8-year-olds[1]). In some hospitals abused children arrive almost daily. Abuse may be physical, sexual, emotional, or by neglect. *Risk factors:* Birth weight <2500g; mother <30yrs; unwanted pregnancy; stress; social class IV and V

Suspect abuse if: ●Disclosure by the child. ●Inconsistent story.
●Late presentation after the injury, often to an unknown doctor.
●The accompanying adult may not be a parent.
●Efforts to avoid full examination, eg after an immersion burn.
●Unexplained fracture; injury to buttocks, perineum or face. Ruptured tongue frenulum results from shaking; also look for vitreous/retinal bleeds, hyphaema (p 514), lens dislocation, bulging fontanelle, head circumference ↑ ± xanthochromia; ►if in doubt, do CT.[2] *Other telling signs:* Cigarette burns; linear whip marks (look for outline of a belt buckle, or the loop of a double electric flex); bruised *non-mobile* baby; suffocation.

Imponderable questions Although non-medical, these need addressing because we are not simply technicians, and they influence what we do, how we react to child abuse, and how the rest of society perceives our rôle.
●Could *proving* of abuse be more destructive than the abuse itself?[3]
●Is it better for him to be loved and battered than neither?
●Is help from the extended family more desirable than the law?
●Is it possible that the parents will grow through crisis, as battering is brought to light, and help given?

Remember that the first aim is to prevent organ damage and murder. If this is a real danger ►*contact the duty social worker today*—eg for an emergency protection order. Offer help to the parents. Learn to listen, leaving blame and punishment to judges. Find out about local policies and referral routes. Remember that very often your duty is not to diagnose child abuse, but to recognize *possible* abuse, and then to get help.[4]

Sexual misuse This may be prevented by pre-school teaching about 'personal safety' and how to say 'No'. Once a child's claim is made, believe it (usually). Know your local guidelines. Follow them. Inform Social Services. If you do not, ask yourself with whom you are colluding. Forensic specimens (eg pubic hair, vaginal swabs) should be taken by an expert who knows how to be gentle, and to avoid a 'second rape'. Prepubertal venereal disease means abuse until proven otherwise. Does abuse cause psychological harm? It is impossible to be sure: although morbidity is increased, this may reflect antecedent events which themselves lead to the abuse (ie confounding variables, eg absent mother, stepfather, which themselves predict later morbidity).[5]

Repertoire of actions by the GP ●Liaise with health visitor or the National Society for the Prevention of Cruelty to Children and have the child put on a Social Services 'at risk' register.
●Admission to a place of safety (eg hospital or foster home).
●Continuing support for parents and protection for siblings.
●Prevention: encourage impulses to be shared, and not acted on.
●Attend a 'case conference' (eg with social worker, health visitor, paediatrician and police) to help decide what action to take. Note: it is not known if such conferences are the best way of making such decisions. Not all members will know the family.

1 *The Times* 18 March 1861 2 H Carty 1995 *BMJ* i 344 3 K Hulme 1984 *Bone People*, Spiral
4 R Morton and B Phillips 1992 *Accidents & Emergencies in Children*, OUP 5 A Weiner 1991 *BMJ* ii 415

A possible sequence of events might be:[1]

Unexplained signs (or disclosure, or allegations), eg odd bruising

↓

'Testing of professional hypotheses' ≈ weighing it up in your own mind

↓

'Clarification by discussion with an experienced colleague' ≈ tell your boss

↓

'Reach a critical threshold of professional concern' ≈ you're both worried

↓

'Weigh the pros and cons of breaking confidentiality' ≈ try to do your best

↓

'Sharing concerns with statutory agencies' ≈ phone Social Services/police

↓

'Act within a timeframe not detrimental to the child' ≈ do it now

↓

'Contemporaneous records detailing all your sources' ≈ write it down now

↓

Preliminary consultation with all concerned: don't promise to keep secrets

↓

'Strategic multidisciplinary discussion' ≈ is an abuse investigation needed?

↓

'Instigation of child abuse investigation' ≈ plan a case conference

↓

'Must parents/child be present?' ≈ bend the ear of the conference chairman

↓

Tell parents & child (if appropriate) what your report to conference will be

↓

Case conference timed to let doctors fulfil their major rôle ≈ ?get a locum

↓

Register your dissent (if any) to the conference conclusions in its minutes

↓

Child is placed on a Register indicating that questions of abuse are unresolved

↓

'Establish networks for information exchange, discussion & advice' ≈ follow-up by Social Services, or a national society protecting children from cruelty (NSPCC)

↓

Second (review) conference to weigh new evidence

↓

Death of a child

↓

Agencies must issue reports to the Area Child Protection Committee (in UK)

↓

Judge issues a life sentence ≈ male breadwinner removed from family for ~10yrs

↓

No ♂ rôle model for siblings ≈ perpetuation of a cycle of poverty and abuse

↓

Unexplained bruising in a member of the next generation

1 DoH 1994 *Child Protection: Medical Responsibilities* and 1991 *Working Together Under the Children Act 1989*, HMSO **NB**: not all our efforts to protect children end thus. Successes are frequent

Sudden infant death syndrome (SIDS)

Definition° 'Death of an infant or young child, which is unexpected by history and in whom a thorough necropsy fails to reveal an adequate cause of death'.[1] SIDS is the leading 'cause' of death in infants aged over 1 week old. NB: minor histological abnormalities are often seen at necropsy.

Epidemiology Peak incidence: 1–4 months; higher incidence in lower socio-economic classes, passive smokers, males, premature babies and during winter; co-existing minor upper respiratory infection is common.

Theories These deaths were often attributed to direct mechanical suffocation, eg from a pillow, or from the family cat misplacing itself. It is now recognized that these deaths are just as common in families without cats, and in cots without pillows. There is no single cause which yet explains all cot deaths, but the following have been proposed:

Obstructive apnoea:	Central apnoea:	Others: Viruses
Inhalation of milk	Faulty CO_2 drive	Overheating/heat stroke
Airways oedema	Prematurity	Immature diaphragm
Pharyngeal collapse	Brainstem gliosis	Passive smoking

Preventing overheating and exposure to parental cigarette smoke are the main areas to concentrate on in prevention. The risk from passive smoking is dose-dependent, and often at least doubles risk. The face is an important platform for heat loss—and it is known that the incidence of SIDS is ~5–10-fold higher among infants usually sleeping prone (17-fold higher if sleeping in a room separated from parents)[2]: ►so recommend sleeping supine (unless there is gastro-oesophageal reflux, Pierre Robin syndrome or scoliosis). Advise as follows:[3]

● Do not overheat the baby's bedroom. Aim for a temperature of 16–20°C.
● Do not use too much bedding, and avoid duvets if less than 1yr old.
● If ill or feverish, consult a GP—do not increase the amount of bedding.
● Have feet come down to the cot's end to avoid migrations under blankets.
● While sleeping, avoid heaters, hot water bottles, electric blankets and hats unless ambient T° is very low. Do wrap up if going out in winter.
● Babies >1month do not need to be kept as warm as in hospital nurseries.

Autopsy is unrevealing. Petechial haemorrhages over pleura, pericardium or thymus, and vomit in the trachea may be agonal events. Causes to exclude: meningococcaemia; epiglottitis; heart defects.

How the GP can help the family on the first day
● A prompt visit to express sympathy emphasizing that no one is to blame.
● Explain about necessary post-mortems and coroner's inquests. The parents may be called upon to identify the body.
● Bedding may be needed to help find the cause of death.
● Don't automatically suppress lactation (bromocriptine 2.5mg/12h PO for 1 week), as continued lactation may be an important way of grieving.
● Many parents will not want anxiolytics, but may want hypnotics.
● Admit a twin sibling to hospital (their risk is increased).

Subsequent help Advise the parents of likely grief reactions (guilt, anger, loss of appetite, hearing the baby cry). Make sure that the coroner informs you of the necropsy result; take some trouble to explain these to the parents. Offer them a chance to talk to a consultant paediatrician. This can provide helpful reinforcement and encouragement to the parents and yourself. The parents may find an electronic apnoea alarm reassuring in caring for later infants. Ask if they would like to join a self-help group.[4]

1 EA Mitchell 1994 *BMJ* ii 607   2 P Pharoah 1996 *Lancet* 347 2   3 DoH advice 1993 PL/COM (93) 4
4 Cot Death Support & Research, 4 Grosvenor Pl, London SW1X 7HD (tel. 0171 235 1721)

Denver developmental screening test

Younger children may be examined while sitting on their mother's lap. Explain to her that the child is not expected to pass every test. Begin with a few very easy tests. Go slowly.

Test materials Ball of red wool, box of raisins, rattle with handle, small bottle, bell, tennis ball, 8 blocks (1 inch cubes).

Administering the test ●Draw a vertical line at the child's chronological age on the charts on p 290-1. If premature, subtract the months premature from the chronological age.
●The items to be administered are those through which the line passes. A parent may administer a test if the child wishes.
●Failure to pass an item passed by ≥90% of children may or may not be abnormal. Note how the child feels during the test.

Test footnotes (See top corner of test boxes on p 290-1.)
1 When prone lifts head up, using forearm support (± hands).
2 No head lag as you pull by the hand from supine to sitting.
3 Child may use a wall or rail for help (but not a person).
4 Throws ball overhand 3 feet to within your reach.
5 Jumps over a distance (eg over an A4 piece of paper).
6 Walk forward, heel within 1 inch of toe.
7 Bounce a ball. He must catch it. Allow up to 3 trials.
8 Walk backwards, toe within 1 inch of heel.
9 Wave wool slowly before eyes: do eyes move 90° to midline?
10 Grasps rattle when it touches his finger tips.
11 Looks for ball of wool dropped out of sight over table's edge.
12 Child grasps raisin between thumb and index finger.
13 Overhand grasp of raisin between thumb and index finger.
14 'Copy this' (circle). Do not name or demonstrate.
15 'Which line is longer?'; turn upside down and repeat.
16 'Copy this' (cross). Pass crossing lines at any angle.
17 'Copy this' (square). You may demonstrate if he fails.
18 Two arms or two eyes or two legs only count as one part.
19 Name the pictures at the bottom of the chart.
20 'Give the block to Mum'. 'Put it on the table'. No gestures.
21 Answer ⅔ of 'What do you do if you are cold/hungry/tired?'
22 Put this on/under/in front of/ behind the chair. No gestures.
23 Answer ⅔ 'Fire is hot, ice is- '. 'Mum is a woman, Dad is a-'. 'A horse is big, a mouse is- '.
24 Define 6/8 of ball, lake, desk, house, banana, curtain, hedge, pavement. An verbal indication of understanding is passed.
25 Answer 3/3: 'What is a spoon/shoe/door made of?' (No others):
26 Smile, talk or wave, to elicit smile (3 tries). Do not touch.
27 While he plays with a toy, pull it away. Pass if he resists.
28 Child need not be able to tie shoes or button at the back.

▶This test is poor at picking up articulatory or mild linguistic deficits—if this is suspected, liaise with speech experts. NB An updated version of this test is being evaluated (Denver II), which contains more language items, new age scales, new categories of item interpretation and a behaviour rating scale.[1]

1 WK Frankenburg 1992 **89** *Paediatrics* 91

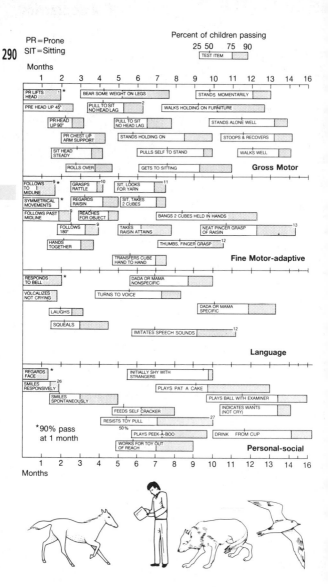

Denver developmental screening test. (After H Silver, C Kempe & H Bruyn 1977 *Handbook of Paediatrics*, Lange.)

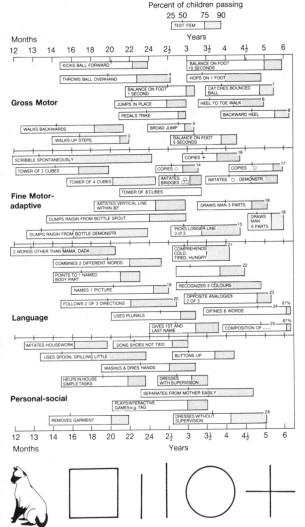

Percent of children passing

25 50 75 90

TEST ITEM

Months · Years

| 12 | 13 | 14 | 16 | 18 | 20 | 22 | 24 | 2½ | 3 | 3½ | 4 | 4½ | 5 | 6 |

Gross Motor

- KICKS BALL FORWARD
- THROWS BALL OVERHAND
- BALANCE ON FOOT 1 SECOND
- JUMPS IN PLACE
- PEDALS TRIKE
- WALKS BACKWARDS
- WALKS UP STEPS
- SCRIBBLE SPONTANEOUSLY
- BALANCE ON FOOT 10 SECONDS
- HOPS ON 1 FOOT
- CATCHES BOUNCED BALL
- HEEL TO TOE WALK
- BACKWARD HEEL
- BROAD JUMP
- BALANCE ON FOOT 5 SECONDS

Fine Motor-adaptive

- TOWER OF 2 CUBES
- TOWER OF 4 CUBES
- TOWER OF 8 CUBES
- IMITATES VERTICAL LINE WITHIN 30°
- DUMPS RAISIN FROM BOTTLE SPOUT.
- DUMPS RAISIN FROM BOTTLE DEMONSTR.
- COPIES +
- COPIES O
- COPIES □
- IMITATES BRIDGES
- IMITATES □ DEMONSTR.
- DRAWS MAN 3 PARTS
- DRAWS MAN 6 PARTS
- PICKS LONGER LINE 3 of 3

Language

- 2 WORDS OTHER THAN MAMA, DADA
- COMBINES 2 DIFFERENT WORDS
- POINTS TO 1 NAMED BODY PART
- NAMES 1 PICTURE
- FOLLOWS 2 OF 3 DIRECTIONS
- USES PLURALS
- COMPREHENDS COLD, TIRED, HUNGRY
- RECOGNIZES 3 COLOURS
- OPPOSITE ANALOGIES 2 OF 3
- DEFINES 6 WORDS
- GIVES 1ST AND LAST NAME
- COMPOSITION OF . . .

Personal-social

- IMITATES HOUSEWORK
- USES SPOON, SPILLING LITTLE
- HELPS IN HOUSE SIMPLE TASKS
- REMOVES GARMENT
- DONS SHOES NOT TIED
- WASHES & DRIES HANDS
- PLAYS INTERACTIVE GAMES e.g. TAG
- BUTTONS UP
- DRESSES WITH SUPERVISION
- SEPARATES FROM MOTHER EASILY
- DRESSES WITHOUT SUPERVISION

| 12 | 13 | 14 | 16 | 18 | 20 | 22 | 24 | 2½ | 3 | 3½ | 4 | 4½ | 5 | 6 |

Months · Years

Paediatric reference intervals

P = plasma; S = serum F = fasting

Biochemistry (1mmol = 1mEq/l)

Albumin[P]	36–48g/l
Alk phos[P]	see below
(depends on age)	
α1-antitrypsin[P]	1.3–3.4g/l
Ammonium[P]	2–25μml/l; 3–35μg/dl
Amylase[P]	70–300u/l
Aspartate aminotransferase[P]	<40u/l
Bilirubin[P]	2–16μmol/l;0.1–0.8mg/dl
Blood gases, arterial	pH 7.36–7.42
PaCO2	4.3–6.1 kPa; 32–46mmHg
PaO2	11.3–14.0 kPa; 85–105 mmHg
Bicarbonate	21–25mmol/l
Base excess	–2 to +2mmol/l
Calcium[P]	2.25–2.75mmol/l;9–11mg/dl
Neonates:	1.72–2.47; 6.9–9.9mg/dl
Chloride[P]	98–105mmol/l
Cholesterol[P,F]	≤5.7mmol/l;100–200mg/dl
Creatine kinase[P]	< 80u/l
Creatinine[P]	25–115μmol/l; 0.3–1.3mg/dl
Glucose	2.5–5.3 mmol/l; 45–95mg/dl
(lower in newborn.	Fluoride tube)
IgA[S]	0.8–4.5g/l (low at birth, rising to adult levels slowly)
IgG[S]	5–18g/l (high at birth, falls and then rises slowly to adult level)
IgM[S]	0.2–2.0g/l (low at birth, rises to adult level by one year)
IgE[S]	< 500u/ml

Iron[S]	9–36 μmol/l; 50–200μg/dl
Lead[EDTA]	<1.75μmol/l; <36μg/dl
Mg[2+ P]	0.6–1.0 mmol/l
Osmolality[P]	275–295mosmol/l
Phenylalanine[P]	0.04–0.21mmol/l
Potassium[P] mean mmol/l:	Day 1: 6.4
Day 2: 6; Day 3: 5.9 (later 4–5.5)	
Protein[P]	63–81g/l; 6.3–8.1g/dl
Sodium[P]	136–145mmol/l
Transferrin[S]	2.5–4.5g/l
Triglyceride[F,S]	0.34–1.92mmol/l (≡30–170mg/dl)
Urate[P]	0.12–0.36mmol/l; 2–6mg/dl
Urea[P]	2.5–6.6mmol/l; 15–40mg/dl
Gamma-glutamyl transferase[P]	<20U/l

Hormones—a guide. ►Consult lab

Cortisol[P]	9am 200–700nmol/l
	midnight <140nmol/l (mean)
Dehydroepiandrosterone sulphate[P]:	
Day 5–11	0.8–2.8μmol/l (range)
5–11yrs	0.1–3.6μmol/l
17α-Hydroxyprogesterone[P]	
Days 5–11	1.6–7.5nmol/l (range)
4–15yrs	0.4–4.2nmol/l
T4[P]	60–135nmol/l(not neonates)
TSH [P]	<5mu/l (higher on Day 1-4)

B=boy; [EDTA]=edetic acid; [F]=fasting
G=girl; [P]=plasma; [S]=serum.

Alk phos u/l: 0–½yr 150–600; ½–2yr 250–1000; 2–5yr 250–850; 6–7yr 250–1000; 8–9yr 250–750; 10–11yr G=259–950, B ≤730; 12–13yr G=200–750, B ≤ 785; 14–15yr G=170–460, B=170–970; 16–18yr G=75–270, B=125–720; >18yr G=60–250, B=50–200.

Haematology mean ± ~1 s.d. Range × 10⁹/l (median in brackets)

Day	Hb g/dl	MCV fl	MCHC%	Retic%	WCC	Neutrophils	Eosins	Lymphs	Monos
1	19.0 ± 2	119 ± 9	31.6 ± 2	3.2 ± 1	9–30	6–26 (11)	.02–.8	2–11	0.4–3.1
4	18.6 ± 2	114 ± 7	32.6 ± 2	1.8 ± 1	9–40				
5	17.6 ± 1	114 ± 9	30.9 ± 2	1.2 ± .2					
Weeks									
1–2	17.3 ± 2	112 ± 19	32.1 ± 3	0.5 ± .03	5–21	1.5–10 (5)	.07–.1	2–17	0.3–2.7
2–3	15.6 ± 3	111 ± 8	33.9 ± 2	0.8 ± 0.6	6–15	1–9.5 (4)	.7–.1	2–17	0.2–2.4
4–5	12.7 ± 2	101 ± 8	34.9 ± 2	0.9 ± 0.8	6–15	(4)		(6)	
6–7	12.0 ± 2	105 ± 12	33.8 ± 2	1.2 ± 0.7	6–15	(4)		(6)	
8–9	10.7 ± 1	93 ± 12	34.1 ± 2	1.8 ± 1	6–15	(4)		(6)	
Months—all the following Hb values below are medians/lower limit of normal									
3	11.5/9	88/80			6–15	(3)		(6)	
6	11.5/9	77/70			6–15	(3)		(6)	
12	11.5/9	78/72			6–15	(3)		(5)	
Years									
2	11.5/9	78/74			6–15	(3)		(5)	
4	12.2/10	80/75			6–15	(4)		(4)	
6	13/10.4	82/75			5–15	(4.2)		(3.8)	
12	13.8/11	83/76			4–13	(4.9)		(3.1)	
14B	14.2/12	84/77			4–13	(5)		(3)	
14G	11.5								
16B	14.8/12	85/78	30–36	0.8–2	4–13	2–7.5 (5)	.04–.4	1.3–3.5	.2–.8
16G	14/11.5								
18B	15/13								

Note *Basophil* range: 0–0.1 × 10⁹/l B[½]₁₂, ≥ 150ng/l.
Red cell folate [EDTA] 100–640ng/ml. B=boys; G=girls.

Platelet counts do not vary with age; range: 150–400×10⁹/l.
Principal source: Musgrove Park Hospital, Taunton, Somerset.

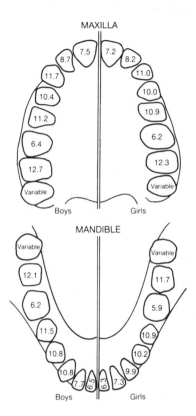

Mean times of eruption (in years) of the permanent teeth. (After D Sinclair 1985 *Human Growth after Birth*, OUP.)

Deciduous teeth	months		months
Lower central incisors	5–9	1st molars	10–16
Upper central incisors	8–12	Canines	16–20
Upper lateral incisors	10–12	2nd molars	20–30
Lower lateral incisors	12–15		

A 1-year-old has ~6 teeth; 1½ yrs ~12 teeth; 2 yrs ~16 teeth; 2½ yrs ~20.

Centile tables

Age Centile—Boys	Weight kg			Height cm			Skull circumference cm		
	3	50	97	3	50	97	3	50	97
Birth term	2.5	3.5	4.4	–	50	–	30	35	38
3 months	4.4	5.7	7.2	55	60	65	38	41	43
6 months	6.2	7.8	9.8	62	66	71	41	44	46
9 months	7.6	9.3	11.6	66	71	76	43	46	47
12 months	8.4	10.3	12.8	70	75	80	44	47	49
18 months	9.4	11.7	14.2	75	81	87	46	49	51
2 years	10.2	12.7	15.7	80	87	93	47	50	52
3 years	11.6	14.7	17.8	86	95	102	48	50	53
4 years	13	15	21	94	101	110			
5 years	14	19	23	100	108	117	49	51	54
6 years	16	21	27	105	114	124			
7 years	17	23	30	110	120	130			
8 years	19	25	34	115	126	137	50	52	55
9 years	21	28	39	120	132	143			
10 years	23	30	44	125	137	148			
11 years	25	34	50	129	142	154			
12 years	27	38	58	133	147	160	51	54	56
13 years	30	43	64	138	153	168			
14 years	33	49	71	144	160	176	53	56	58
15 years	39	55	76	152	167	182			
16 years	46	60	79	158	172	185			
17 years	49	62	80	162	174	187			
18 years	50	64	82	162	175	187			

Centile tables—Girls

	Weight kg			Height cm			Skull circumference cm		
	3	50	97	3	50	97	3	50	97
Birth term	2.5	3.4	4.4	–	50	–	30	35	39
3 months	4.2	5.2	7.0	55	58	62	37	40	43
6 months	5.9	7.3	9.4	61	65	69	40	43	45
9 months	7.0	8.7	10.9	65	70	74	42	44	47
12 months	7.6	9.6	12.0	69	74	78	43	46	48
18 months	8.8	10.9	13.6	75	80	85	45	47	50
2 years	9.6	12.0	14.9	79	85	91	46	48	51
3 years	11.2	14.1	17.4	86	93	100	47	49	52
4 years	13	16	20	92	100	109			
5 years	15	18	23	98	107	116	48	50	53
6 years	16	20	27	104	114	123			
7 years	18	23	30	109	120	130			
8 years	19	25	35	114	125	136	50	52	54
9 years	21	28	40	120	130	142			
10 years	23	31	48	125	136	148			
11 years	25	35	56	130	143	155			
12 years	28	40	64	135	149	164	51	53	56
13 years	32	46	70	142	156	168			
14 years	37	51	73	148	160	172	52	54	57
15 years	42	54	74	150	162	173			
16 years	45	56	75	115	162	174			
18 years	46	57	75	–	–	–			

Measure height exactly: ●Feet together ●Straight legs; loose arms
●Buttocks, shoulders and heels touching wall ●Hair firmly compressed
●Ear canal level with infraorbital margin ●Take reading in expiration.

More up-to-date charts are being designed to take into account new growth data, and to amalgamate American, WHO, and European centile divisions.[1]

1 The 1990 nine centile UK charts are available from Harlow Publishing, Maxwell St, South Shields, NE33 4PU Tel. 0191 427 0195. See TJ Cole 1994 *BMJ* i 641 & DMB Hall 1995 *BMJ* ii 583. NB: the 1995 Castlemead (Buckler) charts have not met with approval from paediatric growth experts because the upper centile on the weight chart is too low, classifying too many children as obese—but these charts have the (unproven) advantage of being based on serial measurements, not cross-sectional data.

Centile growth charts[1]

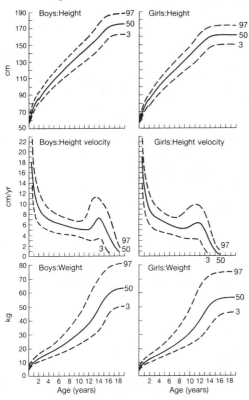

▶It is worthwhile to plot all children on growth charts. See p 185 for ethnic growth charts.

Note: the figures at the end of each of the curves refer to centiles. If a child is 'growing along the third centile' this means that only 3% of children from a healthy population have a growth rate as low or lower than such a child. Not all human populations are the same, so that while such a child might be said to be abnormally small if he was a well-fed Caucasian, this need by no means be the case if he is of Chinese or Bangladeshi extraction. Serial growth measurements tend to correlate well with each other when plotted on simple charts—ie children tend to stay within a given centile. However, measurements on velocity charts do not correlate with each other and this makes it easier to give firm guidelines on when to say that a child is abnormal, based on a single reading. Immediate action needs to be taken if a child is below the third centile on the velocity chart—even on one reading.[2] Another action point is being below the 0.4 centile for height (only 1 in 250 will be below this line).

1 After J Tanner, *Arch Dis Childh* 41 613-35   2 C Brook 1986 *BMJ* ii 1186

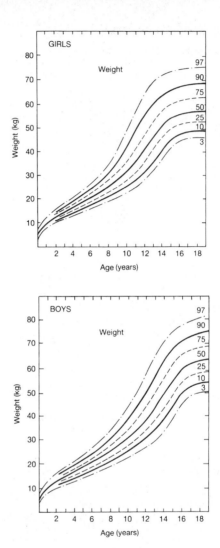

Cross-sectional standards for weight and height attained at each age (English girls and boys). The central line represents the mean, or 50th centile. The two dashed lines above and below it represents the 75th and 25th centiles respectively; ie 25% of the sample fell below the lower line and 25% above the upper line. Other centile lines are also provided. (From D Sinclair 1985 *Human Growth after Birth*, oup, modified from JM Tanner.)

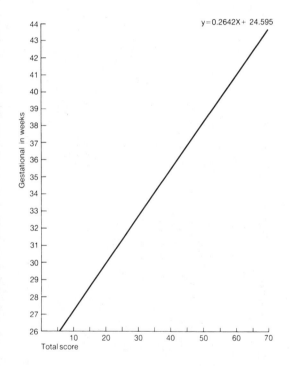

Assessment of gestational age: Dubowtiz system. Graph for reading gestational age from total score. (By kind permission of the *Journal of Paediatrics*.)

Assessment of gestational age: Dubowitz system

Physical (external) criteria

External sign	Score				
	0	1	2	3	4
Oedema	Obvious oedema hands and feet: pitting over tibia	No obvious oedema hands and feet: pitting over tibia	No oedema		
Skin texture	Very thin, gelatinous	Thin and smooth	Smooth: medium thickness Rash or superficial peeling	Slight thickening Superficial cracking and peeling, especially hands and feet	Thick and parchment-like: superficial or deep cracking
Skin colour (infant not crying)	Dark red	Uniformly pink	Pale pink: variable over body	Pale Only pink over ears, lips, palms, or soles	

Skin opacity (trunk)	Numerous veins and venules clearly seen, especially over abdomen	Veins and tributaries seen	A few large vessels clearly seen over abdomen	A few large vessels seen indistinctly over abdomen	No blood vessels seen
Lanugo (over back)	No lanugo	Abundant long and thick over whole back	Hair thinning especially over lower back	Small amount of lanugo and bald areas	At least half of back devoid of lanugo
Plantar creases	No skin creases	Faint red marks over anterior half of sole	Definite red marks over more than anterior half; indentations over less than anterior third	Indentations over more than anterior third	Definite deep indentations over more than anterior third
Nipple formation	Nipple barely visible: no areola	Nipple well defined: areola smooth and flat diameter <0.75 cm	Areola stippled, edge not raised: diameter <0.75 cm	Areola stippled, edge raised diameter >0.75 cm	
Breast size	No breast tissue palpable	Breast tissue on one or both sides <0.5 cm diameter	Breast tissue both sides: one or both 0.5–1.0 cm	Breast tissue both sides: one or both >1 cm	

Assessment of gestational age: Dubowitz system

Physical (external) criteria

External sign	Score				
	0	1	2	3	4
Ear form	Pinna flat and shapeless, little or no incurving of edge	Incurving of part of edge of pinna	Partial incurving whole of upper pinna	Well-defined incurving whole of upper pinna	
Ear firmness	Pinna soft, easily folded, no recoil	Pinna, soft, easily folded, slow recoil places, ready recoil	Cartilage to edge of pinna, but soft in recoil	Pinna firm, cartilage to edge, instant recoil	
Genitalia • Male	Neither testis in scrotum	At least one testis high in scrotum	At least one testis right down		
• Female (with hips half abducted)	Labia majora widely separated, labia minora protruding	Labia majora almost cover labia minora	Labia majora completely cover labia minora		

Nomogram for calculating the body surface area of children

Height	Surface area	Weight

▶▶IM/IV adrenaline in anaphylaxis^{APLS}

Age	ml of 1:10,000*	Age	ml of 1:1000
3–5 months	0.5ml	2yrs**	0.2ml
6–11 months	0.75ml	3–4yrs**	0.3ml
12 months	1.0ml	5yrs**	0.4ml
		6–12yrs**	0.5ml
		Adult	0.5-1ml

Doses may need repeating every 15min, until improvement occurs. Also:
●O_2 (± IPPV) ●Hydrocortisone (4mg/kg IV) ●Chlorpheniramine (0.2mg/kg)
●Colloid (20ml/kg IVI). NB Weight (kg)≈2(Age in yrs+4). OK if 1–10yrs.
*Eg as Min-I-Jet Adrenaline® (10ml 1:10,000 adrenaline).
**For underweight children, halve the dose, or, if weight known, give
10μg/kg (=0.01ml/kg of 1:1000). Adrenaline dose by endotracheal tube:
100μg/kg. If bronchospasm is a feature, give salbutamol 2mg nebulized too.

Blood pressure in childhood[1]

Age	Significant hypertension*		Severe hypertension	
(Years)	*Systolic*	*Diastolic*	*Systolic*	*Diastolic*
At birth	≥96mmHg**		≥106mmHg**	
<2	≥112mmHg	≥74mmHg[k4]	≥118mmHg	≥82mmHg[k4]
3–5	≥116mmHg	≥76mmHg[k4]	≥124mmHg	≥84mmHg[k4]
6–9	≥122mmHg	≥78mmHg[k4]	≥130mmHg	≥86mmHg[k4]
10–12	≥126mmHg	≥82mmHg[k4]	≥134mmHg	≥90mmHg[k4]
13–15	≥136mmHg	≥86mmHg[k5]	≥144mmHg	≥92mmHg[k5]
16–18	≥142mmHg	≥92mmHg[k5]	≥150mmHg	≥98mmHg[k5]

*ie >95th centile. **Measurement requires the use of Doppler devices—it is often only possible to record systolic pressure.

Note 3 or so recordings should be taken (using a snugly fitted cuff whose bladder width is ≥75% of upper arm length) some weeks apart (in general) before hypertension is diagnosed. Because the fifth Korotkoff sound (k5) is often not heard in childhood, the fourth sound (k4) is used for diastolic measurement, until adolescence, when k5 is used.

Commoner causes of hypertension *Newborn infants:* Renal artery stenosis (or thrombosis), congenital renal malformations, coarctation of the aorta, bronchopulmonary dysplasia.
Infants: Renal parenchymal diseases, coarctation of the aorta, renal artery stenosis (OHCM p 398).
6–10 years old: Renal artery stenosis, renal parenchymal diseases, primary hypertension.
Adolescence: Primary hypertension, renal parenchymal diseases.

Clinical assessment *Ask about:* Family history (phaeochromocytoma, OHCM p 554 & 742); GU symptoms. Examine for abdominal masses, bruits, endocrine diseases, coarctation and Turner's syndrome (p 758).

Tests *Urine:* MSU, creatinine clearance, catecholamines.
Blood: FBC, U&E, lipids, as suggested by the physical examination.
Others: Echocardiography; renal ultrasound, isotope scans (p 188) ± IVU.

Drug treatment Indications: severe hypertension (as defined above); end organ damage. The following is a suggested sequence: move down the sequence (using drugs in combination, as required) if BP remains raised, if the previous drug is being taken in a suitable dose, and compliance is assured. Enlist expert advice.
1 *Diuretics:* Eg chlorthalidone 0.5–2mg/kg/day PO.
2 *β-adrenergic antagonists:* Eg atenolol 1–2mg/kg/day PO.
3 *Vasodilators:* Eg hydralazine 200μg/kg/6h PO.
4 *ACE inhibitors:* Eg captopril 0.5mg/kg/8h PO if >6 months old. CI: renal artery stenosis. ►See OHCM p 299. Check U&E 1–2 weeks after starting.

Hypertensive emergencies (In glomerulonephritis, haemolytic uraemic syndrome, or head injury, for instance.) Use sodium nitroprusside 1–8μg/kg/min IVI by pump to allow precise control of IVI rate. Cover the IVI with foil to prevent photodeactivation. Monitor BP continuously, increasing the dose slowly to the required level. CI: severe hepatic impairment. Withdraw over >10–30min to prevent rebound hypertension. In prolonged use (>1 day), monitor blood and plasma cyanide levels (keep to <38μmol/l and <3μmol/l respectively). Note: labetalol is an easier to use alternative, but the manufacturers do not (yet) recommend it for children.

1 Task force on blood pressure control in children. Report of the second task force on blood pressure control in children, *Paediatrics* 1987 **79** 1–25

Milks for preterm babies[1]

If the mother's own breast milk is not available, or if weight gain is not satisfactory on human milk (p 178), a formula milk is needed.

Water This is the most important nutrient. Water comprises 50-70% of weight gain (eg of 15g/day) in preterm babies.

Insensible water loss (IWL) falls with increasing body weight, gestational age and postnatal age; it increases with $\uparrow T°$ (ambient and body) and low humidity. In a single walled, thermoneutral incubator with a humidity of 50-80% IWL≈30-60ml/kg/day. This may double in infants on phototherapy under a radiant heater.

Faecal water loss ≈ 5-10ml/kg (except during diarrhoea).

Urine loss: if ~90ml/kg/day, there is no excessive renal stress.

An intake of 180ml/kg/day (range 150-200ml) of human or formula milk meets the water needs of very low birth weight infants (VLBW; <1500g) under normal circumstances. In infants with heart failure water restriction is necessary (eg 130ml/kg/day).

Energy 130kcal/kg/day (range 110-165) meets the needs of the LBW infant in normal circumstances, and can be provided by formulas with similar energy density to human milk (65-70kcal/dl) in a volume of 180-200ml/kg/day. If a higher energy density is required keep it <85kcal/dl. The problem with energy densities above this is fat lactobezoars and U&E imbalance.

Protein Aim for between 2.25g/100kcal (2.9g/kg/day when fed at 130kcal/kg) and 3.1g/100kcal (4g/kg/day). Lysine should be as high as possible. Precise guidelines on taurine and whey:casein ratios cannot be given. At present LBW formulas are whey predominant. Signs of protein deficiency: a low plasma urea and prealbumin.

Fat Aim for 4.7-9g/kg (fat density 3.6-7.0g/100kcal). Longer chain unsaturated fatty acids (>C12) are better absorbed than saturated fatty acids. Aim to have ≥4.5% of total calories as the essential fatty acid linoleic acid (500mg/100kcal). No recommendations can be made about cholesterol, carnitine or choline.

Carbohydrates Aim for 7-14g/100kcal (<11g/100dl), with lactose contributing 3.2-12g/100kcal (<11g/100dl). Lactose is not essential; substitutes are glucose (but high osmolality may cause diarrhoea) or sucrose (± starch hydrolysates, eg corn syrup oils).

Vitamins A reasonable daily supplement is—Vitamin A: 300µg (1µg = 3.33U). Vitamin D: 20µg (1µg cholecalciferol=40U vitamin D3). Vitamin E: 5mg. Vitamin K: 3µg. Vitamin B1: 50µg. Vitamin B2: 200µg. Vitamin B6: 100µg. Vitamin C: 20mg. Folic acid: 60µg. Niacin, biotin and pantothenic acid need not be added. Giving the above supplement would not be toxic, even if the milk being given contained the maximum permitted limit of vitamins. When the baby has reached 2kg the vitamin intake should be reconsidered.

Elements Na+: 6.5-15mmol/l. K+: 15-25.5mmol/l. Ca^{2+} 1.75-3.5mmol/100kcal. PO_4^{3-}: 1.6-2.9mmol/100kcal. $Ca^{2+}:PO_4^{3-}$ ratio: 1.4-2.0:1. Magnesium: 0.25-0.5mmol/100kcal. Iron: if breast-fed, give 2-2.5mg Fe/kg/day (the recommended total intake). Formula feed infants may need a supplement to achieve this. Iodine: 10-45µg/100kcal. Manganese: 2.1µmol/100kcal. ►1cal=4.18 joules.

1 H Bremer 1987 *Nutrition and Feeding of the Preterm Infant*, Blackwell

Talking to parents about their child's mortal illness

Impending death is always a hard subject to approach—especially, it might be thought, when the death concerned is that of a child. Although death in childhood is now a rare event at least in Britain, there will be few mothers who have not thought of the death of their child—probably from the moment of its first stirrings within her. This is important to appreciate as the mother's and father's preparedness for a child's death is a key factor in determining the pacing of interviews in which mortal diagnoses are imparted. Pacing is important because it acknowledges that the acceptance of a mortal diagnosis is not an event, but an unfolding process of the mind. This process may be marked by uncontrollable outbursts, weeping, anger and reproach. These reactions need saluting as they arise, and the interviewer should resist impulses to suppress or turn away from these negative emotions. There is also an additional temptation when faced in medicine by a difficult and uncertain task: it is to lessen one's insecurity by surrounding oneself with professional talismans: the white coat, medical terminology, scientific objectivity, brief and evasive prognostications and an attitude of lofty condescension. Divesting yourself of these things takes practice. We are all at times frightened of appearing naked before our patients—with nothing to offer and with nothing to shield ourselves against impertinent realities.

A criticism of the above is that it consists of unsubstantiated platitudes. However, the tentacles of clinical science are now embracing even this most difficult and sensitive area. Carefully structured, tested interviews yield the following guidelines:[1]

- Ask both parents to be present (plus a nurse who they trust).
- Take steps to avoid being interrupted. Allow sufficient time.
- Call the family by name.
- Do not avoid looking at the family (mutual gaze promotes trust).
- Name the illness concerned. Write it down.
- Give the address and telephone number of a patient's group where further support can be obtained. Find this by telephoning a central directory of groups (the UK the number is: 0171 240 0671).
- Outline the support available throughout the illness.
- Elicit what the parents now know. Clarify or repeat as needed.
- Answer any questions. Acknowledge that it is difficult to absorb all the information so arrange early follow up.
- Explain that you will be telling the child's GP straight away that you have had a full discussion with them. (Parents are likely to seek his advice: make sure he is fully and promptly informed.)

What about the children themselves? Studies have indicated that children's insight into their impending death may be greater than adults'. So do not assume that a child under your care cannot understand. But expect this understanding to be expressed in different ways. We know that children learn to behave as society expects them to behave. Thus a dying child will play with toys while parents are around, but then engage in frank discussion when alone with peers. The child may have learned that speaking about death makes his parents cry—so he or she will avoid the topic.

1 H Woolley 1989 *BMJ* i 1623

Intraosseus transfusion

Immediate vascular access is required in paediatric practice in the follow-ing circumstances: cardiopulmonary arrest, severe burns, prolonged status epilepticus, hypovolaemic and septic shock. In many cases rapid intra-venous access is not easily obtained, and intraosseous infusion is a rela-tively safe, easy and effective means of obtaining vascular access, and is recommended for life-threatening emergencies in young children in whom other methods of access have failed.

Contra-indications: Osteoporosis, osteogenesis imperfecta, and infection or fracture at the site of insertion.

Preparation: Set trolley: Dressing pack, Betadine®, needles, 10ml syringe, lignocaine 1% (5ml), scalpel, intraosseous needle, paediatric infusion set, 10ml 0.9% saline, adhesive tape.

Choosing the site of insertion: The proximal tibia is the best site. Other sites are the distal tibia, or distal femur). Choose a point in the midline on the flat anteromedial surface of the tibia, two finger-breadths below the tibial tuberosity. The patient's leg should be restrained, with a small support placed behind the knee.

Procedure:
- Sterilize the skin with antiseptic, infiltrate with lignocaine as necessary. (Puncturing the skin with the scalpel is not usually necessary.)
- Insert the intraosseous needle at an angle of 60–90 degrees away from the growth plate, and advance with a boring or screwing motion into the marrow cavity. Correct location of the needle is signified by a decrease in resistance on entering the marrow cavity.
- Stabilize the needle in the cortex, and verify the position by aspirating marrow, or by the easy flushing of 5–10ml of 0.9% saline, without any infiltration of surrounding tissue. The needle should stand upright with-out support, but should be secured with tape.
- Flushing with heparin-treated saline may prevent clotting.
- Connect to IV infusion—but better flow rates are often achieved by syringing in boluses of fluid (standard bolus is 20ml/kg of crystalloid or colloid).

Complications: These are infrequent, but there may be extravasation of fluid, or cellulitis, fractures, osteomyelitis, pain, and fat or bone microemboli. These are more common with prolonged use—so intraosseous infusion should be discontinued as soon as conventional IV access is attained.

NB: Intraosseous delivery may also be used in adults.[2]

1 DH Fisher 1990 *NEJM* **322** 1579-81   2 G Bratteb 1993 *BMJ* **ii** 627. This page kindly supplied by Dr RJ Evans via the *BMJ*

Causes ●Choking ●Trauma/Drowning ●Poisons ●Asthma
●Status epilepticus ●Epiglottitis/Croup ●Pneumothorax ●Sepsis

Basic life support Shout 'help!'. Remove obvious dangers ●Clear airway of obvious obstructions (NB: *blind* finger sweeps may impact foreign body).
●Tilt head and lift chin to slightly extend head, to *open the airway.*
●Assess breathing. Is it audible? Any abdominal or chest movement? Feel with your cheek for expired air—if none, but respiratory movements are present, the airway is obstructed: hold face-down on your lap, with his head lower than his chest, and start the *choking cycle:* 5 back blows → 5 sharp chest thrusts → check mouth → open airway → 5 breaths into mouth; include nose if <1yr old; take a breath yourself to optimize gas before each slow, steady inflation (1-1.5 sec) → Repeat cycle. After 2nd set of back blows, do Heimlich manoeuvre × 5 (p 785; *not* if <1yr, as this may cause internal trauma) → repeat until foreign body is cleared.
●Check pulses: brachial is easiest (mid-humerus, medially—hook your finger over the abducted, externally rotated arm).
●If pulse <60 in infants, or absent if older, start cardiac massage with chest compressions—using 2 fingers on the sternum, 1 finger's breadth below a line connecting the 2 nipples, depressing the sternum 2cm. If >1yr, the heel of the hand is used; the depth of compression is ≥3cm.
●Do chest compressions at 100/min: 5 compressions to each ventilation.
●Get an ambulance. If possible, carry child to phone, to carry on CPR.

Advanced life support[1] When help arrives, attach a monitor.
Asystole: Guedel airway support—or endotracheal (ET) tube, size ≈ the size of the nostril (during intubation, do not interrupt compressions for >30sec) with bag-and-mask ventilation with 100% O_2 at ≥15 litres/min. The right sized airway is one that extends from the mouth to the jaw angle, when laid against the child's face. If <1yr insert it convex side *up,* using a tongue depressor or laryngoscope to guide the tongue out of the way. Once the airway is secure, proceed as follows:
●Get IV access or intraosseous access (p 309) within 90sec.
●Give adrenaline 10µg/kg.
●Carry on chest compressions/ventilations for 3min.
●If still pulseless: adrenaline 100µg/kg IV (a desperate measure). If no parenteral access, give 1st adrenaline dose by ET tube (IV dose × 10).
●Consider need for $NaHCO_3$ (below) and crystalloid IVI (eg 20ml/kg).
●If vagal tone has contributed to arrest consider atropine (0.02mg/kg, up to 1mg in children, 2mg in adolescents).
●If plasma Ca^{2+}↑, Mg^{2+}↑, or K^+↑ give calcium chloride 10-30mg/kg IV.
●If hypoglycaemic give glucose 0.5g/kg (10 or 25% solution). Avoid over-enthusiastic correction (harmful in animal models of asystole).

If ventricular fibrillation or pulseless VT (both rare) the top priority is to give DC shocks: 2J/kg → 2J/kg → 4J/kg. (A precordial thump may work if the arrhythmia was witnessed, eg on ITU.) Proceed with ventilatory support, circulatory access, and repeated doses of adrenaline as above.
3 more DC shocks (4J/kg) are given 1min after each dose of adrenaline. Once this cycle has been completed 3 times it *may* be worth considering bicarbonate (1mmol/kg, by slow IV bolus). Lignocaine (1mg/kg IV) *prevents* (but does not reverse) VF. There are few data on bretylium.

Electromechanical dissociation: Treat this as for asystole. The cause is often exsanguination, so fluid replacement is essential. Treat pneumothorax, cardiac tamponade, poisoning and any U&E imbalance.

1 *APLS*/Europ. Resus. Council 1994 *BMJ* i 1349. Neonates, p 230; adults *OHCM* p 725

▶▶Paediatric cardiac arrest protocols

Ventricular fibrillation
- Precordial thump
- Defibrillate 2J/kg
- Defibrillate 2J/kg
- Defibrillate 4J/kg
- Ventilate & intubate 100% O_2
- IV or intraosseous access (p 309)
- Adrenaline 10μg/kg
- Start CPR/defibrillation/adrenaline cycle:-
  - Cardiopulmonary resuscitation for 1 min
  - Defibrillate 4J/kg
  - Defibrillate 4J/kg
  - Defibrillate 4J/kg
  - Consider and correct problems due to U&E imbalance, drugs, hypothermia
  - Adrenaline 100μg/kg
- Repeat CPR/defibrillation/adrenaline cycle
- Repeat CPR/defibrillation/adrenaline cycle again
- Consider need for alkalysing agent or antiarrhythmic

Asystole
- Ventilate & intubate 100% O_2
- IV or intraosseous access (p 309)
- Adrenaline 10μg/kg
- Start CPR/adrenaline: cycle:-
  - CPR for 3 min.
  - Consider need for alkalising agent and/or fluids
  - Adrenaline 100μg/kg
- Continue looping round CPR/adrenaline cycle

Electromechanical dissociation[1]
- Ventilate & intubate 100% O_2
- IV or intraosseous access (p 309)
- Adrenaline 10μg/kg
- Fluids 20ml/kg
- Start CPR/adrenaline cycle:-
  - CPR for 3 min
  - Treat any: hypovolaemia, tension pneumothorax cardiac tamponade drug overdose hypothermia U&E imbalance
  - Adrenaline 100μg/kg
- Continue looping round the CPR/adrenaline cycle

▶NB: if IV/intraosseous access not established within 90 seconds of trying, give adrenaline at ten times the dose via the endotracheal tube.

1 European Resus. Council & APLS *Guidelines for Paediatric Life Support, BMJ* i 1349 incorporating correction made by APLS in 1994 *BMJ* i 1563

4. Psychiatry

Relevant pages in other chapters Puerperal psychosis p 148; Asperger's syndrome and autism p 742; the consultation and consultation analysis p 422–4; social and ethnic matters p 462 & p 436; prescribing and compliance p 460; behavioural questions in childhood p 206.

The alternative table of contents

Arranged according to the (modified) International Classification of Diseases.

The major adult psychiatric diagnoses are:

Problems with classificatory systems:

- No one system can be equally successful in guiding treatment and in illustrating basic biological distinctions. For example, depression spans the neurosis/psychosis division, and mania has both affective and psychotic components.

- There is no such thing as a correct classificatory system—only more or less useful ones. It is instructive to bear in mind that in ancient times (as reported by Borges) one approved classification of animals had as its first 6 categories: 1 Those animals that belong to the Emperor. 2 Embalmed ones. 3 Those that are trained. 4 Suckling pigs. 5 Mermaids. 6 Fabulous ones. We do not know how many mermaids and other fabulous beings are included in the above classificatory system.

- Not all the diagnostic categories have the same value. If a patient with some features of schizophrenia (eg hallucinations) is subsequently found to have an organic psychosis (eg caused by a brain tumour) the schizophrenic element is usually subsumed under the more powerful diagnostic category of organic illness. This is an example of the 'hierarchy of diagnosis'—in which organic illness 'trumps' schizophrenia and affective disorders, which in turn 'trumps' neurotic disorders.

- Each patient is unique—and is not simply a member of a class, and it is a fundamental error to appear to abandon our patients, once he or she has been categorized. Because of its tragic essence, it is possible for mental illness to be a humanizing experience for those affected (patient, family, and friends)—as well as being intensely destructive one, and there is much to be said for the idea that we should embrace our patients, rather than analyse them. Nevertheless, as ever, the holistic must, on occasion, give ground to reductionist analysis, for diagnosis is the single best tool we have to help our patients.

Mental health and mental illness

The essence of mental health Ideally, healthy humans have:

- An ability to love and be loved. Without this cardinal asset, human beings, more than all other mammals, fail to thrive.
- The power to embrace change and uncertainty without fear—and to face fear rationally and in a spirit of practical optimism.
- A gift for risk-taking free from endless worst-case-scenario-gazing.
- Stores of spontaneous *joie de vivre*, and a wide range of emotional responses (including negative emotions, such as anger; these may be important for motivation, as well as being a natural antidote to pain).
- Efficient contact with reality: not too little; not too much. (As T S Eliot said, human kind cannot bear very much reality.)
- A rich fantasy world enabling hope and creativity to flourish.
- A degree of self-knowledge to encourage the humane exercising of the skill of repairing the self and others following harm.
- The strength to say 'I am wrong', and to learn from experience.
- An adequate feeling of security and status within society.
- The ability to satisfy the requirements of the group, combined with a freedom to choose whether to exercise this ability.
- Freedom of self-expression in whatever way he or she wants.
- The ability to risk enchantment and to feel a sense of awe.
- The ability to gratify his own and others' bodily desires.
- A sense of humour to compensate if the above are unavailable.

Happiness need not be an ingredient of mental health; indeed the merely happy may be supremely vulnerable. All that is needed is for their happiness to be removed. The above are important as they are what a person needs should this misfortune befall him. The above may also be seen as a sort of blue-print for our species' survival.

The essence of mental illness Whenever a person's abnormal thoughts, feelings or sensory impressions cause him objective or subjective harm which is more than transitory, a mental illness may be said to be present. Very often the harm is to society, but this should not be part of the definition of mental illness, as to include it would open the door to saying that, for example, all rapists or all those opposing society's aims are mentally ill. If a person is manic, and is not complaining of anything, this only becomes a mental illness (on the above definition) if it causes him harm as judged by his peers (in the widest sense of the term). One feature of mental illness is that one cannot always rely on patients' judgment, and one has to bring in the judgment of others, eg family, GP or psychiatrist. If there is disparity of judgment, there is much to be said for adopting the principle of 'one person one vote', provided it can be shown that the voters are acting solely in the interests of the person concerned. The psychiatrist has no special voting rights here—otherwise the concepts of mental health and illness become dangerously medicalized. Just because the psychiatrist or GP is not allowed more than one vote, this does not stop them from transilluminating the debate by virtue of their special knowledge.

For convenience, English law saves others from the bother of specifying who has a mental illness by authorizing doctors to act for them. This is a healthy state of affairs only in so far as doctors remember that they have only a small duty to society, but a larger duty to their patient (not, we grant, an overriding duty in all instances—eg when murder is in prospect).

Learning disabilities (mental impairment) This is a condition of arrested or incomplete development of mind which is especially characterized by subnormality of intelligence.

Odd ideas

It is important to decide whether a patient has delusions, hallucinations or a major thought disorder, because if these are present the diagnosis must be: schizophrenia; affective disorder; an organic disorder; or a paranoid state—and it is *not* a neurosis or a personality disorder.

Patients may be reluctant to reveal odd ideas. Ask gently: 'Have you ever had any thoughts which now seem odd—perhaps that there is a conspiracy against you, or that you are controlled by an outside voice or by the radio?'

Hallucinations are a sensory experience (auditory, visual, gustatory or tactile) in the absence of a stimulus. The often-encountered pre- or post-sleep (hypnagogic/hypnopompic) hallucinations are not indicative of pathology. A pseudo-hallucination is one in which the patient knows that the stimulus is in his mind (such as a voice heard within himself, rather than over the left shoulder). They are more common, and need not indicate any mental illness, although they may be an important sign—for example that a genuine hallucination is being internalized before being disbanded altogether. Visual and tactile hallucinations suggest an organic disorder, while auditory hallucinations suggest psychosis.

NB: 2–4% of the general population experience auditory hallucinations, but only ~30% of these have a mental illness.[1]

Delusions These are beliefs which are held unshakably, irrespective of any counter-argument, and which are unexpected, given the patient's cultural background. The belief is usually (but not always) false. If the belief arrives in the mind fully formed, and with no antecedent events or experiences which account for it, it is said to be *primary*, and is strongly suggestive of schizophrenia. Such delusions form around a 'delusional perception', as illustrated by the patient who, on seeing the traffic lights go green (the delusional perception) knew that he had been sent to rid his home town of materialism. Careful history-taking will usually reveal that the delusion is secondary—for example, a person who is depressed may come to think of himself as being literally worthless.

Ideas of reference The patient cannot help feeling that other people are taking notice of the very thing he feels ashamed of. What distinguishes this from a delusion is that he knows the thoughts come from within himself, and are excessive. It is important to distinguish delusions and hallucinations from obsessional thoughts. With any odd idea, get a clear description so that you can decide whether the person hears the thought as a voice (hallucination); whether they believe that the voice is being put into their head—thought insertion (a delusion), or whether they know that the thought really comes from themselves, although it is intrusive (obsessional). An obsessional thought is a sign of neurosis, not psychosis.

Major thought disorder This is recognized from the person's speech. He jumps from one idea to another in a bizarre way (see mania—flight of ideas, p 354—and schizophrenia, p 358).

Note On establishing that an hallucination or a delusion is present, ask yourself: 1 What other evidence there is of psychopathology. Hearing the voice of a recently died spouse is a common hallucination, which does not indicate mental illness. 2 Could the odd ideas be adaptive—and the patient be better off ill? The answer is usually 'No', but not always: a woman once believed she saw aeroplanes flying over her house, and that this information was taken from her head by the British Ministry of Defence. She 'knew' she was playing a key rôle in the defence of Britain. When she was cured of these delusions she committed suicide.

1 M Romme 1994 *BMJ* ii 670

Introduction: psychiatric skills

On beginning psychiatry you are likely to feel unskilled. A 'medical' problem will come as a relief—you know what to do. Do not be discouraged: you already have plenty of skills (which you will take for granted). The aim of this chapter is to build on these. No one can live in the world very long without observing mood swings in both himself or his fellows, and without devising ways to minimize what is uncomfortable, and maximize what is desirable. Anyone who has ever sat an important exam knows what anxiety is like, and anyone who has ever passed one knows how to master anxiety, at least to some extent. We have all of us survived periods of being 'down', and it is interesting to ask how we have done this. The first element is time. Simply waiting for time to go by is an important psychotherapeutic principle. Much of what passes as successful treatment in psychiatry is little more than protecting the patient, while time goes by, and his (and his family's) natural processes of recovery and regeneration bring about improvement. Of course, there are instances when to wait for time to go by could lead to fatal consequences. But this does not prevent the principle from being a useful one.

Another skill with which we are all more or less adept is *listening*. One of the central tenets of psychiatry is that it helps our patients just to be listened to. Just as we all are helped by talking and sharing our problems, so this may in itself be of immense help to our patients, especially if they have been isolated, and felt alone—which is a very common experience.

Just as spontaneous regeneration and improvement are common occurrences in psychiatry, so is relapse. Looking through the admissions register of any acute psychiatric ward is likely to show that the same people keep on being readmitted. In one sense this is a failure of the processes of psychiatry, but in another sense each (carefully planned) discharge is a success, and a complex infrastructure often exists for maintaining the patient in the community. These include group support meetings, group therapy sessions, social trips out of the hospital. We all of us have skills in the simple aspects of daily living, and in re-teaching these skills to our patients we may enable them to to take the first steps in rebuilding their lives after a serious mental illness.

So *time*, *listening* and the *skills of daily living* are our chief tools, and with these simple devices much can be done to rebuild the bridges between the patient and his outside world. These skills are simple compared with some of the highly elaborate skills such as psychoanalysis and hypnosis for which psychiatry is famous. The point of bringing them to the fore is so that the newcomer to psychiatry need not feel that there is a great weight of theoretical work to get through before he starts doing psychiatry. On the contrary, you can engage in the central process of psychiatry from your first day on the wards.

Listening

▶Don't just do something—*Listen!*[1]

One of us once had the good fortune to work on an acute psychiatric ward with a surgeon who before he accustomed himself to the ways of psychiatry would pace aimlessly up and down the ward after he had clerked in his patients, wondering when the main action of the day would start, impatient to get his teeth into the business of curing people. What he was expecting, no doubt, was some sort of equivalent to an operating list, and not knowing where to find one he was at a loss, until he gradually realised that taking a history from a psychiatric patient is not a 'pre-op assessment', but the start of the operation itself. Even quite advanced textbooks of psychiatry appear to have missed the insight of this surgeon, describing psychotherapy as something which should only happen after 'a full psychiatric history' and after assessing what aspects of his problems are to be dealt with in treatment.[2] Of course, there is no such thing as a full psychiatric history. In describing the salient psychological events of a single day even the best authors (eg James Joyce in *Ulysses*) require substantial volumes.

In taking the history and agreeing with the patient what tasks should be attempted, treatment has already started—and indeed for some patients this process is all that is required.

Taking a history sounds like an active process, and suggests a list of questions, and the format of our page on this process (p 322) would seem to perpetuate this error. It is not a question of taking anything. It is more a question of receiving the history, and of allowing it to unfold. If you only ask questions, you will get only answers as replies.

As the history unfolds, sit back and listen. Avoid interruptions for the 1st 3min. Be prepared for periods of silence. If prompts are needed try '... and then how did you feel?' or just '... and then...'; or simply repeat the last words the patient spoke. Do not be over-anxious if the patient appears not to be covering major areas in the history. Lead on to these later, by taking more control as the interview unfolds. At first you will have to ask the relevant questions (detailed on p 322) in a rather bald way (if the information is not forthcoming during the initial unstructured minutes), but it is important to go through this stage as a prelude to gaining information by less intrusive means.

What is the point of all this listening? Listening enables patients to start to *trust* us. Depressed patients, for example, often believe they will never get better. To believe that they *can* get better, patients need to trust us, and this trust is often the start of the therapeutic process. In general, the more we listen, the more we will be trusted. Our patients' trust in us can be one of our chief motivations, at best inspiring us to pursue their benefit with all vigour. A story bears this out. One summer's day, in 334 BC, Alexander the Great fell ill with fever. He saw his doctor, who gave him a medicine. Later he received a letter saying his doctor was poisoning him as part of a plot (it was an age of frequent, and frequently fatal intrigue). Alexander went to his doctor and drank the medicine—then gave him the letter. His confidence was rewarded by a speedy recovery. We think it is unreasonable to expect *quite* this much trust from our patients, and one wonders what can have led Alexander to such undying trust in his doctor. We strongly suspect that such a doctor, above all else, must have been a good listener.

Note that listening, as with any interaction with patients, may have side-effects. It is certainly possible for listening to open up a can of worms (p 347), and then the difficulty is in deciding whether to dissect out each worm in turn, or hurriedly to try to put the lid back on.

1 An aphorism created by Dr A Storr 2 M Gelder 1989 *Oxford Textbook of Psychiatry*, 577–8, OUP

Eliciting the history

Introduce yourself, and explain how long the interview will take. Describe its purpose. Find out how the patient came to be referred, and what his expectations are (eg about treatment). If the patient denies having any problems or is reluctant to start talking about himself, do not hurry him. Try asking 'How are you?' or 'What has been happening to you?' or 'What are the most important things?' Another approach for hospital patients is to indicate why the GP referred the patient and then ask what the patient thinks about this. Sit back and listen, without interrupting, noting exact examples of what he is saying. Take more control after ~3 min, to cover the following topics.

Presenting symptoms Agree a problem list with the patient early on, and be sure that this is comprehensive by asking 'If we were able to deal with all these, would things then be all right?' Then take each problem in turn and find out about onset, duration, effects on life and family; events coinciding with onset; solutions tried; reasons why they failed. The next step is to enquire about mood and beliefs during the last weeks (this is different from the mental state examination, p 324, which refers to the mental state at the time of interview). Specifically check for: suicidal thoughts, plans or actions—the more specific these are, the greater the danger. Discussing suicide does not increase the danger. Questions to consider: 'Have you ever felt so low that you have considered harming yourself?' 'Have you ever actually harmed yourself?' 'What stopped you harming yourself any more than this?' 'Have you made any detailed suicide plans?' 'Have you bought tablets for that purpose?'

Depression—ie low mood, anhedonia, self-denigration ('I am worthless'; 'Oh that I had not been born!'), guilt ('It's all my fault'), lack of interest in hobbies and friends plus biological markers of depression (early morning waking, decreased appetite and sexual activity, weight loss); mania (p 354); symptoms of psychosis (persecutory beliefs, delusions, hallucinations, p 316); drug and alcohol ingestion; obsessional thoughts; anxiety; eating disorders (eg in young women—often not volunteered, and important). Note compulsive behaviour (eg excessive hand-washing).

Present circumstances Housing, finance, work, marriage, friends.

Family history Physical and mental health, job and personality of members. Who is closest to whom? Any stillbirths, abortions?

Birth, growth and development How has he spent his life? Ask about school, play (alone? with friends?) hobbies, further education, religion, job, sex, marriage. Has he always been shy and lonely, or does he make friends well? Has he been in trouble with the law? What stress has he had and how has he coped with it? (Noting early neurotic traits—nail-biting, thumb-sucking, food fads, stammering—is rarely helpful.)

Premorbid personality Before all this happened, how were you? Happy-go-lucky, tense, often depressed? Impulsive, selfish, shy, fussy, irritable, rigid, insecure? ►Talk to whomever accompanies him (eg wife), to shed light on premorbid personality (as well as on current problems).

Next examine the mental state (p 324). You may now make a diagnosis. Ensure that the areas above have been covered in the light of this diagnosis so that the questions 'Why did he get ill in this way at this time?' and 'What are the consequences of the illness?' are answered.

The mental state examination

This assesses the state of mind at the time the interview is being conducted. Make records under the following headings.

- *Observable behaviour:* Eg excessive slowness, signs of anxiety.
- *Mode of speech:* Include the rate of speech, eg retarded or gabbling (pressure of speech). Note its content.
- *Mood:* Note thoughts about harming self or others. Gauge your own responses to the patient. The laughter and grandiose ideas of manic patients are contagious, as to a lesser extent is the expression of thoughts from a depressed person.
- *Beliefs:* Eg about himself, his own body, about other people and the future. Note abnormal beliefs (delusions), eg that thoughts are overheard and ideas (eg persecutory, grandiose). See p 316.
- *Unusual experiences or hallucinations:* Note modality, eg visual.
- *Orientation:* in time, place and person. What is the date? What time of day is it? Where are you? What is your name?
- *Short-term memory:* Give a name and address, and test recall after 5 minutes. Make sure that he has got the address clear in his head before waiting for the 5 minutes to elapse.
- *Long-term memory:* Current affairs recall. Who is the Monarch? (This tests many other CNS functions, not just memory.)
- *Concentration:* Months of the year backwards; see p 353.
- Note the patient's *insight* and the degree of your *rapport*.

Non-verbal behaviour *Why are we annoyed when we blush, yet love it when our friends do so?* Part of the answer to this question is that non-verbal communication is less well controlled than verbal behaviour. This is why its study can yield valuable insights into our patients' minds, particularly when analysis of their spoken words has been unrevealing. For example, if a patient who consistently denies being depressed sits hugging himself in an attitude of self-pity, remaining in a glum silence for long periods of the interview, and when he does speak, uses a monotonous slow whisper unadorned by the slightest flicker of a gesticulation or eye contact—we are likely to believe what we see and not what our patient would seem to be telling us.

Items of non-verbal behaviour:
- Gaze and mutual gaze
- Facial expression
- Smiling, blushing
- Body attitude (eg 'defensive')

- Dress (eg all in black)
- Hair style
- Make-up
- Body ornament (eg ear-rings, tattoos)

Signs of auditory hallucinations:
- Inexplicable laughter
- Silent and distracted while listening to 'voices' (but could be an 'absence' seizure, p 266)
- Random, meaningless gestures

Anxious behaviour:
- Fidgeting, trembling
- Nail-biting
- Shuffling feet
- Squirming in the chair
- Sits on edge of chair

Signs of a depressed mood:
- Hunched, self-hugging posture
- Little eye contact

- Downcast eyes; tears
- Slowness of: thought; speech; movement

Helping patients avoid doctor-dependency

A patient may become over-dependent on his or her doctor in many spheres of medicine. This is a particular danger in psychiatry because of the intimate and intense rather one-sided or asymmetrical relationship which may be built up between the patient and his psychiatrist—who is likely to come to know far more about the patient's innermost hopes and fears than many of his close friends. This encourages the patient to transfer to the therapist thoughts and attitudes which would more naturally be directed to a parent-figure. This process (known as transference) is a powerful force in stimulating doctor-dependency.

Signs of non-therapeutic dependency Repeated telephonings for advice, the inability to initiate any plan without help from the therapist, and the patient's disallowing of attempts by the therapist to terminate treatment (eg by threatening relapse) are all signs that non-therapeutic doctor-dependency is occurring.

Assessing whether dependency is a problem Clearly, in the examples above the patient's dependency on his doctor is non-therapeutic. At other times, for example early on in treatment, doctor-dependency may be quite helpful. In these circumstances the danger is that the doctor will be flattered by his patient's dependency on him. Most therapists either want to be loved by their patients or want to dominate them (or both), and it is important to know, in each session with each patient, just where you lie within the space marked out by these axes. Ask yourself 'Why do I look forward to seeing this patient?' 'Why do I dread seeing Mr X?' 'Why do I mind if this patient likes me?'

▶When you feel good after seeing a patient always ask yourself why (it is so often because he is becoming dependent on you).

Avoiding dependency Planning and agreeing specific, limited goals with the patient is one way of limiting dependency. If the patient agrees from the outset that it is not your job to provide him with a new job, wife, or family the patient is more likely to have realistic expectations about therapy.

Planning discharge from the beginning of therapy is another important step in limiting doctor-dependency. Discharge is easy to arrange from the hospital outpatient department, but for the general practitioner the concept of discharge is diluted by the fact of his contractual obligation to the patient—who is perfectly within his rights to turn up the day after being 'discharged' and demand that therapy be started all over again. He must have more subtle methods at his disposal to encourage the patient to discharge himself. For example, he can learn to appear completely ineffective, so that the dependency cycle (patient presents problem → doctor presents solution → patient sabotages solution → doctor presents new solution) is never started. Another approach is to bore your patient by endlessly going over the same ground, so that the patient seizes control and walks out as if to say 'I've had enough of this!'.

The forgoing makes patients out to be perpetual seekers after succour and emotional support—and so it can seem at times; but a great mystery of clinical medicine is that, spontaneously and miraculously, these patients *can* change, and start leading mature and independent lives. So do not be downcast when you are looking after these poor people: there is much to be said for simply offering a sympathetic ear, staying with your patient through thick and thin, and waiting for time to go by, and for the wind to change. Of course, the wind may change back again, but, if it does, you will not be back at square 1, for you will be able to inject the proceedings with the most powerful psychotherapeutic agent of all: namely hope.

How to identify and use the full range of psychiatric services

Any good psychiatric service will have much more to it than out-patient clinics and hospital wards. The latter are very important, but it is not true to say that the most significant psychiatric exercises go on in this location. Alternatives to the cycle of out-patient/in-patient doctor-led treatments include:

- Responsive community psychiatric teams
- On-call community psychiatric nurses (CPNs)
- Crisis intervention teams
- Drop-in services during weekends and unsocial hours
- Telephone help-lines
- Self-help groups
- Mental illness hostels
- Occupation centres (voluntary or statutory)
- Sheltered work
- Supervised living quarters so patients are not alone
- Self-running group homes

These facilities are not only important in keeping patients out of hospital (a prime aim of those who are purchasing care); they are also important in gradually normalizing the patient's relationship with his environment.

How to help patients not to be manipulative

Fireships on the lagoon We have all been manipulated by our patients, and it is wrong to encourage in ourselves such stiffness of character and inflexibility of mind that all attempts by our patients to manipulate us inevitably fail. Nevertheless, a patient's manipulative behaviour is often counter-productive, and reinforces maladaptive behaviour. A small minority of patients are *very* manipulative, and take a disproportionate toll on your resources, and those of their family, friends and work-fellows. We are all familiar with those patients who Ford Madox Ford describes as being like fireships on a crowded lagoon, causing conflagration in their wake.[1] After destroying their family and their home we watch these people cruise down the ward or into our surgeries with some trepidation. Can we stop them losing control, and causing melt-down of our own and our staff's equanimity? The first thing to appreciate is that, unlike an unmanned ship, these people *can* be communicated with, and you *can* help them without resorting to hosing them down with cold water.

Setting limits One way of avoiding becoming caught up in this web of maladaptive behaviour is to set limits, as soon as this behaviour starts.[2] In a small minority of patients, the therapist may recognize that their needs for time, attention, sedation and protection are, for all practical purposes, insatiable. Whatever the therapist gives, such patients come back for more and more, and yet in spite of all this 'input' they do not appear to get any better. The next step is to realize that if inappropriate demands are not met the patient will not become sicker (although there may be vociferous complaints). This realization paves the way for setting limits on patients' behaviour, specifying exactly what is and is not acceptable.

Take for example the patient who demands sedation, threatening to 'lose control' if it is not given immediately, stating that he cannot bear living another day without sedation, and that the therapist will be responsible for any damage which ensues. If it is decided that drugs do not have a part to play in treatment, and that the long-term aim is for the patient to learn to be responsible for himself, then it can be simply stated to the patient that medication will not be given, and that he is free to engage in destructive acts, and that if he does so this is his responsibility.

The therapist explains that in demanding instant sedation he usurps her professional rôle, which is to decide these matters according to her own expert judgment, and that such usurpation will not be tolerated. If there is serious risk of real harm, admission to hospital may be indicated, where further limits may be set. If necessary, he is told that if he insists on 'going crazy' he will be put in a seclusion room, to protect others.

Another method is to abdicate the power you have to bring about the end the patient or his family desires. So if, for example, the father of a patient who has been labelled 'psychotic' is very keen for repeated admissions for his son, while at the same time hotly resenting the suggestion that the family is in some way exacerbating the condition, an experienced therapist might make progress along these rather dangerous lines: 'It is clear that your wife cannot cope with your son when he is hearing his voices. Although we know that they are quite harmless, they are obviously more upsetting to her than to your son. So just ring the ward when you cannot cope, and send him in—or your wife. It doesn't matter who comes in'. Such strategies (which may be very effective) clearly need careful planning and agreement from all concerned in caring for the family.

---

1 Ford Madox Ford 1915 *The Good Soldier*, Penguin 2 G Murphy 1960 *Am J Psychotherapy* 14 30–47

How to recognize and treat your own mental illnesses

Doctors have a higher than average incidence of suicide and alcoholism, and we must all be prepared to face (and try to prevent) these and other health risks of our professional and private lives. Our skill at looking after ourselves has never been as good as our skill at looking after others, but when the healer himself is wounded, is it clear that his ability to help others will be correspondingly reduced? Our own illnesses are invaluable in allowing us to understand our patients, what makes people go to the doctor (or avoid going to the doctor), and the barriers we may erect to resist his advice. But the idea of an ailing physician remains a paradox to the average mind, so that we may ask: ▶Can true spiritual mastery over a power ever be won by someone who is counted among her slaves?[1] If the time comes when our mental state seriously reduces our ability to work, we must be able to recognize this and take appropriate action. The following may indicate that this point is approaching:

- Drinking alcohol before ward rounds or surgeries.
- The minimizing of every contact with patients, so that the doctor does the bare minimum which will suffice.
- Inability to concentrate on the matter in hand. Your thoughts are entirely taken up with the work load ahead.
- Irritability (defined as disagreeing with >1 nurse/24h).
- Inability to take time off without feeling guilty.
- Feelings of excessive shame or anger when reviewing past débâcles. To avoid mistakes it would be necessary for us all to give up medicine.
- Emotional exhaustion (defined as knowing that you should be feeling pleased or cross with yourself or others, but on summoning up the affairs of your heart, you draw a blank).
- Prospective studies suggest that introversion, masochism, and isolation are important risk factors for doctors' impairment.

The first step in countering these unfavourable states of mind is to recognize that one is present. The next step is to confide in someone you trust. Give your mind time to rejuvenate itself.

If these steps fail, try prescribing the symptom. For example, if you are plagued by recurring thoughts about how inadequately you treated a patient, set time aside to deliberately ruminate on the affair, avoiding distractions. This is the first step in gaining control. You initiate the thought, rather than the thought initiating itself. The next step is to interpose some neutral topic, once the 'bad' series of thoughts is under way. After repeated practice, the mind will automatically flow into the neutral channel, once the bad thoughts begin, and the cycle of shame and rumination will be broken.

If no progress is made, the time has come to consult an expert, such as your general practitioner. Our own confidential self-help group for addiction and other problems is the British Doctors' and Dentists' Group and may be contacted via the Medical Council on Alcoholism (UK tel. 0171 487 4445). If you are the expert that another doctor has approached, do not be deceived by this honour into thinking that you must treat your new patient in any special way. Special treatment leads to special mistakes, and it is far better for doctor-patients to tread well-worn paths of referral, investigation and treatment than to try illusory short cuts.

1 Thomas Mann 1924 *The Magic Mountain*, page 132–3, Penguin

The patient in disgrace

One of the rôles of the psychiatrist is to act as the terminus for patients who have been shunted from hospital department to hospital department. These patients offer their doctor an 'organ recital' of incapacitating physical symptoms for which no physical cause can be found (somatization). She knows that 'there must be something wrong'. In the end, the patient is sent in disgrace to the psychiatrist because the doctor declares that there is 'nothing wrong' and that the patient is 'just a hypochondriac'. The psychiatrist then has to do much work undoing the cycle of symptom-offering and investigation, by acknowledging that there *is* something wrong. The aim is then to reframe symptom-offering into problems which need to be solved. This may start with an acceptance by the doctor that the patient is distressed and looking for helpful responses from her carers which are yet to be defined. After establishing rapport, the psychiatrist will agree a contract[1] with the patient—that he will provide regular consultations for listening to how the patient feels and will try to help, on condition that she acknowledges that past investigations have not helped, and that she agrees not to consult other doctors until a specified number of sessions have elapsed. In addition, cognitive therapy (p 370) which examines the way that conscious thoughts and beliefs perpetuate disability, *can* lead to meaningful symptom reduction. ◁▷

Hysteria

This is a mental disorder in which there are signs or symptoms of disease (in the absence of pathology) which are produced *unconsciously* (so hysteria is different to malingering or lying, and does not, contrary to lay use, mean histrionic). Examples: paralysis, blindness, amnesia—eg in a patient who was homosexually raped: he had no conscious memory for the event, but felt an irresistible urge to write insulting letters about the person who had raped him. In deciding if phenomena are hysterical, ask:
- Has a physical cause been carefully discounted?
- Is the patient young? Beware making the first diagnosis if >40yrs old.
- Have the symptoms been provoked by stress? Ask the family.
- Is there secondary gain?—ie is it worthwhile perpetuating symptoms? Many physical illnesses have gain for the patient, so this sign is only useful if it is absent (so ruling out hysteria).
- Do the symptoms 'make sense' (eg aphonia in a news-reader)?
- Indifference to what should constitute a major handicap (*la belle indifférence*) is of only a little diagnostic use.

Most hysterical conversion reactions subside spontaneously, but if they do not it is important to refer early to a psychiatrist, before the hysterical phenomenon becomes habitual. Epidemic hysteria may occur, in which whole communities become involved, starting with the more suggestible members. In an hysterical fugue the patient disappears for a few days, making pointless journeys about which he has no recollection.

Prognosis in hysteria Follow-up over 7–11 years shows that ~20% (22/99) of those referred to a neurological hospital who had hysteria diagnosed turned out to have organic illness.[3] The question is not so much whether we should, therefore, abandon the diagnosis of hysteria, but why the neurologists were so bad at diagnosing it.

Treatment of hysteria Exploring the stresses in the patient's life may help, as on p 388–96. Hypnosis (or amylobarbitone IV) may be used to enable the patient to relive the stressful events (abreaction).

1 C Deighton 1985 *Psy Med* **15** 515 ◁▷ A Speckens 1996 *EBM* **1** 80 3 E Slater 1965 *J Psychosom Res* **9** 9

Depression

The spectrum of depression Each year 40% of the population have quite
severe feelings of depressed mood, unhappiness, and disappointment. Of
these, ~20% will experience a clinical depression, in which low mood is
accompanied by sleep difficulty, change in appetite, hopelessness, pes-
simism, or thoughts of suicide. *Diagnosis* of major depression:
- Loss of interest or pleasure—*anhedonia* in daily life with dysphoric mood
 (ie 'down in the dumps') plus ≥4 of the following (the first 5 are 'biological' symptoms)—present nearly every day for at least 2 weeks:
- *Poor appetite* with *weight loss* (or, rarely, increased appetite).
- *Early waking*—with *diurnal mood variation* (worse in mornings).
- *Psychomotor retardation* (ie a paucity of spontaneous movement, or slug-
 gish thought processes), or *psychomotor agitation*.
- Decrease in *sexual drive* and other appetites.
- Evidence of (or complaints of) reduced *ability to concentrate*.
- Ideas of *worthlessness*, inappropriate *guilt* or *self-reproach*.
- Recurrent *thoughts of death* and *suicide*, or *suicide attempts*.

Why is major depression so often missed? ●Lack of knowledge.
●Preoccupation with physical disease. ●Stigma of psychiatric labels. ●A
tendency to underrate severity. ●It is not so easy to spot depression if it
coexists with other illness (so often the case).[1]

Classification Classify as: ●Mild; moderate; severe. ●With/without biological features. ●With/without delusions or hallucinations. ●With/with-
out manic episodes (the bipolar/unipolar distinction). These 3 axes
replace the old reactive/endogenous classification.

Why do people get depressed? ●Genetics: identical twins reared apart
show 60% more concordance for depression than dizygotic twins.
●Biochemistry: there are excess 5-hydroxytryptamine (5-HT2) receptors in
the frontal cortex of brains taken from suicide victims. See OHCM p 406.
●Endocrinology: cortisol suppression (dexamethasone suppression test,
OHCM p 550) is abnormal in ~30%. Melatonin is also implicated (p 404).
●Stressful events (eg new births, job loss, divorce, illness): present in 40%.
●Freudian reasons: depression mirrors bereavement—but the loss is of a
valued 'object', not a person. There is ambivalence in an important relationship, with hostility turned inwards. Another idea which some animal
work supports is learned helplessness: if punishment is not contingent
on actions of the person, but is perceived as random, the response is the
helplessness which is, some believe, the hallmark of depression.
●Vulnerability factors: physical illness, pain, and lack of intimate relationships may allow depression to arise and be perpetuated.

Management There is no clear distinction between the low moods we all
get and illness needing vigorous treatment, but the lower the mood and
the more marked the slowness, the more vigorous the treatment.
- Presence of biological features or stressful life events suggests a good
 response to *antidepressants* (p 340)—particularly if symptoms are severe.[2]
- Delusions or hallucinations prompt a physical treatment, ie ECT (p 342)
 or *drugs* (antidepressants ± antipsychotic drugs, p 360).
- The depressive phase of bipolar illness is treated as above, but note:
 1 Be cautious with the physical treatments: they may induce mania.
 2 Lithium prophylaxis may be needed (p 354).
- *Psychological treatment* (p 388–96) is part of the treatment of *all* depression;
 it may be all that is needed in milder depressions.
- *Reasons to admit:* Poor compliance; high suicide drive; isolation.

1 AT Tylee 1993 *Br J Gen Prac* **43** 327   2 ES Paykel 1988 *Pharmacopsychiatry* **21** 15

Suicide and attempted suicide

As we all know, Nature has implanted in us a dominant will to live; what the average mind may be quite unprepared for is the person who with equal but opposite energy pursues his own destruction, urgently discarding all that is partial and provisional, to gain the ultimate goal. He differs from many self-poisoners—who are a very common problem. In girls aged 15–19 the incidence is >1%/yr in some areas. Most self-poisoning does not end in death—but in some places suicide accounts for ½ of adolescent deaths, and is a growing problem in the elderly.[1][2] 'Overdoses' account for 4.7% of all general hospital admissions in those aged 12-20yrs.

Understanding suicide behaviour Suicide behaviour may be a way of:
- Keeping honour and autonomy, and avoiding shame (the Roman view).
- Avoiding something ghastly, eg pain, or another bout of depression.
- Controlling change in families, eg when a member is about to leave.
- Communicating important messages in a not-to-be ignored way.
- Gaining power in family transactions, by escalating conflict.

Causes of suicide behaviour Depression is a leading cause. A common antecedent is an argument with a boyfriend. Emotional immaturity, inability to cope with stress, weakening of religious ties, unemployment, and availability of guns and drugs (psychotropics are the most popular poison) are also important, as is 'copy-cat' behaviour: when a celebrity tries suicide, others will follow. Children who are bullied or placed under intolerable stress to succeed may opt for suicide, eg in Japan—where completed suicide is the cause of death of >600 children/yr. Common reasons are falling behind with homework—and if this reflects your own state after trudging through too many OHCS pages, shut this book, and have a good holiday. Australia has the highest suicide rates in 15-24-yr-olds: 16/100,000.

There are 6 stages in trying to help the survivors:[1]
- Assessment. ●Agreement of a contract offering help (p 339).
- Discussion with the family as to how problems might be tackled.
- Problem-solving by facilitating the patient's understanding of her predicament, and by pointing out how she has coped with problems in the past. The aim is to engender a greater ability to cope in the future as well as helping immediate personal and social problems.
- Prevention: open-access, walk-in clinics or 24h phone service.
- Follow-up—either with family or with the patient alone.

Assessment Think of a target with 3 concentric rings. *The inner ring* is the circumstances of the attempt: what happened that day; were things normal to start with? When did the feelings and events leading up to the act start? Get descriptions of these in detail. Was there any last act (eg a suicide note)? What happened after the event? Was this what did he/she expected? *The middle ring* is the background to the attempt: how things have been over the preceding months. Might the attempt have been made at any time over the last months? What relationships were important over this time? *The outer ring* is the family and personal history (p 322).
After the above, come to the *bulls-eye*—the intentions lying behind the attempt, and the present feelings and intentions. Does the attempt represent a wish to die (a grave, not-to-be-ignored sign); a wish to send a message to somebody; or a wish to change intolerable circumstances? Ask: 'If you were to leave hospital today, how would you cope?' Examine the mental state (p 324) to find out if there is any mental illness. *Summary:* ●Any plan? What? When? Where? ●Are the means available? ●Ever tried before? How seriously? ●Preparations (making a will, giving things away).[3]
   Before arranging hospital admission, ask what this is *for*. Is it only to make you feel happier?—or to gain something that cannot be gained outside hospital. Ask: *Why will discharge be safer in a few weeks rather than now?*

1 D Aldridge 1992 *BJGP* 42 482 2 N Retterstol 1993 *Suicide*, CUP 3 R Vinning 1995 *BMJ* i 126

How to cope with threats of suicide

Suicide behaviour is very powerful. Boyfriend: 'After last night, I think we should separate—we'd better just be good friends. . .' Girlfriend thinks: 'You are not going to treat me like that', but says: 'So we are not getting married after all?' 'Well not for a little while, maybe.' 'So you don't love me . . . I knew you loved Amanda much more than me.' Exit into the bathroom, where she swallows several handfuls of her mother's tranquillizers. Next day in hospital the boyfriend says 'I'm so sorry—I didn't know you loved me that much.' He thinks: 'No one has ever thought that much of me. . . Fancy wanting to kill herself because I wouldn't marry her! I feel guilty; I should not have led her to this. . .' Whereupon he proposes marriage. This disastrous ending may be complicated by threats of suicide from jilted Amanda.

339

The psychiatrist may become enmeshed in these webs of suicide threats, and may wrongly assume that because someone threatens suicide, they should be admitted to hospital (compulsorily if necessary) so that they can be kept under constant surveillance, and suicide prevented. This reasoning has three faults. One is the idea that it is possible to prevent suicide by admission to hospital. There is no such thing as constant surveillance. Secondly, hospital admission may achieve nothing if it simply removes the patient from the circumstances which he must learn to cope with. The third reason is that it is necessary to distinguish between suicide gestures, which have the object of influencing others' behaviour, and a genuine wish to die.

Before completing suicide many will see a GP, and many deaths could be preventable. It is important to ask patients unambiguously about suicide plans (p 322). On deciding that a suicide threat is more manipulative than genuine, very experienced therapists may try influencing the person's use of suicide behaviour by forcing him to face the reality of his suicide talk, eg by asking: 'When will you kill yourself?' 'How will you do it?' 'Who will discover the corpse?' 'What sort of funeral do you want? Cremation, burial, with or without flowers?' 'Who will come?'

Those who die:	Those who do not die:
Often older and male	Young and often female
Often lives alone; unemployed	Lives with family or friends
Mental or painful illness common	These illnesses are rarer
Drug and alcohol abuse common	Drug and alcohol abuse rarer

The above distinction should not be taken to mean that those with features in the right-hand column need less active help than those with features on the left. ►*All suicide threats must be taken seriously*—but the emphasis will differ depending into which group the patient falls. In both groups it may be helpful to form a *contract* with the patient, eg:

● The therapist will listen and help if the patient agrees to be frank, and to tell the therapist of any suicide thoughts or plans.
● Agreement about which problems are to be tackled is made explicit.
● The type of change to aim for is agreed.
● Specification of who else will be involved in treatment (eg other members of the family, friends, the patient's GP).
● An agreement about the timing and place of sessions.
● An agreement about the patient's responsibility to work effectively with the therapist, and to carry out any 'homework'.

Antidepressant drugs

These improve mood and ↑synaptic availability of noradrenaline or 5-HT. The 3 kinds: tricyclic and related drugs; MAOI (below); SSRIs (p 341).

Explain to sceptical patients that drugs *do* work; also discuss side effects. *In agitated patients, offer a sedative antidepressant, eg:*

Amitriptyline (50mg/8-24h PO. Starting dose example: 50mg at night).
Dothiepin (50mg/8-24h PO. Start with 50-75mg at night).
Doxepin (75mg/8-12h PO. Start with 10-50mg at night).
Trimipramine (25-50mg/8 PO. Start with 50mg 2h before bedtime).

Less sedative antidepressants include the following:

Clomipramine (50mg/8-24h PO. Start with 10mg/24h). This drug is particularly helpful if there are phobic or obsessional features.
Imipramine (10-25mg/8-24h PO, up to eight 25mg tablets/day).
Lofepramine (70mg/12h PO. Start with 70mg/24h; few anticholinergic SE).
Nortriptyline (25mg/6-24h PO. Start with 10mg/12h).
Protriptyline (5-10mg PO morning, noon and at 4pm—to avoid insomnia, up to six 10mg tablets/day). This is a stimulant.

▶In the elderly use the smaller doses indicated above. Build up dose over ~10 days (eg to amitriptyline 150mg/24h PO); maintain this if side-effects allow for ≥1 month, before deciding if ineffective. After 2-5 months it may be appropriate to reduce the dose (eg amitriptyline 75-100mg/24h).

Unwanted effects Convulsions (dose related), arrhythmias, and heart block occur (particularly with amitriptyline, which is contraindicated in the weeks after a myocardial infarct; it is particularly dangerous in overdose—so prescribe in small volumes; review suicidal drive often).

Anticholinergic effects (dry mouth, blurred vision, constipation, urinary retention, drowsiness, and sweating) may occur with any tricyclic or related drug, particularly nortriptyline, amitriptyline, imipramine. Explain these to the patient. Indicate that these effects will diminish with time. Until they do, advise against driving and operating machinery.

Hepatic and haematological reactions may also occur, particularly with mianserin. Agranulocytosis may occur up to 3 months after treatment is started (do monthly FBCs during this period).

Interactions The Pill may reduce the effect of tricyclics. Side effects may be worse if phenothiazines are used concurrently. The effect of some hypotensives (eg clonidine but not β-blockers) may be reduced.

Lack of response[1] ●Is the diagnosis correct? ●Are there other diagnoses (eg cancer or CNS disease)? ●Is he taking the tablets?—at full dose for >1 month? Do not expect a response before this. Consider an SSRI (p 341), ECT (p 336)—or try low-dose flupenthixol (0.5-1mg PO in the morning) or a monoamine oxidase inhibitor (MAOI)—but *not with* a tricyclic (tricyclics mustn't be given within 21 days of MAOIs). Try phenelzine 15-30mg/8h PO. The danger is hypertensive crises caused by certain foods and drugs, eg cheese, yeast (Marmite®), pickled herring, narcotics, common cold remedies, levodopa, tricyclics. This may occur up to 2 weeks after stopping MAOIs. Patients must carry a card declaring use of MAOIs, and foods to be avoided. But do not be put off their use: the incidence of crises is only ~17/98,000 patient years.[2] Benefits may be great, eg if there is oversensitivity to rejection by friends, some lightening of mood related to circumstances, hostility, overeating, oversleeping, marked fatigue, panic, irritability, or hypochondriasis.[2] Moclobemide (150-300mg/12h PO pc) is a new MAO-A inhibitor said to have few of the problems of older MAOIs; CI: confusion, phaeochromocytoma); hypertension has sometimes occurred.[3]

Antidepressants in adolescent depression Careful meta-analyses of 12 papers indicates that tricyclics are no more effective than placebos in this group.

1 M Hornig-Rohan 1994 *Psychiatric Ann* 24 220   2 C Bass 1989 *BMJ* ii 345   3 *Drug Ther Bul* 1994 32 6

Selective serotonin reuptake inhibitors (SSRI)

Mode of action 5-HT (serotonin) re-uptake inhibition (OHCM p 406). NB: Serotoninergic system dysfunction is a feature of drug-free depressed people (CSF and brain 5-HT metabolite levels↓; ↑CNS density of binding sites).
There is evidence that the *STin2.9* allele of the serotonin transporter gene confers susceptibility to unipolar depression in >10% of patients. ⬚

Effects of SSRIs SSRIs are effective antidepressants (response rate ≈ that of tricyclics, ie 60-70%). Chief *advantages* over old tricyclics such as amitriptyline are that they are less toxic in overdose and less sedating; the main *disadvantage* is that they are ~30 times the price. SE (all SSRI): nausea (20%), dry mouth (10%) blurred vision (10%), seizures, and anorgasmia. 300-400 people die each year from poisoning with tricyclic related compounds.
In spite of >60 trials of SSRIs and their detailed meta-analyses,[1] no clear advantages have been shown, with apparent equality in drop-out rates, suggesting that from patients' view points they are little different. One problem is that transferring treatment drop-out rates from a trial environment to routine consulting room or ward life is rash as in trials there is much pressure and organization to ensure that the patient 'carries on taking the tablets'—for the short duration of a trial.[2] Although there is also evidence (at least for fluoxetine) that if the lower doses now recommended are used (20mg/day) drop-outs are less than with low-dose régimes of tricyclics,[3] the effect is small, and the data are suspect because of the anomaly of finding that if the dose of fluoxetine is increased, the drop-out rate also falls. Another problem is that in many studies standardized interviews were not used (it is hard to assess trials where inclusion depends on impressions).
If the risk of suicide is high (p 338-9) it might be prudent to use SSRIs in preference to tricyclics. Although some suicidal patients on SSRIs will find other ways to die than via their prescribed drugs, it is nevertheless likely that lives would be saved by using SSRIs in place of traditional tricyclics such as amitriptyline and dothiepin (which account for 70% of deaths by antidepressant poisoning).[4] But, it could be argued that the conclusion from this is to avoid the older, more toxic tricyclics: this does not entail the use of SSRIs (lofepramine, for example, is an alternative).
If patients need to drive there is some evidence that this is safer with paroxetine than with tricyclics.[5]

Treatment régimes ▶Monitor closely: suicide risk is greater early on.

Sertraline: 50mg/24h PO, ↑ if needed by 50mg steps every few weeks to a maximum of 200mg/day; within 8 weeks reduce to a maintenance dose of 50-100mg/day PO. SE: dyspepsia, sweating, AST↑ (if so, discontinue). It is the SSRI least likely to cause agitation or sedation.

Fluoxetine: 20mg/24h PO. SE: anxiety, insomnia, drowsiness, confusion, fever, weight↓, Na+↓, platelets↓, prolactin↑, vaginal bleeding on stopping treatment. If rash or lung fibrosis, discontinue. $t_{1/2} \approx 3$ days.

Fluvoxamine maleate: 100mg/12-24h PO (max 300mg/day). SE: agitation, headache, tremor, fits, pulse↓, AST↑ (discontinue). CI: epilepsy.

Paroxetine: 20mg each morning (↑in 10mg steps to 50mg—40mg if old). SE: dystonia, Na+↓. Anorgasmia rare. Withdrawal signs: sweats, tremor.

Cautions Epilepsy, pregnancy/lactation, ECT. Toxic interactions: MAOI, sumatriptan, Li+, theophylline (fluvoxamine), haloperidol (fluoxetine), β-blockers (fluvoxamine), warfarin. Anticonvulsant antagonism.

1 F Song 1993 *BMJ* **306** 683   2 T Hale 1993 *BMJ* **306** 1124   3 M Robertson 1993 *BMJ* **306** 1125
4 S Milne 1993 *BMJ* **306** 1126   5 I Hindmarch 1992 *Int Clin Psychopharmacol* **6** suppl 4 65-7

Electroconvulsive therapy

Indications *Depression* if: ●Not responding to antidepressants, or ●Psychotic features are present, or ●Needing to be controlled fast (eg patient refusing to drink, high suicide risk, or post-partum): electroconvulsive therapy (ECT) is the fastest acting physical treatment.
Schizoaffective depression: (depression with signs of schizophrenia).
Mania: If not responding to drug treatment.

Contraindications Recent subdural or subarachnoid haemorrhage. Cautions: recent stroke, myocardial infarct, arrhythmias. SE: anterograde amnesia; anaesthesia complications. **Consent**, see p 400.

Technique ●Check the patient's identity. Give calm reassurance away from the site where electroconvulsive therapy (ECT) is going on. (It may be distressing to watch ECT actually happening.)
●Check that the patient has been nil by mouth for >8h.
●Are the consent forms in order (p 400)? In the rare instances in the UK where ECT is given without consent, a second opinion from the Mental Health Commission must state that the treatment is necessary (p 400).
●Ensure that the defibrillator and suction apparatus work satisfactorily.
●Ensure that the anaesthetist knows about allergies or drugs interfering with anaesthesia. For countering ECT-induced vagal stimulation, the anaesthetist is most likely to use atropine or methscopolamine before using an ultra-short-acting anaesthetic agent (eg methohexitone) with muscle relaxation (eg with suxamethonium) to minimize the seizure.
●Study the manufacturer's information on the ECT machine, the charge and energy to be given, and what wave-form is employed (bidirectional or modified sinusoidal, or unidirectional). Has the machine been checked recently? Is a reserve to hand?

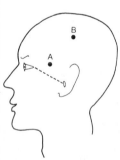

●Establish if unilateral (on side of the non-dominant hemisphere) or bilateral positioning of electrodes is to be used. Unilateral ECT is associated with less anterograde amnesia. If dominance is uncertain (eg in left-handed people) use bilateral positioning.
●In unilateral ECT, position one electrode (A) 4cm above the midpoint between the auditory meatus and the lateral angle of the eye, and the other (B) 10cm away from A, above the ear.
●In bilateral ECT, place electrodes over the same spot (A) on opposite sides ofe head.
●Put electrode jelly on the chosen sites (not so liberally as to allow electrode shorting).
●When the anaesthetist gives the word, give the shock. Be prepared to restrain the patient if paralysis is incomplete. While the current passes, the muscles will contract. This will cease as the current ceases. After ~10sec, further clonic spasms occur, lasting about 1min. The only sign may be lid fluttering. Clonus is probably needed for ECT to be fully effective.
●Put the patient in the coma position. Observe until conscious.

Note: ECT may be frightening for patients and relatives—partly because it has had a bad press; so explain, step-by-step, what will happen; and patients should not witness other patients having ECT. The author (JML) also finds it helpful to tell patients that he would opt for ECT, if the right indications pertained, on a psychiatrist's recommendation, because it can be so effective—indeed life-saving. ECT is not recommended for children.[1]

1 T Baker 1995 *Lancet* 345 65

Anxiety neurosis

Neurosis is the major psychological condition of our age—affecting at least one person in seven. It produces enormous suffering, and staggering losses to economies, in the UK estimated to reach £5.6 billion/yr (⅓ the cost of the NHS).[1] Neurosis refers to maladaptive psychological symptoms not due to organic causes or psychosis, and usually precipitated by stress. Apart from free-floating anxiety and depression, such symptoms are: fatigue (27%), insomnia (25%), irritability (22%), worry (20%), obsessions, compulsions, and somatization (p 744).[2] These are judged to be more intense than normal—ie out of proportion to the stress which precipitates them. Symptoms are not just part of a patient's normal personality, but they may be an exaggeration of personality: a generally anxious person may become even more so—ie develop an anxiety neurosis—as a result of job loss. The *type* of neurosis is defined by the chief symptom (eg anxiety, obsessional, depressive). Before diagnosing neurosis, consider carefully if there is underlying depression which needs treating with antidepressants.

Symptoms of anxiety:[3] Tension, agitation; feelings of impending doom, trembling; a sense of collapse; insomnia; poor concentration; 'goose flesh'; 'butterflies in the stomach'; hyperventilation (so tinnitus, tetany, tingling, chest pains); headaches; sweating; palpitations; poor appetite; nausea; 'lump in the throat'; unrelated to swallowing (globus hystericus); difficulty in getting to sleep; excessive concern about self and bodily functions; repetitive thoughts and activities (p 346). Children's symptoms: thumb-sucking; nail-biting; bed-wetting; food-fads; stammering.

Types ●Generalized anxiety disorder ●Panic disorder ●Simple phobia
●Post-traumatic stress disorder ●Agoraphobia ●Social phobia
●Obsessive-compulsive disorder

Causes ●Stress (eg work—or lack of it, noise, a hostile home).
●Life events (p 336, such as gaining a spouse or losing a job).
●Intrapsychic theories (eg anxiety is excess psychic energy and is a manifestation of repressed hostility or conflicting drives). Neurotic behaviour is seen as a way of reducing excessive energy, and, according to psychoanalytic theory, is more likely to occur if the person has failed to pass normally through the oral, anal, and genital stages of development.

Treatment *Symptom control:* Simple listening is a potent way of reducing anxiety. Explain that headaches are not due to a brain tumour, and that palpitations are harmless. Anything you can do to enrich the patient's relationship with others is likely to help. *Graded exposure* to the anxiety-provoking stimulus is a well-validated therapy (behaviour therapy, p 370).[4] *Anxiolytics* may be indicated to enable effective work to be done with the patient. *Dose example:* diazepam 5mg/8h PO for <6 weeks. The problems of benzodiazepine treatment (see p 366), severely limit their usefulness.

Progressive relaxation training: The patient is taught to tense and relax groups of muscles in an orderly way—eg starting from the toes and working up the body. By concentrating on this, anxiety and muscle tone are reduced. Deep breathing exercises bring about similar changes. The patient must practice often before benefit occurs. Commercial cassette tapes are available for repeated use, to help the learning process.[5]

Hypnosis: This is a powerful mechanism for reducing anxiety. Initially the therapist induces progressively deeper trances using such techniques as guided fantasy and concentration on various bodily sensations, such as breathing. Later, some patients will be able to induce their own trances.

1 J Holmes 1994 *BMJ* ii 1070 2 Off Pop Cens 1996 *B J Psych* 271 70-1 3 M Lader 1994 *BMJ* ii 321
4 I Marks 1994 *BMJ* ii 1071 5 Adelph One, High St, Bottisham, Cambridge, UK (tel. 01223 811679)

Other neurotic disorders

Anxiety neurosis has 4 subgroups: ●Panic disorder (episodes of extreme anxiety and impending calamity) ●Generalized anxiety disorder ●Mixed anxiety and depressive disorder ●Depressive disorder.

Phobic disorders These involve symptoms of anxiety occurring in specific situations only, and leading to their avoidance. These are labelled according to specific circumstance: agoraphobia (*agora*, Greek for market place) is fear of crowds, travel, or situations away from home; social phobias (where he might be minutely observed, eg small dinner parties); simple phobias, eg to dentists, intercourse, Friday the 13th (triskaidecophobia), spiders (arachnophobia), beetles (paint them red, and put black spots on and they become charming ladybirds). There may also be free-floating 'fear of fear', or fear of disgracing oneself by uncontrollable screaming.

It is important to find out exactly what the phobic stimulus is. It may be very specific, eg travelling by car, not bicycle. Why are some situations avoided? If because the patient is deluded that he is being followed or persecuted, paranoia rather than phobia is suggested. Panic attacks are lessened by paroxetine (p 341) ± cognitive therapy (p 370). ◄1►

Obsessive-compulsive disorder Compulsions are senseless, repeated rituals. Obsessions are stereotyped, purposeless words or phrases which come into the mind. They are perceived by the patient as nonsensical (unlike delusional beliefs), and, although out of character, as originating from themselves (unlike hallucinations or thought insertion). They are often resisted by the patient, but if longstanding, the patient may have given up resisting them. An example of non-verbal compulsive behaviour is the rambler who can never do a long walk because every few paces he wonders if he has really locked the car, and has to return repeatedly to ensure that this has, in fact, been done. Cleaning (eg hand-washing), counting, and dressing rituals are other examples. *Pathophysiology:* Positron emission tomography (OHCM p 794) reveals increased blood flow to the orbitofrontal cortex, and reduced blood flow to the caudate nucleus. Successful treatment is reflected by some normalization of metabolism in these areas.[2] *Treatment:* Behaviour therapy (p 370; see also *Neurosis*, p 344). Clomipramine (start with 25mg/day PO) and SSRIs (eg fluoxetine, start with 20mg per day PO) really can help (even if the patient is not depressed).[3] See p 340.

Depersonalization This means an unpleasant state of disturbed perception in which objects (eg parts of the body) are experienced as being changed ('as if made of cotton wool'), becoming unreal, remote, or automatized. There is insight into its subjective nature, and it is not a feature of psychosis. Depersonalization may be primary, or part of another neurosis.

Hysteria See p 334.

Treatment: Behaviour therapy (p 370) is often suitable, provided the patient really wants to change. Antidepressants (p 340) may also help, if combined with behavioural therapy.

◄1► S Oehrberg 1996 *E&M* i 81   **2** R Ramsay 1995 *BMJ* ii 167   **3** M Piccinelli 1995 *Br J Psych* 1995 **166** 424

Stress and post-traumatic stress disorder[1]

Presentation Symptoms of too much stress: intrusive thoughts, dyspepsia, chest pain, fainting, palpitations, transitory (but probably not sustained) hypertension, hyperventilation (OHCM p 48), abdominal distension, nausea, irritability, insomnia, over- or under-eating, depression, poor concentration, poor decision-making, and pre-occupation with trivia.

There are many biological consequences of stress, eg depression; increased weight, BP, and alcohol intake; peptic ulcers; irritable bowel syndrome; delayed wound healing (interleukin-1β mRNA↓)▣; and migraine.

Relieving stress Smoking, alcohol and chattering are the most popular methods. If drugs must be given, and there is no asthma, heart failure or heart block, propranolol (dose example: 10mg/8h PO) may be appropriate to reduce autonomic symptoms—but side effects are common. Safer alternatives which some people try include exercise, singing, progressive relaxation (p 344), and counselling (p 376).

Post-traumatic stress disorder

After great psychological trauma (eg near-death, shell-shock, etc) there may be severe anxiety, depression, alcohol abuse, and irritability. In children, there may be bed wetting and school refusal. Adult symptoms may take a long time to come to light. For example, in the Lockerbie community in the UK over which a jumbo jet disintegrated, men were still presenting (for the first time) with stress-related symptoms related to the event more than 2 years after the crash.[2]

Treatment may focus on psychotherapeutic reliving of the emotional trauma, with skilled interpretation by the therapist. However, it must be stressed that there are no hard and fast rules which guarantee success. There are examples (eg in war veterans) in which the 'talking cures' of psychotherapy may cause more harm than good: here the best advice may be to try to forget the experience.[3]

1 D Wilkinson 1991 *BMJ* ii 191   2 L Dillner 1992 *BMJ* i 1073   3 C Krasucki 1995 *Lancet* 345 1240

Anorexia and bulimia nervosa[1]

Anorexia epidemiology: Incidence: 4–16/10⁶/yr. ♀/♂ ≈ 20. Prevalence: 1–2% of schoolgirls (prevalence is higher in upper socio-economic classes). It is the third most common chronic illness in teenage girls.

Definition of anorexia: There are 4 diagnostic criteria:
1 The person chooses not to eat, leading to potentially dangerous weight loss interfering with normal functioning (eg BMI <17, or, more strictly, a weight less than 85% of weight expected, taking into account height, sex, and population to which the patient belongs—see p 473).
2 Intense fear of becoming obese, even when underweight.
3 Disturbance of weight perception (ie feeling fat when thin).
4 Amenorrhoea: ≥3 consecutive cycles absent (when not on the Pill).

Symptoms: These usually start at ~16–17yrs (12 in boys) and may be precipitated by dieting. She attaches excessive importance to weight reduction (an over-valued idea), and she may become deluded that she is repulsively fat when she is, in fact, very malnourished. Excessive exercise, and induced vomiting may occur; there may be abuse of laxatives, diuretics and appetite suppressants. The patient sees her self-worth as being embodied in her shape and weight. Many patients may also have episodes of binge eating, followed by remorse, vomiting, and concealment. Depression and suicide attempts commonly coexist.

Physical complications of weight reduction: Sensitivity to cold, constipation, amenorrhoea, faints, weakness, fatigue, BP↓, K⁺↓, glucose↓.

Causes: Single factor theories are not favoured. 10% of ♀ siblings may be affected, but this is more likely to show the importance of family behaviour in maintaining symptoms than any direct genetic effect. Rarely, hypothalamic tumours turn up in people originally diagnosed as having anorexia (amenorrhoea *before* weight loss is suspicious). Some workers see the relentless pursuit of thinness as a struggle for control and to gain an identity. Dietary problems in early life, parental preoccupation with food, and family relationships which leave the person without a sense of identity may also be important. There is little foundation for the view that the chief problem is psychosexual immaturity.

Treatment: What is the urgency? If extreme weakness, fainting, K⁺↓ or glucose ↓, then prompt admission to hospital, possibly under the Mental Health Act, may be needed for feeding (IV if needed). In other circumstances work on building up a trusting relationship with the patient. Work on several fronts simultaneously: ●Gaining of weight (agreed targets; aim for weight gain of 1.5kg/week, and a final body mass index of 20–25—special foods are not required).[2] ●Treat the medical complications of starvation ●Helping with relationships (eg family therapy) ●Cognitive-behavioural therapy (p 370). ◁3▷ On admission privileges are removed, and restored as weight rises. *Prognosis:* 2% of anorectic patients die (by starvation), and 16% remain seriously underweight after 4–8yr. Cardiac complications are the commonest cause of death (K⁺↓ and prolonged Q–T interval predispose to arrhythmia).

Bulimia may affect ~10% of young women. Diagnostic criteria are:
●Recurrent episodes of binge eating ●During binging a lack of feeling of
control ●Regular use of mechanisms to overcome the fattening effects of
binges (eg vomit-induction, laxatives, excessive exercise) ●Persistent over-
concern with body weight ●Body weight higher than required for the
diagnosis of anorexia (anorexia trumps the diagnosis of bulimia).

Physical complications: Stomach rupture, haematemesis and metabolic complications following excessive (eg self-induced) vomiting. There may be
painless enlargement of the salivary glands, tetany and seizures. Russell's
sign: calluses form on the back of the hand, following its repeated abrasion against incisors during inducement of vomiting.[1]

Management: Consider behavioural or cognitive techniques (p 368–82). For
example, the patient may agree to limit eating to one room in the house,
and to eat only at meal times. She may agree not to have food in the
house, and when shopping, she may agree to buy only those items on a
shopping list which she has made when satisfied after a meal. It may help
to shop with a friend, and with only sufficient money to buy the items on
the list. *Drugs:* These are rarely used, but a small randomized trial shows
that fluoxetine (p 341) aids weight reduction, and reduces craving for
carbohydrates.[4]

1 M Gelder 1989 *Oxford Textbook of Psychiatry*, 2nd Ed, OUP. & *Drug Ther Bul* 1992 **13**   **2** P Beumont 1993 *Lancet* **341** 1635   ◁3▷ CG Fairburn 1995 *Arc Gen Psy* **52** 304 **4** J Edwards 1992 *BMJ* **i** 1644

Organic reactions

Acute organic reactions (Acute confusional state, delirium) The central feature is impaired conscious level with onset over hours or days. This is difficult to describe; take any opportunity to be shown it. You have the sense when trying to communicate that he is not really with you. He is likely to be disoriented in time (does not know day or year) and, with greater impairment, in place. Sometimes he is unusually quiet, and is found to be drowsy; sometimes he appears agitated, and you are called because he is disrupting the ward. On other occasions the patient appears deluded (for example accusing staff of plotting against him/her) or to be hallucinating. If there is no past psychiatric history, and in the setting of a general hospital (ie the patient is physically ill or has had recent surgery), mental illness is rare and acute confusional state is common. A confusional state is particularly likely if the symptoms are worse at the end of the day.

Differential diagnosis: If agitated, consider anxiety (usually readily distinguished on history-taking). If onset uncertain consider dementia.

Causes: Infection; infarction; drugs (benzodiazepines, opiates, anticonvulsants, digoxin, L-dopa); U&E imbalance; hypoglycaemia; hypoxia; epilepsy; alcohol withdrawal; trauma. For a full list, see p 351.

Investigations: U&E, FBC, blood gases, blood glucose, appropriate cultures, LFT, ECG, CXR. Consider skull X-ray, LP and CT scan.

Management:
1 Find the cause. Examine the patient with above causes in mind; do necessary tests. Start relevant treatment—eg O₂.
2 If agitated and disruptive, sedation may be needed before examination and tests are possible. Use major tranquillizers (eg haloperidol 5–10mg IM/PO; or, chlorpromazine 50–100mg IM/PO). Wait 20mins to judge effect. Give further dose as necessary. *Note:* in alcohol withdrawal use chlormethiazole (OHCM p 682).
3 Nurse ideally in *moderately lit* quiet room with same staff in attendance (to minimize confusion). Repeated reassurance and orientation. In practice a compromise between a quiet room and a place where staff can keep under close surveillance has to be made.

Chronic organic reactions (dementia) *Prevalence:* ~6% of those >65yrs.

Cardinal features: Worsening memory and global intellectual deterioration without impairment of consciousness. Get a good history from friends/relatives. Exclude depression (may need a drug trial).
- Behaviour: restless; no initiative; repetitive, purposeless; sexual disinhibition; social gaffes; shoplifting; rigid routines.
- Speech: syntax errors; aphasia; mutism.
- Thinking: slow, muddled; delusions. Poor memory. No insight.
- Perception: illusions, hallucinations (often visual).
- Mood: irritability, depression (early); emotional blunting with sudden mood changes and emotional incontinence (later).

Identifying treatable causes: ● Haematology: FBC, B₁₂, red cell folate (macrocytosis suggests alcoholism, or B₁₂/folate↓). ESR (malignancy).
- Biochemistry: U&E, LFT, GGT, Ca²⁺ (renal/hepatic failure, alcoholism, malignancy, endocrine causes leading to Ca²⁺↓). T₄ (hypothyroidism).
- Serology: for syphilis (OHCM p 218). After counselling carers, test for HIV if at risk.
- Radiology: CXR (evidence of malignancy). CT scan (hydrocephalus, tumours, subdural haematoma, CNS cysterticosis).

Management: If treatable causes have been excluded, explain this to the relatives, who may feel unable to look after the immobile, incontinent, aggressive patient. A walking frame, an indwelling catheter, promazine syrup (25–50mg/8h PO), day care, holiday admission, and an attendance allowance can transform this picture. If not, long-stay institutional care may be needed. Consider arranging power of attorney.

Causes of organic reactions:

	Acute organic reactions:	Chronic organic reactions:
Degenerative		*Senile dementia, *Alzheimer's Jakob-Creutzfeldt, Pick's or Huntington's diseases
Other CNS	Cerebral tumour or abscess, subdural haematoma, epilepsy, acute post-trauma psychosis	Brain tumour, subdural haematoma, multiple sclerosis, Parkinson's, normal pressure hydrocephalus
*Infective	Many eg meningoencephalitis, septicaemia, cerebral malaria, trypanosomiasis	Late syphilis, chronic or subacute encephalitis, CNS cysticercosis, AIDS
Vascular	Stroke (or TIA), hypertensive encephalopathy, SLE	Thromboembolic multi-infarct (arteriosclerotic) dementia, anaemia
Metabolic	*U&E imbalance, *hypoxia, *liver and kidney failure, non-metastatic cancer, porphyria, *alcohol withdrawal	Liver and kidney failure non-metastatic or metastatic cancer
Endocrine	Hyperthyroid or Addisonian crisis, diabetic pre-coma, hypoglycaemia, hypo/hyperparathyroidism	T4↓, Addison's, hypoglycaemia hypopituitarism, hypo-/hyperparathyroidism
Toxic	*Alcohol, many drugs (check *Data Sheet Compendium*), lead, arsenic, mercury	*'Alcoholic dementia', barbiturate or bromide abuse, too much manganese or carbon disulphide
Deficiency	Thiamine↓ B₁₂↓ folate↓ nicotinic acid↓	Thiamine↓ B₁₂↓ folate↓ nicotinic acid↓ (pallegra)

*=leading causes

Help for relatives of demented patients

- Alzheimer's disease is usually progressive, with mental functioning getting steadily worse; some problems such as aggression *may* improve over time. Both the rate of change and the length of life vary greatly.
- Take every opportunity you can to talk about your predicament with other people in the same position. This is often just as useful as talking to professionals. The Alzheimer's Disease Society exists to put you in touch (UK phone number: 0181 675 6557).
- Accept any offer of day-care for your relative. It will give you much-needed respite from the thankless task of looking after him or her.
- Lock up any rooms in the house which you do not use. Your relative will not notice this restriction—and this may make your life much easier.
- Lock any drawers which contain important papers or easily-spoiled items. This will prevent the patient storing inappropriate things in them, such as compost from the garden, or worse.
- Remove locks from the lavatory—so he/she cannot get locked in.
- Normal sexual relationships will probably stop. Spouses should try not to fall into the trap of asking 'What's the matter with me?'. If sexual behaviour is inappropriate (hypersexuality) drugs are available.**
- Prepare yourself psychologically for the day when he/she no longer recognizes you. This can be a great blow, unless you prepare for it.
- In the UK, apply to Social Services for an Attendance Allowance.

**Consider medroxyprogesterone acetate 100-200mg/2 weeks IM (M Weiner 1992 *Lancet* i 1121)

Alzheimer's dementia (AD)

This is the major neuropsychiatric disorder of our times, dominating not just psychogeriatric wards, but the lives of many thousand sons, daughters and spouses who have given up work, friends, and all their accustomed ways of life to support relatives through the last long years. The struggle of caring for loved ones through terminal illness always puts us on our metal: never more so than when that loved one's personality disintegrates, and the person who is loved is gone long before their eventual death.

Suspect Alzheimer's disease in adults with any enduring, acquired deficit of memory and cognition—eg as revealed in the mental test score and other neuropsychometric tests (p 353). Onset may be from 40yrs (or earlier, eg in Down's syndrome)—so the notions of 'senile' and 'pre-senile' dementia are blurred (and irrelevant). $\female/\male \approx 0.7$.

Genetics This is complex (defective genes on chromosomes 1, 14, 19, 21; the apoE4 variant is a major risk-factor, also bringing forward age of onset).

Histology (Needed for definite diagnosis—but rarely employed.)
- Deposition of β-amyloid protein in cortex (a few patients have mutations in the amyloid precursor protein).
- Neurofibrillary tangles and an increased number of senile plaques.
- A deficit of neurotransmitter acetylcholine from damage to an ascending forebrain projection (nucleus basalis of Meynert; connects with cortex)—amenable in theory to tacrine therapy. The slow, steady decline in AD distinguishes it from the step-wise deterioration of multi-infarct dementia.

Presentation In stage I of AD there is failing memory and spatial disorientation. In stage II (which follows after some years) personality disintegrates (eg with increased aggression and focal parietal signs, eg: dysphasia, apraxia, agnosia and acalculia). Parkinsonism may occur. She may use her mouth to examine objects (hyperorality). In stage III she is apathetic (or ceaselessly active–akathisia), wasted, bedridden and incontinent. Seizures and spasticity are common.

Mean survival 7yrs from onset.

Management *Theoretical issues:* Potential strategies—the MRC menu:
- Preventing the breakdown of acetylcholine, eg tacrine, below.
- Augmenting nerve growth factor (NGF), which is taken up at nerve endings, and promotes nerve cell repair.
- Stimulating nicotinic receptors which may protect nerve cells.
- Inhibition of the enzymes that snip out β-amyloid peptide from APP, (amyloid precursor protein), so preventing fibrils and plaques.
- Anti-inflammatories to prevent activation of microglial cells to secrete neurotoxins (eg glutamate, cytokines) which stimulate formation of APP.
- Regulation of calcium entry (mediates the damage of neurotoxins).
- Preventing oxidative damage by free radicals.

Practical issues: ▶See *Help for relatives of the demented*, p351. Exclude treatable dementias (B₁₂, folate, syphilis serology, T4, ?HIV).
Treat concurrent illnesses (they contribute significantly to confusion). In most people the dementia remains and will progress. The approach to management is that of any chronic illness (OHCM p 66). Involve relatives and relevant agencies. Alzheimer's Disease Society (OHCM p 804).

Liaise with a psychogeriatrician, ask if there is a rôle of tacrine (licensed in the USA—but a licence has been refused in UK) to increase acetylcholine by inhibiting the enzyme for its breakdown. It is often disappointing in practice.[1] It has a serious interaction with suxamethonium.

1 M Rossor 1993 *BMJ* ii 779

Mental test score

This is a quantifiable, standardized way of measuring someone's mental functioning. It is most useful for serial measurements. Maximum score 34; normal ≥29; a change of ≥5 is meaningful.

Knows his name	score 0 or 1	Recognizes 2 people	score 0, 1 or 2
Knows his age	score 0 or 1	Birthday	0 or 1
Time, to nearest h	score 0 or 1	Town of birth	0 or 1
Time of day	score 0 or 1	School	0 or 1
Teach name and		Former occupation	0 or 1
address: test recall after 5min:-		Name wife, sibling or	
Mr John Brown	score 0, 1 or 2	next of kin	0 or 1
42 West Street	score 0, 1 or 2	Monarch's name	0 or 1
Gateshead	score 0, 1 or 2	Prime Minister's	
Day of the week	score 0 or 1	name	0 or 1
Date of month	score 0 or 1	Give years of:	
Current year	score 0 or 1	World War I	0 or 1
		World War II	0 or 1
Place: eg hospital	score 0 or 1	Months of year	
name of hospital	score 0 or 1	backwards	0, 1 or 2
name of ward	score 0 or 1	Count 1-20	0, 1 or 2
name of town	score 0 or 1	Count 20-1	0, 1 or 2

Psychosis (including mania)

►Beware labelling people: and remember that even during the best of times, there is only a thin veil separating us from insanity. In its most florid form, psychosis is the archetype of the layman's 'madness', and it signifies a state of mind in which contact with reality is lost and all the landmarks of our normal mental processes have been suspended and turned awry, with an abandoning of constraints imposed by reason and morality. However, the *usual* picture is much less obvious: the patient may be sitting alone, quietly attending to his or her voices.

Key features suggesting psychosis are: hallucinations, delusions, and thought disorder—defined on p 316 & 358. If one of these features is present, the diagnosis is limited to 4 entities: schizophrenia (p 358), a disorder of affect (ie mania or depression, p 336), a paranoid state, or an organic, ie physical, disease. So the term psychosis is not in itself a diagnosis, but is a useful term to employ, while the underlying diagnosis is being formulated.

Mania and hypomania Humans normally adapt their pace according to the obstacles encountered—but in mania we march round in circles, like a tornado picking up internal speed as we go, destroying everything beneath the conquering tread of an unfettered will, while exhibiting these signs:

- ●Euphoria ●Hyperactivity   ●Hallucinations   ●Over-assertiveness
- ●Appetite↑ ●Pressure of speech   ●Disinhibition   ●Grandiose delusions
- ●Insight↓ ●Sexual desire↑   ●Spending sprees   ●Self-important ideas

Less severe states are termed *hypomania*. If depression alternates with manic features, the term *manic depressive psychosis* is used—particularly if there is a history of this in the close family. Cyclical mood swings without the more florid features (listed above) is termed *cyclothymia*.

Treatment and prophylaxis: Sedation may be required, eg chlorpromazine 25-150mg/4-8h PO, droperidol 5-15mg/4h IV, or 1 dose of zuclopenthixol acetate, 50-100mg IM (into a buttock; it lasts 2-3 days).[1] If compliance is good, and U&E, ECG, and T4 normal, give lithium carbonate 125mg-1g/12h PO as follows. (Note: intracellular levels take time to build up.)

- ●Adjust the dose to achieve a plasma level of 0.8-1.0mmol/l Li+ as measured on the 4th or 7th day of treatment, 12h (± ½h) after the preceding dose. Note: a lower range—0.4-0.6mmol/l—was once used but this is now known to be associated with a 2-3-fold rise in relapse rates,[1] and the level to recommend is controversial. Discuss with colleagues and your lab.
- ●Check Li+ levels weekly until the dose has been constant for 4 weeks; then monthly for 6 months, and then every 3 months, if stable. Do more often if on diuretic, low salt diet or pregnant (avoid Li+ if possible).
- ●If levels are progressively rising, suspect progressive nephrotoxicity.
- ●Do plasma creatinine and TSH every 6 months; Li+ irreversibly affects thyroid and kidney (hypothyroidism; nephrogenic diabetes insipidus).
- ●Avoid changing proprietary, brands as Li+ bioavailability varies.
- ●Make sure you know how to contact the patient by phone if Li+ levels are ↑ (>1.4mmol/l prompts urgent contact with patient to adjust dose).
- ●Toxic effects: blurred vision, D&V, K+↓, drowsiness, ataxia, coarse tremor, dysarthria, hyperextension, seizures, psychosis, coma, and shock.

Refractory mania: consider carbamazepine (OHCM p 450), valproate, ECT.[2]

Organic psychoses These follow some physical stimulus: narcotics; amphetmines; cocaine; lysergic acid; alcohol; glue or solvent abuse; CNS trauma or tumour; epilepsy; puerperal psychosis (see p 148, the stimulus is birth); associated stress, life events, and tiredness are also important.

Paranoid states This implies paranoid symptoms (delusions or hallucinations) which concern a patient's relationships with others, and in whom using one of the other 4 categories of psychosis (above) is unjustified. A common form is a persecutory delusion ('there is a conspiracy against me'). Morbid jealousy (the Othello syndrome—p 754) is a rarer example.

1 G Goodwin 1994 *Prescrib J* 34 19   2 N Coxhead 1992 *Acta Psych Scand* 85 114-8

Schizophrenia: management

Aims These are to treat the patient and the illness as well as possible—and to provide support and advice to the relatives. *Is hospital admission needed?* Yes, if the patient is a danger to himself or others, or hospital is the only place where you can be sure the patient will get the treatment he needs. Treat acute symptoms with major tranquillizers (neuroleptics, eg chlorpromazine 25mg/8h PO, up to 1g/day (500mg if elderly); child: 0.5mg/kg/4–6h PO, if >6yrs ⅓–½ adult dose, max 75mg daily). These usually reduce hallucinations and delusions within 3 weeks (but may not make the patient *feel* better), and have their full effect by 3 months. Longer-term management includes helping the patient engage in as normal a life as possible, in a planned way. Aim to reduce relapse risk by continuing neuroleptics, and trying to ensure a stress-free environment. Relatives need help to be neither critical nor over-protective. Cognitive therapy (p 370) reduces psychotic symptoms. Sessions are devoted to systematically reviewing the start of symptoms. Patients are questioned in detail to reveal faulty reasoning, and then encouraged to monitor psychotic symptoms, and develop coping strategies, eg: ●Changing the focus of attention ●Relaxation techniques ●Modifying exacerbating behaviours. The use of these have been confirmed in trials,[1] and note how their effects may be more humanizing than the ready resort to the 'liquid cosh' (below).

The acutely psychotic patient who is violent (a rare event):
- ●Recognize early warning signs: tachypnoea, clenched fists, shouting, chanting, restlessness, repetitive movements, pacing, gesticulations. Your own intuition may be helpful here. At the 1st hint of violence, get help. If alone, make sure you are nearer the door than the patient.
- ●Do not be alone with the patient; summon the police if needed.
- ●Try calming and talking with the patient. Do not touch him. Use your body language to reassure (sitting back, open palms, attentive).
- ●Get his consent. If he does not consent to treatment, emergency treatment can still be given to save life, or serious worsening of the patient's condition—on the verbal authority of any doctor.
- ●Use minimum force to achieve his welfare (but may entail 6 strong men).
- ●If one of the following IM drugs is needed, monitor BP every 4h.
- ●Later-on liaise with colleagues about long-term care. Will the outside world be at risk if community care is used? If so (or if the person is at danger from suicide or self-neglect) the UK government requires entry of the patient into a *supervision register*, with a key worker (eg a community psychiatric nurse) trained in risk-management and assessment, who can co-ordinate multidisciplinary review meetings to monitor progress. These requirements may be counter-productive,[2] if no extra funds are available and the key worker's chief aim is to keep people out of hospital.

Usual adult ORAL doses	**Max licensed IM dose; halve in elderly**
Group 1 phenothiazines:Sed	
●Chlorpromazine 25–100mg/8h	50mg IM
Group 2 phenothiazines:Ach	
●Thioridazine 25–100mg/6h	No injection available
Group 3 phenothiazines:Ex	
●Trifluoperazine 5–10mg/12h	1–2mg/8h IM (child: 50μg/kg/day)
The butyrophones:Ex	
●Droperidol 5–20mg/4–8h	5–10mg/4h IM/IV (child: 0.5–1mg/day)
●Haloperidol 0.5–5mg/12–8h	2–30mg IM, then 5mg/1h, max total: 60mg
(rarely adults need 120mg/day and adolescents 30–60mg/day)	▶Move to PO route as soon as you can
Initial child's dose is 25–50μg/kg/day.	

NB: doses are difficult, some patients require more than the licensed dose;◻ if so, liaise with a colleague, and follow this advice from the Royal

College of Psychiatrists: ●Be cautious in those over 70yrs ●Do an ECG; beware a long Q–T interval ●Increase doses slowly (less than once weekly), and stop increased dose if no benefit after 3 months ●Review often ●Consider alternatives eg clozapine (below).

Sed = sedative side-effects predominate. *Ach* = anticholinergic side-effects predominate (blurred vision, retention of urine, dry mouth, BP low, precipitation of glaucoma). *Ex* = Extrapyramidal side-effects predominate: ●Akathisia (restlessness) and parkinsonism (use orphenadrine 50 mg/8h PO or benzhexol 1–4mg/6h PO—only if needed). ●Dystonic reactions (eg torticollis, opisthotonus, often occurring in the first weeks of treatment, and responding to biperiden 2–5mg IM or IV slowly, up to 4 times daily. ●Tardive dyskinesia (chewing, grimaces, choreoathetosis)—may be irreversible, but try tetrabenazine 50–200mg/day PO (usually given only after waiting to see what happens for ~6 months after stopping the drug). Neuroleptic poisoning: OHCM p 749.

Other SEs: BP↓, ejaculation delay, depression, fits, jaundice, corneal opacities, rash, WCC↓, and the rare (but fatal) neuroleptic malignant syndrome (hyperthermia, rigidity, dystonia, consciousness disturbance).

Maintenance treatment with drugs Up to ⅔ will relapse if antipsychotic medication is discontinued within the 1st 5 years of treatment.[2] As there is no way of predicting who needs prophylaxis, and because a psychotic relapse is likely to be devastating, the best plan is to offer prophylaxis to everyone, in the lowest effective dose—and by depot injections if compliance is a problem. Low-dose fluphenazine decanoate (5mg IM every 2 weeks) is probably as effective as 25mg/2 weeks IM, and causes less anxiety and depression.[3] On either dosage, ~40% of patients may be expected to show evidence of a mild relapse over the first year. At the first sign of relapse, double the dose.[3] Continuous treatment with oral chlorpromazine[2] may also be effective. Aim to keep the dose less than 100mg/day.

Treatment failures: 30–50% may benefit greatly from clozapine. SE: agranulocytosis in 0.5% (fatal in ~10% of these despite the mandatory WCC monitoring service); fits (in 14% if dose ≥600mg/day PO); myocarditis; sedation—but *not* tardive dyskinesia: its antagonism of striatal D_2 receptors is weak compared with S_2 and S_{1C} antagonism (OHCM p 407).[4,5] Risperidone (blocks D_2, 5-HT$_2$, α_1 and α_2 adrenoceptors) is a novel agent which may improve both positive and negative symptoms.[6]

1 R Ramsay 1995 *BMJ* ii 167 2 IM Marks 1994 Brit J Psychiatry 165 179–94
3 S Hirsch 1986 *BMJ* ii 515 4 S Hirsch 1993 *BMJ* i 1427
5 S Dursun 1993 *BMJ* ii 200 6 JG Edwards 1994 *BMJ* i 1311

Drug problems and addiction (▶See OHCM p 184)

Definitions: *Tolerance:* a drug's early effects are later achievable only by using higher doses. *Withdrawal syndrome:* the physical effects experienced when a tolerance-inducing drug is withdrawn. *Dependence:* only continued doses prevent physical or psychological withdrawal (compulsive activity to achieve this is called *drug addiction*). The drugs involved are the opiate derivatives, amphetamines, cocaine, lysergic acid, hydrocarbons ('glue sniffing'), barbiturates and, to a lesser extent, cannabis. Addiction may be psychological (restlessness and craving when the drug is withdrawn) or physical, such as the 'cold turkey' which occurs in narcotic withdrawal—dilated pupils, D&V, tachycardia, sweating, cramps, and pilo-erection. This may be ameliorated by giving methadone, eg 20–70mg/12h PO, reducing by 20% every 2 days), ideally as part of a régime in which a contract is made with the patient (p 338). Narcotic abusers must, by UK law, be notified to the Home Office. Barbiturate withdrawal may cause seizures and death, and withdrawal should be as an in-patient, giving ⅓ of the previous daily dose as phenobarbitone. Lower the dose over 14 days.

When to suspect drug addiction ●Convictions for crime—to buy drugs.
●Any odd behaviour, eg with visual hallucinations, elation or mania.
●Unexplained nasal discharge (cocaine sniffing).
●The results of injections: marked veins; abscesses; hepatitis; AIDS.
●Repeated requests for pain killers—with only opiates being acceptable.

Counselling parents about how they can counsel their children
●If possible, father and mother should agree a plan together, along with someone who knows the child (eg teacher or GP).
●Accept the child; and try not to seem authoritarian or bossy.
●Find out what the child's attitudes are to drugs. He may think that hard drugs are no more harmful than alcohol and tobacco.
●Bring the subject up with the younger siblings so that they know how to say no to offers of drugs.
●Confidential phone advice is always available (in the UK dial 100 and ask for Freefone Drug Problems).

The dangers of HIV infection (*OHCM* p 212) if drugs are injected has altered the approach to long-term management. It is important to try to maintain a relationship between the health team and the addict, and to educate the patient about avoiding high-risk behaviour. There is controversy over the extent to which addicts should be supplied with clean needles and controlled doses of addictive drugs. See *OHCM* p 184–5.

Alcohol-related problems

Repeated drinking harms a person's work or social life. Addiction implies:[1]

Difficulty or failure of abstinence	Often aware of compulsion to drink
Narrowing of drinking repertoire	Priority is to maintain alcohol intake
Increased tolerance to alcohol	Sweats, nausea, or tremor on withdrawal

Questions ●Can you control your drinking? ●Have you thought you should cut down? ●Do friends comment on your drinking, or ask you to cut down? ●Has alcohol led you to neglect family or work? ●What time do you start drinking? Do you drink before this? ●Go through an average day's drinking. ●Have you been in trouble with the law (violent crime)?

Alcohol and organ damage ●Liver: (↔ in 50% of alcoholics). Fatty liver: acute, reversible; hepatitis—80% progress to cirrhosis (hepatic failure in 10%); cirrhosis: 5yr survival 48% if alcohol intake continues (if it stops, 77%). ●CNS: poor memory/cognition→may respond to multiple high-potency vitamins IM; cortical/cerebellar atrophy; retrobulbar neuropathy; fits; falls; neuropathy; Korsakoff's/Wernicke's encephalopathy (OHCM p 712). ●Gut: D&V; peptic ulcer; erosions; varices; pancreatitis. ●Marrow: Hb↓; haemolysis; MCV↑. ●Heart: arrhythmias; cardiomyopathy.

Withdrawal signs (Delirium tremens) Pulse↑; BP↓; tremor; fits; visual or tactile hallucinations, eg of animals crawling all over one. *Treatment:* ●Admit; monitor vital signs (beware BP↓). ●If vomiting, set up chlormethiazole IVI (OHCM p 442). Chlormethiazole capsules: (each is 192mg chlormethiazole base, an addictive drug). Give 3 capsules/6h PO for 2 days, then 2 capsules/6h for the next 3 days, then 1 capsule/6h for a final 4 days. Diazepam IVI is an alternative if chlormethiazole-allergic.

Treatment (Only feasible if the patient really wants to change.) Group psychotherapy in self-help organizations such as 'Alcoholics Anonymous' may be appropriate, ± agents which produce an unpleasant reaction if alcohol is taken (eg disulfiram 200mg/24h PO). Reducing the pleasure that alcohol brings (and the craving when it is withdrawn) with naltrexone 25-50mg/24h PO (an opioid receptor antagonist sometimes used in opiate addictions)▣ can halve relapse rates.[2] SE: vomiting, drowsiness, dizziness, cramps, joint and muscle pain. CI: hepatitis; liver failure. *Cost:* ~£3.20/day.

Prevention See p 454.

1 RE Meyer 1996 *Lancet* 347 162 2 JP Volpicelli 1995 *Lancet* 346 456

Personality disorders and psychopathy

Personality comprises lasting characteristics which make us the sort of people we are: easy-going or anxious; placid or histrionic; ambitious or stay-at-home; fearless or timid; self-satisfied or doubting; optimistic or pessimistic. (Personality *can* change and develop, and it may even change quickly, eg after religious conversion in which a timid man is remoulded into a fearless activist.) Personality is a spectrum lying between these opposites. Statistical analysis reveals that all these distinctions may in fact overlap, and are describable in terms of a few orthogonal dimensions (eg neuroticism/psychoticism; introvert/extrovert). Those with abnormal personalities may then be defined as occupying the extremes of the spectrum. Abnormal personality only matters if it is maladaptive, causing suffering either to its possessor or his associates. In general, psychological symptoms which are part of a personality disorder are harder to treat than those arising from other causes.

Psychopathy refers to a persistent personality disorder characterized by antisocial behaviour, inability to make loving relationships, and lack of guilt. This is a difficult definition to sustain in any detached way, especially in the light of our comment on p 356—that those whom society wishes to forget, it first makes mad—or bad. Using synonyms such as 'sociopathy' does not get round this difficulty. The patient is impulsive, and regards his closest associates without affection. He does not know how to like or love them. Rather than leading to shame and guilt, his irresponsible actions are more likely to lead to prison. Reason is intact (but is used for selfish ends). Note that psychopathy has forensic overtones. If a criminal act is seen as being part of a psychopathy then the Courts have the option of specifying psychiatric treatment (in a secure hospital if necessary) rather than just prison, so raising difficult questions about psychopathy and responsibility.

Causes: Brain damage, social factors, parental psychiatric illness or laxity have been suggested, but none is pre-eminent.

Treatment is problematic because patients seldom want to change. Peer group pressure (as may occur in group therapy, p 372) may be a motivating force. It is rarely wise to use drugs.

Other personalities *Obsessional personality:* The rigid, obstinate bigot who is preoccupied with unimportant (or vital) details.

Histrionic personality: The self-centred, sexually provocative (but frigid) person who enjoys (but does not feel) angry scenes.

Schizoid personality: the cold, aloof, introspective misanthrope.

Mental handicap and learning disabilities

Those with IQs of 50–70 account for 80% of people with learning disabilities. There is useful development of language, and learning difficulty only emerges as schooling gets under way. Most can lead an independent life. If IQ 35–49, then most can talk and find their way about. In severe learning disability (IQ 20–34) limited social activity is possible. If less than this, very simple speech may be achieved. Special schooling and medical services are needed, as are adequate care and counselling for the families involved. Further information is available from MENCAP.[1] For the causes of retardation see p 204.

---

1 123 Golden Lane, London EC1Y 0RT, UK, tel. 0171 253 9433

Withdrawal of psychotropic drugs

Withdrawing benzodiazepines ►*The withdrawal syndrome may well be worse than the condition for which the drug was originally prescribed.* So try to avoid their use, eg relaxation techniques for anxiety, or, for insomnia, a dull book, sexual intercourse and avoiding night-time coffee may facilitate sleep. If not, limit hypnotics to alternate nights.

One third of those taking benzodiazepines for 6 months experience withdrawal symptoms if treatment is stopped, and some will do so after only a few weeks of treatment. Symptoms appear sooner with rapidly eliminated benzodiazepines (eg lorazepam compared to diazepam or chlordiazepoxide). It is not possible to predict which patients will become dependent, but having a 'passive dependent' or neurotic personality appear to be partially predictive. Symptoms often start with acute anxiety or psychotic symptoms 1–2 weeks after withdrawal, followed by many months of gradually decreasing symptoms, such as insomnia, hyperactivity, panic attacks, agoraphobia and depression. Irritability, rage, feelings of unreality and depersonalization are common, but hallucinations less so. Multiple sclerosis is sometimes misdiagnosed as these patients may report diplopia, paraesthesiae, fasciculation and ataxia. Gut symptoms include D&V, abdominal pain and dysphagia. There may also be palpitations, flushing and hyperventilation symptoms. The problem is not so much how to stop benzodiazepine treatment, but how to avoid being manipulated into prescribing them unnecessarily. This issue is addressed on p 330.

Method of withdrawal: ●Augment the patient's will to give up by elaborating the disadvantages of continuous prescribing. ●Withdrawal is harder for short-acting benzodiazepines—so change to diazepam. ●Agree a contract with the patient—that you will prescribe a weekly supply, but will not add to this if the patient uses up his supply early. ●Withdraw slowly (eg by ~2mg/week of diazepam). Warn the patient to expect withdrawal symptoms, and not to be alarmed.

Withdrawing monoamine oxidase inhibitors (MAOI) Abrupt withdrawal of phenelzine leads to panic, shaking, sweats and nausea in ~¼ of patients. Tranylcypromine has amphetamine-like properties and may be addictive.

Withdrawing tricyclic drugs 4 withdrawal syndromes (all rare):
●Cholinergic activation: abdominal cramps, D&V, dehydration.
●Insomnia, followed by excessively vivid dreaming on falling asleep.
●Extrapyramidal symptoms, eg restlessness and akathisia.
●Psychiatric symptoms: anxiety, psychosis, delirium or mania.

Keep up treatment for ≥6 months post-recovery. Rate of withdrawal: 25mg/5 days (amitriptyline). NB: there are no reports of these withdrawal symptoms from non-tricyclic antidepressants (eg mianserin, trazodone).

Because of withdrawal symptoms it is tempting to stop treatment at the earliest opportunity—but this is a mistake. The severely depressed continue antidepressants for ~6 months after recovery. For amitriptyline, the maintenance dose during this phase of treatment is ~75mg/24h PO. Further slow reductions should be made before stopping the drug.

Phenothiazines Withdrawal symptoms have not been reliably reported.

Introduction to the psychotherapies

'As usual, it was dialogue that combed out my muddle . . .'[1]

Medicine has three great branches: prevention, curing, and healing—and psychotherapy is the embodiment of healing: a holistic approach in which human dialogue and the relationship between doctor and patient is used in a systematic way not only to relieve stress and suffering, but to augment self-esteem and a sense of life's meaning, by producing changes in cognition, feelings, and behaviour.[2] It stands in stark contrast to the most visibly successful but increasingly questioned technical, machine-based sphere of medicine, and we accord it great prominence, in the hope that the explicit descriptions here, and their reverberations throughout our books will produce corresponding reverberations in our minds as we set about our daily work in any branch of medicine, to remind us that we are not machines delivering care according to automated formulae, but humans dealing with other humans. So, taken in this way, psychotherapy is the *essence* of psychiatry—and the essence of all psychotherapy is communication. The first step in communication is to open a channel. The vital rôle that listening plays here has already been emphasized (p 320).

Psychotherapy cannot achieve all things—but not much can be achieved without it. We may prescribe pills for our patients, but if we cannot communicate and share our aims with the patient, those pills are unlikely to be swallowed. And if they are, there is no way of telling that they are working if we have not opened a channel of communication with the patient.

It is not possible to teach all the skills required for psychotherapy in a book, any more than it is possible to teach the art of painting in oils from a book. So what follows in the rest of this section (p 370-96) is a highly selective tour round the gallery of psychotherapy, in an attempt to show the range of skills needed, and to whet the doctor's appetite. It is not envisaged that the doctor will try out the more complicated techniques without appropriate supervision. The psychotherapies may be classified first in terms of *who is involved* in the treatment sessions: an individual, a couple, a family, or a whole group; and secondly they may be classified by their *content and methods* used: analytic, interpersonal, cognitive, behavioural. All psychotherapies are aimed at changing aspects of the patient.

Behavioural therapies aim to change behaviour. If someone is avoiding crowded shops (agoraphobia) a behavioural approach will focus on the avoidance-behaviour. Typically, such approaches will define behavioural tasks which the patient is expected to carry out between sessions.

Cognitive therapy focuses on people's thoughts, and their assumptions. So in the above example of agoraphobia, the therapist would encourage the articulation of the thoughts associated with going into crowded shops. The patient might report that she becomes anxious that she might be about to faint—and then fear that everyone will think her a fool. These thoughts would be looked at using a Socratic approach: 'Have you in fact ever fainted? How likely would you be to faint? If someone fainted in front of you in a shop, what would you think? That they were a fool?'

Psychoanalytical therapies are concerned with the origin of symptoms. They are based on the view that symptoms arise from unresolved issues from childhood. The therapist adopts a non-dominant stance, encouraging the patient to talk without inhibitions. The therapist encourages change by suggesting interpretations for the content of the patient's talk.

The problem of which psychotherapy is the most successful (and in what circumstances) is tackled on p 386.

1 A Miller 1990 *Timebends*, Octopus, London (page 83)   2 J Holmes 1994 *BMJ* ii 1070

Behavioural and cognitive therapies

Behavioural therapy is designed to treat symptoms such as phobias, obsessions, eating and sexual disorders, as well as more general anxiety and mild depression. In those with obsessive-compulsive disorders, rituals (eg hand-washing) respond better than ruminations. The therapy is not suitable for those with schizophrenia, dementia or severe depression.

The first step is to analyse the behaviour that the patient wants to change in great detail. If the patient has agoraphobia (say), find out what somatic and psychological symptoms he has, such as palpitations, and fear of fainting ('fear of fear'), and then relate these to exact points in the behaviour—eg on merely planning to go out, or on entering a particular shop. Next ask the patient to make a hierarchy of such situations, starting with the least threatening. Treatment starts with graded exposure to the least threatening situation (systematic desensitization) combined with teaching a method of anxiety reduction. This might be systematic relaxation, in which groups of muscles are stressed and relaxed in an orderly sequence. For example, while sitting in a chair, the toes are pressed firmly into the ground, and then the knees are pressed firmly together, and then the hands on the knees, and then the fingers of one hand into the opposite palm, each finger in turn. Each pressing lasts for about 30sec and is followed by a period of relaxation, and it is this which counteracts the feelings of anxiety as systematic desensitization through the hierarchy gets under way.

Compliance needs fostering by providing written information about the reasons for practising the new behaviours, and by explaining that the major benefits of treatment only come after hard work. It is wise to enlist the help of the spouse to give encouragement at each achievement.

Rituals such as hand-washing is provoked by a particular stimulus are treated by exposure to the stimulus for some hours while encouraging avoiding the ritual. The therapist illustrates what the patient has to do ('modelling'). Patients may be taught to interrupt obsessional thoughts, eg by doing a particular activity each time they start.

Cognitive therapy The fundamental idea is that mood and thoughts can form a vicious cycle. Using the example of depression: low mood leads to gloomy thoughts, and memories (eg dwelling on exams you did badly at, rather than those in which you performed well). These gloomy thoughts make you feel more depressed (mood) and this lowering of mood makes your thoughts even more gloomy. Cognitive therapy tackles this vicious circle by tackling the thoughts. Take for example the thought: 'I'm a failure, and all my friends are avoiding me'. In cognitive therapy the process is to: ●Clarify exactly what the thought is (do not let it be just a vague negative belief) ●Look for evidence for and against the proposition in the thought ●Look for other perspectives ●Come to a conclusion.

The therapist encourages him to find other explanations by challenging the patient, eg by examining what 'I'm a failure' really means. 'What are the important areas in your life?' 'What do you count as success or failure?' The patient may cite lack of promotion as evidence of failure—countered by the therapist pointing out that failing to achieve some goal is different from being a global failure. There are many kinds of biased thinking which cognitive therapy helps patients to recognize: eg 'black and white thinking', over-generalizing (as in the case of treating one failure as a symbol of everything). Then the therapist might help the patient to look at the judgment of failure within the work area.

Group psychotherapy

The rationale underlying group psychotherapy is that the group provides an interactive microcosm in which the patient can be confronted by the effect his behaviour has on others, and be protected during his first attempts to change. This implies that group psychotherapy is only practical for those who want to change. It has also been found that the most suitable patients are: 1 Those who enter into the group voluntarily, and not as a result of pressure from relatives or therapists; 2 Those who have a high expectation from the group, and do not view it as inferior to individual therapy; 3 Those who have enough verbal and conceptual skills. Patients who are unlikely to benefit include those with severe depression, acute schizophrenia, or extreme schizoid personality (the aloof, cold, hypersensitive introvert); hypochondriacs; sociopathic types (they have low thresholds for frustration and no sense of responsibility to anything); and narcissistic (very self-centred) or paranoid (suspicious and pessimistic about the rôle of others) patients.

Clearly the selection procedure needs to be carried out by someone with much experience. He will aim for a group of, say, 6-8 members balanced for sex, and avoiding mixing the extremes of age. He will decide if the group is to be 'closed', or whether it will accept new patients during its life. He will usually take on a co-therapist, and he will prepare the patients in detail before the group starts. The life of the group (eg 18 months) will develop through a number of phases: a settling-in period when members seem to be on their best behaviour, seeking to be loved by the therapist, and looking to him for directive counselling (which he rarely provides). Next is the stage of conflict, as the patient strives to find his place in the group other than through dependency on the leader. Frustration, anger and other negative feelings are helpful in allowing the patient to test the group's trustworthiness. It is worth learning that expressing negative feelings need not lead to rejection—and this is a vital prelude to the next stage of intimacy, in which the group starts working together.

The therapist will need to steer the group away from outside crises and searches for antecedent causes towards the here and now—eg by asking 'Who do you feel closest to in the group?' or 'Who in the group is most like you?' 'Who would you say is as passive (or aggressive) as you are?' But the therapist must avoid sacrificing spontaneity. He learns to use what the group gives him, eg: 'You seem very angry that John stormed out just now'. He avoids asking unanswerable questions, especially those beginning with 'Why'. His task is to encourage interaction between members and to facilitate learning and observation by members. Special methods used to augment this process include written summaries of group activities, video and psychodrama.

Crisis intervention

This offers immediate short-term help in resolving crises and restoring the ability of the patient to cope. The heroic policeman who climbs on to a roof top to handcuff himself to a man about to cast himself to the pavement hundreds of feet below is engaging in the first step of crisis intervention (rescue), but will not have fulfilled the potential of crisis intervention if he does not go on to address the following issues:

● What events have led to the person's difficulties? Concentrate on his thoughts and actions in the last few days.
● What is his mental state at the moment (p 324)? It is vital to know about depression, suicidal ideas and psychosis.
● When his mental state allows, get to grips with methods he has used in the past to combat stress and resolve crises.
● Who are the significant people in the persons's life?
● What help can the person rely upon from family and friends?
● What solutions has the person (or his or her family, school, or employer) tried in the present crisis? How have they failed?
● If the person has been very severely affected by the crisis, it will be appropriate to offer, and sometimes insist upon, his abandoning of his normal obligations and responsibilities. Temporarily relieving the patient of his responsibilities is necessary to allow concentrated contact ('intensive care') with a therapeutic environment—eg a hospital or crisis unit.
● Ensure by taking practical steps that the patient's commitments are adequately looked after (eg arrange transport of children to foster parents).
● Decide on the best way of lowering arousal (time spent talking is often preferable to administering anxiolytics, which may only serve to delay the natural process of adaptation). If the patient is shocked, stunned or mute, take time to establish the normal channels of communication.
● As soon as the person is receptive, promote a sense of hope about the outcome of the crisis. If there is no hope (a mother, consumed by grief, after losing all her children in a fire), then this too must be addressed.
● The next step is to encourage creative thinking about ways whereby the patient might solve the problems. Start by helping him think through the consequences of all options open to him. Then help compartmentalize his proposed solutions into small, easily executed items of behaviour.

Once the immediate crisis is passed, and the patient has been restored to a reasonable level of psychological functioning, it will be necessary to put him back in charge of his own life. A period of counselling is likely to be appropriate. This is described on p 376. The process of making a contract about therapy is particularly important in encouraging the patient to transfer from the 'sick rôle' to a self-dependent, adult rôle.

Crisis intervention often focuses on loss of face, loss of identity, or loss of faith—in oneself, in one's religion, one's goals or one's roots. 'Perhaps there is an advantage in being born. . . without roots, for one accepts more easily what comes. The rootless have experienced. . . the temptation of sharing the security of a religious creed or a political faith, and for some reason we have turned the temptation down. We [the counsellors] are the faithless; we admire the dedicated, the Doctor Magiots and the Mr Smiths for their courage and their integrity, for their fidelity to a cause, but through timidity, or through lack of sufficient zest, we find ourselves the only ones truly committed—committed to the whole world of evil and of good, to the wise and to the foolish, to the indifferent and to the mistaken. We have chosen nothing except to go on living, 'rolled round on Earth's diurnal course, With rocks and stones and trees'.'[1]

1 Graham Greene *The Comedians* Penguin, 279 and William Wordsworth *Lucy* (given in full in OHCM p 568)

Supportive psychotherapy

There are many people who need continuous psychotherapy, as they find daily activities pose unending stress. The smallest decisions are insurmountable problems, and the patient, lacking even a glimmer of insight, seeks support at every turn. What can we offer here?[1]
- Reassurance—eg that his problems are not unique; he is not mad.
- Clarifying problems; instituting practical help (eg practising for a job interview); explanation (eg about claiming benefits).
- Teaching how to recognize that stress is building up, and how to take the first small steps to reduce anxiety (p 344).
- Positive feedback when he moves away from passive dependency.
- Encouragement in taking even the smallest step towards clearly defined goals—which should be set within his limitations.
- Help to change the patient's social world (job changes, protection from a hostile family, reducing social isolation).
- Counselling for relatives and friends. (Get permission first!)

Counselling

Good novelists (and counsellors) are somehow large enough to embody the world—so their characters (clients) are not just recreated in their own image. *Nothing* human is alien to them. Such exercise of the imagination is what enables virgins to counsel prostitutes—which they *can*, if they are submerged in and are fully aware of human affairs outside themselves.

In its basic (non-directive) form this involves focusing on helping the person clarify what it is they want to do, and what would be needed to achieve it. It does not entail advice, which would run counter to the process of empowering the patient. Fundamental counselling skills consist in listening, understanding and reflecting ($\pm$ summarizing, the use of silences, interpretation—p 378—and confrontation).[2] Other points:
- History-taking: note how previous stress has been coped with.
- Production of an agreed full inventory of problems.
- Redefining problems in terms of attainable goals.
- Use of therapeutic contracts to negotiate small behaviour changes.
- Aim for adult relationships between patient, family, and therapist, so that duties are agreed, eg about frequency and duration of therapy, and what will be expected of the patient by way of 'homework', eg learning anxiety-reducing techniques, and carrying out 'rewards' such as cooking an extra-nice meal with the family if the patient achieves an anxiety-provoking task such as shopping.
- Allowing the expression of negative feelings such as anger—not acted out in reality, but talked about in a cathartic way.
- Reassurance. This is not so simple as it sounds. The therapist must not only give overt reassurance, but also by his demeanour he must reassure the patient that *whatever* he reveals (eg incest or baby battering), he will not be condemned, but accepted. Note: this is one advantage of computer-based psychiatric history taking—the computer is not shocked when the patient tells of drinking two bottles of gin a day: it simply goes on to ask: 'What about whisky?'

Questioning non-directive counselling: Is being non-directive a way of evading responsibility? Is it dishonest to hide our attitudes? What do we lose by being dispassionate? Compassion, perhaps?

1 S Bloch 1986 *Introduction to the Psychotherapies*, OUP
2 N Rowland 1989 *J Roy Col Gen Prac* 39 118

Strategies in long-term psychotherapy[1]

Free association One of Freud's central strategies in psychoanalysis was that the patient should be encouraged to verbalize whatever comes into his mind, without censure, while lying suspended in the limbo world of the psychiatrist's couch. This takes some practice. The therapist uses the information later as raw data, which, after various linking operations and interpretations, can be used to throw light on the patient.

Linking The therapist draws the patient's attention to links between the distant past, the present and future aspirations that he, an outsider, has noted, but which the patient has missed. These links may be between attitudes, eg towards the patient's parents and children. This allows the patient to make some sort of sense of his life, and so to gain insight.

Reflecting The therapist sifts the problems presented to him, reflecting them back to the patient in a way that clarifies them, but without adding any new material or interpretation.

Interpretation An interpretation is a hypothesis to account for a patient's current attitude or emotion. The therapist aims to bring subconscious motives and feelings to the surface, so as to provide the patient with new insights into his life. He notes any resistance (or abnormally flaccid compliance) with which the patient greets such interpretations. The resistance itself will need interpretation, eg with the therapist explaining about how the patient is reacting to him as if he were a parent (transference). The patient's defence mechanisms (eg projecting feelings on to others, denial of feelings) will also need interpretation. Excessive guilt and other super-ego (conscience) manifestations can also be interpreted. The analysis of dreams is another exercise in interpretation. The therapist teaches the patient how to 'decode' a dream's manifest content, in order to expose its more revealing latent content.

The therapist should avoid the irritating and compulsive habit of interpreting everything the patient says or does. The circumstances under which transference, for example, should be interpreted have been fairly clearly defined: 1 When the patient repeatedly shows undue emotion while in therapy. 2 When the flow of the patient's associations becomes blocked. 3 When the therapist believes that interpretation is a necessary step to enable insight to occur. 4 When the patient is coming very close to recognizing the phenomenon of transference for himself.

Confrontation If the patient fails to face up to his psychological problems, and the therapist's interpretation of this fails to lead to the desired change, confrontation may be indicated. This entails a direct statement by the therapist eg 'You are so often late for therapy, that I begin to wonder if you want it to be successful. I might as well spend my time elsewhere.'

Why was Freud so influential? One reason is that his mind is the nearest psychiatrists can attain to one that is continuously scintillating. Who else can make faeces awesome? Try this brilliant, fascinating, useless, sentence: 'Faeces are the child's first gift, the first sacrifice on behalf of his affection, a portion of his own body he is ready to part with, but only for the sake of someone he loves.'[1]

1 S Freud 1955 *History of Infantile Neurosis*, Complete Works **XVII**, Hogarth

Sex therapy for couples

Start with a full (joint) description of the problem. This may be premature (or delayed) ejaculation, female frigidity (anorgasmia), impotence or dyspareunia (eg from spasm—vaginismus—or other physical causes). How did the problem start (eg after childbirth)? Was there ever a time when sex occurred as desired? Is the problem part of some wider problem? What does your partner expect from you? Are you self-conscious or anxious during sex?

Sexual history Early experiences; present practices; orientation to either or both sexes. Has there been difficulty with other partners? When did you meet? What attracted you to each other?

Drugs Alcohol, hypotensives (impotence); tricyclics (delayed ejaculation); β-blockers, stilboestrol, finasteride, contraceptive steroids and phenothiazines (loss of libido). Note: other causes of impotence include diabetes mellitus, cord pathology, hyperprolactinaemia.

The principles of behavioural therapy comprise:
1 Defining the task which the couple wish to accomplish.
2 Reducing the task to a number of small, attainable steps.
3 Asking the couple to practise each small step in turn.
4 At the next session, discussing difficulties encountered.
5 Ameliorating maladaptive attitudes.
6 Setting the next task.

How do these principles work eg when the problem is premature ejaculation and vaginismus? (Both are related to performance anxiety.) One sequence (to agree with the couple) might be:[1]
1 A ban on attempted sexual intercourse (to remove fear of performance failure). Education and 'permission' giving (ie to talk about and engage in 'safe' sexual fantasies) is vital.
2 Touching without genital contact, 'for your own pleasure', initially, with any non-genital part of the body, to explore the range of what is pleasurable, and then to concentrate on whatever erogenous zones are discovered ('sensate focus').
3 Touching as above 'for your own and your partner's pleasure'.
4 'Homework' using a vaginal dilator and lubricating jelly.
5 Touching with genital contact, first in turn, later together. Problems in taking the initiative may surface at this stage. If premature ejaculation is the problem, the partner stimulates the penis, and as orgasm approaches the man signals to his partner—who inhibits the reflex by squeezing his penis at the level of the frenulum.
6 Concentrate on playing down the distinction between foreplay and intercourse, so that anxiety at penetration is reduced.
7 Vaginal containment in the female superior position so that she can stop or withdraw whenever she wants. She concentrates on the sensation of the vagina being filled.
8 Periods of pelvic thrusting, eg with a 'stop/start' technique.

1 J Bancroft 1986 in S Bloch, *Introduction to the Psychotherapies*, OUP

Brief psychotherapy with families[1]

This is a problem-orientated, goal-directed psychotherapy which stresses the importance of social interactions, so treatment aims to include other members of the referred patient's family. Therapy is based on the assumption that people have the resources and potential for growth and the resolution of life's difficulties. When someone's attempted solutions to these difficulties are consistently failing, the therapist's task is to step in and unblock the person's repeated application of these failed solutions, by prescribing new behaviours based on shared insights occurring during the process of family interaction.

The mechanisms of family interaction need to be studied carefully, and with four or more people in the room it is hard for the therapist to keep track of all verbal and non-verbal behaviour going on. To aid him a team of helpers observes the interaction through a one-way mirror, and using an earphone they communicate instructions to him, which may be both verbal (eg *ask the husband* 'What do you do when your wife cries?') and non-verbal (eg *Now turn to look at his wife*). The interaction is videotaped if the family gives its consent, and before the end of the session the therapist retires behind the screen to formulate the insights gained, share the team's hypothesis for the mechanism of the psychopathology, and to check this by reviewing salient points on the videotape. A prescription is formulated and the therapist relays this to the family. This prescription is usually a simple item of behaviour, and it may be negotiated between the members of the family. It is based on the answers family members give to the following sorts of questions, which allow problems to be defined in concrete terms and define the sequences and circularities of family behaviour:

- What does he do (or say) that makes you say he is depressed?
- When he said that, what did you reply?
- What do you do when he behaves like that? Does it help?
- What does your mother do when your father shouts at you?
- Does this happen more indoors, or away from home?
- How long does it last? How do you try to get over it?
- How is this a problem to you? (Depressed? So what!)
- What have other people suggested? What does your mother say?
- Who is the first to see the depression? Your mother or father?
- How do you see the relationship between your sister and mother? How does your mother see it? And your sister?
- If I asked her if she's feeling better, what would she say?
- Of the four of you, who can cheer her up the most?
- What do you want therapy to do? When do you want to change?
- What is the smallest step towards this goal?
- Who wants this change most—you, your husband or your son?

Change often occurs fairly quickly, eg requiring 1–12 sessions.

1 Based on HG Procter's Southwood House approach (Bridgwater, Somerset); M Palazzoli 1980 *Family Proc* **19** 3-12 and M Palazzoli 1978 *Paradox and Counter-paradox*, Jason Aronson (family therapy for schizophrenic families)

Play therapy[1] [2]

The people who have most experience in psychotherapy with children are parents. They hold the key cards for influencing a child's behaviour. These comprise love, mutually understood channels of communication, systems of valid rewards, and a shared knowledge of right and wrong. It is the families without these which are most likely to need the help of professionals. The following guidelines are offered[1] to help these children:

- Take time early on to make friends with the child. Don't rush.
- Accept the child on his own terms—exactly as he is.
- Avoid questioning, praising or blaming. Be totally permissive.
- Don't say 'Don't', and only restrain if about to harm himself.
- Show the child that he is free to express any feeling openly.
- The responsibility to make choices is the child's alone.
- Follow where the child leads: avoid directing the conversation.
- Use whatever he gives you. Reflect his feelings back to him.
- Encourage the child to move from acting out his feelings in the real world, to expressing them in words and play.
- Prepare the parents for change in the child.

In play therapy the child (who may bring his friends) and the therapist play together with toys which give the child an opportunity to verbalize his innermost fantasies. As Virginia Axline explained to one of her 5-year-olds,[2] play therapy is 'a time when you can be the way you want to be. A time you can use any way you want to use it. A time when you can be you.'

Verbatim example of play therapy in action with 6-year-old Dibs[2] He walked around the playroom. Then he picked up a doll. 'Well, here is sister. I'll get her to eat some nice rice pudding, only I'll poison her, and she'll go away. . . forever and ever.' 'You want to get rid of the sister?' the therapist remarks. 'Sometimes she screams and scratches, and hurts me and I'm afraid of her. . . She's five now . . . She'll go to the same school as I go to next year.' 'And how do you feel about that?' 'Well, I don't care. . . she doesn't bother me like she used to.' He went over to the easel and picked up a jar of paint, adding further colours and stirring it well. 'This is poison for the sister. She'll think it is cereal and she'll eat it and then that will be the end of her.' 'So that is poison for the sister and after she eats it, then that will be the end of her?' Dibs nodded 'I won't give it to her yet a while, I'll wait and think it over.'

Before play therapy Dibs had been sullen, mute and unco-operative for years on end. No one had ever been able to engage him in normal family or school activities—an impossible child—until play therapy, by allowing him to express negative feelings, unlocked those reserves of resilience, creativity and zest for life which lie dormant in the heart of every child.

1 V Axline 1969 *Play Therapy*, Ballantine   2 V Axline 1966 *Dibs, in Search of Self: Personality Development in Play Therapy*, Gollancz

385

Comparing the psychotherapies

►Is it true that 'Everyone has won and all must have prizes'?[1]

One reason why it is hard to demonstrate that one form of psychotherapy is better than another is the high proportion of patients (~80%) who benefit from almost any intervention. A small (but perhaps useful) difference between 2 treatments is likely to be swamped by the large proportion of patients who will benefit from either technique.

The methodological problems of comparing different studies of varying validity have now been addressed in a systematic way (using techniques which themselves are not beyond criticism) and yield the following 'conclusions' from research which has met most of the ideal design criteria[1]: randomness; independent assessment; therapy being given by comparable, experienced therapists working for equal amounts of time on their respective patients; well-matched groups; and adequate sample size.

	Which is better?		
	1st item	2nd	Tie
Psychotherapy *vs* minimal or no treatment	9	0	5
Psychotherapy *vs* group psychotherapy	2	2	9
Time limited *vs* time unlimited psychotherapy	2	1	5
Non-directive *vs* other psychotherapies	0	1	4
Behaviour therapy *vs* psychotherapy	6	0	13
Drugs *vs* psychotherapy	7	0	1
Drugs *vs* psychotherapy and drugs	0	6	5
Medical régime alone (eg for asthma) *vs* psychotherapy plus medical régime	1	9	1

The chief finding is the large number of ties, except when psychotherapy is compared with minimal treatment (it is better) and when it is combined with drugs, which confer a more-or-less consistent increased benefit.

The diagnoses in the above studies were mixed (ie depressed, neurotic or schizophrenic, with rather over-inclusive USA criteria), and so the question arises as to whether patients with different diagnoses respond differently to psychotherapy. The few consistent findings are that behaviour therapy is especially suited for treating circumscribed phobias, obsessive compulsive, sexual, and habit disorders, and that cognitive therapy is good for depression and may be more effective than antidepressants in preventing relapse.[2] It may also be more helpful than behaviour therapy in bulimia.[3] Family intervention is known to be particularly helpful in preventing relapse in emotional families with a schizophrenic member.

It is very hard to quantify improvement with psychodynamic therapy (psychoanalysis and its descendants)—but there *are* convincing trials of its use in helping with chronic disease, eg irritable bowel syndrome[4]—but in other cases, therapy may last for decades, and often it is not clear what the exact goal of therapy is, or who should have it. So purchasers of health care (p 435) are reluctant to fund this type of care[5]—but this should not blind purchasers to the very real and large benefits randomized trials have 'proved' in the well-focused, usually brief psychotherapies above.[6]

It is said that the most suitable patients for psychotherapy are the YAVIS people (young, attractive, verbal, intelligent and successful). Research bears this out to some extent. Low socio-economic status, seclusiveness, and having a repressive morality are all known to be negatively correlated with improvement from traditional psychotherapies.

1 The dodo's verdict for the race in *The Adventures of Alice in Wonderland*: see L Luborsky 1975 *Arch Gen Psych* **32** 995   2 T Fahy 1993 *BMJ* ii 576   3 C Fairburn 1993 *Arch Gen Psy* **50** 419   4 E Guthrie 1993 *Br J Psych* **163** 315   5 G Andrews 1993 *Br J Psychiatry* **162** 447-51   6 J Holmes 1994 *BMJ* ii 1070

Excerpts from the diary of a psychotherapist 1: treatment goals*

Formulation My patient, Mr X Smith, is a previously well 68-year-old farmer with one son. His wife died a year ago after a long illness in which her left leg was amputated. He is very sad, feeling low and depressed whatever the external environment is like. He feels guilty about the care he was able to offer his wife in her last illness. He feels a burden to his son, and has restricted all his contacts with the outside world because he feels that people dislike him. He is somewhat suicidal, and this is why he was referred to me by his general practitioner (who had started him on an appropriate dose of amitriptyline). He fulfills the DSM-III criteria for major depression (p 336).

'Thank you for telling me in so much detail about your thoughts and feelings, Mr Smith. In this, the first of our 12 sessions, I would like to start by encouraging you to talk about your wife, although this may be painful at times.' My first goal is to help the patient overcome the guilt surrounding his wife's death. I wouldn't be surprised if he *wanted* her to die, if the illness was long and painful; it may take a few sessions before he feels ready to tell me about this. 'What did you feel like when you took your wife to the ward for the last time, knowing that she wouldn't be coming home again? . . . Was there anything else you could have done for her, in the circumstances?'

My second goal is to help Mr Smith reinstate himself in his social world, and I can start this process right now by getting him to go over his previous contacts with friends and family before his wife's illness, and ways he could go about reawakening these contacts. Mr Smith said he could not get in touch with his previous best friend—it had been so long since they had been together, because of his wife's illness, and it would seem rude now to have neglected him so long. 'Did he know that your wife was ill?' 'Yes.' 'Well, I wonder if he might understand if you explained to him. Might it be worth a try?'

My third goal is to reassure him and to instil a sense of hope. 'One of the reasons why it's hard to get going after such a deep loss as yours is because its hard to face up to the loss and allow yourself to feel all the painful feelings that go with such a loss. Here today you have started to let yourself feel these things. This is good. It will get easier as time goes by, and I'm sure that together we will be able to get over this depression. Yes, it is depression you are in the throes of now—but it doesn't go on for ever. You *will* get better.'

---

*This example is adapted from G Klerman *et al* 1984 *Interpersonal Psychotherapy of Depression*, Basic Books, New York. Our psychotherapist belongs to the school of interpersonal psychotherapy (IPP)—chosen because it offers an internally consistent approach, and its methods have been vindicated in fairly well-controlled 'outcome' research in which IPP has been assessed alone and in conjunction with amitriptyline, and minimal, 'placebo', psychotherapy.

Excerpts from the diary of a psychotherapist 2: other people

Mr Smith seemed reluctant to talk about his wife, so we spent the first half of the session talking about his son and how he was coping with the loss of his mother, and finding out whether he had been able to talk to his son about his loss. We surveyed previous important relationships that he had had in years gone by, with both family and friends. In the second half of the interview I addressed his reluctance to talk about his wife, and he admitted that he found it very painful to do so. He admitted that somewhere deep down he was pleased that it wasn't him that had died, and that such feelings made him feel bad and guilty. He then began to bring forward disturbing material about his wife's last months—how she became demented, did not recognize him, and became abusive.

I let him talk about this for some time, but not so long as to miss the opportunity to explore how he was getting on with recreating his social life. He felt that his depression would make him an unattractive companion in his local pub and that people would ignore him. I confronted him on this, saying gently that if he had not tried to go to the pub, how could he know that people would treat him like that? We talked about the possibility of doing some voluntary work for his favourite charity at weekends.

He said he had developed a troublesome limp in his left leg, and that his doctor had been doing X-rays to find out the cause.

I checked compliance with his amitriptyline by taking a look at the number of tablets in his bottle (which I had asked him to bring with him), and I increased the dose to 75mg at night. He had not found the side effects of a dry mouth and slightly blurred vision too troublesome. He was not surprised that the tablets were not having much effect yet as I had warned him that they take a few weeks to start to work.

I stressed that throughout therapy I would be expecting frank discussion about his thoughts and feelings, and that he should feel free to bring forward topics which affected him greatly.

Excerpts from the diary of a psychotherapist 3–5: change begins

Session 3 Mr Smith reported that he had started to lead a more active life. He had joined a club and visited the pub once, where he had had a short conversation with an old acquaintance. He told of how much his wife, with her restricted personality, had confined his life over decades, and how now, being free from the constraints she placed upon him, he was feeling guilty. This guilt was enhanced whenever he did something of which his wife would have disapproved, for example, driving his car at night.

I encouraged him to talk about his everyday lifestyle, and he described how he still behaved as if his wife was still alive, carefully leaving room for her on 'her' side of the double bed, and moving about the house as if to expect her in the next room. He admitted that he heard his wife's voice calling him from time to time, and he needed reassurance that this hallucination was not a sign that he was 'going mad'.

He told of how he had come to the realization that he, not she, could choose either to restrict himself, or to allow himself to do as he pleased. He was frightened by the new degree of freedom his wife's death had given him, and he was uncertain as to how he was going to use this new freedom.

Session 4 He came in limping, explaining that his doctor had had all the test results back, and he had been to a neurologist—and that no cause could be found for the weakness. He described how the limp had been inhibiting his plans for socializing, but as he spoke he smiled, and seemed quite unconcerned by it. *La belle indifférence* of an hysterical reaction?—I wondered to myself, remembering his description of his late wife's leg amputation. Perhaps he was not ready to socialize yet, and this limp was useful to him as it allowed him to get out of his commitment to himself to re-establish his links with old friends. He is a bit old for an hysterical reaction (but only if this is his first). We went on to cover the same ground as earlier, going over his wife's death, and his reactions to it. I slipped in a few questions about his war experience as a radio operator, and he told me of how he had had to leave the front because he lost his voice.

This session seemed to meander over the same ground as earlier. I thought that maybe there was something else he wanted to bring up, but if there was, there was no sign of it—yet.

Session 5 He started off by saying that he was much improved, and that he was concerned that he might be wasting my time. He went on to describe how he had begun to learn new rules to live by. I suggested that sessions need not just be concerned with his symptoms alone—especially as these were now improving. We could also concentrate on his experience of learning to live according to a new lifestyle. He was pleased at this suggestion, and went on to talk of plans for creating his new life.

Excerpts from the diary of a psychotherapist 6–8: revelations

Session 6 Mr Smith described persisting, frenzied feelings, as if he had to get everything done before something dreadful happened. He had been doing more and more socializing, and was preparing to go on holiday for a weekend with a friend, while his son looked after the farm. This would be his first holiday without his wife. What was the meaning of these frenzied feelings? He related it to a fear of his own death, as a punishment, just when his new life was starting to look promising. He talked about his feelings about not having provided his wife's grave with a tombstone—as if he could not leave his wife in her new resting place while he went off enjoying himself.

In this session I encouraged him to talk at length about his new plans. I was also on the look out for setbacks due to unrealistic assumptions about his own death, his need for self-punishment, and his wife's place in his new life. Instead of introducing these topics myself, I allowed the patient complete control of the topics brought forward, and used the salient features as they were presented to me. As in the earlier sessions, I took the opportunity to counter any false logic inherent in these cognitions.

Our rapport had been steadily growing, but none the less there was an atmosphere of anxiety in our dialogue. He never fully relaxed, except for when he was talking about his limp.

Session 7 He arrived in a much more confident mood, having had several chats with a recently widowed lady who was still very depressed. In talking to her he realized just how much he had improved. This insight gave him a feeling as of rebirth, and he was redoubling his efforts to socialize (so far as his weak leg would allow), as if to make up for lost time. He had started taking evening classes in navigation and these were going well. He was interested in the subject, and was meeting new people.

Session 8 We went over the same old ground of his wife's death and his feelings about it. We'd been over it so much before that I thought there would be no further information to be had. But as he talked, he made progressively deeper revelations about why he was feeling guilty. On the night before his wife's final admission to hospital she had become unbalanced and had accused him of having secret lovers. He lost his temper at this, and found himself with his hands round her throat, wanting to strangle her. He only stopped when she spluttered and became blue. After he had made this revelation all the tension that had been present in our earlier sessions evaporated.

He was able to leave this session without limping.

Excerpts from the diary of a psychotherapist 9–12: termination

Session 9 I decided to make the timing of the end of therapy explicit: 'There are only three more sessions to go after to-day, and I thought this would be a good opportunity to discuss how much progress you think you have made.' The advantage of making termination explicit early is that it provides a definite goal, and that if Mr Smith is going to relapse because he is unready to accept the end of termination, this has a chance of occurring when there are still one or two more sessions to go. He appeared to accept that therapy should end after another three sessions. I explored his reactions to this carefully, giving him an opportunity—if he felt like it—to say how he felt let down or abandoned by myself.

He went on to give details of his navigation classes. He has been doing well—sometimes being the first person in the class to come up with the correct answer. He described how at first he had been reluctant to make friends in the class: 'They all seem so *old!*' He said that he was now able to accept that he himself was getting old. He has active plans to try out his navigation skills in coastal waters with one of the class members who owns a small boat.

Session 10 Mr Smith ran through a number of areas where he believes he has made significant progress: how he is now able to live his life without making 'space' for his wife, and how he is able to enjoy things without feeling guilty, and sometimes without even thinking of his wife at all, for brief periods. He seems to be fairly pleased with the new life he has constructed for himself. He is confident that his improvement will continue, but, rather disappointingly, he did not go on to say that even if he did have a relapse, he would know how to deal with it, and it would not be the end of the world. He wanted to end this session early, and we agreed to this proposal together.

Session 11 He discussed his attitude to therapy, describing how at first he was anxious and suspicious, and then how later, when he began to get better, he thought he should be feeling guilty about getting better too fast. But he did not feel guilty—he just felt pleased. We talked of the depression as of something in the past, dwelling on his improved mood, his increased sense of comfort with himself, his improved relationship with his son, his widening range of activities and his new friends.

Session 12 No new material was brought forward—perhaps a good prognostic sign indicating that it really was the appropriate time to end therapy. We reviewed Mr Smith's progress again, and made arrangements that he should see his own doctor for maintaining and then tailing off his drug therapy. He thanked me, and we parted.

Compulsory hospitalization

▶The patient must have a mental disorder, ie mental illness (p 314), mental impairment (p 314) or psychopathy (p 364), and need detention for treatment of it, or to protect himself or others, before compulsion may be used (if voluntary means have failed).

Admission for assessment (Mental Health Act* 1983, section 2)
●The period of assessment (and treatment) lapses after 28 days.
●Patient's appeals must be sent within 14 days to the mental health tribunal (composed of a doctor, lay person and lawyer).
●An approved social worker (or the nearest relative) makes the application on the recommendation of 2 doctors (not from the same hospital), one of whom is 'approved' under the Act (in practice a psychiatric consultant or senior registrar).

Section 3: admission for treatment (≤6 months)
●The exact mental disorder must be stated.
●Detention is renewable for a further 6 months (annually thereafter).
●2 doctors must sign the appropriate forms and know why treatment in the community is contraindicated. They must have seen the patient within 24h. They must state that treatment is likely to benefit the patient, or prevent deterioration.

Section 4: emergency treatment (for ≤72h)
●The admission to hospital must be an urgent necessity.
●May be used if admission under section 2 would cause undesirable delay (admission must follow the recommendation rapidly).
●An approved social worker or the nearest relative makes the application after recommendation from one doctor (eg the GP).
●The GP should keep a supply of the relevant forms, as the social worker may be unobtainable (eg with another emergency).
●It is usually converted to a section 2 on arrival in hospital.

Detention of a patient already in hospital: section 5(2) (≤72h)
●The doctor in charge (or, in the case of a consultant psychiatrist, his or her deputy) applies to the hospital administrator, day or night—so it is often helpful to obtain early joint care for these patients with a consultant psychiatrist.
●A patient in an A&E department is not in a ward, so cannot be detained under this section. Common law is all that is available, to provide temporary restraint 'on a lunatic who has run amok and is a manifest danger either to himself or to others'[1] while awaiting an assessment by a psychiatrist.[2]
●Plan where the patient is to go before the 72h has elapsed, eg by liaising with psychiatrists for admission under section 2.

Nurses' holding-powers: section 5(4) (for ≤6h)
●Any authorized psychiatric nurse may use force to detain a voluntary 'mental' patient who is taking his own discharge against medical advice, if such a discharge would be likely to involve serious harm to the patient (eg suicide) or others.
●During the 6h the nurse must find the necessary personnel to sign a section 5 application or allow the patient's discharge.

Renewal of compulsory detention in hospital: section 20(4)
●The patient continues to suffer from a mental disorder and would benefit from continued hospital treatment.
●Further admission is needed for the health or safety of the patient—which cannot be achieved except by forced detention.

Section 136 (for ≥72h) allows police to arrest a person 'in a place to which the public have access' and is who believed to be suffering from a mental disorder. The patient must be conveyed to a 'place of safety' (usually a designated A&E department) for assessment by a doctor (usually a psychiatrist) and an approved social worker. The patient must be discharged after assessment or detained under section 2 or 3.

Section 135 This empowers an approved social worker who believes that someone is being ill-treated or is neglecting himself to apply to a magistrate to search for and admit such patients.

---

*This act operates in England and Wales; arrangements in Scotland are slightly different.
1 Lord Justice Keith, 1988 *The Times* May 28   2 R Jones 1991 *Mental Health Act Manual*, 3edn, London, Sweet and Maxwell

Consent to treatment (Mental Health Act)

▶Emergency treatment to save life or to prevent serious harm to the patient can (must) always be given, overriding all the safeguards below.

Background In general a patient must give consent for any procedure (any touching), otherwise the doctor is liable under the UK common law of battery. This means that a competent person can refuse any treatment, *however* dire the consequences. Within common law there are defences to battery other than consent: if a patient is not competent to give or withhold consent, and if treatment is in their best interests, then treatment may be undertaken under 'necessity'. The defence of 'emergency' is to allow restraint where you must act quickly to prevent the patient from harming themselves or others (or committing a crime). For example, you could restrain a patient running amok on a ward before you have a chance to assess the situation. The Mental Health Act enables treatment of someone suffering from a mental disorder *for that mental disorder* under certain circumstances (p 398) and sets down some conditions for consent for treating in certain circumstances, as follows.

Consent is required *and* a second opinion is needed (section 57) All forms of psychosurgery. If the patient is incapable of giving informed consent or declines consent, no treatment can be given.

The Mental Health Act Commission must validate the informed consent. It does this by sending 3 members to interview the patient and review the notes. One member is a doctor (who decides if the treatment is appropriate); others are non-medical. The patient must understand the nature, purpose and likely consequences of recommended treatment for the conditions of informed consent to hold. If the patient is confused, demented, or would consent to *any* treatment, consent is not informed. Signing of an ordinary consent form is needed as well as the certificate from the Mental Health Act Commission. Although grades of informed consent are probably more realistic, the law is black and white: either the consent is informed, or it is not.

Consent is required or a second opinion (section 58)
Note: no treatment can be given to a voluntary patient without his informed consent. If the responsible medical officer (RMO) feels that electroconvulsive or drug treatment for ≥3 months is necessary, and the patient is incapable of giving informed consent, or withholds consent, these treatments may be given provided a second opinion from the Mental Health Commission states that the treatment is necessary. Section 3 of the Mental Health Act is then used. The advantage of doing this is that the patient then acquires well-defined rights for review and appeal. If informed consent is given, the RMO must sign Form 38 (as well as the ordinary consent form) stating that the patient understands the purpose, need for, and likely effects of the treatment.

Children under 16 require the same standard of informed consent from a parent or guardian for prolonged psychotropic treatment. If parents are unwilling or unable to consent, a Care Order may be necessary– under which the local authority social services department takes over responsibility for the child. It must be shown that the child's proper development or health is being avoidably prevented or neglected; or that he is exposed to a moral danger; or that he is beyond the control of his parent or guardian.

Withdrawal of consent Treatment must cease immediately, unless it is necessary to save life. If the treatment is being carried out under section 58, the RMO must contact the Mental Health Act Commission, so that the requirements of this section can be met.

The sections referred to above are those of the 1983 Mental Health Act (England and Wales)

Autism

This neurodevelopmental disorder is, if severe, the antithesis of all that defines mental health (p 314). Prevalence up to 90/10,000 of those <16yrs old—estimates vary considerably.[1] Sex ratio: (♂/♀≈3). Autism is a triad of:

1 Impaired reciprocal social interaction (A symptoms).
2 Impaired imagination, associated with abnormal verbal and nonverbal communication (B symptoms).
3 Restricted repertoires of activities and interests (C symptoms).

Diagnosis of an autistic disorder depends on identifying at least 8 of the following symptoms—and these should include at least 2 'A' symptoms; 1 'B' symptom and 1 'C' symptom.[1]

Impaired reciprocal social interaction (A symptoms)
- Marked unawareness of the existence and feelings of others (eg treating a person as a piece of furniture; being oblivious to others' distress—and to their need for privacy).
- Abnormal response to being hurt (he does not come for comfort; or makes a stereotyped response, eg just saying 'Cheese, cheese, cheese').
- Impaired imitation (eg does not wave 'bye-bye').
- Abnormal play: eg solitary, or using others as mechanical aids.
- Gross impairment in making peer friendships. If he tries at all, the effort will lack the conventions of social interaction (eg reading from the telephone directory to uninterested peer).

Impaired imagination associated with abnormal verbal and non-verbal communication (B symptoms)
- No babbling, facial expressions or gestures in infancy.
- Avoids mutual gaze; no smiles when making a social approach; does not greet his parents; stiffens when held.
- Does not act adult rôles; no interest in stories; no fantasy.
- Abnormal speech production (eg echolalia, ie repetitions); idiosyncratic use of words ('Go on green riding' for 'I want a go on the swing'); misuse of pronouns ('You' instead of 'I'); irrelevances (eg starts talking of train schedules during a conversation about sport).
- Difficulty in initiating or sustaining conversations.

Poor range of activities and interests (C symptoms)
- Stereotyped movements (hand-flicking, spinning, head banging).
- Preoccupation with parts of objects (sniffing objects, repetitive feeling of a textured object, spinning wheels of toys) or unusual attachments (eg to a piece of coal).
- Marked distress over changes in trivia (eg a vase's place).
- Insists on following routines in precise detail.
- Markedly narrow range of interests, eg preoccupied with lining up objects; or in amassing facts about the weather.

Causes: (Mostly unknown) Prenatal rubella, tuberous sclerosis. MRI has shown neocerebellar hypoplasia (vermal lobules VI and VII).[2]

Treatment: This is not effective. Behaviour therapy may be tried.
▶ *A good teacher is more helpful than a good doctor.* Special schooling may be needed (most have a low IQ). Children may learn better by overhearing, than by direct methods. Parents may gain valuable support from self-help groups.[3] Encourage parents to give more attention to the child for 'good' behaviour rather than 'bad', and to have unwavering rules for behaviour. 70% remain severely handicapped. 50% develop useful speech; 20% will develop seizures in adolescence. 15% will lead an independent life. Apply for benefits (disability living allowance if in UK).

1 L Wing 1996 *BMJ* i 327   2 E Courchesne 1988 *NEJM* 318 1349   3 UK tel. 0171 451 3844

Seasonal affective disorder

. . .Lying thus in the sun one is liberated from doubts and from misgivings; it is not that problems and difficulties are resolved, it is that they are banished. The sun's radiation penetrates the mind as well as the body, anaesthetizing thought. . .[1]

Some patients find that symptoms of depression start in the winter months, and remit in spring or summer. It has been postulated that disordered secretion of the indole melatonin (OHCM p 568) from the pineal gland is to blame in some of these patients with 'SAD'. Melatonin, the hormone of darkness[2], is secreted by the pineal only at night—eg at 30μg/night; Evidence supporting the rôle of melatonin in depression includes the following:

- Bright light morning phototherapy (6am–8am) is effective in reducing depression ratings in the winter in those with SAD. This was shown to be associated with delay in the onset of melatonin secretion.[3]
- A dose-response relationship has been shown between the amount of light administered in phototherapy and the degree of improvement in depression (as measured on Hamilton ratings). 6h/day of increased light brought about a 53% decrease in score, whereas treatments of 2h (or red-light treatment) produced only a 25% reduction. These effects were correlated with suppressed plasma melatonin concentrations at 23.00h.[4]

However, evidence in this field of research is often contradictory, and it is probably unwise to rush into recommending light therapy for all patients whose recurrent depressions start in the autumn or winter. This might have the undesirable effect of enticing such patients to book unaffordable winter holidays to exotic locations—with inevitable disappointments and recriminations.

Burnout

Signs of this syndrome of overwork-engendered emotional exhaustion are:
- Stress   •Depression •Libido↓ •Insomnia   •Guilt   •Feeling isolated
- Apathy •Paranoia   •Amnesia •Indecision •Denial •Temper tantrums

Management This is difficult. Some may respond to plans such as these:
- Diagnose and treat any depression (p 336–42 & p 388–396).
- Allow time for the person to recognise that there is a problem.
- More hobbies, and more nice holidays.
- Advice from wise colleagues in the same profession (regular follow up).
- Learn new professional skills.
- Set achievable goals in work and leisure (eg protected time with family).
- Early retirement.

1 AE Ellis 1958 & 1979 *The Rack* Penguin   2 RV Short 1993 *BMJ* ii 952   3 HS Yu 1993 *Melatonin Biosynthesis, Physiological Effects, and Clinical Applications*, CRC Press, Boca Raton, Florida, USA
4 F Winton 1989 *Psychol Med* 19 585

Community care

In the UK, as in many other Western countries, over the last decade, thousands of patients with mental illnesses have had the focus of their care moved from within hospital to the community. Has this been an advantage to them?

Four questions keep recurring, each (ominously) prefixed by a 'Surely. . .'

1 *Surely hospitals will always be needed for severely affected people?* In general, the problem is not the severity of the mental illness, but the social context in which it occurs which determines if community care is appropriate.

2 *Surely community care, if it is done properly, will be more expensive than hospital care, where resources can be concentrated?* Not so—at least not necessarily so. Some concentration of resources *can* take place in the community—for example, in day hospitals and mental illness hostels (p 326-7). It is also true that the 'bed and breakfast' element of in-patient care is likely to be quite expensive, particularly if the running and maintenance costs associated with deploying in-patient psychiatric services are taken into account. In most studies, the cost of each type of service does not differ widely—and in some studies good community care turns out to be ~25% cheaper.[1]

3 *Surely there will be more homicides and suicides if disturbed patients are not kept in hospital?* There are certainly anecdotal reports of these occurrences—and it is clear that however careful selection of patients for community care may be, this could still be a problem.

4 *Surely if in-patient psychiatric beds are not available, however good the daytime team is in the community, some patients will still need somewhere to go at night?* The implication here is that the skills available in bed-and-breakfast accommodation may be inadequate at times of day when there is no other support, other than the general practitioner. Studies which have looked at this have certainly found an increase in non-hospital residential care in those selected for community care, and this increase may be as much as 280% over 5yrs.[2]

Advantages reported for community care are: better social functioning, satisfaction with life, employment, and drug compliance[3]—but in randomized studies in the UK these advantages were not manifest.[4] Furthermore, trends have been repeatedly found indicating that the longer studies go on for, the harder it is to maintain the initial advantages of community care. If it is hard for teams to keep up their enthusiasm during a trial, it will probably be even harder when the trial period has ended[5]—or when team members are ill or away. These considerations may in part explain the observation that with inadequately funded and supervised community care, patients can fail to get the services they require, and when hospitals are being run down, and a patient's condition worsens, so that 'sectioning' followed by admission becomes impossible, the patient is left in the community 'rotting with his rights on'.

1 J Hoult 1984 *Acta Psy Scand* **69** 359   2 A Borland 1989 *Hosp Comm Psy* **40** 369-74   3 LI Stein 1980 *Arch Gen Psy* **37** 392-7   4 M Muijen 1992 *BMJ* i 749   5 P Dedman 1993 *BMJ* i 1359

Poverty and mental health

Social deprivation is positively associated with premature mortality,[1,2] and poverty makes almost *all* diseases more likely (but not Hodgkin's disease, eczema, or melanoma). See *Health and social class*, p 464. In the UK, the numbers of homeless people ranges from 1-2 million.[3] One-third to a half of these suffer from mental illness, and two-thirds do not know where to go for help, and most have no doctor. A 3-tier strategy may help. ●Emergency shelters. ●Transitional accommodation ●Long-term housing.

The cost in health terms to society and the individual is enormous. Illnesses and symptoms such as diarrhoea which may pass as a minor inconveniences to the well-housed may be a major hurdle for the homeless, with severe social and psychological effects. In the UK, government policy (*Working for Patients*) requires each Health Authority to determine the size and morbidity of its resident and its homeless populations. Capture-recapture techniques show that the *unobserved* population of the homeless is about twice that observed. This method of enumeration collects samples (lists) and looks for tags (duplicates) in subsequent counts, and from this determines the degree of under-counting. If all in the subsequent count are duplicates, then there is no underestimate of the original count. Statistical techniques can allow for migration in and out of the population area.[4] These studies show that psychiatric morbidity is greatest in the observed homeless populations: the implication is that the psychiatric illness makes these people more 'visible'.

In the UK, as in many other Western countries, what started out as an enlightened policy of looking after people with mental health problems in the community has resulted in large numbers of psychiatric patients living on the street in great poverty—relieved by occasional admissions to acute units. This 'revolving door' model of care has failed many patients, not least because continuity of care has been compromised.

One way to stop poverty is to pay people not to be poor—specifically not to have more than 1-2 children. For example, the Singapore government is offering poor families $26,400 over 20yrs if they have ≤2 children.

In the UK the 'poverty line' weekly income is calculated on an individual basis depending on circumstances, so for an unmarried mother with 2 young children the figure is £93.85. In fact, state benefits may not amount to this sum. She will receive Income support personal allowance of £46.50 plus £15.95 per child, plus £10.25 family premium with a £5.20 single parent premium (total £93.85). Further benefits such as child benefit or maintenance will lead to a proportionate reduction in this total. From this she might typically pay £3 for water bills, £2.20 for council tax, and £10/week for fuel (in winter). In council flats there are no furnishings. Grants (eg £500) are scarce, so DoH can make a loan for this, deducted at source (eg £13/week)—so the amount some families are living on is less than £60.

Note that there is no evidence that simply living in a deprived area makes a person more prone to illness and death. All the excess mortality and morbidity is explained by the person him- or her-self being poor.[2] Their immediate neighbours who are not poor do not share the same risk. The implication is that targeted health care should be directed towards poor people *wherever* they live, not at poor areas.

1 P Townsend 1989 *Health & Deprivation*, London, Routledge   2 A Sloggett 1994 *BMJ* ii 1470   3 *Br J Psych* 1993 **162** 314–24 See also R Burridge 1993 *Unhealthy Housing: Research, Remedies and Reform* Spon/Chapman and Hall, ISBN 0-419-154410-8   4 N Retterstol 1993 *Suicide: a European Perspective*, CUP

Postnatal depression

The psychiatric and psychological phenomena of pregnancy and birth include the very common, and usually mild, 'blues'; puerperal psychosis; and post-natal depression (the risk of developing this latter is 3-fold that of those with no recent pregnancy). There are many causes, individual and social circumstances, genetic, and hormonal change (this latter 'old favourite' has now been given added veracity by a reasonably successful randomized trial of oestrogen in severely depressed postnatal patients).[1]

Natural history Reviews state that most post-natal depression resolves within 6 months, so we find ourselves thinking: "Good! Those trusty psychiatric tools of 'wait-and-see', and the passage of time will be my main tools in this illness: nothing depends critically on what I do now, to this poor depressed new mother with whom I am now confronted". **NOT SO!**

Consider these facts: ●For the patient, 6 months is a long, long time.
●For the infant, 6 months is more than a long time: it's literally an age.
●Suicide is a waste, but for a young family, a mother's suicide is especially destructive—unthinkable, indeed, for those who have not experienced it.
●There is evidence of impaired cognition and social functioning in the offspring of mothers with postnatal depression.[2]

We therefore submit that our actions may spawn critical outcomes. So what should our actions be? The first step is to try not to be caught unawares by a major depression which apparently strikes like a bolt from the blue, but which, in reality, has been building up over time. Pregnancy and infant-motherhood is supposed to be time of unclouded joy. We professionals often collude with this view. We are always hearing ourselves saying: "Oh Mrs Salt, what a lovely baby! You must be so pleased—and you always wanted a little boy, didn't you? We are so delighted for you. . ." But what if *she* is not delighted? She hardly dares confess her traitorous thoughts that she is unaccountably sad, that she spends the nights crying, and that her exhausted days are filled with a sense of foreboding that she or some other agency will harm the child. The place to start to pre-empt these feelings is in the antenatal clinic. By addressing these issues, and, later, in the puerperium, by expressly asking after them, you give permission for the new mother to tell her woe. When this is revealed, counselling, and in-put from a health visitor and a psychiatrist is appropriate, as is particularly close follow-up. You may need to arrange emergency admission under the Mental Health Act: but the whole point of being prepared for post-natal depression is to avoid things getting this bad.

Pharmacology Tricyclics (p340) and progesterones are often used, but there is little evidence that they are effective.[1] Adding lithium (p 354) may help (and ECT may also be indicated). A single randomized trial ($N=61$) has concluded that transdermal oestrogen is effective in severe postnatal depression. The Edinburgh Postnatal Depression Scale (EPDS) was used to assess outcome (along with other measures). The score started off at 21.8, falling within a month to 13.3—taking some patients out of the 'major depression' category. The control group also improved over time (from 21.3 to 16.5)—but still remained severely depressed.[1] The dose régime was 3 months of transdermal 17β-oestradiol (200µg/day) for 3 months on its own, then with added dydrogesterone 10mg/day for 12 days each month for a further 3 months. CI: uterine, cervical or breast neoplasia; previous thrombo-embolism or thrombophlebitis—and breast-feeding.

As ever, we recommend caution in applying results of a single trial. The key issues are prevention, support, and frequent follow-up, with admission (with the baby) as needed. Before starting oestrogens, try to get a joint decision, with the family, the health visitor, the psychiatrist, and the general practitioner verified by your local drug information service.

1 AJP Gregoire 1996 *Lancet* **347** 930   2 EM Cummings 1994 *J Child Psychol Psychiat* **35** 73–112

5. General practice

Relevant pages in other chapters ►*Every page in all chapters*. All diseases are relevant. This is a key feature of general practice, which is the axis around which all our medical events revolve, every day drawing millions of patients to it before scattering a few handfuls to every corner of the medical establishment. Only in general practice is the full gamut of what can go wrong with people's lives made manifest. Because of the huge and undefined range of general practice, it has been said (not just by GPs themselves) that the general practitioner needs to be the most comprehensively educated of any kind of doctor.[1]

1 D Hill 1969 *Psychiatry in Medicine, Retrospect and Prospect*, London, Nuffield Provincial Hospitals Trust

Index on prevention

Introduction

General practice is not a specialty in the usual sense of the term. General practitioners have no special knowledge of disease (except for diseases in their earliest stages), and they have no unique ways of treating people. If they are specialists at all, then they are specialists not in diseases, but in their patients. They can answer questions like *'Why doesn't Miss Phelps ever attend the hospital antenatal clinic? What would have to change for her to do so?'* There are few medical instances in which GPs are in a position to offer more skill and

than all other doctors—but there are many circumstances when the GP's understanding of his patient is what counts, and for which no amount of expertise can ever be a substitute. So when, for example, Miss Phelps's baby dies, to whom does she turn in her distress? She is not a medical problem; she is not an obstetric problem; she is not even a psychiatric problem—she is Miss Phelps. And the doctor who specializes in Miss Phelps is (or ought to be) her GP. He is the one who knows what to do and what to say; when to be quiet and when to explain; and when to appear and when to disappear, humbled as he is by having witnessed a wealth of such grief.

After perusing the contents page (p 410), the newcomer to general practice might conclude that a GP spends most of his time setting up primary health care teams, conducting audits, loading programs of prevention into his computer, engaging in performance review, sitting in on mother-and-baby or patient participation groups, delegating to his team, and, in those few cases where he actually consults himself, he then follows this with a long period of videotaped consultation analysis. General practice is the oldest branch of medicine, and until very recently it has not felt the need for any of the above. What has brought these activities to the forefront is that we are awakening to a world where there are limits to the success of technological medicine. Even if all our scientists had unlimited abilities and facilities, there would still be a limit to technological medicine. Even if we could turn the whole world into a hospital so skilled as to be able to manipulate the totality of the *mileau intérieur* there would still be a vast abyss between the possibility of health and its attainment. On the other side of the abyss is the patient, surrounded by unknown (but not unknowable) myths, hopes, fears, ideas, expectations and fantasies directing him away from the very things that could help him most. It is the GP's rôle to bridge this abyss, not so much to haul the patient to safety, but rather to be the platform on which the patient crosses of his own (enlightened) free will. In bridging this gap, the GP needs sure foundations on both of its sides, and this is the great challenge of general practice—to foster in oneself an equal love of, and an excellence in, both the technological and the personal realm.

Some definitions and measures of health

Primary care and distributive justice The WHO Alma-Ata statement declares that primary care should 'be made universally accessible to individuals and families in the community, by means acceptable to them, through their full participation, and at the cost that the community and country can afford to maintain in the spirit of self-reliance. . . [and] addresses the main health problems in the community, providing promotive, preventative, curative and rehabilitative services accordingly.' Factors affecting access to health include finance, ideology, and education.

A job description A GP is a licensed medical graduate giving personal, primary, and continuing care to individuals irrespective of age, sex, and illness—attending patients at home, or in consulting rooms, clinics or hospitals. Her aim is to make early diagnosis. These diagnoses will be composed in physical, psychological, and social terms. She will make an initial decision about every problem presented to her as a doctor. She will undertake the continuing management of her patients with chronic, recurrent, or terminal illness. . . She will practise in co-operation with other colleagues, medical and non-medical. She will know how and when to intervene through treatment, prevention, and education to promote the health of her patients and their families. She will recognize that she also has a professional responsibility to the community.

Health 3 definitions to compare: Health is: I The absence of disease.
II A state of complete physical, mental, and social wellbeing.[2]
III A process of adaptation, to changing environments, to growing up and ageing, to healing when damaged, to suffering, and to the peaceful expectation of death. Health embraces the future [so] includes anguish and the inner resources to live with it.[3]

All the above have limitations, but (I) seems least counter-intuitive. Consider the following: ●Was Charles I healthy as he laid his head on the executioner's block? ●Was Ghandi healthy at the end of a hunger strike? ●Can an old person be healthy now if he lost a finger years ago? ●Can animals or babies be healthy? The above give different answers:

Healthy according to definition:

	I	II	III
King Charles	Yes	No	Yes
Fasting Ghandi	Yes	No	Yes
Fingerless old person	Yes	No	Yes
Babies and animals	Yes	Yes	No

Measures of health Scores on the health survey Short Form 36 (SF36) are reproducible and quantifiable, related to patients' clinical state, the GP's decision to refer, and GPs views on severity. It is valid when combined with a patient-generated index of quality of life (which asks patients to name the 5 most important activities/areas affected by their condition, and to value importance of improvements to them) and a daily time trade-off calculation (how much time would you give up to be in perfect health?).[4] By combining instruments, defects in one can be mitigated (eg the SF36 asks if health limits your ability to walk a mile—irrelevant if the patient does not need or want to walk much). *Health need* is the difference between the state now and a goal.[5] Needs may be ranked by the distances between states and goals. Health indicators measure whether needs are met. Abortion rates can be used to measure the need for sex education. Surveys may measure health, but not necessarily needs.

1 WHO 1978 *Primary Care Report* 2 WHO definition 3 I Illich 1976 & 1995 *Medial Nemesis*, Boyars 4 D Rupta 1993 *BMJ* ii 448 5 Per-Erik Liss 1993 *Health Care Need: Meaning & Measurement*, Avebury

What are the determinants of health?

The answer is **wealth**. With wealth comes more stable political systems, and these are what are necessary for literacy and education to flourish, which in turn leads to easy access to clean water (the key issue, as more than 1 billion people have no access to clean water[1]), and the possibility of developing equitable health delivery systems.

A consensus is emerging in some public health units in the UK on what have been the most important influences on health in the last 20 years.[2]
- Improvements in living standards from greater disposable income.
- Improvements in the environment, eg housing and road safety.
- Reduction in tobacco use (but the young continue to take up smoking).
- Improvement in education and communication.
- Better, more effective health services, with easier access to them.

Other influences include nutritional improvement, smaller families, exercise, community self-help, and the prevention of HIV infection.

Future determinants of health are thought to rest on:
- Reducing inequalities in access to health care, and in its content.
- Reducing unemployment (a potent cause of ill health, see p 406).
- Decline in tobacco consumption in all age groups.
- Better health services with more effective, more acceptable treatments.
- Education capable of influencing behaviour so that exposure to identified risk factors is reduced.
- Better protection of the environment, and better housing.
- Applying existing knowledge in a better, more systematic, way.
- Introducing a more patient-centred style of health care, in which patients are not the passive recipients of care, but well-educated partners in the struggle against disease and its causes.

1 J Mackay 1993 *The State of Health*, ISBN 0-671-71147-4   2 R Bhopal 1994 *BMJ* ii 1156

Primary health care

The philosophy of primary health care Primary medical care is the system of care used by people as their first contact with health services. Where needed, referrals are made (eg in ~10% of UK patients) to centres of secondary care—usually district hospitals. Further referral may be necessary to regional 'super-specialists', so constituting tertiary medical care. This seductively simple model of medical provision misses out entirely on the corner-stone of primary health care—the responsibilities that individuals and families have for their own physical and mental wellbeing. ▶*Up to 90% of all health problems are taken care of outside the official health care system.* Unless individuals and families act on their own initiative to promote their health, no amount of medical care is going to make them healthy; coronary artery bypass grafting may be helpful, but it is no substitute for families taking the initiative in promoting their own health by trying to live sensibly (eg *vis-à-vis* diet and smoking) and by learning to look after their own health. In fostering and enlightening this sense of initiative, a large number of groups are involved—not just doctors but also teachers, health educators, politicians and, above all, the families themselves. So in assessing how good a community is at primary health care, one needs to look not just at medical care, but also at social, political, and cultural aspects as well.[1] One must ask questions such as: 'Is society making it easy for individuals to choose a healthy lifestyle?' and 'How is society promulgating health education?' and 'When does this education become indoctrination?'

Primary health care defined as a set of activities
●Providing safe food and water. ●Ensuring freedom from want. ●The basic treatment of illness. ●Provision of necessary drugs. ●Preventive care throughout the lifecycle (p 426).

Primary health care defined as a strategy. No country in the world is rich enough to provide its citizens with everything that medical care can offer. This fact pinpoints the need for the efficient use of limited resources—and this presupposes an effective system of primary health care. To be effective, this must be accessible; relevant to the population's needs; properly integrated; have full community participation; be cost-effective, and characterized by collaboration between sectors of society.[1]

Barriers to primary health care
●People who are rendered helpless and hopeless by the great scourges of our time: unemployment, poverty, and family strife.
●Professionals who want to monopolize and medicalize health.
●Nations which are keener to take up arms than to vaccinate.
●A world which behaves as if it does not know the meaning of social justice and equality, and in which rich and poor fail to share common objectives—or simply fail to share anything.

Some facts ●90% of ill health episodes reported to the NHS are handled in general practice.
●GP services account for <10% of the cost of the NHS.
●The number of unrestricted principals in the UK is rising by ~1.8%/year, and is now 30,320 with 2250 trainees, 250 assistants and 36,000 supporting ancillary staff.

1 H Vuori 1986 *J Roy Col Gen Prac* **36** 398

The primary health care team

Whenever a task can be successfully delegated, delegate it. The antithesis is: *If you want a job done properly, do it yourself.* Nature favours the 1st maxim: when we die all our tasks are either forgotten or delegated, often by default. So the question is not *whether* to delegate, but *when*, and *to whom*. The principle of team work is: *No member is indispensable; all can contribute.*

Doctors As general practice develops it becomes less true that all doctors are 'totipotent'—ie equally able to do any task. Within any group practice, large or small, rôles can compliment each other. Some may specialize in a topic or a procedure, others may lead in information technology, and others see to staff relationships, management, or preventive initiatives. Other rôles include skills in purchasing care (fund-holding, p 466) and practice research. A good question for any partnership to ask itself is what is its range of skills, and is postgraduate training being arranged to fill in any lacunae? These specialisms are helpful in furthering a career in general practice[1] and need not undermine the central rôle of the GP as a generalist.

The community (district) nurse She is usually employed by the local health authority and attached to a practice or to a sector. Sectorization has the advantage of making all her many home visits near to each other but has the disadvantage that she has to relate to many GPs. Her activities include visits to post-op patients for dressings and the removal of sutures, dressing leg ulcers, and giving 'all care' to the elderly housebound, giving injections (eg to blind diabetic patients), and supplying incontinence and other aids. Male nurses may be available for catheterizations.

The community midwife Conducts her own antenatal classes and clinics—often with a GP next door, so that they can advise each other about difficulties as they arise. She will also visit at home. Many do planned home deliveries (where appropriate) and all will do so in emergency. She has a statutory obligation to visit in the puerperium for the 1st 10 days (she has right of access). At 10 days the health visitor takes over (p 466).

The health visitor She has a nursing and midwifery background, and has a further qualification in health visiting. Rôles: developmental testing of children; advice on immunizations, breast feeding, minor illness in children, and handicap; advice to adults about healthy eating and giving up smoking; implementing the strategies of the health education officer (p 438), screening of the elderly in their homes, bereavement visits.

The practice nurse is employed by GPs (some of her salary may be reimbursed by FHSAs). Activities include *tests* (urine, taking blood, audiometry, ECGs, peak flow, prick tests); *advice* (eg about diets); treatment—eg ear syringing (93% of nurses), injections (93%), BP (52%), UTI (30%), otitis media (19%), asthma (16%); *prevention* and *audit*—eg immunizations, BP, cervical smears and family planning (eg if she holds the English National Board Certificate of Competence[2]); *aiding the GP* (diabetes or asthma clinic, chaperoning, IUCD insertion). *Nurse practitioners* have an extended diagnostic rôle as trained by the GP—if happy to take extra responsibility.

Receptionists, secretaries, practice managers These are employed by GPs and salaries are reimbursed as above. Just because they are 'non-medical' it does not follow that they do no medicine—eg a depressed patient may find quiet times to come in and have a chat with a receptionist, and may find this better than official visits. These staff may also be trained to take blood, test urines, do blood pressures, capillary glucose, ECGs, or audits.

Others: Social worker, psychologist, counsellor or facilitator (to encourage preventive activities), physiotherapist. *Beyond the surgery:* Community pharmacists, health education officers (p 438), and community physicians.

1 S Handysides 1994 *BMJ* i 253   2 S Haslett 1987 *J Roy Col Gen Prac* 37 561

Time and the general practitioner

How do general practitioners spend their time? A typical GP with a list of ≤2000 (mean=1902) patients of whom ~ 15% are over 65yrs might see 50 patients in a day with 5 home visits (mean=9745 consultations/GP/yr costing £12.77 each). The major categories of care are for self-limiting illnesses (65% of consultations), chronic illness (20%), and acute major illness (15%). The most common conditions are listed below (persons/yr/GP).[1]

Acute illnesses	Persons/yr	Chronic	Persons/yr
Upper respiratory infections	600	Mental illness	460
Skin disorders	300	'Rheumatism'	100
Emotional disorders	300	High blood pressure	100
Gastrointestinal disease	200	Cardiovascular	50
Lacerations	100	Obesity	40
Acute otitis media	75	Bronchitis	35
Removal of ear wax	50	Anaemia	35
Urinary infections	50	Cancer: old/new	30/5
Acute back syndrome	50	Asthma	30
Menstrual disorders	50	Peptic ulcers	30
Vaginal discharge	30	Varicose veins	30
Migraine	25	Strokes: old/new	20/5
Hernia	20	Diabetes	20
Piles	20	Epilepsy	10
Myocardial infarction	8	Thyroid disease	7
		Parkinsonism	3

Time and the consultation rate ►*Does heavy demand produce short consultations, or do short consultations produce heavy demand by failing to meet patients' needs?* GPs average consultation time is 6.6mins (with some consultations lasting ~½h). The consultation time influences the degree of patient satisfaction,[2] and may influence the consultation rate (2.5–6/patient/yr), with lower return visit rates for longer consultations, (not shown in all studies)[3], and lower rates of prescription issue—especially for antibiotics, and more preventive activities occurring. Mean face to face consultation time is 8min for 10min appointments but only 9.2min for 15min appointments suggesting extra time may not be well used by doctors when booking interval exceeds 10min.[3] Running late is stressful for doctors[3] (and patients): it is easier keeping to time for 10min (rather than shorter) bookings. (Why have appointments at all, since balloted patients prefer non-appointments, and, although they reduce waiting times, exasperation is increased as appointments engender high expectations?) Other factors which increase (↑) or decrease (↓) consultation rates:
- List size (↑,↓), and having personal lists (↓ by 7%).
- Not prescribing for minor ailments—see p 458 (?↓)
- New patients (first year), and patients over 65yrs (↑).
- Time of year (epidemics), geography, and social setting (↓,↑).
- If the GP is extrovert he recalls more, and his rate is higher than others (eg 6/yr *vs* 2/yr). GP age and sex also influences rates.
- High latitudes—within UK (↑). The South-East has lowest rates.[4]
- Social deprivation (↑) and morbidity (↑). Rates fell in the 1960s–70s, perhaps because the burden of respiratory disease lessened.
- Preventive activities (↑). Non-attenders are now invited to clinics.

1 J Fry 1978 *A New Approach to Medicine*, MTP 2 D Morrell 1986 *BMJ* i 869 3 R Potter 1989 *JRCGP* 39 485 4 T Carney 1989 *BMJ* ii 753

Classification and the James Read codes

The problem: my secretary in Worthing wants to find out from your secretary in Bali whether you are seeing more people with myocardial infarction this year compared with last year. She looks through your cherished manual system and does not find very much, because you usually code myocardial infarction as MI, or coronary thrombosis, or acute MI, or acute coronary insufficiency. She concludes that myocardial infarction is very rare in Bali, and advises all her friends to move there at once. How can a mass exodus be prevented? The first step is to produce a database of *preferred terms* (MI=G32) and their *true synonyms* (a dash followed by a synonym code, eg G30-2), along with *misnomers, obsolete terms* (asterisked in Read 3.1) and *homonyms*—whose meaning depends on context—eg 'section', which is legal to psychiatrists, uterine to obstetricians, and frozen to pathologists.

A system which aims to code general practice must be: *comprehensive, computerized, hierarchical, cross-referenced, dynamic,* and *multi-axial.*

Comprehensive		
History/symptoms	Operative procedures	Education
Examination/signs	Drugs and appliances	Administration
Diagnosis	Psychotherapy	Health status facts
Diagnostic procedures	Physical treatments	Occupation/social milieu
Preventive procedures	Alternative therapy	Diagnostic related groups
	Other treatments	Natural language usage

It must facilitate transfer of information between primary, secondary, and tertiary care, and be easy to use by clinical staff, researchers and planners.

Hierarchical Read uses stepwise ranking of classes into subclasses so that it is easy to quantify all operations (7. . . .), all appendix operations (770. .), and all emergency excisions of a normal appendix (77002). There is a *concept* (operations), *subclasses* of this, and a *terminal class*. The code is *alphanumeric* (0–9; A–Z; a–z—with i and o excluded to avoid confusion with 1 and 0, so there are 58 options at each junction in each of the 5 levels of code hierarchy). So there are 656,356,768 (58^5) codes available (enough for all diseases but not for every symptom cluster), and all descriptions are reduced to 5 easily stored and retrieved digits or letters.

Dynamic 656,356,768 codes offers enough *redundancy* to accommodate new ideas. Our own synaptic connexity confers *unlimited* redundancy. This book has 2 kinds of *limited* redundancy: the blank pages offer a strictly limited redundancy, so that a few new ideas can be accommodated in the right place, and a less strictly limited redundancy is provided by the addition of new pages in the wrong place at the end of the text. The Read codes are limited in theory, but in practice almost infinitely expandable (ie dynamic).

Cross-referenced Read codes aim to be superset of all common classificatory systems (ICD-9; ICD-9-CM; ICD-10; OPCS-4, CPT4, BNF), and at least as detailed as any, and to be compatible with them all. *Internal mapping* allows automatic grouping of DRGs (diagnostic related groups) from clinical data.

Multi-axial Read v3.1 is partly multi-axial in that each code remains a unique identifier but no longer acts acts as a fixed address within a branching hierarchy. Relational database techniques are used, so that meningococcal meningitis appears as a subset both of bacterial meningitis, and of meningococcal disease. There exist core terms with any number of linked near-synonyms, and linked qualifiers, such as *Right/Left*, or *Severe/Mild*.▣

So, despite conceptual flaws, and practical limitations and omissions, Read codes are becoming *the* standard for data recording and exchange.

The consultation

'The essential unit of medical practice is the occasion when in the intimacy of the consulting room or the sick room, a person who is ill, or believes himself to be ill, seeks the advice of a doctor whom he trusts. This is a consultation and all else in medicine derives from it.' When medical errors occur, the reason usually lies in a failure of some process of the consultation. We must acquire flair for telling which part of which model is vital at any time, so that in busy surgeries with urgent visits mounting up, both the doctor and his or her patients can survive.

A medical model This is a way of describing what happens in a consultation in purely medical terms: History → examination → investigation → diagnosis → treatment → follow-up.

A task-orientated model[2] This comprises 7 stages:
1 Define the reasons for the patient's attendance in social, psychological, and physical dimensions. Elucidate the nature, history, cause, and effects of the stated problems, along with the patient's ideas, concerns and expectations.
2 Consider other problems and at-risk factors (eg smoking).
3 Choose appropriate action (or inaction) for each problem.
4 Share understanding of the problems with the patient.
5 Involve the patient in his own management, and encourage him to accept responsibilty for promoting his own health.
6 Use time and resources appropriately.
7 Establish or consolidate a good relationship with the patient.

How to achieve these aims is not stated. Presumably healing is amenable to analysis, but such analyses have yet to be propounded.

The problem-solving model Problem stated → problem examined → problem defined → problem agreed. Solutions generated → solutions examined → solutions selected → solutions agreed.

The Stott & Davis model[3]

Management of the presenting problems	Modification of help-seeking behaviour
Management of continuing problems	Opportunistic health promotion

A hypothesis-testing model Information is collected and its validity is ascertained by generating and testing hypotheses.

Goal models (ie ends matter, not means) Aim to:
● Cure; comfort; calm; counsel; prevent; anticipate; explain.
● Enable the patient to put himself back in control of his life.
● Manipulate society to the patient's advantage.
● Facilitate change where change is what the patient desires.
● Increase patients' stature—by tapping the sources of richness in their lives, so freeing them from the shadow of insoluble problems.

In consultations which are going wrong, ask yourself:[4]
'Am I granting as much space to the patient's agenda as to mine?'
'Have I discovered his hopes and expectations—and his fears?'
'Am I negotiating openly with the patient over our clashing ideas?'
'What are my feelings, and how can they be used positively?'
Try saying: 'Things aren't going very well. Can we start again?'

1 J Spence 1960 *The Purpose & Practice of Medicine*, OUP 2 D Pendleton 1983 *Doctor Patient Communication*, Academic Press 3 *J Roy Col Gen Pr* 1979 **29** 201 4 J Middleton 1989 *J Roy Col Gen Pr* **39** 383

Consultation analysis

It is a sad fact that people lose some of their innate skills in communicating while at medical school. Consultation analysis is aimed at reviving and extending this art, and it is known to bring about permanent improvement in those who participate.[1]

Methods The first step is to gain the patient's consent. The method which gives the most information and the most scope for learning employs an observer/director sitting behind a 2-way mirror, who can pass verbal instructions to the doctor through an earphone which is worn unobtrusively. The activity is videotaped for later analysis. By directing the verbal and non-verbal behaviour, the observer can demonstrate the potential of a consultation in ways that the doctor may not have imagined possible. Other methods include simple videotaping, audiotaping, and joint consultations, in which the second doctor either participates in or observes the first doctor's consultations.

Consultation analysis is likely to be a somewhat threatening activity, so rules have been evolved to minimize this.[2] For example, facts are discussed before opinions, the consulting doctor says what he did well, and then the group discusses what he did well. Then the consulting doctor says what he thinks he could have done better, and finally the group says what he could have done better. In practice, these constraints are occasionally stultifying, but it is better to be stultified than hurt.

Mapping the consultation and scoring its effectiveness In the consultation mapped below, the patient's inferior myocardial infarction (sudden chest pains on swallowing hot fluids) was mistaken by the doctor (JML) for indigestion, illustrating that there is no point in being a good communicator if you communicate the wrong message. It also shows how misleading it is to add the scores (50/84, but the patient nearly died).

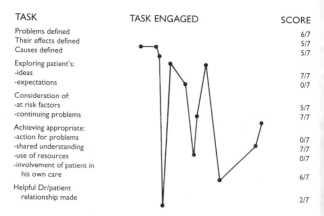

TASK	TASK ENGAGED	SCORE
Problems defined		6/7
Their effects defined		5/7
Causes defined		5/7
Exploring patient's:		
-ideas		7/7
-expectations		0/7
Consideration of:		
-at risk factors		5/7
-continuing problems		7/7
Achieving appropriate:		
-action for problems		0/7
-shared understanding		7/7
-use of resources		0/7
-involvement of patient in his own care		6/7
Helpful Dr/patient relationship made		2/7

1 P Maguire 1986 *BMJ* i 1573   2 D Pendleton 1984 *The Consultation*, OUP

Prevention

▶In all disease the goal is prevention. This is not to say that preventive activities are without side effects.

Classification Preventing a disease (eg by vaccination) is *primary prevention.* Controlling disease in an early form (eg carcinoma *in situ*) is *secondary prevention.* Preventing complications in those already symptomatic is *tertiary prevention.*

Prevention through the human life cycle Preconception (p 94): Is she rubella immune? If not, vaccinate (p 214—ensure that she is using effective contraception for 3 months after vaccination). Is she diabetic? If so, it is wise to optimize glycaemic control as early as possible (p 156).

The child: Vaccination (p 208) and developmental tests (p 288).

Preventing myocardial infarction: Cigarette smoking trebles the risk above the rate for men who have never smoked. A systolic BP >148mmHg (40% of men) doubles the risk, and if the serum cholesterol is in the top fifth of the observed range, the risk trebles. Helping to stop smoking, and treating marked arterial hypertension are the only uncontroversial interventions (p 452). It is probably not worth treating a mildly raised BP (diastolic <110mmHg).[1,2] In America, intensive dietary advice is recommended for those in the top ¼ of the distribution of cholesterol values (this would include 50% of all British men), using drugs for those in the top 10th who do not respond to diet alone. The UK government recommends reductions in total fat and saturated fat to 35% and 15%, respectively, of total energy.[3] GPs and practice nurses have a central rôle in preventing cardiac deaths,[4] eg by identifying and reducing risks in young, first degree relatives of those who have had early coronary events (<50yrs), screening for hypertension, and encouraging less smoking. The fact that the optimum diet for reducing mortality has yet to be declared does not justify inaction with regard to the above preventative measures.

Preventing deaths from breast cancer: Education and self-examination. Well woman clinics. Mammography (using negligible radiation)—cancer pick-up rate: 5 per 1000 'healthy' women screened. Yearly 2-view mammograms in post-menopausal women might reduce mortality by 40%—but the price is the serious but needless alarm caused: there are ~10 false +ve mammograms for each true +ve result. The UK national health service offers 3-yearly single views to those between 55 and 64 years old (older women may be screened if they want, but are not sent for).

Prevention in the reproductive years (p 432): Safer sex education starting in adolescence (teaching about the use condoms need not increase rates of sexual activity[1]); family planning (p 60-8), antenatal/prenatal care (p 94), screening for cervical cancer (p 34), blood pressure, rubella serology.

Old age and prevention: 'Keep fit', pre-retirement lessons, bereavement counselling and occasional visits from the health visitor may help in preventing disease. But the main aim is to adopt the measures outlined above to ensure that there is an old age in which to prevent disease.

Side effects No intervention is without possible side effects (eg those of antihypertensives), and when carried out in large populations the problems may outweigh the benefits.[2]

Smoking and prevention: p 452. **Seat belts and prevention:** p 678. For a fuller list of preventable conditions, see p 411.

1 B Herrmann 1995 *Lancet* **345** 869   2 MRC trial: *BMJ* 1985 **ii** 97   3 DoH 1984 *Diet & CV Disease,* HMSO 4 E Fullard 1984 *BMJ* **ii** 1585. See reports on prevention No 18–21 RCGP publications dept.

Barriers to prevention

Cognitive barriers People often have little thought for the future, and may feel no more attachment to themselves 20 years hence than to a stranger. So the *you* in sentences like: 'Stopping smoking now may stop you getting cancer in 20 years' is not perceived as referring to the same person as the smoking teenager to whom it is addressed. This is a more subtle and insidious version of the '*It won't happen to me*' philosophy.

To some people, over-zealous and sanctimonious-sounding hectoring from bodies such as the UK Health Education Authority creates barriers to prevention, inciting anger and rejection by those who resent their taxes being spent by some State Nanny who assumes that all her charges are 'backward 5-year-olds' who cannot be trusted to think for themselves[1]—so people are now proud to announce that '. . . I eat everything, as much butter and fried foods as I can get. . . I smoke 40-60 cigarettes a day. . . To eat cornflakes, you've got to have sugar on them, and lots of cream, otherwise there is no point in eating them. . . As long as you keep smoking cigarettes, and drink plenty of whisky, you'll go on for ever.'

Psychological barriers All of us at times are prone to promote our own destruction as keenly as we promote our own survival. Knowing that alcohol may bring about our own destruction gives the substance a certain appeal, when we are in certain frames of mind—particularly if we do not know the sordid details of what death by alcohol entails. It provides an alluring means of escape without entailing too headlong a rush into the seductive arms of death. Gambling and taking risks is all part of this ethos.

Logistic barriers A general practice needs to be highly organized to be in a state of perpetual readiness to answer questions like 'Who has not had their blood pressure checked for 5 years?' or 'Who has not turned up to their request to attend for screening'? or 'Who has stopped sending in for their repeat prescriptions for antihypertensives'? On a different front, providing a sequence of working fridges in the distribution of vaccines to rural tropical areas poses major logistic problems.

Political barriers It is not unknown for governments to back out silently of preventive obligations as if influenced by pressure groups who would lose if prevention were successful. Some countries are keener to buy tanks than vaccines.

Ethical barriers If child benefits were available only to those children who had had whooping cough vaccine, much whooping cough would be prevented. This approach is not popular.

Financial barriers A practice may have to pay for extra staff to run an effective screening programme. Coronary artery bypass grafts may prevent some consequences of ischaemic heart disease, but they are too expensive to use on everybody whom they might benefit.

Motivation barriers As we rush out of morning surgery to attend the latest 'coronary' we use up energy which might have been spent on studying patients' notes in the evening to screen to prevent the next one. Changing from a crisis-led work pattern to strategic prevention is one way that practice nurses can lead the way. There is some evidence that they are particularly successful at the meticulous, repetitive tasks on which all good prevention depends.

1 B Levin & P Barthrop 1990 *The Times* 11/8/90 page 10 column 3

Screening

This entails systematic testing of a population, or a sub-group for signs of illness—which may be of established disease (pre-symptomatic, eg small breast cancers), or symptomatic (eg unreported hearing loss in the elderly).

Modified Wilson criteria for screening (1–10 spells iatrogenic):
1 The condition screened for should be an important one.
2 There should be an acceptable treatment for the disease.
3 Diagnostic and treatment facilities should be available.
4 A recognizable latent or early symptomatic stage is required.
5 Opinions on who to treat as patients must be agreed.
6 The test must be of *high discriminatory power* (see below), *valid* (measuring what it purports to measure, not surrogate markers which might not correlate with reality) and be *reproducible*—with safety guaranteed.
7 The examination must be acceptable to the patient.
8 The untreated natural history of the disease must be known.
9 A simple inexpensive test should be all that is required.
10 Screening must be continuous (ie not a 'one-off' affair).

Summary: screening tests must be cost-effective.

Informed consent: Rees' rule[1] Before offering screening, there is a duty to quantify for patients their chance of being disadvantaged by it—from anxiety (may be devastating, eg while waiting for a false +ve result to be sorted out) and the effects of subsequent tests (eg bleeding after biopsy following an abnormal cervical smear), as well as the chances of benefit.

Comparing a test with some 'gold standard'		Patients with condition	Patients without condition
TEST RESULT	Subjects appear to have the condition	True +ve (a)	False +ve (b)
	Subjects appear not to have condition	False –ve (c)	True –ve (d)

Sensitivity: How reliably is the test +ve in the disease? a/a+c
Specificity: How reliably is the test –ve in health? d/d+b

Examples of effective screening	**Unproven/ineffective screening:***
Cervical smears (if >20yrs, p 34)	Mental test score (dementia, p 352)
Mammography (after menopause)	Urine tests (diabetes; kidney disease)
Finding smokers (+quitting advice)	Antenatal procedures (p 95)
Looking for malignant hypertension	Elderly visiting to detect disease

*There is evidence that some screening causes morbidity (mortality-awareness and hypochondriasis↑)—so why is screening promulgated? One reason is that it is easier for governments to be optimistic than to be rigorous.

Why screen in general practice? If screening is to be done at all, it makes economic sense to do it in general practice. In the UK ≥1 million people see GPs each weekday, providing great facilities for 'case-finding' (90% of patients consult over a 5-yr period). Provided the GP's records are adequate, the remaining 10% may then be sent appointments to attend for special screening sessions—eg at the well-woman clinic (p 432). Private clinics do limited work eg in screening for cervical cancer, but there is no evidence that their multiphasic biochemical analyses are effective procedures.

1 *The Times* 9/6/88 2 H Stoate 1989 *J Roy Col Gen Prac* **39** 193 3 T Marteau 1989 *BMJ* ii 527

Problems with screening

All these have affected UK screening programmes, reducing their benefit.

1 Those most at risk do not present for screening, thus increasing the gap between the healthy and the unhealthy—the *inverse care law*.
2 The 'worried well' overload services by seeking repeat screening.
3 Services for investigating those testing positive are inadequate.
4 Those who are false positives suffer stress while awaiting investigation, and remain anxious about their health despite reassurance.
5 A negative result may be regarded as a licence to take risks.
6 True positives, though treated, may begin to see themselves as of lower worth than hitherto.[3]

►Remember: with some screening programmes of dubious value, *it may be healthier not to know*: see OHCM p 654.

General practitioner and shared-care clinics

The following mini-clinics may be conducted in general practice:

- Well-woman/well-man clinic
- Elderly 'non-attending' patients
- Giving-up-smoking clinic (p 452)
- Joint outreach clinics with a consultant who shares care (eg orthopaedics)
- Antenatal clinic
- Hypertension/Diabetic clinic
- Citizen's advice clinic
- Diabetes mellitus clinic
- Asthma clinic

Advantages of mini-clinics
- Easy to keep to management protocol
- Check-lists prevent omissions
- Co-operation cards allow shared care
- Flow-charts to identify trends
- Help from specialist practice nurse
- Fewer referrals to outpatients (↓by 20%)
- Better co-operation with hospitals
- GPs can improve their clinical skills[1]
- Focused dialogue with on-site specialist *does* improve treatment and referrals[1]

Disadvantages
- Extra time needed
- Extra training needed
- Not holistic
- Not flexible
- Value often unproven
- Access to hospital technology↓
- Travelling time by consultants to outreach clinics is wasteful

It is likely that starting a clinic is a good (but not the only) method of implementing the advantages above. The main disadvantage is the temptation for doctors to manage the disease, and to forget about the patient who suffers it: 'I've had enough, doctor. I want to kill myself.' 'Yes, well, your blood sugar is not too bad. Any spells of hypoglycaemia recently?'

Some activities in a well-woman clinic
- Cervical smear & breast examination
- Mammography (for breast carcinoma)
- Pre-conception counselling (p 94)
- Antenatal and postnatal care
- Rubella and tetanus* vaccination
- Smoking and alcohol advice*
- Safer sex advice for HIV*
- Family planning/sterilization*
- Diet and weight* (OHCM p 490)
- Blood pressure*
- Discussion of HRT issues

Well-men It is paradoxical that well-woman clinics are more popular than well-man clinics, since women live longer than men and are more frequent attenders at the surgery (where they make ideal candidates for opportunistic screening). Nurses can do all the well-woman activities *starred above in well-man clinics. One such clinic yielded ≥25% obese; 14% with sustained diastolic BP ≥100mmHg; 66% needing tetanus vaccination; and 29% needing smoking advice. See the OXCHECK study, p 438.

A diabetic mini-clinic ▶*Education is the single most important activity.* The GP and the patients educate each other. Advantages over hospital clinics: the patient sees the same doctor every time; weekly (or more frequent) appointments are possible during periods of difficult control; telephone advice is easily available. Mini-clinics are cheaper than out-patient clinics. Even young insulin-dependent diabetics can be managed wholly in the mini-clinic from the time of presentation (with extra appointments, as necessary) provided there is no overt ketoacidosis—without ever going to hospital.[2] There are dangers in adhering too closely to protocols (p 434). However, the vital test is to dilate the pupil for fundoscopy (p 476 & p 508). The other important areas are diet, blood pressure control, smoking advice and round-the-clock blood glucose monitoring.

Liaison with community based consultant-services is very effective in keeping patients, including children, out of hospital.[2]

1 WP Vierhout 1995 *Lancet* 346 990   2 P Swift 1993 *BMJ* ii 96

Evidence-based medicine[1]

The problem 2 000 000 papers are published each year. Patients may benefit directly from a tiny fraction of these papers. How do we find them?

A partial solution 50 leading journals are scanned not by experts in neonatal nephrology or the left nostril, but by searchers trained to spot papers which have a direct message for practice, and meet pre-defined criteria of scientific rigour (below). Abstracts for these papers are then re-written to give enough background detail to make their significance apparent, and these are published in the bimonthly journal *Evidence-based Medicine*.[2]

Questions used to evaluate papers: 1 Are the results *valid*? (Randomized? Blinded? Were all patients accounted for who entered the trial? Was follow-up complete? Were the groups similar at the start? Were the groups treated equally, apart from the experimental intervention?) 2 What *are* the results? (How large was the treatment effect? How precise was the treatment effect?) 3 Will the results help *my* patients (cost–benefit equation).[1]

Problems with the solution ●The concept of scientific rigour is opaque. What do we want? The science, the rigour, the truth, or what will be most useful to our patients? These may overlap, but they are not co-extensive.
●Will the best be the enemy of the good? Are useful papers rejected because of some blemish? Answer: *all* the evidence needs apprasing, not just that which is unimpeachable.
●Is the standard the same for the evidence for *all* changes to our practice? For example, we might want to avoid prescribing drug X for constipation if there is a chance of a chance that it might cause colon cancer. There are many other drugs to choose from. We might require far more robust evidence than the chance of a chance to persuade us to do something rather counter-intuitive, such as giving heparin in DIC. How robust does the data need to be? There is no science to tell us the answer to this: we decide off the top of our head (albeit a wise head, we hope).
●What about the correspondence columns of the journals from which the winning papers are extracted? It may take years for some unforeseen but fatal flaw to come to light, and be reported in correspondence columns.
●There is a danger that by always asking 'What is the evidence. . .' we will divert resources from hard-to-prove activities (which may be very valuable) to easy-to-prove services. An example might be physiotherapy for cerebral palsy. The unique personal attributes of the therapist may be as important as the objective régime: she is impossible to quantify. It is a much easier management decision to transfer resources to some easy-to-quantify activity, eg neonatal screening for cystic fibrosis.
●Evidence-based medicine can never be always up to date. Reworking meta-analyses in the light of new trials takes time—if it is ever done at all.

Advantages of evidence-based medicine ●It improves our reading habits.
●It leads us to ask questions, and then to be sceptical of the answers: what better definition is there of science?
●As tax-payers, we should like it (wasteful practices can be abandoned).
●Evidence-based medicine presupposes that we keep up-to-date, and makes it worthwhile to take trips around the perimeter of our knowledge.
●Evidence-based medicine opens decision making processes to patients.

Conclusion There is little doubt that, *where available*, evidence-based medicine can be better than what it is superseding. It may not have as much impact as we hope, as gaining unimpeachable evidence is time-consuming and expensive—and perhaps impossible. Despite all these caveats, evidence-based medicine is the most exciting medical development of the decade. Let's all join in by subscribing to its ideals and its journal.

1 F Davidoff 1995 *BMJ* i 1085 and W Rosenberg 1995 *BMJ* i 1122 & ii 259   2 *Evidence-based Medicine* BMA, Journal marketing dept, PO box 299, London WC1H 9JP—tel. 0171 383 6270. See L Ridsdale *Evidence-based General Practice*, ISBN 0 7020 1611 X & A Miles 1995 *Effective Clinical Practice*, Blackwell

Protocols and guidelines

▶ Freedom from only doing ordained tasks is essential for mental health.
▶ Beware accepting even the most enticing protocol without asking yourself if it will affect your sympathy at the bedside. Other questions:

434

• Has the protocol been *evaluated* by someone who is not its author? • What is its *objective* (cost-containment, conformity, or care-enhancement?).

Sympathy is a delicate flower which has often withered before the end of morning surgery; it is also the elusive harmony which makes clinical medicine hum. If your new protocol says that you must do 10 things to Mrs James who happens to have diabetes, both of you may become inconsolably irritated by item 5: the doctor is running out time, and the patient is running out of goodwill. She is really worrying about her husband's incipient dementia, having long-since stopped worrying about her own illnesses. She does not really mind being assailed by lights, forks, stix and lancets, if this is the price that must be paid for a portion of her doctor's sympathy. But if she finds that this sympathy has withered, who knows what her feelings may be, and how she will view her doctor?

Guidelines are friendly, flexible and allow for the frailties of clinical science as it meets reality at the bedside; they can also be individualized and interactive, if instituted in a computerized record during the consultation.[1] Protocols, particularly if they have been handed down from some supposedly higher authority, have a reputation for being strict, sinister, and ultimately stultifying instruments for thought-control. How well do these stereotypes stand up in practice? It is known that doctors working in highly regulated environments with strict protocols perform suboptimally.[2][3][4] It is also worth noting that very few laws are flexible: those which are, are dangerous because they invite abuse. (All men are equal, but. . .). But what laws *can* do is to state when they do not apply. This does not necessarily make them mere guidelines. You could say, for example, that patients have a right to be asked if they want to participate in a protocol, and if they do, that it should be done properly. But what if it is the child's birthday today, and he really does not want to have his finger pricked for a blood glucose test, but he is most willing to go along with all other aspects of a protocol? If you are flexible, the price may be ketoacidosis after the party. Herein lies the central paradox of protocols. They are designed to remove the many indefensible inconsistencies found in clinical medicine, yet the protocols depend on the individual doctor's own flair and instinctive judgement to be applied in the best way.

Perhaps the best approach is to welcome the good protocols, and develop some sort of meta-protocol which should be answered whenever (or almost whenever) such protocols are not adhered to. Why did you not adhere to the protocol? Please tick the appropriate box. ☐My own convenience. ☐The patient's stated wish. ☐The patient's stated wish after being given full information. ☐The protocol is contraindicated in this case because my instinctive flair and judgement tells me so.

In conclusion, each protocol should come with a comment on the likely effect on the doctor's sympathy; how often, on average, one would expect the protocol not to be adhered to; and what resources and time will be required to implement the protocol and its matching meta-protocol. There should also be a statement about each element of the protocol, and just what benefit it is likely to confer—not on the ideal patient, but on the ones likely to be found in the population being offered the protocol. If a protocol does not come with these safety features, beware it, unless, that is, you have invented it yourself—for your own protocols will always have a special status, being infallible, ingenious and innocuous.

1 I Purves 1992 *BMJ* ii 1364   2 T O'Dowd 1991 *BMJ* ii 450   3 S Shortell 1988 *NEJM* **318** 1100
4 A Nicoll 1986 *Lancet* i 606 5 See also *Clinical Guidelines* RCGP Occasional Pr 58 (1992), JM Gimshaw 1993 *Lancet* **342** 1317 and Clinical Resource & Audit Group 1993 *Clinical Guidelines*, Scottish Office

The placebo effect

Placebo effects are very powerful and are important not just in research, but also in demonstrating to us how our demeanour may be just as important as the drugs we give. It is often said that if two obstetric units (for instance) do things quite differently (eg for or against artificial rupture of the membranes) then if one unit is vindicated by research, the other unit must be wrong. This need not follow—not just because the population served by the 'errant' unit may be different. The errant unit may believe it is a centre of excellence, and its staff may rupture membranes with joy in their hearts, knowing they are fulfilling their destiny as the best obstetric unit bar none. This feeling may communicate itself to labouring mothers, who, due to an interaction between communication, beliefs, cognition, and contractility, have their babies with few complications—so much so, that if the unit went over to the 'correct' method, their results might come to mirror their plummeting self-confidence. (This is an important reason for the failure of imposed protocols which look good on paper: see p 434.) Research in this area is very difficult to do, because you cannot easily control for 'joy in the heart'—but with care it can be systematically analysed: in a placebo-controlled study of antihypertensives the partners of the enthusiastic doctor broke the code, and told him that his experimental treatment appeared similar to existing treatments without telling him who was having the active drug, and who was having the placebo. ►From this point, there was an immediate, marked increase in BP in both groups, although the difference between the drug and the placebo was maintained.[1] What we believe and how we behave really do matter—sometimes more than the drugs we use.

The mechanism for the placebo-effect is disputed. It is not the case that there is a 'placebo-responsive' personality. Anxiety reduction is important, and endorphins may play a part.

From the above it should not be assumed that we should give placebos to all our patients, and be enthusiastic about their likely effects. Most of this book is founded on the precept that we should be honest and straightforward with our patients. Nevertheless, it may often be unwise to share too many of our doubts with our patients—as demonstrated by the GP who randomly assigned his consulting style in those with apparently minor illness to a 'positive encounter' or a 'negative encounter'. In the former, patients were given a diagnosis and told they would be better in a few days, the latter group were told that it was not certain what the matter was. Just over ⅓ of the patients having negative encounters got better in 2 weeks, but over ⅔ of the positive group did so.

Nocebo effects Pharmacists are in the habit of 'leafleting' customers with long lists of their drugs' side effects, which, like Voodoo death, may become self-fulfilling prophesies. When one of the authors (JML) gives diamorphine IV to patients having a myocardial infarction, he says in a loud voice 'Your pain will end now' (maximizes placebo effect)—and, in a very quiet voice to the spouse, 'Get a bowl: he may be sick.'—to avoid the reverse 'nocebo' effect. We tend not to tell patients about side-effects such as impotence, justifying this dishonesty on the grounds that in such a sensitive area, the mere mention of impotence will bring it about. Central to this debate is the question of how paternalistic we should be. We do not know the answer, but simply recommend getting to know individual patients well, and having a flexible approach: in doing so we may occasionally strike the right balance.

1 JA Turner 1994 *JAMA* **271** 1609-13 See also VM Oh 1994 *BMJ* ii 69

Ethnic matters

How to avoid offending Western manners This is not just a question of 'please', 'thank you', 'after you', and 'I'd love to. . .' rather than 'I don't mind if I do. . .' with feigned enthusiasm—Westerners are notoriously sensitive to gaze and mutual gaze: not enough, and they think you are shifty; too much, and you are making unwelcome advances; the same goes with interpersonal distance: if you get too close, invasion is threatened.

How to avoid offending Eastern manners ●Avoid prolonged eye contact and loud speech—which can indicate lack of respect.
●Control your gesticulations (the 'thumbs up' sign, for example, may be considered very rude).
●Do not expect an Afro-Asian to answer intimate questions, without first explaining about confidentiality.
●A psychiatric referral may destroy eligibility for marriage.

Hindu names ●First names are often male and female, but middle names always denote sex. ●First (eg Lalita) and middle (eg Devi) names are often written together (Lalitadevi). ●When asked his name a Hindu is likely to give only his first 2 names, withholding his family name, to be polite. This can cause great confusion with receptionists, and duplication of notes.

Sikh names Few Sikhs have a family name. Singh and Kaur indicate only sex and religion, so that extra identification is needed on the notes (use the father's or husband's name also).

Muslim names Sometimes the last name is not a family name, and the first name is not the personal name. There may be no shared family name. The first name is often religious (eg Mohammed). Common female second names are Banu, Begum, Bi, Bibi, and Sultana. They are of as little help in identification as the title Ms is in UK names, but because they are given as second names they are often used for filing purposes (so other family members should be entered as well, to give extra identification).

Clinical phenomena ●Centile charts make normal Bangladeshi, South Indian and Chinese babies seem small (below 3rd centile).
●Bilateral cervical and inguinal lymphadenopathy may be normal in Asian and African children (or it may indicate TB).
●Genetic diseases may be common, eg glucose-6-phosphate dehydrogenase ↓ (sex-linked recessive), sickle-cell anaemia in Africans and West Indians; haemoglobin E disease in Chinese.
●Unusual malignancies (oesophagus in Japan, tongue from betel nuts).
●The cow is sacred to Hindus—so do not offer beef insulin (or pork insulin to those of Jewish or Islamic religions).
●Some Asians will consider measles to be a curse from the Goddess Mata and will opt for an offering in the temple, rather than a visit to the GP.
●Some Muslims will avoid all alcohol (so do not offer tonics).
●Taking drugs may be a problem for Muslims during Ramadan.

Muslims and death (Muslims are the largest non-christian UK sect).
●Religious observance requires prompt burial, not cremation.
●Washing and shrouding is an important ceremony done by respected elders of the same sex (they should be immunized against Hepatitis B).
●The dead body is sacred; and never abandoned by relatives; so it is rare for undertakers to be involved. The body may be moved in a family van.
●Bereavement lasts for 3 days, during which prayers in the home are said almost continuously. The wife is expected to stay at home for up to 4½ months after death of a spouse. Plain, but not black clothes are worn.
●Shoes are to be taken off before entering the house of the dead.
●Transplantation may be OK for Arab Muslim, but Indian Muslims (the latter may allow corneas; bone marrow does not count as a transplant).

1 AR Gatrad 1994 *BMJ* ii 521

Health education

What is education?—four incomplete answers

●Education is the system used for passing down, from one generation to the next, society's values, attitudes, and culture. Thus are crime, duplicity and double standards (and, on a good day, idealism) perpetuated in a kind of cultural inheritance.

●Education is an activity carried out on ignoramuses by people who think they know better.

●Education is about changing people. It usually ends up implying 'change your ways . . . or else.' The most extreme form of education is prison.[1]

●Education performed on one mind by another, under duress, is indoctrination. Indoctrination has its uses. Its value is measured by its propensity to encourage *self-education*, through, for example, travel, reading, or dialogue. Self-education is the food of the mind: the procedure by which we can touch the great minds of the dead and know we are not alone in all our confusions and questionings. By standing on their shoulders we can find a new view of our world—sometimes, even, new worlds to view.

Health education messages To be effective, they must be specific and direct. Eg in getting people to sign on for help for drinking problems, it is of little use saying 'If you don't stop drinking you'll get these diseases. . .' (only ~25% respond); saying 'Signing on is good for you because of these benefits. . .' (~50% respond); saying 'If you don't sign here, you've had it. . .' brings the biggest response. A certain amount of 'fear' in the message is not bad: in enlisting patients for a tetanus vaccine a 'low fear' message gets a 30% response, while a more fear gets a 60%. Optimum messages must be very specific about dates, times and places of help. Too high a level of fear is counterproductive. A gruesome film about the worst effects of caries produces petrified immobility, not self-help or trips to dentists.

The messenger Peers may be better than authority-figures (eg in stop smoking messages). A message about breast feeding will come best from a mother. However, if the issues are not well understood, authority may be helpful (the *BMJ* is more effective than *Woman's Own* in suggesting to mothers that a new formulation of aspirin should not be taken).

Changing attitudes This is usually needed if behaviour is to change—according to the following paradigm:

Knowledge → Attitudes → Intentions → Behaviour

As the Chinese thought reformers knew so well, attitude changes depend on a high level of emotional involvement. NB: the arrows are not always in the above direction. If our behaviour is inconsistent with our ideas (cognitive dissonance) it is usually the ideas, not the behaviour which changes.

The health education officer (health authority) These are likely to have a nursing, teaching or health visiting background. They may have a postgraduate diploma in health education. There are ~300 in England and Wales. A team may comprise a technician and a graphic artist, as well as clerical staff. One rôle is to give information and Health Education Authority leaflets. They also liaise with health visitors and the district medical officer, as well as engaging in planning and research.

Examples of health education at work ●Education about 'safer sex' and the prevention of AIDS. ●Leaflets and tape/slide programs can (slightly) increase knowledge of breast self-examination (which is associated with smaller tumours and less spread in those presenting with breast cancer).

Health promotion by nurses Nurses are the experts in this field[2]—but even they are not very effective in reducing coronary risk. In the community-based OXCHECK randomized trial ($N=6124$, aged 35–64) serum cholesterol fell by only 0.08–0.2mmol/l—and there was no significant difference in rates of giving up smoking, or in body mass index. Systolic (and diastolic) BP fell by ~2.5% in the intervention group receiving dietary and lifestyle advice.[3] Blanket health promotion may not be a complete waste of resources, but it is certainly expensive for rather limited gains. Slightly more optimistic results were obtained by the *Family Heart Study Group*.[4]

The conclusion may be that energies are best spent on those with highest risk as determined in routine consultations by a few 'simple questions' about smoking, family history, etc. One trouble is that these questions are not always innocuous. It is not necessarily a good thing to bring up 'strokes and heart attacks in the family' in, for example, consultations about tension headaches. OXCHECK is not the last word—and there is some evidence that if lipid-lowering drugs were used very much more extensively, cholesterol (and cardiac events) could fall by up to 30%.

1 P Theroux *Down the Yangtze*, Penguin ISBN 0-14-600032-3, page 35–6 2 JL Curzio 1990 *J Hum Hyperten* 4 665 3 OXCHECK 1994 *BMJ* i 308 & 1995 BMJ i 1099 4 1994 BMJ i 313

Patient groups

Mother-and-baby groups These are best set up in the first weeks after the birth of four or five babies. The health visitor encourages the group to form. A doctor may attend the group—regularly to start with, then less often as the group becomes self-sufficient. After a year or two a large practice will have a number of groups running. One aim is to increase motivation (through discussion) to enhance the uptake of health education and preventative medicine. Another aim is to ease the stresses involved in becoming a responsible parent by providing a social support network. A mother, noting for the first time her beautiful baby's ability to hate, to destroy and to hurt, may find it a relief to know that other babies are much the same.

Patient participation groups ►Working *with* your patients is as important as working *for* them. The health care team meets with patients' representatives to discuss some of the following:[1]

- Dealing with complaints (less adversarial than with formal methods).
- Harmonizing the 'consumer's' and the 'provider's' aims.
- Feedback to aid planning, implementation and evaluation of services.
- Identifying unmet needs (eg among the isolated elderly).
- Improving links between the practice and other helpers.
- Health promotion in the light of local beliefs (p 416).
- Pressurizing health authorities over inadequate services.

Owing to lack of interest, or to there being no clear leader or task, up to 25% of groups have closed. The complaint that participation mechanisms lead to tokenism (ie the democratic ideal has been exercised, but what has been created is just a platform for validating the *status quo*) does not turn out to be the case when a patient group has power over funds which it has raised. Here, our experience is that analysis may be penetrating and decision swift—in a way which makes even the best-run health authorities look pedestrian.

The Patients' Association This group represents and furthers the interests of patients by giving assistance, advice, and information. It aims to promote understanding between patients and the medical world. Publications: Patient Voice and a directory of self-help organizations. See also the Contact-a-Family Directory.[2]

Self-help organizations Many thousands of these groups have been set up all over the world for sufferers of specific rare or common diseases. They offer information, companionship, comfort and a life-line to patients and their families. They can share techniques and self-remedies—but a danger is that they share nightmares as well, for example, unnecessarily graphic descriptions of their children dying of cystic fibrosis may be spread, causing unneeded despondency. They raise funds for research, providing a 'welcome alternative to the expensive services of paid professionals'. Full directories exist.[3][4]

Community health councils These are independent monitors of health services. A member may be taken along by a patient to hearings of complaints made against doctors held by FHSA service committees.

1 HD Chase 1993 *Br J Gen Prac* **43** 341   2 A Brownlea 1987 *Soc Sci Med* **25** 605   3 CaF Directory 1991 (tel. 0171 222 3969) ISSN 0964 0703   4 Room 33, 18 Charing Cross Road, London WC2H OHR (tel. UK 0171 240 0671)

Dying at home

The UK death rate is ~12/1000/yr, or ~30 deaths/GP/yr. 65% die in hospital, 10% die in hospices, public places, or on the street, and 25% die at home. Of these deaths at home over half will be sudden. In the remainder, the GP has a central rôle to play in enabling the patient to die a dignified death in the way that he or she chooses. Pain relief and symptom control are the central preoccupations of death in hospices, where death has already been somewhat medicalized, but in those who choose to die at home there often runs a fierce streak of independence, so that their main aim is to carry on with the activities of normal living—come what may. This may cause distress to relatives who feel that the dying person is putting up with unnecessary pain. An open discussion is often helpful in harmonizing the family's aims. The important step is to find out what a patient wants, and then to enable him to do it, being aware that his aims may change over time. The next step is to find out about his hopes and fears and how they interact with those of the family.

Pain ▶See OHCM p 766. Diagnose each pain. Bone pain will need opiates, ± NSAIDs, eg naproxen 250mg/8h PO pc or prednisolone 5mg/8h PO; pain from constipation responds to laxatives or enemas. Document each pain's response to treatment. Morphine (oral) is given eg every 4h. Strength examples (/5ml): 10mg, 30mg, 100mg. Once the daily dose which controls pain is known, consider giving the same 24h dose as modified release tablets every 12h (eg MST® 5, 10, 15, 30, 60, 100 or 200mg). Take the 1st MST tablet 4h after the last dose of morphine solution. (Starting dose of MST in children: 200–800μg/kg/12h PO.) If pain breaks through, give extra morphine solution, and ↑MST at the next dose (do not give more than twice daily to avoid pharmacodynamic confusion). If the patient cannot swallow, use diamorphine SC, at ½ the daily oral morphine dose, given eg divided into 4-hourly injections or via a pump. Laxatives and antiemetics will be needed.

Help ●Family ●Community nurse ●Hospice ●Friends/neighbours
●Night nurses ●GP/health visitor ●CRUSE[1] ●Pain clinic

Bereavement ▶*There is in this world in which everything wears out, everything perishes, one thing that crumbles into dust, that destroys itself still more completely, leaving behind still fewer traces of itself than Beauty: namely Grief.*[2]

The normal grieving process: Numbness→denial→yearning→depression →guilt and aggression→reintegration. This process may take years. The process may become pathological if a major depression (p 336) is precipitated. An extended example of such a reaction is described on p 388–396. It is often tempting for the doctor to try and 'do something' by prescribing psychotropics, but it is known that most bereaved people do not want this, and there is no evidence that drugs reduce subsequent problems. Sympathy and helping the patient to shed tears is probably the most valuable approach. Counselling after bereavement is effective.[3]

After bereavement the risk of death in the spouse is increased in the first 6 months (men) or in the second year (women). Men and the younger bereaved are at the greatest risk.[4] It is not known whether this is due to shared unfavourable environments or to psychological causes (eg mediated by the immune system). The main causes of death are vascular, cancer, accidents and suicide.

Activities which we should try to avoid[5]
- Distancing tactics: 'Everyone feels upset when there is bad news, but **443** you'll soon get used to it.'
- False reassurance: 'I am sure you will feel better; we have good anti-emetics these days.'
- Selective attention: 'What is going to happen to me? I'm beginning to think I'm not going to get better this time. The pain in my hip is getting worse.' Doctor: 'Tell me more about your hip.'

Breaking bad news
There are at least 6 central activities to the breaking of bad news.[6]
1 Choosing a quiet place where you will not be disturbed.
2 Finding out what the patient already knows or surmises.
3 Finding out how much the person wants to know. You can be surprisingly direct about this. 'Are you the sort of person who, if anything were amiss, would want to know all the details?'
4 Sharing information about diagnosis, treatments, prognosis, and specifically listing supporting people (eg nurses) and institutions (eg hospices). Try asking 'Is there anything else you want me to explain?'
5 Be responsive, and recognize the patient's feelings.
6 Planning and follow-through. The most important thing here is to leave the patient with the strong impression that, come what may, you are with him or her whatever, and that this unwritten contract will not be broken.

1 CRUSE (Help for the widowed), 126 Sheen Rd, Richmond, Surrey, tel. 0181 940 4818 for list of local helpers. 2 M Proust 1925 *Albertine Disparue* 3 C Parkes 1986 *Bereavement: Studies of Grief in Adult Life*, Penguin 4 B McAvoy 1986 *BMJ* ii 835 5 P Maguire 1985 *BMJ* ii 1711 6 R Buckman 1992 *How to Break Bad News*, PaperMac ISBN 0-333 54564-7

The agents of terminal care[1]

1 Alum irrigation (1%) by catheter for *massive bladder bleeding*.

2 Amitriptyline 25-50mg PO at night for *nerve destruction pain*.

3 Betadine vaginal gel® for foul *rectal* or *vaginal discharges*.

4 Bisacodyl tablets (5mg), 1-2 at night help *opiate-constipation*.

5 Buprenorphine sublingual 0.2-0.4mg/8h—good in pain if there is *dysphagia or vomiting*. It is not a pure agonist. 'Ceiling' effects negate dose increases, compared eg with oxycodone (below).

6 Cholestyramine 4g/6h PO (1h after other drugs): *itch* in jaundice.

7 Cyclizine 50-100mg/4-8h PO, PR, IM/SC is useful for *vomiting*.

8 Diamorphine PO: 2mg≈3mg morphine. SC: 1mg diamorphine≈1.5mg morphine. Sustained release morphine is available, eg MST-30® (30mg/12h PO). Use IV diamorphine in *massive haemoptysis*.

9 Enemas, eg Arachis oil, may help *resistant constipation*.

10 H₂ antagonists (eg cimetidine 400mg/12h PO) help *gastric irritation*—eg associated with gastric carcinoma.

11 Haloperidol (p 360) 5-10mg PO helps *agitation, nightmares, hallucinations* and *vomiting*.

12 Hydrogen peroxide 6% cleans an unpleasant-feeling *coated tongue*.

13 Hyoscine hydrobromide 0.4-0.8mg SC/8h or 0.3mg sublingual: *vomiting* from upper GI obstruction or noisy *bronchial rattles*.

14 Intercostal nerve blocks may lastingly relieve *pleural pain*.

15 Low residue diets may be needed for *post-radiotherapy diarrhoea*.

16 Metronidazole 400mg/8h PO mitigates anaerobic *odours* from tumours; so do charcoal dressings (Actisorb®). Aerosol masking attempts often make equally unpleasant compound smells.

17 Nerve blocks are useful for *resistant pain*.

18 Naproxen 250mg/8h pc: *fevers* caused by malignancy or bone pain from metastases (consider splinting joints if this fails).

19 Metoclopramide 10mg/8h PO or SC: *vomiting-related gastric stasis* (if this fails, try domperidone 60mg/8h rectally).

20 Pineapple chunks release proteolytic enzymes when chewed eg for a *coated tongue*. (Sucking ice or butter also helps the latter.)

21 Spironolactone 100mg/12h PO + bumetanide 1mg/24h PO may help distressing symptoms of *distension* which accompanies *ascites*.

22 Steroids: most useful is dexamethasone: give 8mg IV stat to relieve symptoms of *superior vena cava* or *bronchial obstruction*—or *lymphangitis carcinomatosa*. Tabs are 2mg (≈15mg prednisolone). 4mg/12-24h PO may *stimulate appetite*, or reduce ICP↑ *headache*, or induce, (in some patients) a satisfactory sense of *euphoria*.

23 Table fans ± supplemental humidified oxygen helps hypoxic *dyspnoea*.

24 Chlorpromazine helps air hunger (eg 12.5mg IV, 25mg suppository).[2]

25 Thoracocentesis (± bleomycin pleurodesis) in pleural effusion.

26 Syringe divers (eg for drugs marked 〰〰 above) or suppositories are used when *dysphagia* or *vomiting* make oral drugs useless. For *pain* try oxycodone 30mg suppositories (eg 30mg/8h, ≈30mg morphine). For *vomiting* try prochlorperazine 25mg suppositories (eg 25mg/12h). *Agitation:* try diazepam 10mg suppositories (eg 10mg/8h).

▶List every symptom the patient has. Diagnose each one individually.
▶Make a plan to relieve each symptom. Review progress often.

For example, the patient may have constipation from drugs, a coated tongue from poor fluid intake and 3 sorts of pain—eg from an invasive cancer, bony metastases and abdominal distension from ascites.

---

1 This page reflects adult doses only See C Regnard 1992 *A Guide to Symptom Control in Advanced Cancer*, 2nd ed, Haigh & Hochland 2 D Walsh 1993 *Lancet* ii 450

Records and computers

►We cannot make ourselves better people by using a system or a machine: reflection, dialogue, and action are more likely routes to self-improvement.

►It is the care and interest a doctor takes in his records that matters, not the details of the particular system employed.

Problem-orientated records *The database:* This comprises past medical history (PMH), occupation, marital status, and address.

The problem list: Eg 'violent husband' or 'unexplained breathlessness' (not '?bronchitis'). Problems are grouped as active or inactive.
Subjective interpretation: How the patient sees the problems.
Objective: Physical examination and the results of bedside tests.
Assessment: Social, psychological and physical interpretation.
Plan: Do the following tests. . .' or 'Wait on events'; treatment: eg 'Start psychotherapy' and explanation—note what the patient has been told.

French Weir system This uses FHSA cards (free to all British GPs):
FP5: This is the record of immunizations.
FP9A: Pink (male) PMH summary card. Use for both sexes.
FP9B: Blue (female) summary card is used as a drug card. These are placed in front of the continuation cards (FP 7), which are kept in time order by a treasury tag. A second treasury tag orders hospital letters.

Computers—The big innovation in general practice over recent years. The reason for this large investment in money and time is 2-fold: computers make it easier to fulfil NHS requirements (elderly home visiting; prevention targets). Computers can allow one to answer interesting questions quickly (but don't believe statements that contain the phrase 'at the press of a button' and, on rare occasions, reliably. Vast GP databases ($>8 \times 10^6$ patient-yrs) allow detailed epidemiological questions to be answered. *Other uses:*
●Computerized age/sex and disease registers (p 448).
●Faster communication with laboratory and hospital (Read codes, p 421).
●Better and *more* preventive care, not just in theory, but in practice (eg 8–18% more vaccinations in consultations with a computer on the desk, more BP measurements, with a ~5mmHg fall in BP in those with hypertension; he price is longer consultations—by 48–90sec). [1]
●Audit: eg computerized record of everyone's BP and vaccinations.
●Prescribing: computers save money (8%) by making generic prescribing easy. They are more likely to be associated with complete records (95% *vs* 42%). The electronic BNF allows simultaneous viewing of various sections, and 'hot-links' to drug interactions. Computer assisted prescribing linked to Read codes (eg PRODIGY) can speed up prescribing, and aims to improve quality by automatically linking prescribing to guidelines.[2] Compters can send personal letters to patients, in the light of new drug information.
●Integration of data from multiple sources—eg drug interactions.
●Quick exploration of hypotheses. (Do I see more depression in winter?)
●Assessing eligibility for social security benefits.
●Rule-based systems/neural networks (p 35)—is this ECG/smear abnormal?
●Keeping up-to-date, and decision support. Examples include: the Internet, the *Personalized Medical Record* and *Mentor*,[3,4] the latter uses this text, OHCM, and the *Oxford Handbook of Clinical Rarities*. This database is updated every 2 weeks by modem.[3] *Mentor* contains hundreds of thousands of facts and an intelligent indexing system, and is linked to the patient's medical record via the Read classification (p 421) to enable explanation of apparently unrelated or obscure signs, symptoms and results—eg chest pain, depression, MCV↑ and melaena are explained by alcoholism, with related trauma (fractured ribs) or cardiomyopathy.
NB: there are big problems in realizing the potential of computers: see p 447.

◁1▷ F Sullivan *E-BM* **1** 96 2 R Walton 1995 *BMJ* ii 1181 & RS Evans 1994 *Arch Int Med* **120** 135
3 '*MENTOR*'–from EMIS (UK tel. 01132 582454/01132 591122), Park House Mews, 77 Back Lane, Off Broadway, Horsforth, Leeds (UK) LS18 4RF 4 CG Westerman 1993 *BMJ* ii 679

Paperless medical enterprises?

We have been recording our interactions with patients using computers for near enough a decade—not long in the history of medicine, but at least two generations in the world of computers. We have also been using computers for other activities. We use a sophisticated networked system when we shop (credit cards), bank, and write using typesetting software rather than 1000 hot metal workers. From these vantage points we look back to the age of DOS (Microsoft's old 'Disk Operating System') and other embryonic operating systems as if to a past ice age. Why is medical computing still stuck in this ice age? The reason is that we have not been very good at capturing the consultation in electronic form, and the benefits of doing so are not very great—ie not enough to drive development. Unlike banking, where computers bring sufficient advantages to warrant investment so that systems *have* to work efficiently, in medical computing this imperative is absent. So, unfortunately, we operate both manual and electronic records.

The coding systems for capturing consultations do not work. It's probably not their fault. In the daily tussle of trying to reduce Mrs Salt (OHCM p 478) to a 5-digit code, we can lose what we are trying to record. We are asking too much of the codes: to speed data entry, aid audit, decision support, planning, and delineating and communicating human predicaments.

This is not like banking or typesetting a document: it is ultimately a philosophical, biographical and poetic enterprise. We cannot expect a coding system to do all this, and free text entry does not work very well either. We *have* to have two systems if one of them is electronic. What we put in the electronic record is coded detail, using a shoehorn to force-fit an incongruent match. If we want to know a bit more of the real truth, we have to look in the physical record. Medical computing will not fly until double data entry is a thing of the past—but this cannot be as there is no good electronic way of dealing with ECGs, diagrams, hospital letters, slowly dawning possibilities, subtle uncertainties, and verbatim examples of patients' problems using their own words. What should we do? Give up the unequal struggle? There is a real danger of this. Computers are no longer new. We do not now come to them with a forgiving frame of mind welcoming a new medium to explore, but in the cynical knowledge that compromise and frustration is just round the corner. So to someone who says "Let's put our efforts into a *single* system, but a manual one", we no longer answer from conviction, but from the point of view of "We could hardly fulfil our terms-of-service requirements if we did this". So we limp on. But it's not quite this bad: we have written as if the old system was reasonably OK, if pedestrian. But anyone who has come to an unfamiliar, complicated patient described in a manual system will know how difficult it is to get simple information from the notes. Has the patient ever had codeine, or penicillamine, before? This might take half an hour of textual analysis, which is why few of us ever used to bother than do more than ask the patient (not a bad solution when all else fails).

Now computerized medical environments take the ease of answering this sort of question for granted—and herein lies the problem. Our greed for ever more fancy programs and the runaway success of computing applications in a few simple environments leads us to be over-critical and so pessimistic about our own not-so-simple computer environment. Using a manual system presupposes being able to read and write efficiently. These skills needed years of teaching and practice, starting at an ideal, formative age. This enables us to read, write, talk, and listen all at the same time—which is how most manual records get written, ie in time that isn't there. Real-time computerized data capture needs its own allotted time *at the moment*. But we are not so very far from easier data entry: graphical user interfaces, pen systems for writing on the screen combined with the will to make more space for training may yet bring us out of the ice age.

Audit—two schools of thought

Audit has various meanings. To FHSAs it means checking up that we are fulfilling our terms of service, and advising on cost-effective use of resources via its increasingly powerful supervising financial auditor and its Medical Audit Advisory Group. Here audit is unrelated to education, and has only a tenuous link with quality. Academically and professionally audit means quality control by counting and measurement; it merges with research,[1] but is usually more about aims and changing behaviour, whereas research aims to reveal new facts, theories or relationships. Audit means asking questions like: 'Have we any agreed aims in medical practice?' and 'Are we falling short of these aims?' and 'What can we do to improve performance?' The process of audit can be divided into 6 stages.[2] These are illustrated below, using a practice which wants to know whether something more formal than their ad hoc methods of summarizing notes is needed.

1 **Aim (ie choice of topic)** This might be to have all important information about a patient immediately available in an accessible form.

2 **Setting standards** The practice may decide that notes must be in chronological order, fat notes should have a summary card, and patients on long-term treatment should have a treatment card.

3 **Observing current practice** The practice manager arranges a random sample of the notes for 'quality control' analysis.

4 **Compare performance with targets** A record is made of the proportion of notes in chronological order, and whether summary cards are used.

5 **Implementing changes** If some notes analysed are not up to standard, a method is devised to achieve this. For example, practice staff could place notes in chronological order, and each doctor could aim to summarize a certain number per day to complete the job in a specified time.

6 **Evaluation** Have the plans worked? A later random sample is analysed. This last element is of some importance, as it prevents us deluding ourselves that our present strategies are necessarily effective.

As demonstrated above, other people's audit exercises are very boring. It is only when a practice engages in audit itself that interest is aroused, and it can become quite satisfying to watch one's practice develop through a series of audits. Once an age-sex register exists, or the practice is computerized, it is possible to do audits on many aspects of care, to answer questions such as 'Is our care of diabetics adequate? Would a mini-clinic be effective? Are all our fertile female patients rubella immune?'

Notice that the practice manager can have a central rôle in running an audit exercise—eg by relieving doctors of the burden of data collection, and is able to communicate the results of the audit in a practice's annual report, along with less formal audits of the consultation and home visits rates, referrals to hospital and whatever else the practice does.

Possible dangers of audit (No intervention is without side effects.)

- It takes time away from eye-to-eye contact with patients.
- In becoming the province of professional enthusiasts, it can alienate some practice members, who can then ignore the results of the audit.
- There is no guarantee that audit will improve outcomes.
- It may limit our horizons—from the consideration of the vast imponderables of our patients' lives in a world of death, decay, and rebirth—to a preoccupation with attaining tiny, specific, and very limited goals.
- Some doctors fear that in espousing audit they risk transforming themselves from approachable but rather bumbling carers and curers who perhaps don't know *exactly* where they are going, into minor administrative prophets, with too much of a gleam in their eyes and zeal in their hearts.

1 IK Crombie 1993 *Audit Handbook* Wiley, ISBN 0-471-93766-5 2 M Lawrence 1994 *BMJ* ii 513

Peer review during a practice visit

GPs ask other GPs to place them somewhere between 2 opposed views:[1]

1 Professional values

He balances his own convenience against that of his patients, keeping the interest of the wider community in mind.

He puts his own convenience above the needs of patients, having no concern for wider responsibilities.

He believes in the importance of the continuity of care, and gives as personal and comprehensive a service as possible.

He does not think that this matters. He delegates too much. Clinical interests are dominated by hobby horses.

He subjects his work to critical self scrutiny and review by colleagues. He enjoys being a GP and he accepts the obligation to maintain his own health.

He is complacent about his work and never reviews it.
He has become defeated, or drives himself excessively.

2 Accessibility

It is easy to see him quickly if needed.

You cannot see him quickly.

He does not keep patients and staff waiting unnecessarily.

The doctor is regularly late for appointments.

3 Clinical competence

He consistently gives evidence of taking relevant histories, and appears to listen to what his patient says.

He fails to elicit histories appearing not to listen to what the patient is saying.

His use of drugs is appropriate.

Prescribing is inappropriate

4 Ability to communicate

He creates a calm, receptive atmosphere which encourages patients to talk freely

No communication with patient in defining the reasons for the patient's attendance.

Regular meetings take place with other members of the health care team. He encourages a free exchange of ideas.

He discourages talk among the team. He is insensitive, misunderstanding their rôles.

There are ~70 such statements in the RCGP document (as well as much else), and the main difficulty is not what the scheme leaves out (eg cost containment), but that it is so thorough that some might quail before so much goodness-in-theory because of what actually happens in practice. If this is the case, the best plan is to buy the College document, but only to read it in very short bursts until one's practice gradually comes into shape—when the whole document can be read without inducing a fit of despair about one's own performance.

1 *What sort of Doctor? J Roy Col Gen Prac* 1985 Report 23

Patient satisfaction

▶*The patient is the nearest thing to an infallible judge of what constitutes good medicine.* Patient satisfaction is one of the aims of all our endeavours, and is one of the very few measures of *outcome* (rather than *process*) which is fairly easy to measure, eg by questionnaire agreement with these 13 statements (*Baker's dozen*[1]).

Satisfaction
1 I am totally satisfied with my visit to this doctor.
2 Some things about my visit to the doctor could have been better.
3 I am not completely satisfied with my visit to the doctor.

Professional care:
4 This doctor examined me very thoroughly.
5 This doctor told me everything about my treatment.
6 I thought this doctor took notice of me as a person.
7 I will follow the doctor's advice because I think he/she is right.
8 I understand my illness much better after seeing this doctor.

Relationships:
9 This doctor knows all about me.
10 I felt this doctor really knew what I was thinking.
11 I felt able to tell this doctor about very personal things.

Perceived time:
12 The time I was allowed with the doctor was not long enough to deal with everything I wanted.
13 I wish I could have spent a bit longer with the doctor.

Why do patients change their doctor? The most common reasons are that either the patient has moved, or the doctor has retired or is perceived to be too far away. There are additional reasons:[2]

Patient needs:		Organizational problems:		Problems with doctor:	
One doctor for all the family	5%	Long waits	13%	Lost confidence in	21%
Want woman doctor	4%	No continuity of care	6%	Dr not interested	10%
Want alternative medicine	2%	Rude receptionist	6%	Rude Dr	10%
		Appointments wanted	1%	Prescriptions criticized	5%
Obstetric needs	1%	Open surgeries wanted	1%	Dr hurried	4%
		Other staff rude	1%	Visits problem	4%
				Communication poor	3%
				Referral problem	2%

Another approach to gaining satisfaction is to agree and publish standards of care that patients can expect, along with performance figures for how nearly these standards are achieved in practice. This is the philosophy behind the UK government-led *Patient's Charter* and the British Standards kitemark BS5750—the aims of which are:
●Set standards—eg by agreement with patient participation groups (p 440).
●Monitor progress towards these standards, and publish progress locally.
●Provide information about how services are organized. Maximize choice.
●Let users know who is in charge of what, and what their rôles are.
●Explain to users what is done when things go wrong, and how services are improved, and what the complaints procedure is.
●Show that tax-payers' money is being used efficiently.
●Demonstrate customer satisfaction.

This culture has proved alien to most GPs, perhaps because of a very necessary preoccupation with illness and its curing, rather than service, and its glorification.

1 R Baker 1990 *Br J Gen Prac* **40** 487   2 B Billinghurst 1993 *Br J Gen Prac* **43** 336

Stopping smoking tobacco

Epidemiologists estimate that ~50% of smokers will die of smoking if they do not give up—and it is not the case that they only lose a few years: a quarter of a century is more likely for those dying between the ages of 35 and 69.[1] Stopping smoking diminishes the excess risk from tobacco, so that after 10-15yrs the risk of lung cancer effectively reverts to that of lifelong non-smokers. A similar, but quicker diminution of excess risk (↓ by ~50% in the 1st year) is found for deaths from coronary artery disease and, to a lesser extent, risk of stroke.[2] ► *60% of smokers want to give up.*

Annual UK health costs of smoking ●GP prescriptions: £52 million ●GP consultations: £89 million ●Hospital episodes: £470 million

Advantages of stopping smoking Saving of life (110,000/yr in UK).
●Larger babies (smokers' babies weigh on average 250g less than expected, and their physical and mental development may be impaired).
●Less bronchitis (accounts for millions of lost working days).
●Less risk from the combined contraceptive pill (cardiovascular risk ↑ × 20 in those smoking >30 cigarettes/day).
●Less risk from passive smoking (cot deaths, bronchitis in children and lung cancer in non-smoking spouses).
●Return of the sense of taste and smell.

Helping people who want to stop smoking ●Ask about smoking in consultations—especially those concerned with related diseases.
●Ensure that advice is congruent with beliefs about smoking.
●Concentrate on the benefits of giving up (as above).
●Invite the patient to choose a date (when there will be few stresses) on which he will become a non-smoker.
●Suggest that he throws away all smoking accessories (cigarettes, pipes, ash trays, lighters, matches) in advance.
●Inform friends of the new change.
●Consider offering nicotine chewing gum, chewed intermittently (to limit release of nicotine). ≥Ten 2mg sticks may be needed/day. Results have been rather inconsistent, and transdermal nicotine patches may be more reliable (easier to use, with minimal instruction required), ~doubling the give-up rate (eg from 11.7% to 19.4%,[3] compared with placebo). A dose increase at 1 week is helpful for some who are in difficulties.[4] NB: gum may be preferable to patches in those most addicted (eg craving on waking).[4] Giving additional detailed written advice offers no added benefit to simple advice from nurses.[3] Always offer follow up.[6]

For those who do not want to give up Give them a health education leaflet, record this fact in the record, and try again later.

Cigarette smoking is declining (but not fast enough to reach UK government *health of the nation* targets), but is *increasing* among the young. ►In the UK, ~450 children start smoking per day; and ¼ of school leavers smoke regularly.[6] Numbers are rising. This has prompted the Health Education Council's family smoking education project for schools.

1 R Doll 1994 *BMJ* ii 901 & R Peto 1992 *Lancet* 339 1268   2 SG Wannamethee 1996 *E-BM* 1 95
3 D Mant (ICRF GP research group) 1993 *BMJ* i 1304   4 M Russell 1993 *BMJ* i 1308
5 J Tang 1994 *BMJ* i 21   6 M Russell 1979 *BMJ* ii 231

Reducing alcohol intake

With the toll that excess alcohol takes in terms of personal misery and the national purse (>£1600 million/yr in the UK), the need to reduce alcohol intake should rank as one of the leading aims of preventive care. The reason why alcohol is not at the top of the agenda is not just that doctors are so fond of it (the profession has three times the national rate of cirrhosis), but because there is a powerful and pervasive lobby which ensures that alcohol is cheaper (in relative terms) and more readily available than ever before—so that its use on an individually moderate scale arouses no comment. It is assumed to be safe, provided one is not actually an alcoholic. However, it is more helpful to view alcohol risks and benefits as a spectrum (see the French Paradox, OHCM p 482 for the *benefits* of alcohol). Problems are listed on p 363. A strategy to reduce the bad effects of alcohol in your patients:

- Whenever a patient presents with symptoms or signs related to alcohol, be sure to ask in detail about consumption.
- Question any patient with 'alerting factors'—accidents, driving offences, child neglect, assault, attempted suicide, depression, obesity.
- Question all remaining patients (not infants) as they register, consult or attend for any health check.

Helping people to cut down ● Take more non-alcoholic drinks.
- Limit your drinking to social occasions—and learn to sip, not gulp.
- Reduce the frequency of sips, eg by shadowing a slow drinker in the group. Don't pick up your glass until he does (and don't hold your glass for long: put it down to avoid unconscious sipping).
- Don't buy yourself a drink when it is your turn to buy a drinks' round.
- Reduce the period of drinking—go out to the pub later.
- Take 'days of rest' when no alcohol is drunk.
- Learn graceful ways of refusing: 'No more for me please, I expect I'll have to drive Jack home' or 'I'm seeing what it's like to cut down.'

Maintaining reduced drinking ● Agree goals with the patient.
- Suggest he keeps an alcohol diary in which he records all drinking.
- Teach him to estimate his alcohol intake (u/week, p 362).
- Consider an 'Alcohol Card' in the notes to show: units/week; pattern of drinking; reasons for misuse; each alcohol-related problem (and whether a solution has been agreed and action implemented); job record; family events; biochemical markers (GGT, MCV); weight.
- Give feedback about how he is doing—eg if GTT falls are discussed at feedback, there is much lower mortality, morbidity and hospitalization compared with randomized control subjects.[1]
- Include the family in plans for on-going alcohol reduction. Agree a system of 'rewards' for sobriety.
- Group therapy, self-help groups, disulfiram, local councils on alcohol, community alcohol teams and treatment units may also help (p 362).

Setting limits for low-risk drinking eg ≤20u/week if ♂; ≤ 15u/week if ♀— there are no absolutes: risk is a continuum. NB: higher limits are proposed, on scant evidence[1] (eg 4u/day for men and 3 for women). 1u is 9g ethanol, ie 1 measure of spirits, 1 glass of wine, or a pint of beer.[2,3]
▶ *Primary care is a good setting for prevention*: intervention leads to less alcohol consumption by ~ 15%, reducing the proportion of heavy drinkers by 20%—at one-twentieth the cost of specialist services.[4] There is no evidence that GP intervention has to include more time-consuming advice such as compressed cognitive/behavioural strategies.[5,6] Simple advice works fine as judged by falling GTT levels—at least for men. After interventions, women may report drinking less, but this is not reflected in a falling GTT.[5]

1 M Marmont 1995 *Lancet* 346 1643 2 R Doll 1994 ii *BMJ* 911 3 J Gaziano 1995 *BMJ* ii 3 4 J Volpicelli 1995 *Lancet* i 456 5 P Anderson 1991 *BMJ* i 766 6 P Anderson 1993 *Br J Gen Pract* 43 386

Alternative and holistic medicine

Alternative (complementary/fringe/non-orthodox) medicine These terms are used to describe therapies which are not included in conventional medical/paramedical training. Some are the orthodoxies of a different time (eg herbalism) or place (the Ayurvedic medicine of India), some are mainly diagnostic techniques (iridology), some mainly therapeutic (aromatherapy). Some doctors are suspicious of unorthodox medicine, and feel that its practitioners should not be 'let loose' on patients. However, in many places the legal position is that, however unorthodox a practitioner's treatment may be, he or she cannot be convicted of unethical practice in the absence of demonstrable harm to patients.[1]

Many patients (1 million annually in UK) consult alternative practitioners, usually as a supplement to orthodox treatment and for a limited range of problems—predominantly musculoskeletal. Many will feel unable to tell their doctor.

Modern medicine is criticized (both from within and from without) for sacrificing humanity to technology, and with little benefit for many patients. In contrast to the orthodox doctor, alternative practitioners may be seen as taking time to listen, laying on hands rather than instruments, and giving medicines free (not always!) from side effects.

Many alternative therapies seem scientifically implausible, and controlled trials are difficult to design, as treatment is often individualized and the rôle of the therapist crucial. However, the same comments could be made of much of orthodox practice. GPs are showing increasing interest in what their rivals are offering, and some incorporate alternative therapies into their armamentarium. Some of the therapies commonly available to patients are briefly described below.

Acupuncture: Traditionally used to treat a wide range of conditions, acupuncture is increasingly employed in orthodox practice for pain relief, control of nausea and treatment of addiction. For these, endorphin release provides a scientific rationale.

Homeopathy: This is based on the principles that like cures like, and that remedies are made more efficacious ('potentiated') by infinite dilution. NHS GPs may prescribe homeopathic remedies on an FP10. Randomized trials suggests real (small) benefits, eg in asthma, but nobody knows if the reason *why* they are effective is 'like cures like'.[2][3]

Manipulative therapies (osteopathy and chiropracty): These are widely used and of proven benefit in musculoskeletal problems, but some use them to treat more general conditions such as asthma.[4]

Clinical ecology Starting from the fact that atmospheric pollutants, toxins and xenobiotic chemicals (from other organisms) are known to be harmful, a system is built up around techniques (using intradermal injections) for provoking and neutralizing symptoms related to foods.[1][5]

Holistic medicine[5] The holistic approach means taking a broad view: of the patient as a whole person, of the rôle of the therapist, of the therapies used. The patient's autonomy is encouraged through patient involvement in decision-making and nurturing of self-reliance. Specialism does not exclude holism; nephrologists can be as holistic as naturopaths. As shown on p 422, most models of the GP consultation are based on a patient-centred approach which is essentially holistic. Compare the sequence 'bronchitis → antibiotic' with 'bronchitis → smoker → stressed → redundancy-counselling → ?antibiotic'.

1 K Mumby 1993 *BMJ* ii 1055 2 D Reilly 1994 *Lancet* 344 1601 3 1991 *BMJ* i 316 4 WP King 1988 *Otolaryngol Head Neck Surg* 99 263–71 & 272–7 5 P Pietroni 1988 in *Primary Care* p 114 Heinemann

Living dangerously

Ten years ago a patient had a seminoma treated, apparently successfully, in a well-known London hospital, to which he had been referred from his distant Sussex village. In the year that follow-up stopped, the patient had a major myocardial infarction—again, followed by an apparently reasonable recovery. But the patient became morbid, self-centred and depressed, perhaps because of the dawning appreciation of his mortality, his residual breathlessness, and his inability to carry out his hobby of carpentry. His GP tried hard to cheer him up, and rehabilitate him by encouraging exercise, sex, a positive self-image, and alternative hobbies. Rehabilitation was almost working when he began to develop headaches and kept asking forlornly whether these were a sign that his cancer had spread to his brain. There were no signs of recurrent tumour or raised intracranial pressure. His GP appreciated that there *was* a chance that the tumour was resurfacing, but judged that starting a pointless chain of investigations would be disastrous to the patient's mental health. So instead of arranging CT scans the GP interpreted the patient's forlorn question for him, saying that he was only asking questions like this because he was in a negative frame of mind, and the patient and his GP developed strategies to avoid negative cognitions, in co-operation with the local consultant who had helped to look after his myocardial infarction. The headaches improved, and the pressure to investigate was resisted.

Had a CT scan been done, it would by no means have achieved reassurance if it was negative—in the patient's frame of mind he would be all too willing to ask if the CT scan is 100% reliable, and then to request some other test in addition, and so on, until illness had become a major preoccupation. So in this case, it was rational for the doctor to live dangerously, take risks, and be prepared to take the blame if things had gone wrong.

How do we cope with and thrive on uncertainty? The first step is to get away from the idea that if you do not do all you can to reduce uncertainty, you are somehow being lazy. (The reverse may be true.) The next step is to share the uncertainty with a colleague—to see if he or she agrees with your judgment. From the medicolegal viewpoint it is wise to document your thought processes. Another caution is to follow in the steps of those adventurous but wise mountaineers who never plan a route without also planning an escape route: in the medical sense, this means the triad of follow-up, the taking of the family into one's confidence, and honest reflection on the chances of error and the chances of detecting it. This means that, as far as possible, you will get early warning of error, and then be able to adjust your therapeutic approach in line with the way the illness unfolds.

Minor illness

Many people with apparently minor illnesses visit their GPs, although GPs do not have a sufficient monopoly of this to justify being called 'triviologists'. Most minor illnesses are dealt with from a disease point of view, in the relevant chapter in which they occur. Here the concern is with the study of minor illness itself, and in this context much minor illness does not come to the general practitioner: only *people* come to general practitioners, and it may not be known for some time whether the symptoms are serious or minor (minor to whom?). A GP may not want to spend all his time on minor conditions, but this may become almost unavoidable if he issues a prescription for such complaints. This reinforces attendance at the surgery, as a proportion of patients will come to assume that a prescription is necessary. In current practice, GPs rate about 14% of their consultations as being for minor illness (mild gastroenteritis, upper respiratory problems, presumed viral infections, 'flu, and childhood exanthemata).[1] More than 80% are likely to receive a prescription, and >10% are asked to return for a further consultation. Why does this great investment of time and money occur? Desire to please, genuine concern, prescribing to end a consultation and therapeutic uncertainty may all play a part. Positive correlations with low prescribing rates include a young doctor, practising in affluent areas and long consultation times. Patients in social classes I and II are more likely to get a visit for minor ailments than those in other social classes. Membership of the Royal College of General Practitioners does not influence prescribing rates.[1] Not everyone wants to reduce prescribing, but advice is available for those who do:[2]

- Using a self-care manual explaining about minor illness.[3]
- Using self-medication (eg paracetamol for fever).
- Using the larder (eg lemon and honey for sore throats).
- Using time (eg pink ear drums[4]—follow up).
- Using granny (a more experienced member of the family).
- Pre-empting the patient's request for antibiotics (eg for a sore throat), eg: 'I'll need to examine your throat to see if you need an antibiotic, but first let me ask you some questions. . . From what you say, it sounds as if you are going to get over this on your own, but let me have a look to see.' [GP inspects to exclude a quinsy.] 'Yes, I think you'll get over this on your own. Is that all right?'

1 C Whitehouse 1985 *J Roy Col Gen Prac* **35** 581   2 G Marsh 1977 *BMJ* ii 1267
3 D Morrell 1980 *BMJ* i 769   4 D Brooks 1983 *Update* **26** 1961–4

Prescribing and compliance

On any day ~60% of people take drugs, only half of which are prescribed. The others are sold over the counter (OTC). The commonest OTCs are analgesics, cough medicines and vitamins; for prescribed drugs the common groups are CNS and cardiovascular drugs, and antibiotics. On average, 6–7 NHS prescriptions are issued/person/year (21 in Italy and 11 in France).

GPs account for 75% of NHS annual prescribing costs (>£2000 million, or ~10% of the total cost of the NHS), although many of these 'GP drugs' will have been initiated in hospital. The cost of these prescriptions has risen by a factor of 4.75 since 1949 (after allowance for inflation) and is ~£80,000/GP/year. Positive correlations with low prescribing rates include a young doctor, practising in an affluent area, and a longer consultation time (>7 min). The reason for this may be that if extra time is spent with the patient, more explanation about minor ailments (p 458) may be given, so that a patient's expectation for a prescription is replaced by enlightened self-awareness.

General practice formularies The aim is to reduce the drug 'bill' and to make prescribing more effective, by producing an agreed list of favoured drugs. This voluntary restriction can work in tandem with compulsory NHS restricted lists, and lead to substantial savings[2] (eg 18%). The DoH recommends development at individual practice level but this time-consuming task may be better achieved by adapting an existing local formulary.[2] Unless you wish to reinvent the wheel, this would seem excellent advice.

Dispensing doctors In rural areas where there is no chemist's shop GPs are allowed to dispense to their patients. Their annual prescribing rate is 70% of their non-dispensing fellow GPs.

Compliance (Does the patient take the medicine?)

▶ *There is no point in being a brilliant diagnostician if nobody can be persuaded to take your treatments.*

Even in life-threatening conditions, compliance is a major problem occurring in up to 56% of patients (eg adolescents with acute lymphatic leukaemia). The following have been found to be associated with increased compliance:
- Being able to identify with a personal doctor.
- Patient's overall satisfaction with the doctor.
- Simple therapeutic regimens.
- Supplementary written information[3] (use short words—Flesch formula >70, OHCM p 5).
- Longer consultation times or prescribing on home visits.
- Prescribing in association with giving health education.
- Continuity of care by the GP.
- Short waiting time for appointments.
- The encouragement of self-monitoring by the patient.
- Belief in the efficacy of the treatment.

Monitoring compliance: Monitoring plasma drug levels is the most reliable way of doing this, but it is cheaper to ask patients to return with their tablets, so that you can count them.

1 *Drug Ther Bul* 1991 **29** 25   2 P Green 1985 *J Roy Col Gen Prac* **35** 570–2   3 P Ley 1976 in *Communications Between Doctors and Patients*, Ed A Bennett, OUP

461

Social matters

Unemployment and the family ~50% of children in care have parents who are unemployed. UK data from the 1971 census showed an association between child deaths (0–4yrs old) and unemployment, lower social class and overcrowding. Babies whose fathers are employed tend to be heavier at birth (by 150g) than unemployed fathers' babies, after adjusting for other factors.[1] Accidents[2] and infection are more rife among children of the unemployed compared with carefully selected controls, and their mothers may be more prone to depression. As unemployment rises, so does child abuse.[1] Other factors identified with this rise are marital discord, debt, and parents' lack of self-esteem, as affected families reveal: 'When he lost his job he went absolutely bonkers. He changed completely. He became depressed and snappy. Frustrated.'[1]

Marital breakdown heads the list of problems of women in general practice with neurosis, and comes second (to employment difficulties) in men, and is a leading factor in >60% of suicide attempts.[3] In the USA divorced males have the highest rates of mortality. The greater incidence of cardiac deaths is most marked in young divorced males. Being divorced and a non-smoker is nearly as dangerous as smoking a pack a day and staying married.[3] Marital harmony (eg cuddling) protects from cardiac death, as shown in one prospective study of 10,000 Israeli hearts.

UK social security benefits[4] *The elderly:* If the only source of income is the pension, it may well be worthwhile applying for Income support (formerly supplementary benefit). This provides a basic personal allowance ± a premium (eg £36.80/wk if <25yrs or £46.50 if >25yrs) depending on needs arising out of old age, sickness, disability (or family responsibilities).

The disabled child premium is for a child who is registered blind or who is receiving an attendance or mobility allowance.

The disability premium (£19.80/wk) is for claimants or partners receiving income support, attendance or invalid care allowance, or the severe disablement allowance. It is payable if the claimant (not the carer) has not worked because of ill health for 28 weeks.

The severe disability premium (£35.05/wk) is for those receiving an attendance allowance, if no one is receiving invalid care allowance for looking after him or her and he or she is living alone (defined by DHSS).

This system of a basic allowance plus premiums is used for calculating housing benefit for those who are not entitled to income support. No housing benefit is due if the claimant has more than £16,000 in savings.

Young families: If the bread-winner is low paid consider applying for family credit,[5] which does not need repaying, and a loan from the cash-limited social fund for buying essential equipment (eg a cooker). Family credit (average value £25/week) is a tax-free cash payment to families in which the bread-winner works for ≥16h/week. Those earning ≤£110–£170/week may be eligible (depending on number of children and savings—which, can sometimes be up to £8000).

The disability living allowance: This is a tax-free benefit, not affected by earnings or savings, for people <65yrs old, and is divided into two components. The *Care* component for those needing help with washing, dressing, toilet and (if >16yrs) preparing main meals. This can be claimed even if no-one is actually giving the care (£46.70 high rate, £31.20 middle, £12.40 low). The *Mobility* component is for those aged ≥5yrs who have difficulty walking, or who need help to ensure that walking is not dangerous (unsafe or disorientated patient)—it is £32.65 high rate; £12.40 low. There are special rules for those with terminal illness (<6 months to live) to ensure prompt payment. These require a care worker, often the GP, to complete a DS1500 form (for which a fee may be claimed by the doctor).

The attendance allowance: This is tax-free weekly benefit for people aged 65 or over who need help with personal care because of illness or disability.[7] This is payable to the person needing attending to, and not the person attending. Depending on the disability, the weekly rate is either £31.20 or £46.70. All applications are made to Social Services.

Calculating benefits is a complex task, and benefit eligibility and amounts frequently change. One way of keeping up to date is to use software such as the free *Lisson Grove Welfare Benefits Program* to allow specific amounts to be calculated for individual circumstances.[6] Advice is also available on a free number (0800 666555).

1 R Smith 1985 *BMJ* ii 1707   2 R Alwash 1988 *BMJ* i 1450   3 J Dominian 1985 *Update* **31** 809
4 A Ketley 1988 *BMJ* i 1446   5 Family credit unit, Government buildings, Warbreck Hill, Blackpool FY2 0YF   6 Lisson Grove Health Centre, Gateforth St, London NW8 8EG (tel 0171 724 0480)
7 *Which Benefit?*, FB2, Benefits Agency

Health and social class

Throughout human history there have been inequalities in the health of classes and populations, caused by social factors. With the introduction of the British National Health Service, with its ideal of equal access to medical care for all groups in society it was assumed that differences in the health of different social (occupational) classes would be eliminated. We now know that this has not happened, and this has been amply documented in various reports such as the *Black Report (Inequalities in Health)* and the *Health Divide*.[1]

The Registrar General's scale of 5 social or occupational classes

Class I	Professional	eg lawyer, doctor, accountant
Class II	Intermediate	eg teacher, nurse, manager
Class III^N	Skilled non-manual	eg typist, shop assistant
Class III^M	Skilled manual	eg miner, bus-driver, cook
Class IV	Partly skilled (manual)	eg farmworker, bus-conductor, or packer
Class V	Unskilled manual	eg cleaner, labourer

There is a remarkable concurrence of evidence concerning the factor by which mortality rates are higher in social class V compared with those of social class I (with regular gradations between). For stillbirths, perinatal deaths, infant deaths, deaths in men aged 15–64 and women aged 20–59 this factor is respectively 1.8, 2, 2.1, 2, and 1.95.[1] The same sort of factors hold true for specific diseases such as the standardized mortality ratios (SMR) from lung carcinoma (1.98), coronary heart disease (1.3) and cerebrovascular disease (1.9). Only malignant melanoma, eczema, and Hodgkin's disease in early adulthood[2] show a reverse ('disease of affluence') trend. Note: the SMR is the ratio of mortality rates in one class compared with the average for the whole population. The whole population has an SMR of 1.00.

There is more emotional stress and chronic ill health in the 'lower' occupational classes. Furthermore, people who own their own homes have less ill health than private tenants, especially if they live in residential retirement areas.[1]

Within occupations the effect of social class is seen in a 'purer' way than when groups of many occupations are compared: in a study of >17,000 Whitehall civil servants there was a greater than 3-fold difference in mortality from all causes of death (except genitourinary diseases) comparing those in high grades with those in low grades. Similarly in the army there is a 5-fold difference in mortality rates from heart disease between the highest and the lowest ranks.

It has been pointed out that being ill makes a person 'descend' the social scale, but it has been estimated that this effect is not large enough to account for the observed differences between classes. It is much more likely that the differences are due to factors such as smoking behaviour, education, marital status, poverty and overcrowding.[1]

1 M Whitehead 1987 *The Health Divide*, Health Education Council
2 N Gutensohn 1982 *Cancer Treatment Reports* 66 689–95

Healing

Since neolithic times, healing has had a central place in our culture, and has long been recognized as 'mor bettir and mor precious þan any medicyne'.[1] Recently medicines have improved greatly, so that the rôle of doctors as the purveyor of medicines has eclipsed their more ancient rôles. We all recognise the limits of our rôle as prescribers, and we would all like to heal more and engage in repetitive tasks less often. But what, we might ask, *is* healing? How is it different from curing? Healing is, at one level, something mysterious that happens to wounds,[2] involving inflammation → granulocyte, macrophage, and platelet activation → release of platelet derived growth factor and transforming growth factors α and β → neovascular growth → fibroblast-mediated contraction → proteoglycans and collagen synthesis, lysis, and remodelling.

On another level, healing involves transforming through communication—a kind of hands-on hypnosis. We can cure with scalpels and needles, but these are not instruments of communication. Here is a serendipitous example of healing (an all too rare event in our own practice). On a rainy February evening, after a long surgery, I visited a stooped old man at the fag-end of life, with something the matter with his lung. 'I suppose it's rotting, like the rest of me—it's gradually dying.' I reply: 'Do you think you're dying?' 'Aren't we all?' 'Green and dying' I reply for some reason, half remembering a poem by Dylan Thomas. The patient looks mystified: he thinks he misheard, and asks me to repeat. 'Green *and* dying' I say, feeling rather stupid. There is a pause, and then he rises to his full height, puffs out his chest, and completes, in a magnificent baritone, the lines: '. . . Time held me green and dying, though I sang in my chains like the sea.'[3] By chance I had revealed a new meaning to a favourite poem of his which perhaps he thought was about childhood, not the rigours of his old age. Both our eyes shone more brightly as we passed on to the more prosaic aspects of the visit. This illustrates the nature of healing: its unpredictability, its ability to allow us to rise to our full height, to sing, rather than mumble, and how externally nothing may be changed by healing—just our internal landscape, transformed by a moment of illumination. It also shows how healing depends on communication, and how it is bound up with art. Healing may be mysterious, but it is not rare. We have so often kissed the grazed knees of our daughters that we expect the healing balm of kisses to wear out, but, while they are young, it never will, because children know how to receive but not how to doubt, and the kiss is the paradigm of healing: contact between two humans, wordless service of the lips, which the sterile advice of a microbiologist would never condone, an activity for which there is no Read code, and which we know as healing.

The bread and butter of our work comprises the sifting of symptoms, deciding what is wrong, and prescribing treatment—all tasks which, according to an historic prediction by no less a personage than the editor of the *Lancet* are destined for delegation to microchips.[4] This implies that our chief rôle will be as healers and teachers. There is much to be said for the idea that we should throw away the paraphernalia of mechanistic medicine, our formularies, our computers, and our audits—and insist on returning our primordial rôle. We should, perhaps, pay more than lip-service to our daughter's view of general practice, revealed when she wanted one of us to return home promptly for a family outing, when she said 'Can't you just kiss them all a bit faster, and come home quickly today?'

No doubt there will always be some way to go before healing, the central ideal of medicine, becomes its central activity. After all, the last thing any of us wants, when struck down by appendicitis, is a poet or a healer—but last things will always retain their power to set us thinking.

1 Anon ~ 1400 *Secreta Secret.* Gov. Lordsh. EETS 66 & OED 1ed **V** page 152 col 1 2 *Oxford Textbook of Surgery* 1ed 1994 page 5, OUP 3 Dylan Thomas *Fern Hill* 4 *Lancet* Ed 1995 **345** 1126

Purchasers and providers

Two contrasting principles ►He who pays the piper, calls the tune.
►Priceless therapeutic assets cannot be bought or sold: these include compassion, continuity of care and commitment.

Never just ask how good a structure is without also asking how good it is at transforming itself: that which cannot transform, dies. The UK National Health Service is the largest employer in the Western world and for years the search has been on to find ways to control and transform this dear, mighty thing. The purchaser-provider split is the most powerful lever yet developed for this purpose. *Purchasers* commission care by drawing up contracts with competing *providers*, who deliver the care. The better they deliver care (do *not* pause to ask what 'better' means: speculation on this point might ruin the argument) the more likely they are to get the contract next year. The catch is that all the extra effort the provider makes to out-perform a contract this year will be assumed and taken for granted next year. The same may hold true if purchasing is used for the imposition of guidelines ('evidence based purchasing').[1] What has been created is a treadmill which goes faster and faster, while taking less and less account of individual patients' and doctors' legitimate but varying needs. Unless the market is rigged, natural selection ensures that the fittest and fastest providers survive. Patients and tax-payers benefit—until the point where cynicism and exhaustion set in. There is no evidence that once the purchaser-provider path is chosen, then cynicism and exhaustion *inevitably* follow, and there is evidence at local level that benefits accrue, and services become more tuned to consumers' desires. (Consumers are not infallible judges of what constitutes health—but they are the best judges we have.)

Controlling change—from on top: an example from maternity[2]
1 Government sets up an expert group to change a specific area—eg maternity, containing mothers, midwives, ministers, obstetricians and general practitioners (these are jokers in the pack, because they are simultaneously consumers, purchasers *and* providers).

2 Issuing of objectives and indicators of success—eg by 5 years:
● Each woman should be entitled to carry her own notes.
● She should have a named midwife to ensure continuity of care.
● Women should be able to choose their place of delivery (98% are in consultant units, but surveys show 72% wanted other options—22% saying they would like the choice of home birth). Every effort must be made to achieve the outcome that she believes is best for her baby and herself.
● ≥75% of women should know the person who is to deliver them in labour.
● Midwives should have direct access to some beds in all maternity units.
● ≥30% of women should have a midwife as the lead professional.
● ≥30% of admitted deliveries to be admitted under midwife management.
● All front-line ambulances should have paramedics to support midwives.
● All women should have access to information about local services.

3 The group's attractive-looking report is issued[2] (using tax-payers money) to all groups and personnel involved (except mothers).

4 Debate is stimulated, and progress is reviewed.

The anatomy of change Ideals (woman-centred care)→ Specific policy objective (all women to have the chance to discuss their care)→ Purchasers' action point (set up Maternity Services Liaison Committee with lay chairperson)→ Providers' action point (provide link-workers, and advocacy schemes for women whose first language is not English).

Controlling change—from grass-roots upwards Example: GP fundholders buying on-the-spot consultant services for their patients (p 470).

1 M McKee 1995 *BMJ* i 101 2 DoH 1993 *Changing Childbirth*, HMSO, London, ISBN 0 11 321623 8

Fitness to drive[1] (Ordinary UK licences only)

Ordinary UK driving licences issued by the DVLA (Driver and Vehicle Licensing Agency—formerly DVLC) are inscribed 'You are required by law to inform Drivers Medical Branch, DVLA, Swansea SA99 1AT at once if you have any disability (either physical or medical condition), which is, or may become likely to affect your fitness as a driver, unless you do not expect it to last more than three months.' It is the responsibility of the driver to inform the DVLA. It is the reponsibility of their doctors to advise patients that medical conditions (and drugs) may affect their ability to drive and for which conditions patients should inform the DVLA. Drivers should also inform their insurance company of any condition disclosed to the DVLA.

Driving is prohibited if: Serious CNS or CVS disorders—see below—and if:
- Severe mental disorder (including severe mental impairment).
- Severe behavioural disorders.
- Alcohol dependency (including inability to refrain from drink driving).
- Drug abuse and dependency.
- Psychotic medication taken in quantities to impair driving ability.

Vision Acuity (± spectacles) should be sufficient to read a 79.4mm high number plate at 20.5 metres (~6/10 on Snellen chart).
- Monocular vision is allowed if the visual field is full.
- Binocular field of vision must be >120°.
- Diplopia is not allowable unless mild and correctable (eg by an eye patch).

Cardiovascular conditions[2] People should not drive:
- Within 1 month of myocardial infarction: if uncomplicated recovery there is no need to inform the DVLA (1990 guidelines).
- Within 1 month of angioplasty, pacemaker, heart valve/artery surgery, 2 months of heart ± lung transplant.* (* = inform DVLA.) Defibrillating pacemakers or implanted anti-tachycardia devices preclude future driving.
- Angina provoked by driving.*
- Symptomatic arrythmia/bradycardia.* Asymptomatic atrial fibrillation is OK.
- If medication reduces alertness, or causes vertigo or faintness.
- If unexplained syncope, acquired complete or 2° heart block.*

Diabetes mellitus All on oral hypoglycaemics or insulin must inform the DVLA. If insulin-treated, drivers must demonstrate satisfactory control, and must recognise the onset of any hypoglycaemia—the main risk. Check that vision conforms to required standard (above). Advise to avoid driving if hypoglycaemic risk ↑ (eg meal delay; or after excess exercise). Carry rapidly-absorbed carbohydrate in vehicle and stop, turn off ignition and consume it if any warning signs. A card should be carried to say which medications they are using to aid with resuscitation if needed. If an accident is due to hypoglycaemia a diabetic driver may be charged with driving under the influence of drugs.

CNS disorders Disabling giddiness, untreatable vertigo, and significant problems with movements preclude driving. DVLA need to know about unexplained blackouts, multiple sclerosis, Parkinson's (any 'freezing' or on-off effects), motor neurone disease, recurrent TIAs and strokes. In the latter the licence is usually withheld for 3 months depending upon results of examination by an independent doctor, and sometimes a driving test. Those with dementia should only drive if the condition is mild. Relatives should be encouraged to contact the DVLA if they believe a dementing relative should not be driving and the doctor may feel the need to breach confidentiality and inform the DVLA in good faith for demented or psychotic patients (phone DVLA adviser on 01792 783686). Detailed advice as to when to drive after neurosurgery and with brain tumours is given in reference 1.

Epilepsy and brain surgery: A licence may be granted if:
1 Free from attack for 1 year prior to licence validity date—or
2 For previous 3 years attacks only occur during sleep.

All drivers developing epilepsy must inform DVLA. Those with a single fit, their first fit during sleep, brain surgery, intracranial haematoma removal, or depressed skull fracture must usually wait one attack-free year before driving. If a first attack has been diagnosed and the person has persistent 3-per-second spike and wave activity on the EEG they are still regarded as suffering from epilepsy and are precluded from driving. Epileptic drug withdrawal risks a 40% seizure rate in the 1st year. Those wishing to withdraw from medication should cease driving from the beginnig of withdrawal and not recommence until 6 months after treatment has ceased.

Drugs Driving, or being in charge of a vehicle when under the influence (including side effect) of a drug is an offence under the Road Traffic Act 1988. Many drugs affect alertness and driving ability (check *Data Sheets*), and many are potentiated by alcohol so warn patients not to drive until they are sure of side effects, not to drink and drive, not to drive if feeling unwell, and never to drive within 48h of a general anaesthetic.

1 DF Taylor 1995 *Medical Aspects of Fitness to Drive,* published by the Medical Commission on Accident Prevention. 35–43 Lincoln's Inn Fields, London WC2A 3PN and DVLA 1993 *Guide to the Current Medical Standards of Fitness to Drive* 2 R Gold 1990 *Health Trends* 22 31

Referral statistics

UK GPs are obliged to give Family Health Service Authorities details of their referrals to hospitals. They are also obliged to be prepared to discuss referrals with medical representatives of FHSAs—implying that some GPs over-refer, and waste hospital resources. What is the evidence for this? Why is there a 4-fold difference in referral rates between GPs?

If high-referring GPs refer unnecessarily, then the proportion of their referrals resulting in admission should be smaller than that of practices of similar size with low-referring GPs. Usually, this is not the case. Those with high referral rates have high admission rates.[1] How far does this relation hold? If I refer an ever-increasing number of my patients to a geriatric clinic, must a time come when admissions level off? The idea of a 'levelling-off effect' is important. If the consultant's actions are 'correct', and the GP's expectation as to the outcome of referral are uniform (probably never true) then when a levelling-off effect is observed, it may be true that the *average* referral rate is optimal, and that low-referrers are depriving patients, and high-referrers are wasting resources. In fact, levelling-off effects are rarely found—except in general surgery. What are we to conclude from most of the other specialties where no levelling-off effect is observed? Perhaps specialists admit a fixed proportion of patients referred to them. There is some evidence that this is true for ENT consultants and tonsillectomy. Another possibility is the Coulter-Seagroatt-McPherson hypothesis—that consultants have a threshold of severity for admission (eg a claudication distance of 50 metres) and even the majority of patients from the high-referrers fulfil this criterion. In this case (assuming the consultant is right), even the high-referrers are not referring enough. This may be true for all forms of angiography, for example. However, if the consultant is over-enthusiastic, and over-optimistic about the benefits of treatment, then the lower referrers are to be applauded for limiting the excesses of the consultant.

Overall, referral rates are no more variable than admission rates—even among populations with similar morbidities. The reason is probably that there is still a great deal of uncertainty underlying very many clinical decisions. We do not know who should have knee replacements, coronary angiography, cholecystectomy, aneurysm surgery, transplants, or grommets.

►There is no known relationship between high or low referral rates and quality of care. Here are 3 cautions in interpreting referrals:[2]

1 Individual list size should not be used as a denominator, as it takes no account of differing work-loads within a practice. Consultations per year would be a better denominator.

2 If doctors within the practice have special interests, these must be taken into account in comparing referral patterns.

3 Years of data are needed to compare referrals to rarely-used units.

Referral statistics and fund-holding Since 1991, UK GPs have been able to receive funds for purchasing care for their patients—a controversial idea partly because of fears that in trying to keep to budget, GPs might refer less, or simply opt for the cheapest solutions, or might 'force' patients to use the private sector. Analysis of 28,371 referrals[3] from matched fund-holding and non-fundholding practices has not substantiated any of these fears: indeed referral rates rose in fund-holding practices even in specialties where referral often leads on to significant expense (eg surgery). But as budgets tighten, referrals may mirror prescribing, where fundholding has clearly driven down costs, with >60% saving money;[3] but note that non-fundholders are also underspending their budgets, showing that fund-holding is not the only valid way of addressing prescribing costs.▪

1 A Coulter 1990 BMJ ii 273   2 M Roland 1990 BMJ ii 98   3 A Coulter 1993 BMJ i 433 & ii 1186

Looking at words

Accommodation The active changing of lens shape to focus near objects.

Acuity A measure of how well the eye sees a small or distant object.

Amblyopia Reduced acuity which is not from an anatomic optic defect.

Amsler grid Test chart of intersecting lines used for screening for macular disease. Lines will appear wavy and squares distorted to those with macular disease.

Anisocoria Unequal pupil size.

Anisometropia Having different refractive errors in each eye.

Aphakia The state of having no lens (eg removed because of cataract).

Blepharitis Inflamed lids.

Canthus The medial or lateral angle made by the open lids.

Chemosis Oedema of the conjunctiva.

Choroid Vascular coat between the retina and the outer scleral coat.

Ciliary body Portion of uvea (uveal tract) between iris and choroid, containing the ciliary processes and ciliary muscle (for accommodation).

Conjunctiva Mucous membrane on anterior sclera & posterior lid aspect.

Cycloplegia Ciliary muscle paralysis preventing accommodation.

Dacryocystitis Inflammation of the lacrimal sac.

Diopter Units for measuring refractive power of lenses.

Ectropion The lids evert, (especially lower lid).

Entropion The lids invert, (so that the lashes may irritate the eyeball).

Epiphora Passive overflow of tears onto the cheek.

Fornix Where bulbar (scleral) and palpebral (lid) conjunctivae meet.

Fovea The tiny, vital, cone-rich area of retina capable of 6/6 vision.

Fundus That part of retina normally visible through the ophthalmoscope.

Keratoconus The cornea is shaped like a cone. See p 520.

Keratomalacia The cornea is softened.

Limbus The annular border between clear cornea and opaque sclera.

Macula Rim of avascular retina surrounding the fovea.

Miotic An agent causing pupil constriction (eg pilocarpine).

Mydriatic An agent causing pupil dilatation (eg tropicamide).

Near point Where the eye is looking when maximally accommodated.

Papillitis Inflammation of the optic nerve head.

Optic cup The cup like depression in the centre of the optic disc.

Optic disc The portion of optic nerve seen ophthalmoscopically in the fundus of the eye.

Presbyopia Age-related reduction of near acuity from failing accommodation.

Pterygium Wing shaped degenerative conjunctival condition encroaching on the cornea.

Ptosis Drooping lids

Refraction Ray deviation on passing through media of different density; OR determining refractive errors, and correcting them with lenses.

Retinal detachment The sensory retina separates from the pigmented epithelial layer of retina.

Sclera The whites of the eyes starting from the corneal perimeter.

Scotoma A defect causing a part of the field of view to go missing.

Slitlamp A device which illuminates and magnifies the structures of the eye.

Strabismus (squint) Eyes deviate, so they are not looking at the same thing.

Tarsorrhaphy A surgical procedure for uniting upper and lower lids.

Tonometer A device for measuring intraocular pressure.

Uvea Iris, ciliary body and choroid.

Vitreous Jelly filling the globe behind the lens.

Vitrectomy Surgical removal of vitreous.

Examination of the eye

To assess the optic nerve (nerve of vision), test visual acuity, visual field and colour vision (p 517).

Visual acuity This is a measure of central (macular) vision. Always test acuity carefully as loss of acuity is a grave sign. Record it accurately, especially in a patient with eye injury. Examine the right eye first. Sit the patient 6 metres from the Snellen's chart. Instruct the patient to obscure the left eye with an 'eye paddle' and read the Snellen chart from the top using the right eye. Then, similarly, test acuity of left eye. If the patient wears glasses record acuity with and without glasses. The last line completed accurately indicates the acuity for distant vision.

The chart is designed so that the top line can be read by someone with normal vision at 60 metres, the next at 36 metres, the next at 24, the next at 18, the next at 12, the next at 9, and the next at 6 metres. Acuity is recorded as 6/60, 6/36, 6/24, 6/18, 6/12, 6/9, 6/6 to indicate the last line accurately read (6/6 vision is normal). For acuities of worse than 6/60 the patient can be brought forward to 5, 4, 3, 2 and 1 metre from the chart to read the top line. If he can read it then acuity is expressed as that distance eg 5/60, 4/60, 3/60, 2/60 or 1/60. If the vision is below 1/60 ask the patient to count your fingers at ½ metre distance. This is recorded as CF (count fingers). If they cannot count your fingers move your hand in front of the eye at ¼ metre distance. If the patient can appreciate that your hand moves record HM (hand movement). If the patient cannot appreciate hand movement, dim the light in the examination room and shine a torch light into the eye. If the patient can perceive the light, record PL. If there is no light perception, record 'no PL'. The eye is blind.

If the patient sees less than 6/6 with or without glasses, examine again with a pinhole in front of the eye. A narrow beam of light then enters the eye eliminating the need to focus a beam. If the patient has only refractive error, there will be an improvement in vision as seen through the pin hole. This is an important test as it eliminates ocular pathology as a cause of reduced acuity. You can make a pin hole with a 22G needle in a 10 × 10cm opaque card. Check that you can see through the pin hole before giving it to the patient.

If the patient is older than 40 years and complains of blurring of near vision, the cause may be presbyopia (p 490). Near vision can be tested using a near vision testing card (p 478). If the patient can read N5 at 30cm, near vision is normal.

Visual field This is the area that can be seen with both eyes without shifting the gaze. The uni-ocular field is smaller than the binocular field. When assessing for visual field defects establish whether the defect affects only one eye or both, whether there are clear boundaries to the defect, if the boundaries lie in the vertical or horizontal meridians, and to what degree acuity is affected. For confrontation tests see page 492.

External examination: *Lids:* They should be symmetrical and retract normally on upward gaze (abnormal in dysthyroid disease). Note ptosis (p 480), spasm, inflammation or swellings (p 480). *Conjunctiva:* Look for inflammation, discharge, follicles or cobblestone patterns on tarsal conjunctiva of upper lid, and subconjunctival haemorrhage (p 496). *Cornea:* Examine with a torch for opacity, abrasion, ulcer (the latter two stain green with 1% fluorescein), or oedema. *Anterior chamber:* It is filled with clear aqueous but can be cloudy in anterior uveitis (p 494), may have sterile pus (hypopyon) with corneal ulcer, or blood (hyphaema) after injury. *Pupils:* Pupils should be equal and react to light and accommodation (PERLA). They are small and irregular in anterior uveitis, dilated and fixed in acute glaucoma (p 494). For other pupillary abnormalities see p 488. *Lens:* When the lens is normal the pupil appears black. With mature cataract it may appear white.

Extraocular movements It is particularly important to examine these in those with diplopia. Ask the patient to watch a pencil move horizontally, then vertically. Avoid extremes of movement as inabililty to maintain fixation stimulates nystagmus. For eye movements and squint see page 486.

Ophthalmoscopy This is used to examine for pathology in the lens, vitreous and retina. Start with high + numbers (red). To examine the lens and the vitreous focus the beam of the ophthalmoscope at the pupil at about 1 metre from the eye. In the normal eye there is a red glow from the choroid (called the red reflex). Any lens opacity (cataract) will be seen as a black pattern obstructing the red reflex. Blood or loose floaters in the vitreous will be seen as black floaters. Red reflexes are absent with dense cataract and intraocular haemorrhage. When the retina is in focus examine carefully the optic disc (should have precise boundaries and central cup, p 502). Note pallor or swelling. Examine the radiating vessels and macula (ask the patient to look at the light).

Aids to successful ophthalmoscopy:
- Ensure that the batteries are fully charged.
- Darken the room as much as possible.
- Remove the patient's spectacles and dial up the appropriate lens to correct for the resulting refractive error (– lenses correct myopia, + lenses correct hypermetropia).
- If the patient is very myopic or is aphakic, try examining with his spectacles on. The disc will appear very small.
- If you find ophthalmoscopy difficult using your non-dominant eye, try using your dominant eye for examining *both* fundi—while standing behind the seated patient, whose neck is fully extended.
- Always check the lens for opacities before trying to examine the fundus.
- Always get close enough to the patient—even if one of you has had garlic for lunch.
- Consider using a short-acting mydriatic to dilate the pupil (see p 518).
- Remember that most retinal tears are peripheral and are difficult to see without special equipment. It is not possible to see the periphery of the retina with the ophthalmoscope.

Slit lamp examination This instrument has a bright light source and a horizontally mounted microscope to examine the structures of the living eye. The light source can be converted to a slit (hence the name). Tonometric attachments allow intraocular pressure measurement.

N. 48

He Moved

N. 36

forward a few

N. 24

chattering gems. He

N.18

knew so exactly who would

N. 14

happen to sneeze calmly through
the open door. Had there been

N. 12

another year of peace, the battalion
would have made a half-designed system

N. 10

of drainage. A silent fall of immense snow came
near oily remains of supper on the table.

N. 8

We drove on in our old sunless walnut. Presently
classical eggs ticked in the new afternoon shadows.

H
A L
T N C
O L H A
E C T N O
C L O H N A
A E N L O M C T

N. 6

We were instructed by my cousin Jasper not to exercise by country
house visiting unless accompanied by three geese or gangsters.

N. 5

The modern American did not prevail over the pair of redundant bronze puppies.
The worn-out principle is a bad omen which I am never glad to ransom on purpose.

479

H

T

O Y

H U V

A T Y M

X O W U H

Y U V T X O

A W I M H Y T

Test types (N. 48–N. 5 opposite) should be read at 30cm.

The external eye

Entropion This is inturning of the eyelids. The lower lid is the more commonly affected. Usually due to degenerative changes in the fascial attachments of the lower lids and surrounding musculature, in the UK it is rare below 40yrs of age. As the eyelids turn in, the eyelashes rub the cornea and constantly irritate the eye. Taping the (lower) eyelids to the cheek gives temporary relief, surgery gives more lasting correction.

Ectropion This occurs in the older population, and those with facial nerve palsies. There is eversion of the lower lid resulting in irritation of the eye, watering (drainage punctum malaligned) and occasionally exposure keratitis. Plastic surgery may correct the deformity but the problem is more complex in facial palsy.

Ptosis This is drooping of the upper eyelid. Its lower border usually lies mid-way between the superior limbus and upper pupillary margin. *Causes:* •Congenital (absent nerve to the levator muscle, poorly developed levator); •Mechanical (oedema, xanthelasma or tumour of the upper lids); •Myogenic (muscular dystrophy, myasthenia gravis); •Neurological (third nerve palsy, p 486; Horner's syndrome, p 488—slight ptosis). Congenital ptosis is corrected surgically early if the pupils are covered or if it is unilateral (risk of amblyopia ex anopsia).

Lagophthalmos This is difficulty in complete closure of the eyelids over the eyeball. *Causes:* Proptosis, exophthalmos, mechanical impairment of lid movements (eg injury or burns to the lids), leprosy, paralysed orbicularis oculi giving sagging lower lid. Corneal ulceration and keratitis may follow. Lubricate eyes with liquid paraffin ointment. If corneal ulceration develops, temporary tarsorrhaphy (stitching the lids together) may be needed.

Styes (hordeolum externum) These are abscesses of the glands of Zeis and Moll at eyelash bases. They 'point' outwards and may cause extensive inflammation of the eyelids. Treatment is with local antibiotics (eg fusidic acid). Less common is **hordeolum internum**, abscesses of the meibomiam glands. These 'point' inwards, cause less local reaction but leave a residual swelling called a chalazion or a Meibomiam cyst (tarsal cyst) when they subside. Treatment for these residual swellings is incision and curettage under anaesthesia with topical antibiotic (eg fusidic acid) application several days post-operatively.

Blepharitis This inflammation of the eyelids may be due to local infection (eg staphylococcal) or be associated with seborrheic dermatitis. Eyes have 'burning' itching red margins and may have scales on the lashes. Treat with regular saline bathing and local antibiotic applications (eg fusidic acid). If inflammation remains very persistent, consider adding 1% hydrocortisone cream.

Pinguecula These degenerative yellow nodules appear on the conjunctiva either side of the cornea (typically the nasal side) commonly in adults. If they become inflamed (pingueculitis)—they may respond to topical steroids. If it encroaches on to the cornea, as it may in dusty, wind-blown life-styles, the word pterygium is used (surgery may be needed).

Dendritic ulcers These corneal ulcers are caused by the *Herpes simplex* virus. They produce photophobia and epiphoria (watering). If steroid eye applications are given in their presence there is massive amoeboid ulceration and risk of blindness. *Diagnosis:* 1% fluorescein drops stain the lesion. *Treatment* is with acyclovir 3% eye ointment 5 times per day for at least 3 days after complete healing.

Tears and lacrimation

The lacrimal glands are on the superior temporal side of the orbits. The tear film excreted over the eye drains via the lacrimal puncta (found at the medial side of the upper and lower eyelid) through the lacrimal sac, lacrimal duct and inferior meatus (just lateral to the inferior turbinate) into the nasal passages. Dry eyes may be due to insufficient tear secretion, and watering eyes may be due to blockage of the drainage system.

Acute dacryocystitis This is acute inflammation of the tear sac which is located medial to the medial canthus. This may spread to surrounding tissues and result in systemic upset. Immediate antibiotic therapy may resolve the infection. Failure will lead to local abscess formation which may need drainage.

Infantile dacryocystitis In babies the nasolacrimal duct may not be canalized at birth and may not open fully until the child is three months old. Tear sacs tend to get infected and a sticky discharge is produced. Ask an ophthalmologist to teach the mother to massage the sac to empty the contents four times daily and then apply antibiotic (eg gentamicin) eye drops. Should this fail after several months, probing of the duct under anaesthesia is an option.

Chronic dacryocystitis This tends to occur in the middle-aged and elderly. There is distension of the lacrimal sac, discharge of mucopus into the eye and blockage of the nasolacrimal duct. Syringing the lacrimal drainage system should be performed early to try to clear the system—and may need to be repeated. In some cases dacryocystorhinostomy (a surgical procedure to establish communication between the lacrimal sac and the nasal cavity) may be needed.

Dacroadenitis There is pain and swelling on the temporal side of the upper eyelid. The upper eyelid may appear S-shaped. The cause may be viral (mumps, measles, influenza) although gonococcal organisms can be the cause. Chronic swelling can occur in sarcoid, TB, lymphatic leukaemia or lymphosarcoma.

Tear production The volume of tears normally *in* the eye is 6µl, the turnover *rate* being 1.2µl/min. Tears are similar in electrolyte concentration to plasma, but rich in proteins, especially IgA. They also contain lysozyme and β-lysin which have antibacterial properties.

Dry eye syndrome (keratoconjunctivitis sicca) This may be due to ↓ tear production by the lacrimal glands (Sjögren's syndrome associated with connective tissue disorders (especially rheumatoid arthritis), mumps, sarcoid, amyloid, lymphoma, leukaemia, haemochromatosis, old age); due to excess evaporation of the tears (post-exposure keratitis); or due to mucin deficiency in the tears (avitaminosis A, Stevens–Johnson syndrome, pemphigoid, chemical burns). Schirmer's test (strip of filter paper put overlapping lower lid; tears should soak >15mm in 5min) reveals insufficient production. Artificial tears may be used for symptomatic relief.

Excess lacrimation *Causes:* Emotion, corneal abrasions or foreign body, conjunctivitis, iritis, acute glaucoma.

Epiphora (ie normal volume, but not reabsorbed) *Causes:* Ectropion, entropion, blockage of drainage system.

Lacrimal system

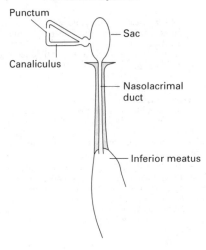

Orbital swellings

Lesions within the bony orbit tend to present with proptosis (ie exophthalmos), whatever the pathological origin. Proptosis (protrusion of the orbital contents) is a cardinal sign of intraorbital problems. If pressure is eccentric within the orbit there will be deviation of the eyeball and diplopia. Pain in the orbit usually arises from neighbouring structures (eg sinusitis).

Orbital cellulitis This usually results from spread of infection from the paranasal air sinuses. There is acute inflammation within the orbit with fever, swelling of the lids, proptosis, and immobility of the eye. Admit to hospital for prompt treatment. Systemic antibiotics, eg cefuroxime 1.5g/8h IV (child 20mg/kg/8h), are needed to prevent extension into the meninges and cavernous sinus thrombosis. Blindness is a risk from pressure on the optic nerve or thrombosis of its vessels.

Carotico-cavernous fistula This is usually due to rupture of a carotid aneurysm with reflux of blood into the cavernous sinus. There is engorgement of the blood vessels of the eye with oedema of eyelids and conjunctiva. Exophthalmos is classically described as pulsating. There is a loud bruit over the eye. Symptoms usually subside spontaneously. Carotid ligation may be necessary.

Orbital tumours Primary neoplasms are rare (angiomas, dermoids, meningiomas, or gliomas of the optic nerve). Secondary tumours are more common. Reticuloses can form orbital deposits (examine liver, spleen, nodes; do FBC). In children unilateral proptosis may be the first sign of a neuroblastoma. Nasopharyngeal tumours occasionally invade the orbit, as may mucocoeles and pyocoeles of the ethmoid and frontal sinuses. CT scan pictures give a clear representation of the orbit.

Hyperthyroidism may cause exophthalmos (p 510).

Ophthalmic shingles

This is zoster of the first (ophthalmic) branch of the trigeminal nerve and accounts for 20% of all shingles with only thoracic nerves being more commonly affected (55%). Pain, tingling or numbness around the eye may precede a blistering rash which is accompanied by much inflammation. In 50% of those with ophthalmic shingles the eye itself is affected with 40-77% having corneal signs and 50-60% iritis. Nose-tip involvement—Hutchison's sign—means involvement of the nasociliary branch of the trigeminal nerve which also supplies the globe and makes it highly likely that the eye will be affected. The eye can be seriously involved with little rash elsewhere. *Presentation:*

- Mucopurulent conjunctivitis
- Limbal lesions ▣
- Preauricular node tenderness
- Scleritis
- Episcleritis
- Visual loss
- V nerve palsy
- Keratitis
- Iritis (± atrophy)
- Pupillary distortion
- Optic atrophy

Treatment: Give acyclovir 800mg five times a day PO for 7 days to reduce viral shedding,▣ accelerate healing time and reduce incidence of new lesions. Start within 4 days of shingles onset. It is advisable for all with ophthalmic shingles to see a specialist within 3 days to exclude iritis with a slit lamp. Prolonged steroid eyedrops may be needed.

Eye movements and squint

To maintain single vision, fine coordination of eye movement of both eyes is necessary. Abnormality of the coordinated movement is called squint. Other names for squint are strabismus and tropia. Exotropia is divergent (one eye turned out) squint: enotropia is (one eye turned in) convergent squint. Prominent epicanthic folds (diagram) may produce pseudosquint.

Non-paralytic squint These squints usually start in childhood. The range of eye movements is full. Squints may be constant or not. All squints need ophthalmological assessment as vision may be damaged if not treated.
Diagnosis: Difficult, eg in uncooperative children. Screening tests:
1 Corneal reflection: reflection from a bright light falls centrally and symmetrically on each cornea if no squint, asymmetrically if squint present.
2 Cover test: movement of the uncovered eye to take up fixation as the other eye is covered demonstrates manifest squint; latent squint is revealed by movement of the covered eye as the cover is removed.

Convergent squint (esotropia) This is the commonest type in children. There may be no cause, or it may be due to hypermetropia (p 490). If the eye is left without focused input, vision may never develop satisfactorily in that eye (it has *amblyopia ex anopsia*).

Divergent squint (exotropia) These tend to occur in older children and are often intermittent. Amblyopia is less commonly a problem.

Management Remember 3 *'O's: Optical; Orthoptic; Operation*. Treatment should start as soon as the squint is noticed. *Optical:* assess the refractive state of the eyes after cyclopentolate 1% eye drops; the cycloplegia allows objective determination of the refractive state of the eyes; the mydriasis allows a good view into the eye to exclude abnormality, eg cataract, macular scarring, retinoblastoma, optic atrophy. Spectacles are then provided to correct refractive errors. *Orthoptic:* the good eye may be patched to encourage use of the one tending to squint. *Operations* (resection and recession of rectus muscles) may help alignment and give good cosmetic results.

Paralytic squint Diplopia is most marked when trying to look in the direction of pull of the paralysed muscle. When the separation between the two images is greatest the image from the paralysed eye is furthest from the midline and faintest.

Third nerve palsy (oculomotor) Ptosis, proptosis (as recti tone ↓), fixed pupil dilatation, with the eye looking down and out. Causes: p 488.

Fourth nerve palsy (trochlear) There is diplopia and the patient may hold his head tilted (ocular torticollis). The eye is elevated in adduction and cannot look down and in (superior oblique paralysed). Causes: trauma 30%, diabetes 30%, tumour, idiopathic.

Sixth nerve palsy (abducens) There is diplopia in the horizontal plane. The eye is medially deviated and cannot move laterally from midline, as the lateral rectus is paralysed. Causes: tumour causing increased intracranial pressure (compresses the nerve on the edge of the petrous temporal bone), trauma to base of skull, vascular.

The cover test

Pseudosquint Wide epicanthic folds give the appearance of squint in the eye looking towards the nose. That the eyes are correctly aligned is confirmed by the corneal reflection.

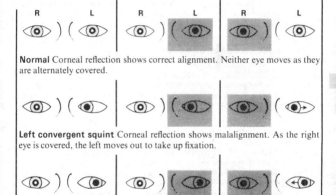

Normal Corneal reflection shows correct alignment. Neither eye moves as they are alternately covered.

Left convergent squint Corneal reflection shows malalignment. As the right eye is covered, the left moves out to take up fixation.

Left divergent squint Corneal reflection shows malalignment. As the right eye is covered, the left moves in to take up fixation.

NB If there is **Eccentric fixation** (ie foveal vision so poor that it is not used for fixation), the deviating eye will not move to take up fixation. Corneal reflection shows that malalignment is present. The cover test relies upon the ability to fixate.

Pupils

Pupil inequality Light detection by the retina is passed to the brain via the optic nerve (afferent pathway) and pupil constriction is mediated by the oculomotor (third) cranial nerve (efferent pathway). The sympathetic nervous system is responsible for pupil dilatation via the ciliary nerves.

Afferent defects (absent direct response) The pupil will not respond to light, but it will constrict if light is shone into the other eye (consensual response). Causes; optic neuritis, optic atrophy, retinal disease. The pupils will appear the same size (consensual response unaffected). If the Marcus Gunn swinging flashlight test is used, the affected eye will appear to dilate when the lamp is swung away from the consensual eye into the affected eye. Constriction to accommodation will still occur.

Efferent defects The third nerve also mediates eye movement and eyelid retraction. With a complete third nerve palsy there is complete ptosis, a fixed dilated pupil, and the eye looks down (superior oblique still acting) and out (lateral rectus acting). Causes: cavernous sinus lesions, superior orbital fissure syndrome, diabetes mellitus, posterior communicating artery aneurysm. The pupil is often spared in vascular causes eg with diabetes and hypertension. Pupillary fibres run in the periphery, and are the first to be involved in compressive lesions by tumour or aneurysm.

Other causes of a fixed dilated pupil Mydriatics, trauma (blow to iris), acute glaucoma, coning (OHCM p 678).

Holmes–Adie pupil This tends to occur in young adult women. Sudden onset of blurring of near vision occurs as accommodation is partially paralysed. The pupil is slightly dilated, there is no response to light and a very slow response to accommodation. If knee and ankle jerks are absent, too, the Holmes–Adie syndrome is present.

Horner's syndrome This occurs on disrupting the sympathetic nerve supply to the iris. The pupil is meiotic (smaller) and there is partial ptosis of the eyelid, but normal range of lid movement (supply to levator palpebrae is by the third nerve). The pupil does not dilate in the dark. Loss of sweating usually indicates a lesion proximal to the carotid plexus—if distal, the sudomotor (*Sudor*=sweat) fibres will have separated, so sweating is intact. Causes: ●Posterior inferior cerebellar artery or basilar artery occlusion ●Multiple sclerosis ●Hypothalamic lesions ●Syringomyelia in the pons ●Cavernous sinus thrombosis ●Cervical cord, mediastinal or Pancoast's tumour ●Aortic aneurysm ●Klumpke's paralysis (p 724) ●Cervical lymphadenopathy.

Argyll Robertson pupil This occurs in neurosyphilis and diabetes mellitus. There is bilateral meiosis with pupil irregularity. There is no response to light but there is response to accommodation (The prostitute's pupil accommodates but does not react.) The iris mesoderm is spongy, the pupils dilate poorly, and there may be ptosis.

Causes of light-near dissociation (-ve to light +ve to accommodation): Argyll Robertson pupil; Holmes–Adie syndrome; meningitis; alcoholism; tectal lesions eg pinealoma; mesencephalic lesions; thalamic lesions. The path from the optic tract to the Edinger–Westphal nucleus is disrupted but deeper cortical connections remain intact, so accommodation is spared.

Identifying the dangerous red eye

Ask yourself the following questions:

1 Is acuity affected? A quick but sensitive test is the ability to read newsprint with refractive errors corrected with glasses or a pinhole. Reduced acuity suggests dangerous pathology.

2 Is the globe painful? Pain is potentially sinister, foreign body sensation may be so, irritation rarely is.

3 Does the pupil respond to light? Absent or sluggish response is sinister.

4 Is the cornea intact? Use fluorescein eye drops, p 496. Corneal damage may be due to trauma or ulcers.

Enquire also about trauma and discharge, general health and drugs, and remember to check for raised pressure.

▶If in doubt, obtain a specialist opinion today.

	Conjunctivitis	Iritis	Acute glaucoma
Pain	±	++	++ to +++
Photophobia	+	++	−
Acuity	normal	↓	↓
Cornea	normal	normal	steamy or hazy
Pupil	normal	small	large
Intraoccular pressure	normal	normal	↑

More red eyes—cornea and conjunctiva

Corneal problems *Keratitis* is corneal inflammation (identified by a white spot on the cornea—indicating a collection of white cells in corneal tissue). *Corneal ulceration* is an epithelial breach; it may occur without keratitis, eg in trauma, when prophylactic antibiotic ointment (eg gentamicin 0.3%) may be used. Ulceration with keratitis is called ulcerative keratitis and must be treated as an emergency—see below. There is pain, photophobia, and sometimes blurred vision. Non-infective corneal ulceration may result from contact lens exposure, trauma, or previous corneal disease.

Ulcerative keratitis: Use fluorescein drops and a bright light (ideally with a blue filter, shone tangentially across the globe) to aid diagnosis. Corneal lesions stain green (the drops are orange and become more yellow on contact with the eye). Ulcers may be bacterial (beware pseudomonas: may be rapidly progressive), herpetic (simplex or zoster), fungal (candida, aspergillus), protozoal (acanthamoeba) or from vasculitis eg in rheumatoid arthritis. ▶Refer the same day to hospital as treatment depends upon the cause and delay may cause loss of sight. Anyone with corneal ulceration or stromal suppuration must have an urgent diagnostic smear (for Gram stain) and scrape by an experienced person. Liaise with microbiologist about samples. *Treatment:* See p 497.

Treat zoster infections with acyclovir (p 484). For *Herpes simplex* dendritic ulcers, see p 480. Cycloplegic drugs (p 518) may relieve pain due to ciliary spasm and prevent iris adhesions.

Episcleritis Inflammation below the conjunctiva in the episclera is often accompanied by an inflammatory nodule. The sclera may look blue below engorged vessels—which can be moved over the area, unlike in scleritis, where the engorged vessels are deeper. The eye aches dully and is tender to the touch, especially over the inflamed area. Usually no underlying cause is found, but it may complicate autoimmune diseases. It responds to steroid eye drops (eg clobetasone butyrate eye drops 0.1%/6h).

Scleritis Rarely, the sclera itself is inflamed. There is more generalized inflammation with oedema of the conjunctiva and thinning of the sclera (if very severe globe perforation is a risk). It may be associated with connective tissue disorders. Refer to a specialist.

Conjunctivitis ▶Conjunctivitis is usually bilateral, if apparently unilateral consider other diagnoses (p 495), eg acute glaucoma.

The conjunctiva is red and inflamed, and the hyperaemic vessels may be moved over the sclera, by gentle pressure on the globe. Acuity, pupillary responses, and corneal lustre are unaffected. Eyes itch, burn and lacrimate. There may be photophobia. Purulent discharge may stick the eyelids together. The cause may be viral (highly infectious adenovirus)—small lymphoid aggregates appear as follicles on conjunctiva, bacterial (purulent discharge more prominent), or allergic. The affliction is usually self-limiting (although allergic responses may be more prolonged). In prolonged conjunctivitis, especially in young adults or those with venereal disease consider chlamydial infection—see ophthalmia neonatorum, p 100.

Treatment: Usually with antibiotics, eg gentamicin 0.3% drops/3h and 0.3% ointment at night. For chlamydia use tetracycline 250mg/6h PO and as 1% ointment 6-hourly for at least one month. For allergic conditions use sodium cromoglycate 2% eye drops/6h.

Subconjunctival haemorrhage This harmless but alarming looking collection of blood behind the conjunctiva from a small vessel bleed requires no treatment. It clears spontaneously. If recurrent, look for a bleeding diathesis and check blood pressure.

Management of corneal ulcers[1]

Smears and cultures Liaise with microbiologist. Take:
1 Smear for Gram stain (if chronic ulcer Giemsa, PAS (Periodic acid Schiff) for fungi, ZN (Ziehl-Neelsen) or auromine for TB).
2 Conjunctival swab to blood agar for tear film contaminants.
3 Corneal scrape (by experienced person) from multiple areas of ulcer edge with needle for direct innoculation.
4 Request the cultures detailed below.

Acute history: Presume bacterial, so culture with blood agar (grows most organisms), chocolate agar (for Haemophilus and Neisseria), nutrient broth (anaerobes), cooked meat broth (aerobes/anaerobes).
Chronic history: Consider rarities; culture as above plus BHI (brain heart infusion) broth (for fastidious organisms and fungi), Sabourauds plate for fungi, anaerobic blood agar (peptococcus, proprionobacteria) + thioglycollate (anaerobes).
Unusual features use also viral transport medium, Lowenstein-Jensen agar slope for TB, *E Coli* seeded agar for acanthamoeba.

Initial eyedrop management Based on urgent Gram stain result.
● Copious bacteria of one type:
   Gram +ve: cefuroxime 50mg/ml (or methicillin 20mg/ml).
   Gram -ve: gentamicin 15mg/ml* (or ticarcillin 10mg/ml).
● Fungal elements only seen: miconazole 1%.
● Acute disease but scanty/multiple or no bacteria: gentamicin 15mg/ml*: + cefuroxime 50mg/ml or + methicillin 20mg/ml.
● No organisms in chronic disease: request PAS for fungi. If -ve await culture result at 24-72h without treatment: consider reculture or biopsy before starting treatment.

*Gentamicin 15mg/ml 'forte' preparation (▶see p 519).

Use drops every 15min day and night. If >1 drug used, alternate them. To prepare drops, see p 519. Vary therapy with severity. Beware drug toxicity causing delayed epithelialization (especially gentamicin). Deep or limbic ulcers may need systemic treatment.

When the organism is known Use cefuroxime 50mg/ml for staphylococci and other Gm +ve organisms; penicillin G 5000u/ml for streptococci; gentamicin 15mg/ml* for pseudomonas, enterobacteriaceae, moraxella; amikacin 10mg/ml for mycobacteria, miconazole 1% for candida and aspergillus; neomycin 0.5% + propamidine isethionate 0.1% (± a steroid) for acanthamoeba.

   Cycloplegia is used as adjunctive therapy. Steroid drops eg prednisolone 0.5%/6h are occasionally needed to help to control vascularization and reduce inflammation. Only use them after positive culture and usually not in the 1st 10 days treatment. Use with great caution if pseudomonas or fungal infection the cause.

   When culture-positive ulcers fail to respond consider whether treatment is sufficient, or too toxic (preventing epithelialization) or whether initial culture may be wrong. Ulcers which perforate may need penetrating grafts to include all infected material or other specialized surgical techniques.

---

1 Information kindly supplied by Mr JKG Dart FRCS, Moorfields Eye Hospital, City Rd, London EC1V 2PD

Sudden painless loss of vision

When an elderly person complains of sudden painless loss of vision, first think of central retinal artery occlusion due to temporal arteritis. Other causes are: amaurosis fugax, ischaemic optic neuropathy, vitreous haemorrhage, retinal detachment, migraine.

The fovea (macula) is the only part of the eye with 6/6 vision. Pathology affecting it causes most drastic visual loss.
▶ The patient wants an answer to the question 'Will I go blind?'
▶ Every patient requires specialist attention unless the cause is unequivocally migraine. ▶Always check the ESR since diagnosis of temporal arteritis can save the sight of the other eye.

Amaurosis fugax is temporary loss of vision 'like a curtain coming down'. It may precede permanent visual loss due to temporal arteritis or embolus, so diagnosis of the cause may save sight. NB: transient visual lasting 30–60 seconds in papilloedema is due to increased intracranial pressure.

Ischaemic optic neuropathy The optic nerve is damaged if the posterior ciliary arteries are occluded by inflammation or arteriosclerosis. Fundoscopy shows a pale, swollen optic disc.

Temporal arteritis (giant cell arteritis): The importance of recognizing this is that the other eye is at risk until treatment is started. There may be general malaise, jaw claudication (pain on chewing) and tenderness of the scalp and temporal arteries (check for pulses). There is an association with polymyalgia rheumatica (OHCM p 674). An ESR >40 is suggestive; temporal artery biopsy may miss the affected section of artery. Start prednisolone 80mg/24h PO promptly. Tailing off steroids as ESR and symptoms settle may take more than a year.

Arteriosclerotic ischaemic optic neuropathy: Hypertension, lipid disorders and diabetes may predispose younger patients to this. Treating these conditions may protect the vision in the other eye.

Occlusion of a central retinal artery There is dramatic visual loss within seconds of occlusion. In 90% acuity varies between light perception and finger counting. An afferent pupil defect (p 488) appears within seconds and may precede retinal changes by 1h. The retina appears very white but with a cherry red spot at the macula. Exclude temporal arteritis. The occlusion is usually due to thrombus or to embolus (listen for carotid bruits). *Treatment:* If you see the patient within 1h of onset apply firm pressure on the globe. Press on the eyeball and increase the pressure until the patient feels pain and then suddenly release the pressure. It may dislodge and drive the embolus into one of the branches. There is no reliable treatment, and if the occlusion lasts much longer than an hour the optic nerve will atrophy, causing blindness. If a single branch of the retinal artery is occluded, the retinal and visual changes relate only to the part of the retina supplied.

Vitreous haemorrhage This is a particularly common cause of visual loss in diabetics with new vessel formation. It may also occur in bleeding disorders, retinal detachment and with central retinal vein or branch vein occlusions. With a large enough bleed to obscure vision, the red reflex is absent and the retina may not be visualized. Vitreous haemorrhages undergo spontaneous absorption and treatment is expectant for the haemorrhage itself and directed against the cause (eg photocoagulation of new vessels). If the vitreous haemorrhage does not resolve within 3 months, vitrectomy may be performed to remove the blood in the vitreous. Small extravasations of blood produce vitreous floaters, (seen by the patient as small black dots or tiny ring-like forms with clear centres) which may not greatly obscure vision.

Central retinal vein occlusion Incidence increases with age. It is commoner than arterial occlusion. Chronic simple glaucoma, arteriosclerosis, hypertension, and polycythaemia are predisposing causes. If the whole central retinal vein is thrombosed, there is sudden visual loss with acuity reduced (eg to finger counting). The fundus is like a 'stormy sunset' (the angry-looking red clouds are haemorrhages alongside engorged veins). There is also hyperaemia. Long-term outcome is variable, with possible improvement for 6 months to one year; peripheral vision tends to improve most, leaving macular vision impaired. The main problems are macular oedema and neovascular glaucoma secondary to iris neovascularization. About ⅓ of eyes show significant non-perfusion on fluorescein angiography, of which one-half would develop neovascular glaucoma (called '100 days glaucoma' as it develops about this time after the occlusion). Panretinal laser photocoagulation is effective in treating and preventing neovascular glaucoma. At present there is no effective treatment for macular oedema from central retinal vein occlusion.

Branch retinal vein occlusion There is unilateral visual loss and fundal appearances in the corresponding area. Retinal capillary non-perfusion can lead to retinal new vessel formation. Treatment of this neovascularization (confirmed by fluorescein angiography) with laser photocoagulation, reduces risk of intraocular haemorrhage by 50%. Macular oedema persisting for months without improvement may be treated with grid pattern argon laser macular photocoagulation.

Other causes of sudden loss of vision in one eye are:
- Retinal detachment (p 506).
- Acute glaucoma (painful—see p 494).
- Migraine.

Stroke patients may complain of monocular blindness but visual field testing will usually reveal a homonymous hemianopia. Sudden bilateral visual loss is unusual (may be CMV infection in HIV patients, p 510).

Subacute loss of vision

Optic neuritis This is inflammation of the optic nerve. Unilateral reduction in acuity occurs over hours or days. Discrimination of colour is affected—reds appear less red, 'red desaturation'—and eye movements may hurt. The pupil shows an afferent defect (p 488). The optic disc may be swollen (papillitis) unless the inflammation is central (retrobulbar neuritis). Recovery is usual over 2–6 weeks, but 45–80% develop multiple sclerosis (MS) over the next 15yrs. Causes other than MS are neurosyphilis, other demyelinating diseases (eg Devic's disease), Leber's optic atrophy, diabetes mellitus and vitamin deficiency.
Treatment: High-dose methylprednisolone for 3 days (250mg/6h IV), then oral prednisolone (1mg/kg/day) for 11 days appears to reduce the risk of developing MS in the next 2yrs, from ~15% to 7%.[1] Seek expert advice.

1 RW Beck 1993 *NEJM* **329** 1764

Gradual loss of vision

The possible causes for gradual loss of vision in one eye are choroiditis, a creeping inferior retinal detachment (p 506), or a choroidal melanoma. If the loss is bilateral (usually asymmetrical), the cause is more likely to be

cataract, chronic glaucoma (p 502), diabetic and hypertensive retinopathy (p 510), senile macular degeneration, or optic atrophy (see below).

Choroiditis (choroidoretinitis) The choroid is part of the uvea (iris, ciliary body + choroid), and inflammatory disorders affecting the uvea may also affect the choroid. The retina may be invaded by organisms which set up a granulomatous reaction (which can be mistaken for a retinoblastoma). Toxoplasmosis and toxocara are now more common than TB. Sarcoid is another cause.

Tests: CXR; Mantoux; serology; Kveim test (the reagent is available in the UK on a named-patient basis from Porton Down). In the acute phase, vision may be blurred, a grey-white raised patch is seen on the retina, vitreous opacities occur, and there may be cells in the anterior chamber. Later, a choroidoretinal scar (white patch with pigmentation around) will be seen, these being symptomless unless involving the macula. Treat the cause.

Malignant melanoma of the choroid is the commonest malignant tumour of the eye. Appearing as mottled grey/black on the fundus, they produce a retinal detachment immediately over the growth. Spread is haematogenous or by local invasion of the orbit. Treatment is usually by enucleation of the affected eye although local treatment is sometimes possible.

Senile macular degeneration This is the commonest cause of registrable blindness in the UK. It occurs in elderly people who complain of deterioration of central vision. There is loss of acuity; visual fields are unaffected. The disc appears normal but there is pigment, fine exudate, and haemorrhage at the macula. Occasionally the macular area is oedematous and lifted by a large mass of exudate—called disciform degeneration. In the majority there is no effective treatment. A few people can be treated by the laser. Use of visual aids may help them to read.

Tobacco amblyopia This is optic atrophy induced by tobacco and is due to cyanide poisoning. There is gradual loss of central vision. Loss of red/green discrimination is an early, permanent sign.

Optic atrophy The optic disc usually appears pale, although the degree of pallor does not correspond to the visual loss. Optic atrophy may be secondary to increased intraocular pressure (in glaucoma), or retinal damage (as in choroiditis, retinitis pigmentosa, cerebromacular degeneration), or be due to ischaemia (as in retinal artery occlusion). Toxic causes other than tobacco are methanol, lead, arsenic, quinine, and carbon bisulphide.

Other causes: Leber's optic atrophy (p 752), multiple sclerosis, syphilis, external pressure on the nerve (intraorbital or intracranial tumours, Paget's disease affecting the skull).

Chronic simple (open angle) glaucoma

▶Simple glaucoma is asymptomatic until visual fields are severely impaired; hence the need for screening—but most people found to have raised intraocular pressures (eg on ~10% of routine NHS sight test in those >40yrs old) do not have glaucoma when their fundi and peripheral fields are charted (this can be a lengthy and tricky business). Once raised pressures are found, lifelong follow-up is needed (at least yearly, and very much more often in the early stages)—which is why glaucoma accounts for such a heavy burden in ophthalmology clinics (eg 25% of the work), and why highstreet optometrists are being encouraged to take over at least part of this monitoring burden. Glaucoma accounts for 7% of new blind registrations.

Pathogenesis Increased intraocular pressure (>21mmHg) causes cupping of optic disc, with capillary closure, hence nerve damage, with sausage-shaped field defects (scotomata) near the blind spot, which may then coalesce to form major defects. The nasal and superior fields are lost first with the last vision remaining in the temporal field.

Normal optic cups are similar (left and right) in shape and occupy <50% of the optic disc. In glaucoma these enlarge, especially along the vertical axis. As damage progresses the optic disc becomes pale (atrophic), and the cup wider and deeper—so that the blood vessels emerging from the disc appear to have breaks in them as they disappear into the cup and are then seen at the base again (see p 503). Since the central field is intact, good acuity is maintained, so patients may only present when there is major irreversible optic nerve damage. Some people develop glaucoma with normal intraocular pressures.

Prevention Those most at risk have a family history of glaucoma (risk ↑ 10-fold, particularly among siblings), are myopic, or have diabetic or thyroid eye disease. Check pressures (optician, using tonometry) in those with a positive family history, regularly from the age of 35 years.

Treatment The aim is to reduce the intraocular pressure to <21 mmHg. Surgery is used if medical treatment fails.

Betaxolol 0.5% drops (or timolol 0.25%–0.5%) Use twice daily. They reduce the production of aqueous. They are β-blockers, so use with caution in asthma and heart failure. (Systemic absorption occurs with no 1st-pass metabolism by the liver). SE: dry eyes, corneal anaesthesia, allergy, subtle changes in exercise tolerance.

Pilocarpine 0.5–4% drops reduce resistance to the outflow of aqueous. It causes miosis and may cause blurring of vision and brow ache due to ciliary muscle spasm. Rarely, retinal detachment can occur. Presbyopes tolerate it better than the young and short sighted. Use 4 times daily. Because of these problems, pilocarpine is second-line treatment, or, with improving surgical options, some would say not even second line.

Adrenaline 1% drops reduce aqueous production and outflow resistance. Caution with heart disease; avoid in closed angle glaucoma. SE: sore red eyes. Use once or twice daily.

Acetazolamide 250mg/24h–500mg/12h PO reduces the production of aqueous (it is a carbonic anhydrase inhibitor). SE: lassitude, dyspepsia, K⁺ depletion, paraesthesia. Side effects limit its use, except perhaps to gain rapid control as treatment starts.

Surgery: Trabeculectomy is the commonest operation. There is strong evidence that it acts as a drainage operation. The effect of laser trabeculoplasty is often short-term.

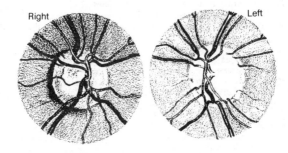

The two optic discs of a patient with open angle glaucoma which has not yet damaged the right optic disc. The left optic disc is grossly cupped and atrophic. (From J Parr *Introduction to Ophthalmology*, OUP.)

Cataract

▶When a cataract is found, measure the blood glucose (to exclude DM).

A cataract is a lens opacity. The 4 major causes of blindness in the world are cataract, vitamin A deficiency, trachoma, and onchocerciasis. The common-est cause of cataract in the West is ageing of the lens. They are found in 75% of over 65s but in only 20% of 45–65-year-olds.

Other causes Diabetes mellitus; galactosaemia; hypocalcaemia; intrauterine rubella or toxoplasmosis (p 98); longstanding posterior uveitis; genetic (Werner's syndrome, dystrophia myotonica); secondary to trauma, electric shock, irradiation (infra-red or X-rays), corticosteroid use.

Ophthalmoscopic classification is by lens appearance. With immature cataracts the red reflex still occurs; if dense cataract there is no red reflex, or visible fundus. *Nuclear* cataracts change the lens refractive index and are common in old age, as are the cortical spoke-like wedge-shaped opacities. Anterior and posterior *polar* cataracts are localized, are commonly inherited, and lie in the visual axis. *Subcapsular* opacities from steroid use are just deep to the lens capsule—in the visual axis. *Dot opacities* are common in normal lenses but are also seen in fast-developing cataracts in diabetes or dystrophia myotonica.

Presentation The main symptom of cataract is blurred vision. Unilateral cataracts are often unnoticed, but loss of stereopsis may affect distance judgement. Bilateral cataracts may cause gradual visual loss ± frequent spec-tacle changes due to the refractive index of the lens changing, may cause daz-zling—especially in sunlight, and may cause monocular diplopia. In children they may present as squint, loss of binocular function, as a white pupil, as nystagmus (infants), or as amblyopia.

Treatment Childhood cataracts require special surgical techniques. Adult cataracts do not necessarily require treatment, and most will never need it. No treatments prevent or arrest their development. Check acuity regularly. Surgical removal of the lens is the only treatment. The main question influ-encing timing of surgery is: 'Does your reduced sight prevent you doing what you want to do?' On this scale, seamstresses will get surgery before wine-tasters.

Extracapsular cataract extraction with posterior chamber lens implant is the surgery of choice. In this procedure the anterior part of the lens capsule is meticulously torn and removed. The posterior part of the lens is left in situ, like a bag, to hold the intraocular lens. The cortex and nucleus of the lens is removed either mechanically through a large incision at the limbus, or after emulsifying the lens (phacoemulsification), through a smaller inci-sion. The intraocular lens is then implanted into the capsular bag.

Day-case surgery (eg with phacoemulsification) is possible using local anaesthetic agents to produce a Tolosa–Hunt syndrome (p 758). The patient can choose to have a lens implanted which will focus on the horizon or near-to—or on an intermediate position. Spectacles will be needed to counteract healing astigmatism—but do not order these for some months after surgery, and warn the patient that further refractive changes may occur until the ini-tial spectacles are ordered.

The posterior capsule may become opaque with time (in 50%), and like cataract, cause a gradual blurring of vision. YAG lasers can be used to make a hole in the capsule (capsulotomy) to restore vision.

Complications of cataract surgery:
- Posterior capsule rupture (4%)
- Section requiring re-suture (0.8%)
- Rubeotic glaucoma (0.3%)
- Posterior capsule thickening (8%)
- Broken or protruding sutures (4%)
- Endophthalmitis (0.6%)
- Vitreous haemorrhage (0.3%)

Phacoemulsification
Small 4mm incision + ultrasonic
  fragmentation of lens, with
  aspiration of fragments
Number of sutures needed: 0–3
Healing 2–4 weeks
Astigmatism less

Extracapsular extraction
Larger incision and manual
  expression of lens

6–8 sutures required
8–12 weeks for healing
Post-op vision may not be so good

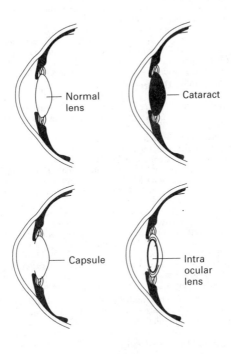

The retina

Anatomy The retina consists of an outer pigmented layer (in contact with the choroid), and an inner sensory layer (in contact with the vitreous). At the centre of the posterior part lies the macula (the centre of which is termed the fovea), appearing yellowish and slightly oval. This has many cones, so acuity is greatest here. ~3mm medial to the fovea is the optic disc, which contains no rods or cones (the visual field's blind spot).

Optic disc Think *colour; contour; cup*. *Colour* should be a pale pink. It is more pallid in optic atrophy (p 500). *Contour*: the disc may appear oval in astigmatic eyes, and appear abnormally large in myopic eyes. Disc margins are blurred in papilloedema (eg from raised intracranial pressure, malignant hypertension, cavernous sinus thrombosis), and with optic neuritis. Blood vessels radiate away from the disc. The normal arterial/venous width ratio is 2:3. Venous engorgement appears in retinal vein thrombosis; abnormal retinal pallor with artery occlusion; and haemorrhages with exudates in hypertension and diabetes. *Cup*: the disc has a physiological cup which lies centrally and should occupy about one-third of the diameter of the disc. Cup widening and deepening occurs in glaucoma (p 502).

Retinal detachment This may be 'simple', idiopathic, secondary to some intraocular problem (such as a melanoma, or fibrous bands in the vitreous in diabetes), occur after cataract operation, or be secondary to trauma. Myopic eyes are more prone to detachment, the higher the myopia, the greater the risk.[1]

In simple detachment, holes in the retina allow fluid to separate the sensory retina from the retinal pigment epithelium.

Detachment may present as painless loss of vision, sometimes described as a curtain falling over the vision (the curtain falls down as the lower half of the retina detaches upwards). 50% of patients developing detachment have premonitory symptoms—flashing lights or the sensation of spots before the eyes as the retina has been abnormally stimulated prior to detachment.[1] Detachment of the lower half of the retina tends not to pull off the macula, whereas upper half detachments do. If the macula becomes detached, central vision is lost and does not recover completely even if the retina is successfully replaced. Rate of detachment varies but upper halves tend to be quicker. Field defects may be detected. Ophthalmoscopy may reveal a grey opalescent retina, ballooning forward. *Treatment:* Urgent referral, eg for scleral silicone implants, cryotherapy, pneumatic retinopexy, or argon or laser coagulation to secure the retina.

Retinitis pigmentosa This familial disorder is a common cause of retinal degeneration. Particles of black pigment fleck mid and peripheral fundus. The earliest symptoms night blindness, often occurring in adolescence. Retinal vessels attenuate, and optic atrophy and blindness ensue.

Retinoblastoma The typical patient is a child <3yrs old. If it is bilateral (30%) the cause is usually hereditary. The 'retinoblastoma' gene, present in everyone, is normally a suppressor gene or anti-oncogene. Those with hereditable retinoblastomas have one altered allele in every cell. If a developing retinal cell undergoes mutation in the other allele, a retinoblastoma results. The retinoblastoma gene is the best characterized tumour suppressor gene (called 'RB1'; 27 exons; >200 kilobases of genomic DNA). Its product is a nuclear phosphoprotein which helps regulate DNA synthesis. *Presentation:* They may grow out (exophytic) or in, ie endophytic—into the vitreous, causing a white pupil (leukocoria), squint, inflammation, and an absent red reflex. Parents may notice this phenomenon when studying flash photographs of their children—in which only one eye is red. *Treatment:* Enucleation or laser ablation. Arrange follow up because of risk of osteosarcoma, typically in the 2nd decade of life.[2]

1 A Chignell 1987 *BMJ* ii 661   2 G Skuse 1995 *Lancet* 345 902

Toxoplasmosis This obligate intracellular protozoan sometimes causes bilateral typical 'punched-out' heavily pigmented chorioretinal scars which can be seen on routine fundal examination. If scars involve the macular area there will be dramatic loss of acuity. Toxoplasmosis is a cause of about 25% of uveitis in the UK.

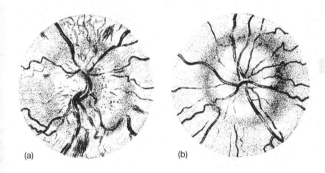

(a) Papilloedema from raised intracranial pressure. The disc is swollen forwards and also outwards into the surrounding retina. The disc margin is completely hidden and in places retinal vessels are concealed, because oedema has impaired the translucency of the disc tissues. The retinal veins are congested and there are a few haemorrhages. (b) Early papilloedema. The vessels crossing the disc are obscured in places due to the loss of translucency of the prelaminar tissue in which they lie. (From J Parr *Introduction to Ophthalmology*, OUP.)

The eye in diabetes mellitus

Diabetes can be bloody and blinding – the leading cause of blindness in those aged 20-65(UK). Almost any part of the eye can be affected: cataract and retinopathy are the chief pathologies. 30% of adults have ocular problems when diabetes presents. At presentation, the lens may have a higher refractive index (possibly due to dehydration) producing relative myopia. On treatment, the refractive index reduces, and vision is more hypermetropic—so do not correct refractive errors until diabetes is controlled.

Structural eye changes Diabetes may cause cataract. Most commonly this is a premature senile cataract, but young diabetics can also be affected at presentation. In this case, the lens has taken up a lot of glucose which is converted by the enzyme aldolase reductase to sorbitol. Sorbitol cannot diffuse through the lens but increases the osmotic pressure within the lens; water is then taken up which can cause the lens fibres to rupture and a cataract thereafter forms. Rarely, the iris may be affected by diabetes, with new blood vessel formation on it (rubeosis), and, if these block the drainage of aqueous fluid, glaucoma can be caused.

Retinopathy *Pathogenesis:* High retinal blood flow induces a microangiopathy in capillaries, precapillary arterioles and venules, causing occlusion and leakage. *Vascular occlusion:* Occlusion produces ischaemia which leads to new vessel formation in the retina, the optic disc and on the iris ie *proliferative retinopathy.* New vessels can bleed and produce vitreous haemorrhage. As new vessels carry along with them fibrous tissue, retraction of the fibrous tissue increases the risk of retinal detachment. Occlusion also causes *cotton wool spots* (ischaemic nerve fibres). *Vascular leakage:* High retinal blood flow caused by hyperglycaemia (and BP↑ and pregnancy) damages capillary pericytes (these cells reinforce endothelial cells).[1] As pericytes are lost, capillaries bulge (microaneurysms) and leak. Leakage produces oedema and hard exudates. Hard exudate is made up of lipoprotein and lipid filled macrophages. The rupture of microaneurysms, when at the nerve fibre level, produces flame shaped haemorrhages; when deep in the retina, *blot haemorrhages* are formed.

▶Pre-symptomatic screening enables laser use of photocoagulation. Screen by regular eye examination or retinal photography. Lesions are mostly at the posterior pole and can be easily seen with the ophthalmoscope. *Background retinopathy* comprises microaneurysms (seen as 'dots'), haemorrhages (flame shaped or 'blots') and hard exudates (yellow patches). Vision is normal. Background retinopathy can progress to sight threatening maculopathy or proliferative retinopathy. *Maculopathy:* Leakage from the vessels close to the macula produce oedema and damage to the macula. *Proliferative retinopathy:* Fine new vessels appear on the retina and optic disc. Engorged tortuous veins, 'cotton wool spots' (ischaemic nerve fibres), large blot haemorrhages, and vitreous haemorrhage may also be seen.

▶Refer those with maculopathy or proliferative retinopathy urgently to an ophthalmologist for treatment to protect vision.

Treatment Good control of diabetes can prevent new vessel formation. Concurrent diseases may accelerate retinopathy (eg hypertension, renal disease, pregnancy and anaemia). Treat these (as appropriate), and hyperlipidaemia. Photocoagulation by laser is used to treat both maculopathy and proliferative retinopathy. Definite indications for photocoagulation are new vessels on the optic disc and vitreous haemorrhage. If vitreous haemorrhage is massive and does not clear, vitrectomy may be needed.

CNS effects Ocular palsies may occur, typically nerves III and VI. In diabetic third nerve palsy the pupil may be spared as fibres to the pupil run peripherally in the nerve, receiving blood supply from the pial vessels. Argyll Robertson pupils and Horner's syndrome may also occur (p 488).

The eye in systemic disease

Systemic disease often manifests itself in the eye and, in some cases, eye examination will first suggest the diagnosis.

Vascular retinopathy This may be *arteriopathic* (arteriovenous *nipping*: arteries nip veins where they cross—they share the same connective tissue sheath) or *hypertensive*–arteriolar vasoconstriction and leakage (producing hard exudates, macular oedema, haemorrhages, and, rarely, papilloedema). Thickened arterial walls appear like wiring (called 'silver' or 'copper'). Narrowing of arterioles leads to localized infarction of the superficial retina seen as cotton wool spots and flame haemorrhages. Leaks from these appear as hard exudates ± macular oedema/papilloedema (rare).

Emboli passing through the retina produce *amaurosis fugax*–seeing 'a curtain passing across the eyes'. They may arise from atheromatous plaques (listen to carotids) or from heart (from valves or infarct site). Treat by aspirin (OHCM p 436). Retinal haemorrhages are seen in leukaemia; comma-shaped conjunctival haemorrhages and retinal new vessel formation may occur in sickle-cell disease; optic atrophy in pernicious anaemia. Note also Roth spots (retinal infarcts) of infective endocarditis (OHCM p 312).

Metabolic disease Diabetes: p 508. Hyperthyroidism and exophthalmos: OHCM p 543. In myxoedema, eyelid and periorbital oedema is quite common. Lens opacities may occur in hypoparathyroidism. Conjunctival and corneal calcification may occur in hyperparathyroidism. In gout, monosodium urate deposited in the conjunctiva may give sore eyes.

Granulomatous disorders (TB, sarcoid, leprosy, brucellosis, and toxoplasmosis) can all produce inflammation in the eye; TB and syphilis producing iritis, the others posterior uveitis. TB, congenital syphilis, sarcoid, CMV and toxoplasmosis may all produce choroidoretinitis. In sarcoid there may be cranial nerve palsies.

Collagen diseases These also cause inflammation. Conjunctivitis is found in SLE and Reiter's syndrome; episcleritis in polyarteritis nodosa and SLE; scleritis in rheumatoid arthritis; and uveitis in ankylosing spondylitis and Reiter's syndrome (OHCM p 669). In dermatomyositis there is orbital oedema with retinal haemorrhages.

Keratoconjunctivitis sicca Sjögren's syndrome, (OHCM p 708) There is reduced tear formation (Schirmer filter paper test), producing a gritty feeling in the eyes. Decreased salivation also gives a dry mouth (xerostomia). It occurs in association with collagen diseases. Treatment is with artificial tears (tears naturale, or hypromellose drops).

AIDS Those who are HIV+ve may develop CMV retinitis, characterized by retinal spots ('pizza pie' fundus, signifying superficial retinal infarction) and flame haemorrhages involving more and more of the retina. This may be asymptomatic but can cause sudden, untreatable, visual loss. If it is present it implies full-blown AIDS, and a low CD4 count. Cotton wool spots on their own indicate HIV retinopathy, and it may present before the full picture of AIDS; it is a microvasculopathy, not a retinitis.▣

Candidiasis of the aqueous and vitreous is hard to treat. Kaposi's sarcoma may affect the lids or conjunctiva.

Tropical eye disease

Trachoma Caused by *Chlamydia trachomatis* (serotypes A, B, and C), this disease is spread by towels, fingers and flies—typically where it is hot, dry and dusty and the people are poor, living near to their cattle. 400 million people are affected (100,000,000 are children).

Diagnosis: Stage 1: There is lacrimation. Follicles under the upper lid give a fine granular appearance. *Stage 2:* There is intense erythema. The follicles are larger and underneath both lids. A fine pannus and capillaries grow down towards the cornea. *Stage 3:* The follicles rupture and are replaced by scar tissue. The pannus is more advanced. The cornea may ulcerate. *Stage 4:* Scar tissue distorts the lids and causes entropion. Eyelashes scratch the cornea, which ulcerates.

Treatment: Mass antitrachoma treatment: tetracycline 1% eye ointment is used 12-hourly for 5 days each month for 6 months. In active disease use 8-hourly for 6 weeks + sulphadimethoxine 1g stat, then 500mg/24h PO for 10 days (erythromycin if a child). Single-dose azithromycin (20mg/kg) is effective.[1][2] Surgery: tarsal plate rotation.

Prevention: Good water and sanitation. Regular washing of faces.

Onchocerciasis This is caused by the microfilariae of the nematode *Onchocerca volvulus*, transmitted by black flies of the Simulium species. Of the 20-50 million people affected, 95% live in Africa. In some areas it may cause blindness in 40% of the population. Unless the eye is affected, problems are mostly confined to the skin. Fly bites result in nodules from which microfilariae are released. These eventually invade the eye, mainly the conjunctiva, cornea, ciliary body and iris, but occasionally the retina or optic nerve. Sometimes they may be seen swimming in the aqueous or lying dead in the anterior chamber. The microfilariae initially excite an inflammatory reaction, then fibrosis occurs around them. Reaction around the dead microfilariae in the cornea causes corneal opacities (nummular keratitis). Chronic iritis causes synechiae formation and may precipitate cataracts. The iris may become totally fixed. *Tests:* Skin snip tests, triple-antigen serology; polymerase chain reaction. *Treatment:* Seek expert help. Ivermectin is the treatment of choice: 150µg/kg PO as a single dose every 6-12 months, until the adult worms die (OHCM p 247). In lightly infected expatriates, the 1st 3 doses are recommended to be monthly, with observation in hospital after the first dose (reactions are common in expatriates).[3]

Xerophthalmia and keratomalacia These are manifestations of vitamin A deficiency, eg if weaned early onto vitamin A deficient milk products, or toddlers who eat few vegetables. Peak incidence: 2-5yrs; 40 million children worldwide. Night blindness and dry conjunctivae (xerosis) occur early. The cornea is unwettable and loses transparency. Small grey plaques (Bitot's spots) are commonly found raised from interpalpebral conjunctiva. Vitamin A reverses these changes. Early corneal xerosis is reversible. Corneal ulceration and perforation can occur. In keratomalacia there is massive softening of the cornea ± perforation and extrusion of the intraocular contents. *Treatment:* For children: retinol palmitate 50,000U IM monthly until the eyes are normal (adult dose 100,000U, weekly to start with); or oral retinyl palmitate 200,000U PO. β-carotene (a provitamin) 1.2×10^6 U PO, is as effective,[4] and cheaper. Avoid vitamin A in pregnancy (vitamin A embryopathy).

1 A Potter 1993 *BMJ* ii 214   2 RL Bailey 1993 *Lancet* 342 453   3 DR Churchill 1994 *Trans Roy Soc Trop Med Hyg* 88 242   4 C Carlier 1993 *BMJ* ii 1106

Eye trauma

▶Prevention is the key. Wear goggles—or plastic glasses when near small moving objects or using tools (avoids metal splinters, fish-hooks, and squash-ball injuries). ▶Always record the acuity of both eyes (if the uninjured one is blind take all injuries very seriously). If the patient is unable to open the injured eye, instill a few drops of local anaesthetic (amethocaine 1% drops) and wait a few minutes after which the patient should be able to comfortably open the eye. Examine lids, conjunctiva, cornea, sclera, anterior chamber, pupil, iris, lens, vitreous, fundus and eye movement. An irregular pupil may mean globe rupture. Afferent pupil defects (OHCM p 488) do not augur well for sight recovery. Note pain, discharge, or squint. CT may be very useful (foreign bodies may be magnetic, so avoid MRI).

Contusions The eyes are well protected by the bony orbital ridges. Severe contusions from large objects may damage the eye, but smaller objects such as champagne corks, squash balls, and airgun pellets (p 679) cause local contusion, eg resulting in lid bruises and subconjunctival haemorrhage (if the posterior limit of such a haemorrhage cannot be seen, consider fracture of the orbit). Both usually settle within 2 weeks. Any injury that has penetrated the eyeball should receive immediate specialist treatment.

Intraocular haemorrhage usually affects the acuity and should receive specialist attention. Blood is often found in the anterior chamber (hyphaema): small amounts usually clear spontaneously but larger amounts filling the anterior chamber may need evacuation. Even a small amount of hyphaema must be carefully evaluated. One of the dangers from hyphaema is the development of secondary glaucoma. Serious secondary haemorrhage may occur within 5 days and may produce sight threatening secondary glaucoma. Sometimes the iris is paralysed and dilated due to injury (called traumatic mydriasis). This usually recovers in a few days but sometimes it is permanent. Vitreous haemorrhage will cause dramatic fall in acuity. There will be no red reflex on ophthalmoscopy. Lens dislocation, tearing of the iris root, splitting of the choroid, detachment of the retina, and damage to the optic nerve may be other sequelae; they are more common if contusion is caused by smaller objects rather than large.

Blows to the orbit may cause blowout fractures with orbital contents herniating into the maxillary sinus. Tethering of the inferior rectus and inferior oblique muscles causes diplopia. Test the sensation over the skin of the lower lid. Loss of sensation indicates injury to the infraorbital nerve, confirming a blowout fracture. Fracture reduction and muscle release is necessary.

Sharp and penetrating wounds Refer urgently—as the longer the delay the more the risk of ocular contents being disturbed, extruded, or infected. With uveal injury there is risk of sympathetic ophthalmia in the other eye. ▶A history of flying objects (eg work with lathes, hammers, and chisels) should prompt careful examination and X-ray to exclude intraocular foreign bodies. ▶▶Do *not* attempt to remove a large foreign body (knife; dart). Support the object with padding. Transport supine. Pad the *unaffected* eye to prevent damage from conjugate movement. Consider skull X-ray or CT to exclude intracranial involvement.

Foreign bodies These may be hard to see: examine all parts of the eye—as they may cause chemosis, subconjunctival haemorrhage, irregular pupils, iris prolapse, hyphaema, vitreous haemorrhage and retinal tears. If you suspect a metal foreign body X-ray the orbit. With high-velocity foreign bodies, consider orbital ultrasound: pickup rate is 90% vs 40% for X-rays—but the technique is difficult, and possibly unsuited to busy A&E departments. Removal of superficial foreign bodies may be possible using a triangle of clean card. Evert the eyelids to check no material underneath. Use gentamicin 0.3% drops afterwards to prevent infection. Have a low threshold for asking for the advice of a senior colleague.

Corneal abrasions These are often inflicted by small fast-moving objects such as children's finger-nails and twigs. They may cause intense pain. Apply a drop of local anaesthetic eg 1% amethocaine before examination. They stain with fluorescein and should heal within 48h. Antibiotic drops should be used until healing occurs. Eye pads should be used initially if local anaesthetic has been instilled to protect the cornea, but otherwise are not needed,[1]

6. Ophthalmology

Burns Treat chemical burns promptly by holding eyelids open and bathing the eyes in copious clean water while the specific antidote is sought. Often the patient will not hold the eye open due to excruciating pain. Amethocaine drops applied every minute for 5 minutes will relieve pain and aid irrigation. All burns may have late serious sequelae, eg corneal scarring, opacification, and lid damage. Alkali burns are more serious than acid.

Arc eye Welders and sunbed users who don't wear protection against uv light may damage the corneal epithelium. There is a foreign body sensation, watering and blepharospasm. Give mydriatics (see p 518), pad the eye, and await recovery. It is a very painful condition so be generous with analgesia.

1 DL Easty 1993 *BMJ* ii 1022

Blindness and partial sight

The pattern of blindness around the world differs considerably, depending on local nutrition and economic factors. 80% of the world's blind live in developing countries. The diseases responsible for most of the blindness in the world are trachoma, cataract (50% of the world's blindness), glaucoma, keratomalacia, onchocerciasis and diabetic retinopathy. In the past smallpox, gonorrhoea, syphilis, and leprosy (10% of those affected were blind) were also common causes of blindness, but are less so now.

Rates of blindness are higher than 10/1,000 in some parts of Africa and Asia, but in the UK and the USA rates are 2/1,000. Blindness may be voluntarily registered in England, registration making one available for certain concessions. Although the word blind suggests inability to perceive light, a person is eligible for registration if their acuity is less than 3/60, or if >3/60 but with substantial visual field loss (as in glaucoma). There are 149,670 registered blind in the UK (1994), and a further 115,710 partially sighted. The criterion for partial sighted registration is that the acuity is <6/60 (or >6/60 with visual field restrictions).

The causes of blindness have changed considerably in the UK over the last 70 years. Whereas in the 1920s ophthalmia neonatorum (p 100) was responsible for 30% of all cases of blindness found in English blind schools, this is now a rare but treatable disease. Retrolental fibroplasia was common in the 1950s, mostly affecting premature infants: monitoring of intra-arterial oxygen in premature babies tries to prevent this. With the increasing population of elderly, the diseases particularly afflicting this population are the common causes of blindness. Nearly two-thirds of the blind population are over 65 years of age, and nearly half over 75. Macular degeneration, cataract, and glaucoma are the three commonest causes of blindness.

In England and Wales the responsibility for blind registration lies with the local authority. Application for registration is made by a consultant ophthalmologist and is voluntary, not statutory. Registration as blind entitles one to extra tax allowances, reduced TV licence fees, some travel concessions, and access to talking books. Special certification from an ophthalmologist is necessary for the partially sighted to receive talking books. At one time it was statutory that the registered blind should receive a visit from a social worker but this is no longer the case, although the social services employ social workers who specialize in care of the blind. The Royal National Institute for the Blind[1] will advise on aids, such as guide dogs (available if required for employment).

Special educational facilities provide for visually handicapped children. Special schools have a higher staff/pupil ratio, specialized equipment, and many have a visiting ophthalmologist. The disadvantage is that the children may not mix much with other children—especially if they board.

---

1 Roy Nat Inst Blind, 224 Great Portland Street, London WIN 6AA (tel. 0171 388 1266)

Colour vision

Colour blindness For normal colour vision we require cone photopigments sensitive to blue, green and red light. The commonest hereditary colour vision defect is X-linked failure of red-green discrimination (8% ♂ and 0.5% ♀ affected—so those with Turner's syndrome have ♂ incidence and those with Klinefelter's have ♀ incidence). Blue–yellow discriminatory failure is more commonly acquired and sexes are affected equally.

Diagnosis: This is by use of coloured pattern discrimination charts eg Ishihara plates.

Depressed colour vision may be a sensitive indicator of acquired macular or optic nerve disease.

Monochromatism This may be due to being born without cones (resulting in low visual acuity, absent colour vision, photophobia and nystagmus), or, very rarely due to cone monochromacy where all cones contain the same visual pigment, when there is only failure to distinguish colour.

Drugs and the eye

Drugs applied locally to the eye may be absorbed through the cornea and produce systemic side effects. For example, there is danger of inducing bronchospasm or bradycardia in susceptible individuals by the use of β-blocking eye drops (such as timolol) for glaucoma.

The eye does not retain drops for as long as ointments and two-hourly applications may be needed. Eye ointments are particularly suitable for use at night and in conditions where crusting and sticking of the lid margins occurs. Allow 5min between doses of drops to prevent overspill. The antibiotics most used are not those generally used systemically. The most commonly used are fusidic acid, neomycin, gentamicin and framycetin. (Chloramphenicol drops should only be used if there is no alternative, as marrow aplasia may occur via conjunctival absorption.[1]) All eye preparations have warnings not to use for more than one month. Contamination of drops is rare, as they contain bacteriostatic drugs, and in practice may be used for longer.

Mydriatics These drugs dilate the pupil. They also cause cycloplegia and hence blur vision (warn not to drive). Pupil dilatation prior to examination is best achieved using 0.5% or 1% tropicamide which lasts for 3h. 1% cyclopentolate has an action of 24 hours and is preferred for producing cycloplegia for refraction of children. These drugs may be used to prevent synechiae formation in iritis. ►In the over-60s with shallow anterior chambers (especially if a family history of glaucoma) they may precipitate acute glaucoma so only use if requested to do so by an ophthalmologist.

Miotics These constrict the pupil and increase drainage of aqueous. They are used in the treatment of glaucoma (p 502). Pilocarpine 1–4% is the most commonly used.

Local anaesthetic Amethocaine (0.5% drops) is an example. This may be used to permit examination of a painful eye where reflex blepharospasm is a problem, and to facilitate removal of a foreign body. ►It abolishes the corneal reflex so use to treat pain is to risk corneal damage. To relieve pain, give an eye pad, and be generous with oral analgesia.

Steroid-containing drops ►These are potentially dangerous as they may induce catastrophic progression of dendritic ulcers (p 480). Ophthalmoscopic examination may miss dendritic ulcers, and slit lamp inspection is essential if steroid drops are being considered, eg for allergy, episcleritis, scleritis or iritis.

Iatrogenic eye disease Glaucoma may not only be precipitated by mydriatics; other drugs may precipitate or aggravate glaucoma. Those particularly implicated are steroids and those with anticholinergic effects (some antiparkinson drugs and tricyclic antidepressants).

A few drugs affect the retina if used in the long-term. Patients taking ethambutol should be warned to report *any* visual side-effects (loss of acuity, colour blindness). Chloroquine and other antimalarials are also implicated and damage may occur with long-term administration, especially if high doses are used.

1 M Donna 1995 *BMJ* i 1217

Obtaining & preparing antibiotic eye drops[1]

Fortified guttate gentamicin is 15mg/ml—(the normal commercial gentamicin is 3mg/ml); penicillin 5,000 units/ml, methicillin 20mg/ml and antifungals can be obtained from the Chief Pharmacist, Moorfields Eye Hospital (tel. 0171-253 3411).

Antibiotics can be home-made as follows:
Gentamicin forte: add 2ml of 40mg/ml IV gentamicin to a 5ml bottle of commercial guttate gentamicin (3mg/ml).

Other antibiotics can be made up using IV preparations to the required concentration using water or normal saline. These are stable for the time recommended for IV solutions in the manufacturers *Data Sheets*. Penicillin G can be used up to 500,000U/ml.

1 Prepared from information supplied by Mr JKG Dart FRCS, Moorfields Eye Hospital, City Rd, London EC1V 2PD

Contact lenses

Contact lenses are commonly worn—and commonly problematic.

80% of contact lenses are worn for cosmetic reasons. Only 20% are worn because lenses are more suitable for the eye condition than spectacles. Among this 20% a minority wear the lenses to hide disfiguring inoperable eye conditions, a greater proportion have them for very high refractive errors. Myopia above -12.ODS and hypermetropia above +10.ODS are indications for lenses because equivalent spectacles produce quite distorted visual fields. Contact lenses after cataract removal come into this category. Lenses are used for ocular reasons for those with ocular conditions such as after corneal ulceration or trauma when a new front surface of the cornea is needed to see through, and in keratoconus. Keratoconus is a rare degenerative, slowly progressive corneal condition with thinning and anterior protrusion of the central cornea. Blurred vision is the only symptom. Contact lens use may compensate for corneal distortion early, but later corneal grafting may be needed.

Types of lens Hard lenses are 8.5-9mm in diameter and are made of polymethylmethacrylate (PMMA). Gas-permeable hard lenses are about 0.5cm larger and are designed to allow gas to permeate through to the underlying cornea. They can only be made to cope with a limited degree of astigmatism and do not wet as well as standard hard lenses so may mist up in the day. With the advent of the larger (13-15mm diameter) soft contact lenses it was hoped that many of the problems with hard lenses could be circumvented. Soft disposable lenses can be worn for 2-4 weeks and then disposed of. With astigmatism adequate correction is not achieved because the lens fits the astigmatic cornea too well and is flexible. A conventional hard or gas permeable lens can usually correct a small amount of astigmatism but high astigmatism requires a toric lens (a soft lens made especially to correct astigmatism). They are more delicate than hard lenses and need meticulous cleaning. Extended wear lenses can be worn for up to 4 months. Some people even wear coloured lenses designed just to change eye colour.

Patients may suffer from keratoconjunctivitis or giant papillary change in the upper tarsal conjunctiva, possibly due to sensitization to the cleansing materials used or to the mucus which forms on the lens.

Cleaning lenses Different cleaning solutions made by different manufacturers should not be mixed.[1] With hard lenses 2 solutions are usually used, one for rinsing and cleaning, and one for storage. The storage solution should be washed off before the lens is inserted. Soft contact lenses, being permeable, tend to absorb the chemicals, so weaker solutions for cleaning are used. In addition, the lenses are usually intermittently cleaned with another system (eg enzyme tablets) to remove mucoprotein on their surface. Sensitivity to cleaning agents usually presents as redness, stinging, increased lens movement, increased mucus production and thickened eyelids. It may be necessary to stop wearing lenses for several months. When restarting use a preservative-free cleansing system.

Complications
1 Contact lenses are the most important risk factor for corneal ulceration. Corneal abrasion is common early while adjusting to wear. Pain ± lacrimation occurs some hours after removing the lens.
2 Sensitization to cleaning agents.
3 Losing the lens within the eye. Hard lenses may be lost in any fornix, soft lenses are usually in upper outer fornix.
4 Keratitis especially for extended-wear (>6 days) soft lenses.[1]
5 Staining by rifampicin or fluorescein.

Laser photorefractive keratotomy (PRK)

This novel way of correcting myopia without necessary recourse to spectacles entails removing microscopic portions of tissue from the front of the cornea, so altering its shape and refractive properties. In general, one may expect treatment to be accurate to within 1 diopter of expectation in eyes with a myopic refraction of 1-6 diopters, and an astigmatism of *no more* than 1.5 diopters. Note the following points:

●PRK is irreversible.
●The greater the myopia, the more the laser needs to penetrate, and the greater the risk of corneal scarring—with subsequent opacification.
●Intermediate myopia (7-12 diopters) *may* be treatable but regression is more likely.
●There is no reliable evidence that astigmatism can be reduced or eliminated permanantly.
●Presbyopia, not being the 'fault' of the cornea, is not amenable to PRK.
●Be critical of reports claiming that excimer or holmium lasers can treat hypermetropia reliably.
●Longterm follow-up information is lacking. PRK only started in 1988.

Contraindications ●Unstable myopia (eg patients under 21 years old) ●Corneal infection ●Keratoconus ●Pregnancy (unpredictable healing) ●Those on steroids ●Those with unrealistic expectations.

What to tell the patient PRK is only suitable for some eyes, and some patients. It is an outpatient procedure, using eyedrops as local anaesthesia. Application of the laser lasts only 15-90 seconds, depending on the degree of short-sightedness. Immediately after there may be quite severe pain (eye drops, oral analgesia, and an eye patch will be given to minimize this). 2-4 days off work is usually necessary. In the few weeks following PRK, vision may be blurred. The healing process tends to produce over-correction of refractive errors. A difference in focus of the two eyes will occur when the first eye has PRK; this can be ameliorated by wearing a contact lens on the other eye. Stress the need for regular post-operative examinations by the ophthalmologist.

Warn the patient that even after the post-operative period there may be a noticeable (but generally subjectively unimportant) visual haze.

Outcome ~80% of eyes heal in a predictable way—so the result is less certain in 20%, with over-correction, under-correction, or regression of myopia being problems. Some patients may also experience glare (a particular problem during night-driving)—but is more of a feature after radial keratotomy rather than after PRK.[1]

1 JF Taylor 1995 *Medical Aspects of Fitness to Drive*, Medical Commission on Accident Prevention, page 122. We thank Mr John Dixon Salt for providing material on which this page is based.

7. Ear, nose and throat diseases

7. Ear, nose and throat diseases

ENT examination

Doctors working in ENT departments are readily distinguishable by their concave mirrors with central hole worn over the eye. These reflect light from a bright independent source in a concentrated beam, giving good illumination and stereopsis, leaving both hands free for manipulations. Fibre-optic devices for nasendoscopy obviate the need for headgear.

Examination of the ear Examine the external *auditory meatus*, *pinna*, and its environs for swelling and for inflammation. Swab any *discharge*, and remove any *wax* (p 530). Attach the largest comfortable earpiece to the otoscope, and examine the full length of the auditory meatus and *drum*. Seeing the drum is the first skill to master, and can be achieved quite readily: but the real skill lies in seeing what lies *behind* the drum. This takes many years of practice. Begin by pulling the pinna up and backward to straighten the external auditory canal. (In infants, pull the pinna backwards and *down*.) The handle of the malleus makes a definite landmark behind the tympanic membrane. Anteroinferiorly is the light reflex which is reflected due to the degree of concavity of the membrane in this region. To left-right orientate yourself (eg when looking at pictures of drums) know that the light reflex points to the toes (ie anteroinferior). Try to get a good look at the posteriorsuperior quadrant of the drum: it is behind this section that there is the most important clockwork: the *posterior mallear fold*, the long crus of the *incus*, and the *facial nerve*. The colour, translucency and any bulging of the membrane are noted together with any perforations. What is their position, and does the margin extend to the periphery? *Perforations* of the pars flaccida may indicate serious pathology (p 536). Mobility of the membrane can be demonstrated using an aural speculum with a sealed glass in front and teat on the side to attach a little rubber balloon. As the balloon is squeezed the drum should move. A patent *Eustachian tube* is indicated by movement of the drum during a Valsalva manoeuvre.

The nose and throat In order to examine the throat and nasal passages the parts needing to be examined are moved into the beam of light reflected from the mirror. First ask the patient to open his mouth without protruding the tongue. Apply a tongue depressor just beyond the highest part of the tongue and ask the patient to say 'ah'. The *palate* rises opening the *oropharynx* so *tonsils*, *posterior wall of pharynx* and anterior and posterior pillars of the *fauces* may be examined. In some people, saying 'aye' (as is in '*a*ble') or simply watching while the patient breathes (with the tongue depressed), is more revealing.

A warmed mirror held below the gap behind the soft palate then reveals the *nasal septum*, the posterior end of superior, middle and inferior *turbinates*, the lateral walls of the *nasopharynx*, the openings of *Eustachian tubes*, the *pharyngeal recess*, the roof of the *nasopharynx* and *adenoids*. Anterior rhinoscopy (ie holding open the nares with a special Thudiculum speculum) enables the front of the inferior and middle turbinates to be seen.

Indirect laryngoscopy is performed, with the tongue protruded, using a large warmed mirror. As the patient breathes, the *epiglottis* and the posterior part of the inlet of the *larynx* is seen. As the patient says 'ee', the interior of the larynx and *vocal cords* may be seen. Flexible fibre-optic endoscopy, using a 3mm nasendoscope, is used for more detailed examinations. Admission to hospital is now mainly for taking of biopsies rather than for examination under anaesthesia.

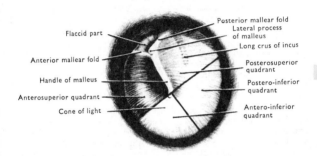

Left tympanic membranes seen from the lateral side. The 4 arbitrary quadrants are indicated by solid lines and by the handle of the malleus. (From *Cunningham's Manual*, Vol. 3, OUP.)

Hearing tests

When assessing deafness, determine the site of the disability, the severity, and the cause. The aim of the examination is to find out if the deafness is treatable, and whether it is part of some other process (eg acoustic neuroma). The first part of the examination should be to remove any wax that is occluding the external auditory meatus.

Tuning fork tests Use a tuning fork with a frequency of 512Hz.

Rinne test: With normal hearing, air conduction (tuning fork held lateral to the external auditory meatus with prongs aligned parallel to the meatus) is better than bone conduction (base of tuning fork placed on mastoid process). When air conduction is greater than bone, the result is termed Rinne positive. Positive responses occur with normal ears and sensorineural (perceptive) hearing loss. Rinne negative response (bone conduction>air) occurs with conductive deafness. If one ear has severe or complete sensorineural deafness, a false negative response may be elicited because the cochlea of the other ear may pick up the sound by bone conduction. Use of a Barany noise box, supplying distracting noise to the other ear during the test, prevents this.

Weber test: The foot of the vibrating tuning fork is placed on the patient's forehead and he is asked in which ear the sound is heard. Sound localizes to the affected ear with conductive deafness, to the contralateral ear in sensorineural deafness, and does not localize to either ear if both ears are normal.

Audiometric tests These quantify loss and determine the site.

Pure tone audiometry uses electronic equipment which emits tones at different strengths over the frequency range of 250Hz to 8000Hz in a sound-proofed room. The patient registers when he first hears the sound and the intensity is recorded in decibels. This records air conduction. A bone conduction threshold can also be obtained by using a transducer over the mastoid process.

Tympanometry (acoustic impedance) This detects fluid in the middle ear (which causes low pressures, p 538), with a sensitivity of 90% and a specificity of 75%. It measures the proportion of an acoustic signal which is transferred from the external to the internal ear, and compares the absorbed and reflected components to compute middle ear pressure.[1] A probe seals the ear canal, varying pressure is introduced into the canal and the compliance of the ear drum recorded on a graph. A normal ear shows a smooth bell-shaped compliance curve. Fluid in the middle ear flattens the curve. High intensity sound introduced into the ear (>85dB) produces a notch on the graph as the stapedius muscle contracts (mediated by the seventh nerve). 5% of the population have absent stapedius reflexes in otherwise normal ears.

Speech audiometry This examines discrimination of speech above the threshold. It indicates whether sensorineural defects lie in the cochlea or the auditory nerve. It can be used to predict whether a hearing aid would benefit a patient.

Assessing hearing in children Co-operation is necessary for the above tests. In babies under 6 months hearing is assessed from the startle response or blinking. At 6 months they should turn their head to sounds. From 3 years audiometry is possible.

1 A Maw 1992 *BMJ* i 67

Referrals for speech therapy

For a full analysis of the problems of *delay in talking*, see p 203.

1 in 7 preschool children may have transient speech and language problems. Many resolve spontaneously or with professional help but serious disorders affect 1 in 20 primary and 1 in 80 secondary school children. 110,000 UK children are so seriously affected that they cannot be understood outside the immediate family.

Refer for speech therapy assessment if:

0-18 months:
- Feeding difficulties from anatomical or neurological disorder, eg cleft lip, or cerebral palsy.

2 years:
- If vocabulary is <30 words or no phrases (but not if good communication skills without speech, and he or she seems on the edge of talking).

3 years:
- Speech is unintelligible.
- Using sentences of 2 words only.
- No descriptive words or pronouns used.
- Limited comprehension (eg cannot identify scissors or pen by 'which do we draw with').
- Parental anxiety, if the child is going through a stage of non-fluency.

4 years:
- Speech is not clear. (Problems with 'r', 'th' and lisps can be left until 5 years.)
- Sentences used are less than 3 words; vocabulary is limited.
- Difficulty in carrying out simple commands.

5 years:
- Persisting articulation difficulties.
- Difficulty understanding simple sentences.
- Difficulty in giving direct answers to simple questions.
- Difficulty with sentence structure; immature sentences; word order.

6 years and older:
- Persisting articulation problem.
- Difficulty understanding spoken language.
- Difficulty with verbal expression.
- Stammering.
- Voice problems.

Children with voice problems, eg hoarseness or excessive nasality, are best referred at whatever age they present.

Painful ears

When someone complains of painful ears, examination should include not only the ear for local sources (eg otitis externa p 536, furunculosis p 536, otitis media p 536, or mastoiditis) but sources of referred pain should also be sought.

534

Referred pain If no local pathology, think of referred pain. Five nerves refer pain to the ear. The auricular branch of the trigeminal nerve may refer pain from the sphenoidal sinus or the teeth. The greater auricular nerve (C2,3) may refer pain from wounds or glands in the neck and from cervical disc or arthritic lesions. A sensory branch of the facial nerve refers pain in geniculate herpes (Ramsay Hunt syndrome, p 756). The tympanic branch of the glossopharyngeal nerve and the auricular branch of the vagus refer pain from the throat to the ear, eg in tonsillitis; quinsy; carcinoma of the posterior third of the tongue, pyriform fossa or larynx.

Bullous myringitis Viral infections (influenza), *Haemophilus influenzae* and *Mycoplasma pneumoniae* can cause painful haemorrhagic blisters on the ear drum and in the external ear canal. There may also be haemorrhagic fluid in the middle ear.

Barotrauma (aerotitis) This is damage caused by undergoing changes in atmospheric pressure in the presence of an occluded Eustachian tube. It afflicts aircraft travellers and divers. It can be prevented by repeated Valsalva manoeuvres during aircraft descent and use of decongestants (eg xylometazoline spray every 20 minutes into the nose, starting 1h before landing). Those with middle ear effusions or unresolved otitis media should not fly. Those affected may suffer sensations of pressure, severe pain and deafness. Examination reveals fluid behind the drum and haemorrhagic areas in the drum. There is conductive deafness. The fluid usually clears spontaneously over several weeks.

Mastoiditis This was a sequelae in 1-5% of otitis media sufferers in pre-antibiotic days. Due to impaired drainage from the middle ear, pressure builds up in the mastoid air cells and there is breakdown of the bony partitions between them. The process can take 2-3 weeks. Those affected complain of pain, low grade fever, malaise and hearing loss. There is variable discharge which is commonly foul smelling. It should be suspected in those with discharging ears of greater than 10 days' duration. Classical swelling behind the ear with downward displacement of the pinna implies subperiosteal abscess formation—a particular feature of mastoiditis. Mastoid radiography may exclude the diagnosis by showing normal air cells, but haziness over the cell system can occur in mastoiditis or otitis externa. Initial treatment is intravenous antibiotics (eg ampicillin 500mg/6h IV) and myringotomy (take cultures and adjust antibiotics as necessary). If resolution does not occur, mastoidectomy will be necessary.

Childhood deafness

Although temporary deafness in childhood due to middle-ear effusion and glue ear is a common problem, permanent deafness is much less common (incidence in infancy 1-2/1000). It is, however, essential that deafness is picked up early, so that as much help as possible can be given to restore hearing to enable children to learn to speak.

Causes
- *Hereditary:* These include Wardenburg's, Klippel-Feil, and Treacher Collins' syndromes and mucopolysaccharidoses.
- *Acquired in utero:* Maternal infection (rubella, CMV, influenza, glandular fever, syphilis), ototoxic drugs.
- *Perinatal:* Anoxia, birth trauma, cerebral palsy, kernicterus.
- *Postnatal:* Mumps, meningitis, ototoxic drugs, lead, skull fracture.

Detection Hearing should be assessed before the age of 8 months. Those with a family history of deafness, and those who were exposed to prenatal, perinatal or postnatal hazards should receive particular attention because they are 10 times more likely than the general population to be affected. Younger babies are not at present formally tested for hearing although they may produce a startle response to sounds. From 6 months to 1 year children may be given the distraction test—by sitting them on their mother's lap with a distractor sitting in front of them to attract their attention sporadically (hence keep them facing the midline). The tester stands 3 feet behind the mother and tests each ear in turn. Low frequencies are tested with the spoken voice, high frequencies with a rattle. Rustling paper gives a broad spectrum stimulus. Refer for specialist testing if there is doubt about hearing. From 3 years old, pure tone audiometry may be used.

From 12 months to 2yrs children are unco-operative, hence the need for objective tests (also for neonatal screening): such a test is measurment of otoacoustic emisssions—a microphone in the external meatus detects tiny cochlear sounds produced by movement in its basilar membrane.[1]

Another useful way of achieving objective tests of hearing is the use of tympanometry (p 532)—or evoked response audiometry, in which a recording electrode is applied behind the ear, in the ear canal, or through the tympanic membrane. The ear is stimulated by sound and the responses generated are picked up, amplified, and fed to a computer. (This takes place in an acoustically treated room.)

Treatment Once deafness has been detected treatment aims to provide as good hearing as possible to help children learn to speak and later with education. Teachers of the deaf make arrangements for the fitting of hearing aids and help monitor progress. Children usually need higher gain from their hearing aids than adults. Ear moulds may need frequent changing to maintain a good fit. Parents should be encouraged to talk as much as possible to their deaf children. Children may be educated at normal schools with visits from teachers of the deaf, or, (for the partially hearing) in specialized units in normal schools, or in schools for the deaf, depending upon individual needs. For a few children, a cochlear implant may be suitable (p 452).

Not everyone will agree that a deaf child needs treatment. In some families with deaf parents, the arrival of a deaf child is especially welcomed. Here, deafness may be considered a variant, rather than a handicap, and rich communication is possible by sign language in those born to it. It is certainly possible to imagine a world in which everyone is deaf, and in which deafness does not matter—but, for better or for worse, this will not be the world in which the young child will grow up, and have to relate to.

[1] A Hinton 1994 *BMJ* ii 651

Deafness in adults

Deafness, unlike blindness, is not a quantified disability; *any* degree of hearing loss may be described as deafness. It is frustrating for those affected and for friends and families, as the niceties of communication start to fail, leading to isolation, with no innuendo, no jokes, no asides, and no music. ~3 million adults are hearing-impaired (UK).

Classification *Conductive deafness:* There is impaired transmission of sound through the external canal and middle ear to the foot of the stapes. External canal obstruction (wax, discharge from otitis externa, foreign body, developmental abnormalities); perforation of the ear drum (trauma, barotrauma, infection); problems with the ossicular chain (otosclerosis, infection, trauma); and inadequate Eustachian tube ventilation of the middle ear with effusion present (eg secondary to nasopharyngeal carcinoma) all result in conductive deafness.

Sensorineural deafness: results from defects central to the oval window in the cochlea (sensory), cochlear nerve (neural) or, rarely, more central pathways. Ototoxic drugs, eg streptomycin and aminoglycosides (esp. gentamicin) and most causes of deafness in infancy are sensorineural. Infections (meningitis, measles, mumps, flu, herpes, syphilis), cochlear vascular disease, Ménière's (p 546) and presbyacusis (senile deafness) are all sensorineural. Rare causes: acoustic neuroma, B₁₂ deficiency, multiple sclerosis, secondary carcinoma in the brain.

Management ●Classify the deafness (p 532). ●Find treatable causes. ●Exclude the dangerous: acoustic neuroma, cholesteatoma, effusion from nasopharyngeal carcinoma. Sudden onset sensorineural deafness demands urgent investigation. ●Find out what procedure to advise, eg surgery for perforations, otosclerosis—or simply the administration of the most suitable type of hearing aid. ●Those with sensorineural deafness so bad that they cannot benefit from a hearing aid may benefit from a cochlear implant.[1] This device takes ~2h to fit (under GA); an external device processes sound, transmitting it across the skin to the subcutaneous receiver coil—which supplies an electrode which directly stimulates the auditory nerve, bypassing the cochlear. Rehabillitation is needed to understand the new sounds. Cost (UK) up to £35,000. ▣

Otosclerosis Usually bilateral; ♀/♂≈2; 50% of those affected have a family history. Symptoms usually appear in early adult life and are made worse by pregnancy. Pathology: vascular spongy bone replaces normal bone around the oval window to which there is adherence of the stapes footplate. There is conductive deafness (hearing is better in the presence of background noise), sometimes also tinnitus and vertigo. Replacing the stapes with an implant helps 90%.

Presbyacusis (senile deafness) Loss of acuity for high frequency sounds starts before 30yrs of age and rate of loss for the higher frequencies is progressive thereafter. Deafness is therefore gradual in onset and people do not usually notice it until the lower frequencies involving speech are affected. Hearing is most affected in the presence of background noise. There is no treatment other than hearing aids.

Prevention of deafness[2] The single most successful way of reducing deafness is to limit damaging noise (<85dB/8h day) exposure at work and leisure—as indicated by finding talking difficult, ringing in the ears during exposure, or sounds appearing muffled after exposure.

1 E Douek 1990 *BMJ* ii 74   2 *Lancet* Ed 1991 ii 21 ▣

Tinnitus

Tinnitus (from Latin *tinnire*–to ring) is ringing or buzzing in the ears. Almost all of us have experienced tinnitus, but only 0.5–2% are severely affected. Although children sometimes experience transient tinnitus they often do not mention it. Peak age for onset is at 50–60yrs.

Mechanism Obscure. May be spontaneous otoacoustic emission, or ephatic transmission (cross-talk) between adjacent nerve fibres.

Causes Unknown; hearing loss (20%); wax; viral; presbyacusis; noise (eg gunfire); head injury; suppurative otitis media; post-stapedectomy; Ménière's; anaemia; hypertension; impacted wisdom teeth.
▶Investigate unilateral tinnitus fully to exclude an acoustic neuroma.
Drugs: Aspirin; loop diuretics; aminoglycosides (eg gentamicin).
Objectively detectable tinnitus: Palatal myoclonus (clicking tinnitus); temporomandibular problems; AV fistulae; bruits; glomus jugulare tumours.
Psychological associations: Redundancy, divorce, retirement.

Another group of sufferers are those who have roaring in the ears when breathing through the nose which is abolished when they breathe through the mouth. On examination, the tympanic membrane may be seen to move on respiration. The cause may be a patulous Eustachian tube and relief may be experienced after application of silver nitrate to the Eustachian tube orifice or submucosal injection of Teflon® to narrow the lumen.

History Site of tinnitus (ear or central); character; alleviating and exacerbating factors; otalgia; otorrhoea; vertigo; head injury; family history of deafness or tinnitus; sleep; social surroundings (tinnitus is worse if isolated or depressed); drugs.

Tests and examinations Otoscopy to detect middle-ear disease; hearing tests (tuning fork and audiometry); tympanogram to examine middle-ear function and stapedial reflex thresholds.

Treatment *Psychological support* is very important (eg from a hearing therapist). Exclude serious causes; reassure that tinnitus does not mean madness or serious disease and that it often improves in time. Cognitive therapy is recommended.[1] Patient support groups can help greatly.[2]
Drugs are disappointing. Avoid tranquillizers, particularly if the patient is depressed (nortryptiline is the drug of choice here),[1] but hypnotics at night may be helpful. Carbamazepine has been disappointing; betahistine only helps in some patients in whom the cause of the tinnitus is Ménière's disease.

Masking may give relief. White noise is given via a noise generator worn like a post-aural hearing aid. Hearing aids may also help those with hearing loss by amplifying desirable sound. Section of the cochlear nerve relieves disabling tinnitus in 25% of patients (but deafness results).

Glomus jugulare tumours A vascular neoplasm arising from the jugular body, which may grow to present as a cerebellopontine angle mass, or, in the external auditory canal, as a pulsatile mass. They may secrete 5-HT. It is a rare cause of tinnitus, but a common cause of *pulsatile* tinnitus.[3]

1 L Luxon 1993 *BMJ* i 1490   2 British Tinnitus Assocn, 105 Gower Street, London WIE 6AH
3 C Hawkes 1993 *BMJ* ii 262

Vertigo

Vertigo can be horrible, confining people to home, making them fearful and depressed. When people complain of feeling swimmy, dizzy, unreal, or panicky, find out if an illusion of movement of self or environment is present—ie vertigo. If so, does it last seconds (usually benign position vertigo, below, if there are no CNS signs), hours (eg migraine), or days (look for a central cause, ie involving nerve VIII, brainstem vestibular nuclei, medial longitudinal fasciculus, cerebellum, or vestibulospinal tract—see below).

Vestibular (peripheral) vertigo is often severe, and may be accompanied by nausea, vomiting, hearing loss, tinnitus and nystagmus (usually horizontal). Hearing loss and tinnitus are less common in central vertigo (it is usually less severe). Nystagmus may be horizontal or vertical with central vertigo—and may be different in each eye (eg in the abducting eye).

Causes *Peripheral:* Ménière's; vestibular neuronitis; benign postural vertigo; gentamicin; labyrinthitis; cholesteatoma; trauma; Eustachian obstruction. *Central:* Acoustic neuroma; multiple sclerosis; head injury; epilepsy; migraine (maybe no headache); vertigo; vertebrobasilar ischaemia (other CNS signs will be present); geniculate herpes; syphilis.

Examination and tests Test cranial nerves, cerebellar function, reflexes. Do Romberg's test (+ve if balance is worse when eyes are shut—implying defective joint position sense or vestibular input). Assess nystagmus. Do provocation tests, below. If equivocal, consider audiometry; electronystagmography; brainstem auditory evoked responses; calorimetry (the only way to test each labyrinth separately; irrigate each canal with water 7° above and 7° below body temp—is nystagmus induced?); CT; MRI; EEG; LP.

Benign positional vertigo There are attacks of *sudden-onset* rotational vertigo lasting >30sec provoked by head-turning. Common after head injury, it may be due to damage in the utricle. *Diagnosis:* Provocative test: with the patient supine, the head is lowered ~30° below the level of the couch, and turned 30-40° to one side. If +ve, the patient experiences vertigo and rotary nystagmus towards the undermost ear, after a latent period of a few seconds. This lasts <1min (adaption). On sitting, there is more vertigo (± nystagmus). ►If any of these features are absent, seek a central cause. *Pathogenesis:* Displacement of the otoconia from the maculae (the receptor for sensing acceleration in the utricle and saccule). The otoconia then settle on the lowest part of the labyrinth. *Causes:* Head injury; spontaneous degeneration of the labyrinth, post viral illness or stapes surgery; chronic middle-ear disease. *Treatment:* If not self-limiting within a few months, consider physiotherapy referral for teaching: ●Repeated adoption of the position which causes vertigo (habituation). ●Formal Epley head exercises to disperse otoconia. A last resort is denervating the posterior semicircular canal or obliterating it by laser (transmastoid) as an option, but deafness may follow.

Vestibular neuronitis follows a febrile illness in adults in winter and is probably viral. Sudden vertigo, vomiting and prostration are exacerbated by head movement. *Treatment:* Try cyclizine 50mg/8h PO. Recovery occurs within 2-3 weeks. It is difficult to distinguish from 'viral labyrinthitis'.

Ménière's disease Dilatation of the endolymphatic spaces of the membranous labyrinth causes paroxysmal vertigo lasting up to 12h with prostration, nausea and vomiting. Attacks occur in clusters. Tinnitus occurs and sensorineural deafness is often progressive. Treat acute vertigo symptomatically (cyclizine 50mg/8h PO). Betahistine 8-16mg/8h PO gives unpredictable results but is worth trying—as are diuretics such as chlorthalidone. Operative decompression of the saccus endolymphaticus may relieve vertigo, prevent progress of the disease and also conserve hearing. Labyrinthectomy *may* relieve vertigo but causes total ipsilateral deafness.

1 SW Denholm 1993 *BMJ* ii 1508

Throat infections and tonsillar tumours

In the past GPs readily gave antibiotics for sore throats in case the cause was a β-haemolytic streptococcal infection which may have resulted in rheumatic fever. Rheumatic fever is now rare in the West (p 194). We know that rheumatic fever patients had often had their sore throats treated so antibiotics had not prevented the illness. GPs may not wish to give antibiotics for simple sore throats, as many are caused by viruses—in any case the proportion of those with sore throats consulting may be as low as 1 in 18. Because sore thoat is such a common presentation, treating *everyone* with penicillin is not only expensive, and encouraging of doctor-dependency (p 326), but also risks more deaths from anaphylaxis[1] than would be saved by any possible benefit.●* Throat swabs do not offer much help (sensitivity 26–30%; specificity 73–80%).[2] Numerous antigen detection kits are available (specific, but not sensitive).[3] Other pathogens apart from streps: staphs, *Moraxella catarrhalis*, mycoplasma, chlamydia, haemophilus.

Tonsillitis The young often complain of abdominal pain with tonsillitis. Older children may complain of sore throat, fever, malaise, difficult swallowing and painful neck lymph nodes.

   The appearance of a rash after 48h on neck and upper chest spreading rapidly to abdomen and limbs suggests *scarlet fever*. It spares the mouth area, giving rise to contrasting circumoral pallor—and the tongue is covered with a white 'fur', which, when cleaned off, leaves prominent papillae (the 'strawberry' tongue). Cause: group A streptococcus.

Treatment: The use of antibiotics has been a source of great debate as in 50% the cause is a virus, and antibiotics make no difference. In reality, a variety of factors influence the decision whether to give antibiotics. When antibiotics are given penicillin, 250mg/6h PO is often used for up to 10 days—and this is the treatment of choice in scarlet fever. In spite of a recent study,[4] amoxycillin is contraindicated as it will cause a rash in almost all whose pharyngitis is due to Epstein–Barr virus.

Tonsillectomy indications have been hotly debated and the number of operations performed is declining. Recurrent tonsillitis is still an indication for referral and controlled trials show that the incidence of sore throats after tonsillectomy is reduced. Tonsillectomy is performed after quinsy to prevent recurrence and in the rare event of tonsils being so large as to cause cor pulmonale or obstructive sleep apnoea in those bearing them. Complications of operation include rare but real risk of death from haemorrhage, and behaviour problems in children.

Local complications of tonsillitis Retropharyngeal abscess: This is rare. Presentation is as an unwell child who fails to eat or drink. Lateral X-rays of the neck show soft tissue swelling. *Treatment:* Incision and drainage of pus under general anaesthesia in a head-down position to prevent aspiration. Peritonsillar abscess (quinsy): These usually occur in adults. Treatment of throat cellulitis by high-dose penicillin may be preventive. Once developed, the tonsil is obscured at examination by the soft palate. There is difficulty in swallowing ± lock-jaw. *Treatment:* Incise under local anaesthetic, or do abscess tonsillectomy under general anaesthetic.

Tonsillar tumours Commonest in the elderly, these present with sore throat, dysphagia and otalgia. Unilateral tonsil enlargement is ominous and grounds for excision biopsy. Pathology: squamous carcinomas (70%), reticulum cell sarcoma, lymphosarcoma. *Treatment:* Radiotherapy, surgery and cytotoxics may all be used depending upon the type of tumour.

1 PS Little 1994 *BMJ* ii 1010   2 C Del Mar 1992 *Med J Aust* 156 572
3 H Marcovitch 1990 *Arch Dis Chi* 65 249   4 P Shvartzman 1993 *BMJ* i 1170

ENT tumours

Carcinoma of larynx/hypopharynx Incidence: 2000/yr in England and Wales. Usually squamous cell carcinomas, occurring more in men, they are likely to be caused by smoking. *Sites:* Supraglottic, glottic, subglottic. Glottic tumours have the best prognosis as they cause hoarseness earlier and spread to the nodes of the neck later than other types. *Presentation:* Persistent progressive hoarseness, then stridor, difficulty or pain on swallowing and cause pain in the ear if the pharynx is involved; haemoptyses. *Diagnosis* is made by laryngoscopy and biopsy. *Treatment* may be radiotherapy or total laryngectomy ± block dissection of the neck glands. After laryngectomy patients have a permanent tracheostomy, hence the need to learn oesophageal speech. It is horrible to wake up from surgery with an unsightly hole in the throat and to be unable to express yourself: pre-op counselling is only a partial answer.[1] Patients are usually discharged after 10-14 days with a plastic stent or metal cannula to keep the tracheostomy open—which may be discarded some weeks later. Excessive secretions and crusting around the stoma are common, needing meticulous attention, to avoid obstruction—humidified stomal covers (eg Laryngofoam®) may get round these problems. Advise to take care while having a bath, and to avoid fishing and deep water (unless expert training is to hand). *Late complications:* Stenosis; recurrent pneumonia. Suggest a Laryngectomy Club.[2]

Nasopharyngeal malignancy is much commoner in China than in UK (25% of all malignancy *vs* 1%). Cause (in China) may involve:[3] ●Genetics (abnormal HLA profiles); ●Early infection with Epstein-Barr virus; ●Weaning babies onto salted fish (?*N*-nitroso carcinogens).

Tumours may be carcinomas, lymphoepitheliomas or lymphosarcomas. Lymphatic spread is usually early to deep cervical nodes (between mastoid process and mandible). Local spread may involve cranial nerves via the foramen lacerum or jugular foramen. Symptoms: epistaxis, diplopia, nasal obstruction, neck lumps, conductive deafness (Eustachian tube affected), or cranial nerve palsy (all but I, VII and VIII can be affected).
Diagnosis is by posterior rhinoscopy, inspection palpation and nasopharynx biopsy under general anaesthesia, + skull-base radiology.
Treatment is with radiotherapy. The prognosis is often poor.

Maxillary sinus tumours are usually squamous carcinomas developing in the middle-aged or elderly. Suspect in those developing chronic sinusitis for the first time in later life. Early presentations may be with blood-stained nasal discharge and nasal obstruction. Later there may be swelling of the cheek, swelling or ulceration of the buccoalveolar plate or palate, epiphora due to a blocked nasolacrimal duct, ptosis and diplopia as the floor of the orbit is involved and pain in the distribution of the 2nd branch of the 5th cranial nerve. Local spread may be to cheek, palate, nasal cavity, orbit, and pterygopalantine fossa. *Tests:* Radiology with tomography reveals erosion of the bony walls. A Caldwell Luc approach may be needed for histological diagnosis. *Treatment options:* Radiotherapy; radical surgical removal; cytotoxics.

Acoustic neuroma These slow growing neurofibromas (schwannomas) often arise from the acoustic nerve's vestibular division, giving progressive[4] ipsilateral tinnitus ± sensorineural deafness. Big tumours may give ipsilateral cerebellar signs or ICP↑ signs. Giddiness is common, vertigo is rare. Trigeminal compression above the tumour may give facial numbness. Near-by cranial nerves may be affected (esp. V, VI, VII).
Tests: X-rays may show an enlarged auditory canal. MRI has made earlier diagnosis possible. *Treatment:* Surgical removal may be possible.

7. Ear, nose and throat diseases

Pharyngeal pouch This is a mucosal herniation at Killian's dehiscence of the inferior constrictor. There may be halitosis, regurgitation of food, and a neck lump (usually left side). Diagnose by barium swallow. *Treatment* is usually surgical.

Pharyngeal carcinoma Oropharyngeal tumours are often advanced at presentation. Symptoms include vague sore throat, sensation of a lump, referred otalgia and local irritation by hot or cold foods. Hypopharyngeal tumours may give dysphagia, voice alteration, otalgia, stridor and throat pain. *Treatment* is with a combination of surgery, chemotherapy and radiotherapy.

Plummer–Vinson syndrome (Patterson-Kelly-Brown syndrome) There is an oesophageal web associated with iron-deficient anaemia, and possibly post-cricoid carcinoma.

1 M Gleeson 1994 *BMJ* i 1452 2 UK tel: 0171 381 993 3 *Lancet* Ed 1989 ii 840 4 A Wright 1995 *BMJ* ii 1141&1421

Dysphagia

Dysphagia is difficulty in swallowing food or liquid. Unless it is associated with a transitory sore throat, it is a serious symptom, and merits further investigation, usually by endoscopy, to exclude neoplasia. If the experience is one of a lump in the throat *at times when the patient is not swallowing*, the diagnosis is likely to be anxiety (globus hystericus).

Causes *Malignant:*	Neurological causes:	Others:
Oesophageal cancer	Bulbar palsy (OHCM p 466)	Benign strictures
Gastric cancer	Lateral medullary	Pharyngeal pouch
Pharyngeal cancer	syndrome	Achalasia (below)
Extrinsic pressure	Myasthenia gravis	Systemic sclerosis
(eg lung cancer)	Syringomyelia (OHCM p 476)	Oesophagitis
		Iron-deficient anaemia

Differential diagnosis There are 4 key questions to ask:
1 Can fluid be drunk as fast as usual, except if food is stuck?
   Yes: Suspect a stricture (benign or malignant).
   No: Think of motility disorders (achalasia, neurological causes).
2 Is it difficult to make the swallowing movement?
   Yes: Suspect bulbar palsy, especially if he coughs on swallowing.
3 Is the dysphagia constant and painful?
   Yes (either feature): Suspect a malignant stricture.
4 Does the neck bulge or gurgle on drinking?
   Yes: Suspect a pharyngeal pouch (food may be regurgitated).

Investigations FBC; ESR; barium swallow; endoscopy with biopsy; oesophageal motility studies (this requires swallowing a catheter containing a pressure transducer).

Nutrition Dysphagia may lead to malnutrition. Nutritional support often needed prior to treatment.

Oesophageal carcinoma This is associated with achalasia, Barrett's ulcer (p 694), tylosis (an hereditary condition in which there is hyperkeratosis of the palms), Plummer-Vinson syndrome (below) and smoking. Survival after resection is rare after 5 years (OHCM p 164).

Benign oesophageal stricture *Causes:* oesophageal reflux; swallowing corrosives; foreign body; trauma. *Treatment:* dilatation (endoscopic or with bougies under anaesthesia).

Barrett's ulcer See OHCM p 694.

Achalasia There is failure of oesophageal peristalsis and failure of relaxation of the lower oesophageal sphincter. Liquids and solids are swallowed only slowly. CXR: air/fluid level behind the heart, and double right heart border produced by a grossly expanded oesophagus.

Treatment: Myomectomy cures ~75% of patients. Pneumatic dilatation may also help.

Plummer-Vinson (Paterson-Brown-Kelly) syndrome There is an oesophageal web associated with iron-deficient anaemia, and possibly postcricoid cancer.

Differential dermatological diagnoses

Ring-shaped lesions ●Psoriasis: red areas topped by silver scales, p 586.
●Fungal lesions (often asymmetrical, and spreading from a focus, p 590).
●Discoid eczema (if coin-shaped, the word 'nummular' is used).
●Erythema multiforme: targets with central blister or haemorrhage, p 580.
●Granuloma annulare: purple patch eg on the hand dorsum, with the skin surface remaining intact, unlike fungal infections.
●Basal cell carcinoma: white nodular pearly periphery, p 582.
●Urticaria: red and white maculopapular nettle rash, associated with allergens or cold temperatures. Urticaria is localized skin or mucosal oedema caused by release of vasoactive agents (eg histamine).
●Pityriasis rosea (p 594); lichen planus (p 594).
●Burns (if on a child, consider a cigarette burn from maltreatment).
●Leprosy (anaesthetic white area within the ring) or secondary syphilis.

Linear lesions ●Koebner phenomenon (p 586)—psoriasis or lichen planus.
●Linear urticaria (scratches giving triple response—dermatographia).
●Dermatitis artefacta (self-inflicted trauma, concealed by the patient).
●Psoralen-induced phytophotodermatitis, eg after battling with giant hogweed or rue in the garden on a sunny day.
●Impetigo: a papulopustular staphylococcal (± streptococci) itchy rash, which may spread along scratch marks. Lesions often start around the mouth and nose. It is topped by honey-coloured crusts, and responds well to flucloxacillin 125mg (child) to 250mg (adult) per 6h PO.
●Herpes zoster: linear if the edge of the affected dermatome is linear.

White patches ●Pityriasis alba: post-inflammatory patch on child's face.
●Tinea versicolor: fungal; brown and white 'raindrop' pattern (p 590).
●Vitiligo: symmetrical, sometimes itchy with hyperpigmented borders, associated with autoimmunity (eg pernicious anaemia, thyroid disease, diabetes, Addison's disease, alopecia areata).
●Rarely piebaldism (from birth, with a white forelock), or leprosy.

Brown spots ●Freckles (on sun-exposed areas) or moles are the commonest.
●Lentigos (like freckles, only darker, and not affected by sunlight).
●Melanomas and dysplastic naevea: see p 584.
●Senile lentigo: solitary on an elderly cheek; may become malignant.
●Chloasma (=melasma)—typically a self-limiting, pale, irregular, brown patch darkening in summer on the face of someone on the Pill, or pregnant (melanocyte activity ↑ in response to ♀ hormones and sunlight). Management: avoid sunlight; try camouflage cosmetics.
●Café au lait spots: faint; irregular. If >5, consider neurofibromatosis.
●Seborrhoeic warts: brown, greasy papules. Non-malignant; not true warts.
●Basal cell carcinoma, melanoma and solar keratoses (p 582).
●Systemic diseases (eg haemochromatosis, Addison's disease).

Subcutaneous nodules Rheumatoid nodules, rheumatic fever, PAN, xanthelasma, tuberous sclerosis, neurofibroma, sarcoid, granuloma annulare.

Itching (the desire to scratch); if this produces papules, prurigo is present; if the skin is like Morocco leather, lichenification is present.
Questions: is there itch and blisters (*urticaria*); is there burning and itching (*dermatitis herpetiformis*); is the itch worse at night, and are others affected (scabies); what provokes itch? Following a bath suggests *polycythaemia* or *aquagenic urticaria*. Is there exposure eg to animals or fibre glass (*atopic eczema*)? Is the patient old (*pruritus of old age*)?
Look for local causes: scabies burrows in the finger webs; search hair shafts for lice (p 598); knee and elbow blisters (dermatitis herpetiformis, eg seen with coeliac disease); Wickham's striae (p 594, lichen planus).

Psoriasis

This chronic, remitting papulosquamous inflammatory skin disease—in which the epidermal cell cycle time is much reduced (from ~311h to eg 36h), affects ≥2% of Caucasians (frequent in smokers and/or alcohol users, less frequent in Blacks). Histology: elongation of rete ridges & papillae (± oedema); absence of the granular layer; parakeratosis and the formation of Munro microabscesses by the migration of neutrophils out of dilated papillary capillaries ('squirting papillae').

Presentation May be at any age (nappy rash, p 182)—mean 28yrs. The chief lesion is a red plaque—topped by silvery scales in hairy areas. Plaques are *guttate* (raindrop-like), *geographic* or *rings*. Scales and elevation are absent in intertriginous groins and axillae. Extensor skin on elbows and knees is often affected. If hands and feet are affected, sterile pustules (palmar pustulosis) and fissuring may be nasty. Pustulosis may lead to severe illness with fever, leucocytosis and lymphopenia. It may be triggered by systemic steroid withdrawal. Worsening psoriasis may lead to total erythroderma.

Signs Koebner's phenomenon: physical trauma (eg wounds, hat band pressure, sunburn, scratching) results in linear plaque formation.
Auspitz sign: pinpoint bleeding as scales are picked off plaques.

Psoriasis beyond the skin Nail pitting; corneal nodules; oedema; liver damage; arthropathy, seen in 7% of those with psoriasis in 1 of 5 ways:

1 DIP joint involvement, with nail pitting and onycholysis.
2 Like ankylosing spondylitis.
3 Like rheumatoid arthritis.
4 Asymmetrical large joint polyarthritis, associated with dactylitis.
5 Arthritis mutilans: severe erosive disease with opera-glass hand (*main en lignette*).

Differential diagnosis Fungi (p 590); lichen planus (papules tend to occur on flexor surfaces, having fine striae); mycosis fungoides (may require biopsy differentiation); seborrhoeic dermatitis (no individual plaques, but overlap occurs with some patients' lesions starting out as typical seborrhoeic dermatitis and then turning into psoriasis).

Treatment Explanation,[1] rest and removing triggers (stress, trauma, infection, drugs such as Lithium, chloroquine, or β-blockers) are important. *Steroids* and *coal tar* may be combined (Alphosyl HC® cream = 0.5% hydrocortisone + 5% tar). As coal tar is messy—and as using high potency steroid creams can lead to skin atrophy (especially on the face and flexural surfaces), both are giving way to treatment of typical plaque psoriasis with twice daily *calcipotriol* ointment or cream (vitamin D derivative inhibiting cell proliferation; CI: pregnancy; very extensive psoriasis; avoid face). Alternative: a short-term *dithranol* régime: dithranol 0.1% cream is applied carefully to lesions and covered with dressings to keep the cream in contact with lesions for 30 minutes. The cream is then washed off, and reapplied daily for ~4 weeks. Increase concentration every 48h (0.25%, 0.5%, 1%)—if no burning or irritation occurs. Caution: do not apply to the face or flexures or inflamed areas. Wash out the bath to avoid staining.
Long-wave ultraviolet (UVA) irradiation: (after ingesting a psoralen = PUVA) or short-wave UVB without a psoralen, may help, but long-term problems of skin ageing and neoplasm formation cannot be ignored. Its best indication is for those with severe psoriasis who are over 45 years old.[1]
Cytotoxic drugs: Life-ruining psoriasis may justify *methotrexate* ~10–25mg weekly PO (interacts with co-trimoxazole and NSAIDs; liver biopsy—eg every 1–2yrs, or after 1.5g methotrexate given detects the 5% of long-term users who develop cirrhosis on this drug). Monitor FBC. Other agents: cylosporin, etretinate, hydroxyurea.

1 Psoriasis Association, 7 Milton St, Northampton NN2 7JG (tel. UK 01604 711129) ⌑

Preventing infantile eczema

In susceptible infants, ie those with a strong family history:
- Encourage the mother to breast feed.
- Consider reducing allergen exposure through breast milk by getting the mother to avoid all dairy products, eggs, fish, peanuts and soya-beans in her diet.[1] Give her a daily calcium supplement (1g PO). Ensure that her overall diet is adequate.
- If breast feeding is impossible, many try soya milk (eg Wysoy®)—but this does not work.[1] The most meticulously conducted blind trials show that a much better alternative is a milk in which the protein (casein) is hydrolysed (eg Nutramigen® or Pregestimil®).[1]

Helping children with eczema

Scratching is a constant feature, and a constant threat the integument's integrity—but you cannot tell a child not to scratch. But you can *distract* the child, although this can no doubt seem like a full-time occupation.

Gently rubbing the skin is also said to be a help,[1] as is daily filing of the nails and the wearing of mittens (or cotton tubular bandages) to prevent excoriation—but avoid over-heating. It can be a great relief to everyone when the child goes to sleep (although scratching may continue). To help sleep in a scratching child, experts may recommend higher doses of trimeprazine than are needed for antihistamine effects—eg 7.5-30mg for those aged 6-12 months and 10-50mg for those older than 1 year.[2]

Note also the following, and discuss with child and parents as necessary:
- Infantile eczema gets better with time. Nature provides the light at the end of the tunnel—although asthma is often the next event along the line.
- Elimination diets rarely help, and may cause neurosis in the parents and nutritional deficit in the child. You cannot rely on skin or blood tests to find the minority who will benefit from avoiding dietary triggers.
- Other, possibly more important, triggers are dust mites and pets (does the child improve when on holiday?).

1 R Chandra 1989 *BMJ* ii 228 2 TJ David 1995 *Prescrib J* 35 202

Superficial fungal infections

Fungi are eukaryotes (>1 chromosome) with a differentiated nucleus and rigid polysaccharide cell walls. They are heterotrophs (living off organic matter) and they do not photosynthesize. Cells form hyphae, which interlace to make a mycelium. Arthrospores are the result of fragmenting hyphae. There are 5 genera which are important in skin mycoses: *Microsporum, Epidermophyton, Trichophyton, Candida* and *Malassezia*. The first 3 are moulds, accounting for 90% of all fungal skin infections (also called 'ringworm' or 'tinea'). Proteolytic enzymes allow invasion of the skin, hair and nails. Sites affected: groins (tinea cruris); feet (tinea pedis); nails (tinea unguium) and head—tinea capitis (rare in the UK). *Microsporum canis* (spread mostly by cats(!) to any site) is zoophilic, so tending to cause more acute infections than the other dermatophytes—eg *T rubrum* (which affects all areas); *T mentagrophytes* (feet, nails); *T violaceum* and *T verrucosum* (scalp); *E floccosum* (groins, feet); *M audouinii* (scalp).

The lesions consist of a well-defined red area, with a fine scale, often tending to clear in the middle, to produce the characteristic ring lesion.

Yeasts are single-celled fungi which reproduce by budding. They are dimorphic in that they can also produce hyphae—depending on the environment. *Pityrosporum orbiculare* (also called *Malassezia furfur*) is a lipophilic yeast which causes pityriasis (or tinea) versicolor—producing red-brown, scaly patches on the trunk. Lesions are often white on black people and brown on white people. It is also implicated in seborrhoeic eczema. *Candida albicans* is a skin opportunist: always there, always ready. Favourite places include the mouth, the vagina, around the nail (chronic paronychia), web spaces and inframammary areas. Lesions are red and scaly, often with satellite papules.

Athlete's foot is an itchy, macerating and fissuring lesion between the toes. It may be caused by Candida species or *T mentagrophytes* or *T rubrum*, together with various (mostly Gram +ve) bacteria.

Diagnosis Culture; microscopy of skin scrapings will reveal hyphae. Keratin is dissolved by the use of 20% potassium hydroxide on the slide for 15 minutes (aided by gentle heating). This is the best way of demonstrating *M furfur*, which stains readily with ink. Scalp dermatophytes may present with kerions (pustules) and hairloss—hairs infected with *Microsporum* fluoresce green with Wood's (UV) light.

Treatment Benzoic acid 6% with salicylic acid 3% ointment (Whitfield's ointment) is useful for many fungal infections, but has been superseded by topical imidazoles (eg clotrimazole 1% cream). Griseofulvin is only active against dermatophytes. Give 250-500mg/12h PO for ≥6 months (adults) if nails are affected. If the nail is deformed and painful, urea 40% in aqueous cream may be used under an occlusive dressing for 10 days, allowing the nail to be removed easily. The key factor in treating paronychia is keeping the lesion dry.

Terbinafine is an expensive alternative (eg 250mg/24h PO for 1–3 months) in difficult mycoses, eg where keratin is thick (nails, soles, palms).[1,2] SE: abdominal pain, rash, headaches.

Bacteria and viruses, and the skin

Erythrasma This is caused by *Corynebacterium minutissimum*: a superficial groin infection, not unlike tinea cruris—but it is not raised or vesicular. Treatment: a week's erythromycin (eg 250mg/6h PO) ± clotrimazole cream.

Folliculitis This represents a host of discrete pustules in hair follicles—which may be caused by bacteria or chemicals. Treatment: erythromycin 250mg/6h PO—or a topical antibiotic.

Furunculosis Staphylococcal boils in a hair follicle—on coalescing, they form a carbuncle. Treatment: exclude diabetes. Let the pus out. Antibiotics, such a flucloxacillin 250mg/6h PO ac, may be needed.

Erysipelas This streptococcal cellulitis often affects the face (or leg), causing fever and flu-like symptoms. The skin is red and oedematous (*peau d'orange*). *Treatment:* Co-amoxiclav (250/125) 1 tablet/8h PO—or erythromycin 500mg/8-12h PO (or phenoxymethylpenicillin 500mg/4h PO ± flucloxacillin 250mg/6h PO). For *impetigo*, see p 578.

Viral skin infections Zoster; molluscum contagiosum (clusters of pearly umbilicated papules, caused by a DNA virus—which respond to pricking with a cocktail stick with a little phenol on its point); warts.

1 M Goodfield 1992 *BMJ* i 1151   **2** *Drug Ther Bul* 1992 30 47

Warts

These are caused by infection of epidermal cells by the human papilloma virus (HPV)—from other warts, either from another part of the patient, or from someone else. Warm moist floors or towels may also be a source. Common warts (HPV 2) are round papules with an uneven, horny surface. Plane warts (HPV 3) are smaller and lie flat in the skin. Warts on the plantar surface of the foot are known as verrucas (eg HPV 4—or HPV 1 for deep, painful plantar warts known as myrmecia). Multiple warts may coalesce to form a mosaic wart (HPV 2). Genital warts (HPV 6, 11, 16, 18) are known as condylomata acuminata. They often have peduncles. Many warts regress spontaneously (due to cell-mediated immune mechanisms), and 20% of normal people lose their warts within 3 months. Therefore, many people will not need treatment. Those with sarcoidosis or autoimmune diseases keep their warts for longer than healthy subjects.

Treatment: This may be requested for non-resolving warts. Most respond to salicylic acid, eg as 12% flexible collodion. Apply with care, so that surrounding skin is not damaged. The aim is to remove the overlying hyperkeratotic layers and destroy the underlying epidermis. Lasers, cryotherapy, curettage and electrocautery are alternatives—but not necessarily more effective. Surgical excision should be avoided because this leaves scars. Anogenital warts are often treated with podophyllin as a 15% paint (which must be washed off after ≤6h). If there are many warts, treat them in rotation, to avoid toxic effects from podophyllin absorption. Avoid in pregnancy (it is teratogenic). In pregnancy, liquid nitrogen cryotherapy may be used. Treat he patient's sexual partner too. Women should be followed up with yearly cervical smears because of the known association between genital warts and cervical carcinoma. Plane warts are best left alone, whereas mosaic warts often need prolonged treatment, and even then may be refractory to conventional treatment. Refractory warts should referred to a specialist who may use a retinoid or intralesional interferon or bleomycin.

Acne vulgaris

This is a papulo-pustular/nodulocystic inflammatory rash, usually confined to the face and trunk, which is very common in adolescents (~90% are affected). It is caused by the overproduction of sebum, and sebaceous duct blockage, hence the main lesion—the comedo ('blackhead')—the black top of which is produced by melanin, not dirt. Rarely, there is coexistent seronegative polyarthritis. Acne is androgen-dependent. Colonization by *Propionibacterium acnes* is probably an important factor. Healing by scarring may occur, and if keloid scars (ie with excessive connective tissue) form, great disfigurement may occur. However, due to the age at which acne occurs, even minor disfigurement can be a great handicap—so patients need a sympathetic doctor.

Treatment: A good way to start is to dispel some of the myths about acne—eg that it is caused by being dirty, that sufferers never bother to wash, and that chocolate or greasy foods are to blame. Mild acne may be treated by *benzoyl peroxide 5% lotion* (may be bought over-the-counter). *Topical 1% clindamycin* (3 times the cost, and *not* to be used with benzoyl peroxide) is available as a lotion for dry skins or as a solution with a 'roll-on' applicator (Dalacin T® solution) for more oily skins.[1] Use twice daily, for up to 12 weeks. In the more severely affected, try 6 months *erythromycin* or *oxytetracycline* (500–250mg/12h PO 1h ac—start with the higher dose). *Azelaic acid cream* is an alternative antibacterial/anticomedonal (keratinocyte inhibition) agent which can be used twice daily for up to 6 months.[2] Anti-androgens may also be useful in women (eg *cyproterone acetate* 2mg/24h PO from the 5–26th day of the cycle)—this may be combined with ethinyl oestradiol 35µg as Dianette®. It is worth referring those with severe cystic acne to hospital for treatment with *isotretinoin* (provided liver function tests and lipids are normal). Isotretinoin is teratogenic.

1 *Drug Ther Bul* 1992 **30** 33   2 *Drug Ther Bul* 1993 31 50

Miscellaneous skin complaints

Haemangiomas *Pyogenic granuloma:* This is an acquired haemangioma that presents as a red papule which bleeds very easily. *Treatment:* cautery. The *'strawberry naevus'* is a haemangioma appearing around the time of birth, which may grow to a large size. Treatment is not needed as they regress.

Lichen planus is a papular, often itchy rash which can occur at any age, but is uncommon if very young or old. The typical lesion is a 2–5mm mauve papule whose flat top may reveal white streaks (Wickham's striae). Sites affected: wrists, flexor aspects of forearms, genitals, lumbar region, ankles. It may be linear, following lines of trauma (Koebner phenomenon). On the glans penis, the lesion appears as annular white lesions, and in the mouth as lace-like streaks with a smooth surface. It usually resolves within a year of onset. Associations: immunological diseases, hepatitis C infection. *Treatment:* If needed, try betamethasone valerate cream (or mouth pellets).

Pityriasis rosea The patient is usually a young adult, who presents in the spring or autumn with a pink rash of oval forms on the chest. Lesions show centripetal scaling, and their long axes may be aligned. A key diagnostic point is that it begins with a solitary scaly, indurated patch, called a *herald patch*, and it precedes the rash proper by ~2 weeks. *Treatment:* none: it is a self-limiting condition, and may be of viral origin.

Acne rosacea This is a syndrome of hypertrophied sebaceous glands, with dilatation of skin blood vessels, affecting the face alone. It may start as flushing. Symmetrical telangiectasia, papules and pustules follow. It is not a form of acne. The cause is unknown. In men, skin thickening can cause gross nasal excrescences (rhinophyma) which may respond to peeling away of the skin. Blepharitis, conjunctivitis and corneal vascularization may occur. *Treatment:* oxytetracycline (250mg/12–24h PO 1h ac) for many months. Metronidazole 200mg/12h PO is an alternative (avoid alcohol).

Hair loss (Alopecia) *Alopecia totalis* means loss of all hair in a specified area (eg scalp). *Alopecia totalis et universalis* means the loss of all body hair. If the skin is normal, the cause may be *alopecia areata* (patches of hair loss on the scalp). Hairs are characteristically short and broken ('exclamation mark'). Hair usually regrows in time. Other causes: normal male-pattern baldness (so in females think of ↑androgens—eg Cushing's disease, adrenal tumours or polycystic ovaries), thyroid disease, pernicious anaemia, hypopituitarism, stress, cytotoxics, anticoagulants, iron deficiency.
   If the skin is white and atrophic (scarring alopecia), think of: discoid lupus erythematosus, lichen planus, local scleroderma, trauma, irradiation. *Treatment of alopecia areata* is unsatisfactory. 2% minoxidil solution has been tried, but tends to produce only vellus fluff rather than a good crop of hair—and this falls out when treatment is stopped.

Bullae (blisters) Bullae are large vesicles (>1cm across). *Causes:*
- Insect bites (look for grouped blisters, each with a central punctum).
- Drug-induced photosensitivity (phenothiazines, sulphonamides, thiazides).
- Fixed drug reactions (bullae are always at the same position whenever the drug challenge occurs—eg with salicylates, barbiturates, quinine).
- Pompholyx (itching dermatitis with vesicles on the hand—benign).
- Dermatitis herpetiformis (severe itch; bullae grouped on knees, elbows, scalp, shoulders. Association: coeliac disease. It responds to dapsone).
- Stevens–Johnson syndrome (OHCM p 652)—mucous membrane blisters, plus target lesions with a central bulla on hands. Often systemically ill; oral steroids may be tried. *Causes:* Virus infections, sulphonamides.
- Pemphigus is an autoimmune disease (antibody to epidermal cell membrane). Bullae are superficial and rupture easily. Secondary infection may occur. Tests: biopsy with immunofluorescence. Enlist expert help (± admission). Prednisolone (~100mg/24h PO) ± azathioprine is needed.
- Pemphigoid. Antibody is directed at the basement membrane. Bullae are deep and less readily ruptured. *Treatment:* prednisolone (~60mg/24h).
- Other causes: heat, frostbite, friction, contact dermatitis.

- If the skin is red all over, with spreading bullae, suspect toxic epidermal necrolysis (synonym: scalded skin syndrome). This acute, red, blistering, necrotic condition affects all the skin. If the patient is a child, suspect Staphylococcal infection, eg phage type 71. The split in the skin allowing the formatuion of blisters in the stratum granulosum. In adults, the cause is more likely to be malignancy, or its treatment (or some other drug reaction, eg sulphonamides, or barbiturates).

Gravitational leg ulcers

Gravitational leg ulcers (also called hypostatic venous ulcers) are second-ary to high venous pressure—often due to failure of the muscle pump in the calf. Fibrinogen diffuses out of the vessels, encircling them in a cuff, and preventing adequate tissue oxygenation. The first sign is often a red patch of inflammation, which becomes itchy, scaly, greasy and pigmented (with haemosiderin). The surface then breaks down to form the ulcer. If healing occurs, this is with granulation tissue—which appears as a red, shiny, uneven surface (formed by macrophages, fibroblasts and vessels). Growth of new skin then follows. Prevalence: ~1%. Typical site: around the medial malleolus (toes or heel if arterial). They are chronic, with 40% of ulcers still unhealed after 1yr, and 10% being present for ≥5 years. Popliteal vein incompetence (on ultrasound) indicates that healing will be slow. If foot pulses are not obviously good, bedside Doppler is useful in determining if the cause is arterial, not venous. The 'pressure index' is the ankle systolic BP divided by the brachial BP—normally ~0.9. If 0.8-0.6, only use compression hosiery with utmost caution—and never if <0.6. (The pressure index is hard to interpret in incompressible, calcified arteries).▣

Other causes of ulcers Arterial disease, sickle-cell anaemia, squamous cell carcinoma (p 582), yaws, polyarteritis nodosum and other vasculitides, skin infections, diabetes mellitus, neuropathies, drug effect.

Treatment Relieve hypostasis by exercise and elevation (heel higher than hips). Graduated support stockings also help (p 598), but more effective is the '4-layer bandage system' giving 40mmHg pressure at ankle with 17mmHg at the calf. Use *only* if no arterial problems, ie ankle pressure index <0.8.▣ A moist environment helps, under dressings; but maceration/cellulitis need to be prevented by changing dressings often (eg every 12-24h). Permeable dressings which allow vapour to escape need changing less often. The exu-date on the ulcer bed aids healing. Bacterial colonization is common, but need not impair healing unless the organism is *Pseudomonas aeruginosa*. There is little evidence that incorporating antiseptics into dressings helps, and those which use hypochlorite may be harmful. During debridment at dressing changes care is needed to avoid damaging new epithelium.

Dressings: Simple absorbent dressings (eg non-adherent 'NA' dressings; Melolin®—apply shiny side down) are the mainstay of treatment.
- Semi-permeable dressings (eg OpSite®) are polyurethane dressings, with a backing which adheres to normal surrounding skin. They are permeable to gases but not to bacteria—but exudate accumulates beneath them.
- Medicated semipermeable occlusive dressings (eg Calaband®, Viscopaste PB7®). They contain antiseptics; use with graduated support stockings.
- Debriding agents—eg Dextranomer (Debrisan®). After cleaning, put a piece of gauze on the ulcer bed and cover with a layer of dextranomer beads, and top with a secondary dressing. Replace before the beads become saturated. The beads remove bacteria and exudate by capillary action. Controlled trials find this more effective than other treatments.
- Tulle dressings impregnated with paraffin (Jelonet®) or antiseptic (Bactigras®) offer only non-adherence and moisture. Change often.

If there is much slough, or a heavy exudate, hydrogel may be useful.▣ If the ulcer enlarges, do not blame the dressing (usually), but look to increas-ing elevation, and improving nutrition, and the arterial supply.

Pinch skin grafts: After local anaesthesia, pieces (0.2-0.5cm across) of skin from the anterior thigh are raised on a syringe needle and detached using a scalpel and then transfer to the ulcer and covered with NA dressings.

Aspirin: 300mg PO apparently aids healing (healing rate after 4 months: 30% with aspirin; 0% without, in a small, randomized trial.)▣

Pressure sores

In patients rendered immobile by age, or CNS problems such as stroke, cord lesions or MS, uninterrupted pressure on an area of skin may lead to ulceration, and extensive, painful, subcutaneous destruction—eg on the sacrum, heel, or greater trochanter. Malnutrition, arteriopathy, and old age make the condition more likely to occur—particularly if nursing care is wanting. They are a big problem. Cost: UK—£150 million/yr; USA—$3 billion/yr[1].

Classification/staging
Stage I: Non-blanching erythema over intact skin.
Stage II: Partial thickness skin loss—eg a shallow crater.
Stage III: Full thickness skin loss, extending into fat.
Stage IV: Destruction of muscle, bone, or tendons.

Prevalence ~7% of hospital inpatients have pressure sores; most are over 70 years old. Up to 85% of paraplegic patients have pressure sores.

Complications Osteomyelitis; pyoarthrosis

Treatment
- Prevent the condition getting worse (see below).
- Improve nutrition (eg vitamin C). Insulin if hyperglycaemic.
- Treat systemic infection with antibiotics.
- Dress the area. There is no convincing evidence from randomized trials which favours any one type of dressing. See p 596.
- Vascular reconstruction, if needed, and if practicable.
- Split thickness skin grafts.
- Neurosensory myocutaneous flap surgery.[1]

Prevention
- Find an interested, knowledgeable nurse to educate the patient.
- Proper positioning, with regular turning (eg every 2h, see p 726 & p 732, alternating between supine, and right or left lateral position). Use pillows to separate the legs.
- Functional electrical stimulation can prevent sores in paraplegics, by inducing the buttocks to change shape, and by improving blood flow.[2]

1 RK Vohra 1994 *BMJ* ii 853   2 AC Furguson 1992 *Paraplegia* 30 474

Graduated support hosiery

In the UK, elastic stockings are now prescribed not according to type of yarn but by performance. This falls into 1 of 3 classes.

The aim has been to make NHS hosiery more effective, easier to put on and more acceptable (with fashionable colours and styles).

Class I (old name: light-weight, elastic yarn)
Compression at ankle: 14-17mmHg; at mid-calf: <80% of this; thigh: <85% of calf value. That is, the compression is graduated. Used in venous insufficiency, the aim is to compress the superficial venous system, so forcing blood back into the deep system.

Indications: Early varicose veins; varicose veins in pregnancy.

Styles: Thigh length [T]; below knee [K]; toe open [O]—so foot length matters less—or closed [C]. Open heels are not available.

Colour examples: Tan: [TKC]Duomed®; mink, honey or black: [TKC]Lastosheer Class I Stocking®.

Class II (Ankle 18-24mmHg; calf <70% ankle; thigh <70% calf)
Indications: Varicose veins of medium severity, leg ulcer treatment, mild oedema, varicose veins in pregnancy.

Styles: Thigh length [T]; below knee [K]; toe open [O] or closed [C].

Colour examples: Mink or honey: [TKC]Lastosheer Class II Stocking®.

Class III (Ankle 25-35mmHg; calf <70% ankle; thigh <70% calf)
Indications: Gross varicose veins; post-thrombotic venous insufficiency; gross oedema; leg ulcer treatment.

Styles: Thigh length; below knee; toe open [O]—so foot length matters less, or closed [C]. Heels may be open* or closed.

Colour examples: Tan: [O]Duomed®; light: [C]Eesilite®. These may either be thigh or knee length.

Measurement Measure the thinnest part of the ankle, the fattest part of the calf, and the mid-thigh circumference (cm). 80% of patients will conform to a stock size (see table below).

Sizes (cm) of a representative make of stocking (Duomed®; Medi UK)

Ankle	Calf	Thigh	Ankle	Calf	Thigh
Small size			*Large size*		
19-21	28-34	42-56	25-27	36-42	54-68
Medium size			*Extra large size*		
22-24	32-38	48-62	28-30	40-46	60-74

Made-to-measure stockings are available in all 3 classes (and are the only way to prescribe class I or II stockings with open heels).

Prescribing The doctor specifies the length and number of garments (not pairs), and the class. Other variables may be left to the patient's choice. The pharmacist may do the measuring.

Problems Putting on stockings can be difficult. Wearing rubber gloves helps the process, as does putting on a nylon stocking first. Twice weekly washing (in warm water, not hot) prolongs the life of stockings—do not wring them out. Ideally, there should be no tourniquet effect at the top, and there should be little difference in performance if a particular leg does not match exactly a standard size. In independent trials, 2 brands came closest to these ideals: TX (from Brevet) and Thrombex (from Seton).[1]

▶Limb ischaemia is a contraindication to using support hosiery.

1 *Bandolier* 1995 **18** 2

Photosensitive eruptions

Photosensitivity may be due to visible light, or UVA (320–400 nanometres, accounting for 90% of ultraviolet sun radiation) or UVB (higher energy, so more damaging, 290–320nm). UVC (200–290nm) is filtered out by the ozone layer. Sun-blocking preparations are assigned a factor (Sun Protection Factor, SPF) relating to UVB (which is 1000 times as potent a cause of sunburn as UVA).[1] Skin protected by a blocker of SPF 10 should be able to withstand a given intensity of UVB 10-times longer than the same skin without protection—before it becomes burned. To get good blocking of UVA, it is necessary to use reflective inorganic compounds such as zinc oxide or titanium dioxide—and these may also help in preventing premature ageing of skin, and some skin cancers (even when 'microfined' these leave a visible white film on the skin). Protection against UVA is not quantified as an absolute value on sun-blocking preparations. Instead, a star system may be used. 4 stars indicate that the blocker gives equal protection to UVB and UVA, 1 star indicates that the block is predominantly a block of UVB.

Other effects of ultraviolet: melanomas, and ageing of the skin.

Predominantly UVA photodermatoses Polymorphic light eruption (below), drug-induced photosensitivity, cutaneous porphyrias.

Predominantly UVB photodermatoses Chronic actinic dermatitis, systemic lupus erythematosus, xerodermatitis, vitiligo, albinism, herpes labialis.

Polymorphic light eruption Prevalence: 20% in temperate climes.
Presentation: Typically in a young woman in spring and summer, as 'prickly heat' (which it is not). There are itchy, erythematous papules, vesicles or plaques—induced after minutes' or a few hours' exposure to sun. Window-glass may not afford protection. Not all sun-exposed areas need be affected. The face, for example, may be spared. The rash often fades within 48h, but may last for a week. In most people it is due to sensitivity to UVA.

Treatment and prevention: Regular use of a sunscreen to UVA, eg Uvistat Ultrablock® is as effective and less time-consuming than 'skin hardening' in hospital with the use of PUVA.

Drug-induced photosensitivity In any a photosensitivity, ask what drugs the patient is taking. Culprits may be thiazides, tetracyclines, phenothiazines, NSAIDs, nalidixic acid, or chlorpropamide. UVA is the usual cause.

Sunburn There are 4 types of skin *vis à vis* tanning:
I Always burns, never tans.
II Burns easily, sometimes tans.
III Tans easily, burns rarely.
IV Always tans, never burns. Skin types I and II should be advised to use a sunscreen with SPF of 15 or more.
Treatment: Tap-water cool soak; water-based emollients; aspirin 600mg/6h PO pc limits erythema (and relieves pain)—as do moderately powerful steroids (eg clobetasone butyrate cream), eg for 3 days. If severe, protect the blisters, and give a tapering week's course of prednisolone, starting with 60mg/24h PO.[2] ▶Educate about prevention.
Prevention: Slip, slap, slop (p 584—Uvistat Ultrablock® cream has an SPF of 30 and is water-resistant). Those with type III skins may use screens with SPF of 10. Discourage prolonged sunbathing as ultraviolet light may be immunosuppressive; also, if a UVB screen is used, there may be no warning of excessive exposure to UVA, as the skin will not burn.

1 R Ratnavel 1993 *Prescriber's J* **33** 63   **2** EL Lloyd 1994 *BMJ* ii 587

Infestations

Scabies (*Sarcoptes scabei*) This mite is an arachnid (having 8 legs). Only females make the characteristic itchy skin burrow. Her size is ~0.3×0.4mm; she burrows at ~2mm/day (larvae only embed: they do not burrow). The eggs she leaves behind hatch after 4 days. Spread is common in families (here the secondary attack rate is 38%). Those <45yrs are most susceptible.

Presentation: A papular rash (eg on abdomen or medial thigh—a separate phenomenon to burrows) and nocturnal itch are leading symptoms. *Incubation period:* ~6 weeks (during which time sensitization to the mite's faeces and/or saliva occurs). Finger web spaces and wrists (flexor surface) are preferred sites for burrows (seen in ≥85%). Lesions on the penis produce red nodules. Burrows are non-linear and are 3-15mm long. They are better felt than seen.

Diagnosis of scabies: Trying to tease a mite out of her burrow with a needle for microscopical examination ($\times$ 10) often fails—but if a drop of oil is placed on the lesion, a few firm scrapes with a scalpel will provide microscopically recognizable faeces and eggs.

Differential diagnosis: Any pruritic dermatosis (p 578).

Treatment:[1] ►Treat all the household. Give written advice, eg:
- Take a warm bath and soap the skin all over.
- Scrub the fingers and nails with a firm brush. Dry your body.
- Apply malathion 0.5% liquid (Derbac-M®), or, if pregnant or <6 months old, monosulphiram 25% in methylated spirit (diluted in 3 parts water to one part of the liquid), from the neck down (include the head in those <2 yrs old). Remember to paint *all* parts, including the soles.
- Wash off after 24 hours. If you wash your hands before the 24 hours is up, reapply the liquid to the parts you have washed.
- Use fresh pillow cases and sheets, if you have any.
- Treatment may worsen itch for 2 weeks—so use calamine lotion.
- Warnings: avoid the eyes.

Other treatments include lindane lotion, or permethrin dermal cream. CI: infants <6 months old; pregnancy.

Head lice (*Pediculosis humanus capitis*) These blood-feeding lice are 3mm long, and attach to the base of hairs. Egg containers (nits) are glued firmly to hairs. They are 1-2mm long; the emerging nymph is capable of reproduction (eg 8 eggs/night) in ~10 days. Spread is by head-to-head contact. Secondary infection may cause impetigo or furunculosis.

Treatment: Malathion lotion (0.5% in alcohol eg Suleo M Lotion®): rub into dry hair, comb, allow to dry; wash out after 12h. Avoid alcoholic solutions in those under 3yrs and in asthmatics (try aqueous Derbac M Lotion®). This may not work—if so, personal experience favours a 10 minute application of 1% permethrin (Lyclear®). [2] NB: twice-daily systematic combing (fine tooth, on wet hair) can prevent infestations. Resistance to many drugs occurs (liaise with Public Health).

Crab lice (*Phthiriasis pubis*) Eye-brows and lashes, pubic and axillary hair may be involved. They may be sexually transmitted.

Treatment: Aqueous malathion 0.5% for 12h to *all* hairy areas; repeat after 7 days to kill lice emerging from any surviving eggs.

Flea bites *Pulicidae* spread plague (OHCM p 224), typhus (OHCM p 236), Q fever (OHCM p 236), tape worm (OHCM p 248), and listeria (OHCM p 222). The animal (eg cat or dog) which spreads the flea may not itch or scratch itself. *Treatment:* Aqueous malathion lotion.

---

1 Dermatology dept, Musgrove Park Hospital ◄2►R Vender 1995 *BMJ* i 604 & *Bandolier* 1995 **20** 6

601

Mast cell disorders

Mastocytosis (urticaria pigmentosa) There are too many mast cells in the skin, eg in numerous nests or as a single mastocytoma. These release histamine, leucotrines, and heparin (which may have systemic effects) and proteases (which have a local effect).

Presentation: This is unpredictable, and symptoms need not be progressive—but there may be progressive myeloproliferation/mast-cell leukaemia. Systemic forms can occur with no skin signs.☑ In babies and other patients there may be blisters or pigmented swellings, which swell when scratched (or after a bath, or on exercise). In older patients, there may be flushing or itching. Systemic mast cell infiltration may produce:

- Splenomegaly
- Thrombocytopenia
- Bleeding disorders
- Osteoporosis
- Osteosclerosis
- Colic
- Diarrhoea
- Right heart failure
- Pulmonary hypertension
- Leukaemic mastocytosis

Tests: Urinary histamine may be ↑ (helps if no skin signs). Labs may also report N'-methylhistamine (MH) and N'-methylimidazoleacetic acid (MIMA).

Differential diagnoses: Legion—eg idiopathic facial flushing syndrome.

Treatment: This is difficult. H_1 and H_2 antagonists may be helpful. Single skin lesions can be excised. Disodium cromoglycate may help some patients with systemic symptoms.[1]

Lasers in dermatology*

Lasers (light amplification by stimulated emission of radiation) have an increasing rôle, and it is worth knowing who may benefit from them. The destructive energy of lasers can be concentrated in time and space, and the light's wavelength can be adjusted to match the target lesion. Energy can be selectively taken up by targets as long as the volume delivered is less than the target's capacity to absorb it. If this limit is exceeded, unnecessary destruction is caused. Variables to specify—an example from keloid scar treatment: wavelength 585nm, pulse duration 0.45ms, spot size 5mm, mean fluence per pulse 7 Joules/cm^2, treatment interval: 6 weeks.[2] NB: melanin in dark skins may be a problematic competing chromophore.

Photoacoustic damage is produced by shock waves from very intense energy delivered over picoseconds.

Photocoagulation implies less rapid, and less intense treatment.

Selective photothermolysis is somewhere between the above 2 categories. It may produce minimum scarring. Modification of the duration of delivery is possible (Q-switching and mode-locking).

Examples of laser-treatable skin disorders[3]

Port wine stain: Best treated when young (smaller, smoother lesions).

Tuberous sclerosis: Angiofibromata (mis-named adenoma sebaceum) can cover much of the face.

Strawberry naevi: These do not need treatment, unless they fail to resolve spontaneously. If they ulcerate, get expert advice.[3]

Flat, pigmented lesions: These are problematic because it is essential to know that they are benign.

Tattoos: If multi-coloured, multiple wavelengths must be used.

Keloid (hypertrophic) scars: These raised, red, nodular scars are unsightly, and may itch. They are very difficult to treat, but lasers can help somewhat.[2]

1 J Dermatol Surg 1993 19 295   2 T Alster 1995 Lancet 345 1198   3 Disfigurement Guidance Centre, Cupar, Fife, Scotland (UK tel. 01337 870281). *We thank Dr Ian Comaish for inspiring this topic.

9. Orthopaedics and trauma

Relevant pages in other chapters Any medical or surgical illness from the acute abdomen to zoster may first present to a casualty officer. So almost every page in this book and in its sister volume, the *Oxford Handbook of Clinical Medicine*, has relevance to the casualty officer. ▶See *Pre-hospital care* (p 780–806); *Radiology* (OHCM p 720–94).

Note: for childhood poisoning, see p 252–4; (adults, OHCM p 690–6).

▶Many detailed treatments are described in this chapter, and it is not envisaged that the inexperienced casualty officer will try them out except under appropriate supervision. The importance of enlisting early expert help (either at once or, if appropriate, by calling the patient back to the next morning's registrar clinic) cannot be overemphasized.

Assessing the locomotor system

The aim is to screen efficiently for rheumatological conditions—and also to assess any motor disability. It is based on the GALS locomotor screen.[1] ▶Don't *just* catalogue muscle function. (Details of testing each muscle may be found on p 664.) Find out what a person can do—eg with her arthritic fingers, can she do up zips or buttons? Can she open a tin? Is she using any special devices to assist in daily life? What can't she do that she would like? Does she have any comments on her joints?

Essence Ask questions, look, compare, move and palpate. If a joint feels normal to the patient, looks normal to you, and has a full range of movement, it usually *is* normal.

Valgus or varus? In a valgus deformity, the part of the limb distal to the deformity is angled *away* from the midline. (Varus is the other way.)

3 screening questions ●Are you free of pain or stiffness in your muscles? ●Can you dress all right? ●Can you manage stairs? If 'Yes' to all 3, muscle problems are unlikely. If 'No' to any, go into functional detail.

A screening examination To be done in light underwear. (No corsets!)
Observe from behind: Is muscle bulk OK (buttocks, shoulders)? Is the spine straight? Symmetrical paraspinal muscles? Any swellings or deformities?
Observe from the side: Normal cervical and lumbar lordosis? Kyphosis?
'Touch your toes, please': Are lumbar spine and hip flexion normal?
Observe from in front for the rest of the examination. Ask him to:
'Tilt head towards shoulders': Is lateral cervical flexion normal?
'Put hands behind head': Tests glenohumeral and sternoclavicalar movement.
'Arms straight': Tests elbow extension. Then test supination and pronation.
Examine the hands: See OHCM p25. Any deformity, wasting, or swellings?
'Put index finger on thumb': Tests pincer grip. Assess dexterity.
Observe the legs: Normal quadriceps bulk? Any swelling or deformity?
Find any knee effusion: Sit the patient down, take the leg on your lap, and do the patella tap test (this also tests for patello-femoral tenderness), or, to be more sensitive, watch any fluid moving from compartment to compartment by stroking upwards over the medial side of the knee, and then downwards over the lateral side. Aspirate it. Any blood, crystals or pus?
Observe feet: Any deformity? Are arches high or low? Any callosities? These may reflect an abnormal gait of some chronicity.
'Walk over there please': Is the gait smooth? Is there good arm swing? Is the stride length OK? Normal heel strike and toe off? Can he turn quickly?

Additional palpations ●Press over the midpoint of each supraspinatus muscle, to elicit the tenderness of fibromyalgia (p 675).
●Squeeze across 2nd-5th metacarpals. Any synovitis? Repeat in metatarsals.
●Palpate for crepitus with your palm on his knee (patient supine), while you passively flex knee and hip to the full extent. Is movement limited?
●Internally rotate each hip in flexion.

The GALS system for quickly recording your findings[2]

	Appearance:	Movement:
G (Gait) ✓		
A (Arms)	✓	✓
L (Legs)	✓	✓
S (Spine)	✓	✓

A tick (✓) means normal. If not normal, then substitute a cross with a footnote to explain what the exact problem is.

1 M Doherty 1992 *Annals of the Rheumatic Diseases* **51** 1165-9   2 As approved by the education committees of the Arthritis and Rheumatism Council and the British Society for Rheumatology (1991)

The neck

Examination ►►After trauma, the neck must be immobilized ie with rigid collar, sand bags and tape, until a lateral cervical spine X-ray has been seen (except if very restless—use hard collar only, as otherwise the neck is vulnerable when the body moves on an immobilized head).

The posture of the neck and any bone tenderness are noted. The range of movements: flexion; extension (mainly occipito-atlantoid joint); rotation (mainly atlanto-axial joint); lateral flexion (whole of cervical spine). Rotation is the movement most commonly affected. The arms are examined for weakness that might signify root lesions (shoulder abduction C5; elbow flexion C5,6; elbow extension C6,7; wrist extension C6,7; wrist flexion C7,8; grip and abduction of fingers against resistance T1). Reflexes are examined: biceps C5,6; supinator C5,6; triceps C7. Dermatomes are illustrated on p 665. If cord compression is suspected examine lower limbs for signs of this (see p 608)—eg upgoing plantars, hyperreflexia.

Spasmodic torticollis Sudden onset of a stiff painful neck with torticollis may occur in adults due to spasm of the trapezius and sternocleidomastoid muscles. *Treatment:* The condition is self-limiting but heat, manipulation, wearing of a collar, muscle relaxants and analgesia may ease resolution.

Infantile torticollis Thought to be a result of damage during birth to the sternocleidomastoid, affected children present (between 6 months and 3 years) with a tilted head (ear nearer shoulder on affected side). There is retarded facial growth on the affected side, hence facial asymmetry. Early, there is a tumour-like thickening in the muscle. *Treatment:* If this persists, physiotherapy to lengthen the muscle may help. Later treatment involves division of the muscle at its lower end.

Cervical rib Congenital development of the costal process of the C7 vertebra is often asymptomatic but may cause thoracic outlet compression. Similar symptoms without demonstrable radiological abnormality is called a scalenus or first rib syndrome. Thoracic outlet compression involves the lowest trunk of the brachial plexus (p 724) and the subclavian artery. Pain or numbness may be felt in hand or forearm (often on the ulnar side), there may be hand weakness and muscle wasting (thenar or hypothenar). *Diagnosis:* Radial pulses may be weak and the forearm cyanosed. X-rays may not reveal cervical ribs, as symptoms may be caused by fibrous bands. Arteriography will show subclavian compression. *Treatment:* Physiotherapy to strengthen the shoulder elevators may improve symptoms, but rib removal or band division may be needed.

Prolapsed cervical disc Those between C5,6 and C6,7 are most commonly affected. Central protrusions may give symptoms of spinal cord compression (p 608; refer to neurosurgeon). Posterolateral protrusions may cause a stiff neck, pain radiating to the arm, weakness of muscles affected by the nerve root, and depressed reflexes. *Tests:* X-rays may show disc narrowing; also CT or MRI scan. *Treatment* is with NSAIDs, and a collar. As pain subsides, physiotherapy may help to restore mobility. Surgery may be indicated, in the light of CT and MRI findings.

Cervical spine

In major trauma the first X-ray to be performed after resuscitation is a cross-table lateral of the cervical spine. All seven cervical vertebrae must be seen, along with C7-T1 junction: do not accept an incomplete X-ray—try traction on the arms, or a swimmer's view. Occasionally subluxations do not show up without flexion and extension views (perform only on the advice of a senior colleague).

▶If a cervical spine injury is suspected you must immobilize the neck until it is excluded.

▶A cross-table portable in the best hands will miss up to 15% of injuries.

When examining the film, follow 4 simple steps:

●*Alignment:* Of anterior and posterior vertebral body, posterior spinal canal and spinous processes. A step of <25% of a vertebral body implies unifacet dislocation; if >50% it is bifacetal. 40% of those <7yrs have anterior displacement of C2 on C3 (pseudosubluxation); still present in 20% up to 16yrs. 15% show this with C3 on C4. In this physiological subluxation, the posterior spinal line is maintained. Angulation between vertebrae >10% is abnormal.

●*Bone contour:* Trace around each vertebra individually. Look for avulsion fractures of the body or spinous process. A wedge fracture is present if the anterior height differs from posterior by >3mm. Check odontoid (open mouth view but can be seen on lateral view); don't mistake vertical cleft between incisors for a peg fracture!)—epiphyses in children can be mistaken for fractures; the distance between the odontoid and anterior arch of C1 should be <3mm (may be increased in children). Type 1 fractures are of the odontoid peg, type 2 of its base, and type 3 fractures extend into the body of C2.

●*Cartilages:* Check that intervertebral disc space margins are parallel.

●*Soft tissues:* If the space between the lower anterior border of C3 and the pharyngeal shadow is >5mm suspect retropharyngeal swelling (haemorrhage; abscess)—this is often useful indirect evidence of a C2 fracture. Space between lower cervical vertebrae and trachea should be <1 vertebral body.

The neck and cord compression

Cervical spondylosis (See OHCM p 468.)

Cervical spondylolisthesis

This is displacement of one vertebra upon the one below. *Causes:*
1 Congenital failure of fusion of odontoid process with the axis or fracture of the odontoid process (skull, atlas and odontoid process slip forward on axis).
2 Inflammation softens the transverse ligament of atlas (eg rheumatoid or complicating throat infections), the atlas slips forward upon the axis.
3 Instability after injuries. The most important consequence of spondylolisthesis is the possibility of spinal cord compression. Treatments used include traction, immobilization in plaster jackets and spinal fusion.

Spinal cord compression

Pressure on the cord may be due to bone displacement or collapse, disc prolapse, local tumour or abscess. Root pain (p 618) and lower motor neurone signs occur at the level of the lesion with upper motor neurone signs and sensory changes below the lesion (spastic weakness, brisk reflexes, upgoing plantars, and loss of co-ordination, joint position sense, vibration sense, temperature and pain). Cord anatomy is such that dorsal column sensibilities (light touch, joint position sense, vibration sense) are affected on the same side as the insult, but spinothalamic tract interruption affects pain and temperature sensation for the opposite side of the body 2-3 dermatome levels lower than the affected sensory level. As the cord ends at L1, compression at this vertebral level affects information in the cord relating to a lower dermatome. To determine the cord level affected behind a given vertebra, add the number in red to that of the vertebra concerned, thus: C2-7: +1. T1-6: +2. T7-9: +3. T10 has L1 and L2 levels behind it; T11 has L3 and 4, L1 has sacral and coccygeal segments.

Lower lumbar problems can cause cauda equina compression characterized by muscular pain, dermatomal sensory changes (if the lowest sacral dermatomes are affected the genitals are anaesthetic), and retention of urine and faeces. ▶These signs indicate urgent neurosurgical referral.

The shoulder

Anatomically, this gleno-humeral joint is lax, depending on surrounding rotator cuff muscles for stability—ie the sheath of tendons of supraspinatus (1st 15-30° of abduction), subscapularis (internal rotation) infraspinatus (lateral rotation), and teres major (lateral rotation + extension). Biceps' long head traverses the cuff, attaching to the top of the glenoid cavity.

History Any trauma, arthritis or neck problems? Is pain sharp (eg capsulitis) or dull (eg osteoarthrosis)? Does movement make it worse (eg trauma, impingement, or, eg if young, dislocation). General health OK? (if aches and pains all over, suspect fibromyalgia, or polymyalgia).

Examination Strip to the waist. To assess gleno-humeral movement, hold the lower half of the scapula to estimate the degree of rotation of the scapula over the thorax. Half the range of normal abduction is achieved by scapula movement. *Abduction* is tested by raising the hands from the sides sideways to above the head; *flexion* by raising hands forwards and upwards; *extension* by backward movement of the elbows; *external rotation* by holding the elbows against the sides flexed at 90° and moving the hands outwards (normal 80°); and *internal rotation* by placing the back of the hand against the lumbar spine and moving the elbows forward (an easier test is to assess how far behind his back the patient can reach—'imagine you are doing a bra up at the back').

The muscles used for movement at the shoulder joint
- Flexion (forward movement): pectoralis major, deltoid, coracobrachialis.
- Extension: deltoid (latissimus dorsi, pectoralis major and teres major begin the extension if the shoulder starts out flexed).
- Abduction: supraspinatus (for first 15°); deltoid thereafter.
- Adduction: pec. major, latissimus dorsi, teres major, subscapularis.
- Medial rotation: pec. major, deltoid, latissimus dorsi, teres major, sub-scapularis.
- Lateral rotation: deltoid, teres minor, infraspinatus.

NB: rotator cuff muscles are most important in rotation: subscapularis, teres minor, infraspinatus (supraspinatus is also part of the rotator cuff).

Movement of the scapula on the chest wall
- Elevation: (shrug shoulders) levator scapulae, trapezius
- Depression: serratus anterior, pectoralis minor
- Forward action (eg punch): serratus anterior; pectoralis major
- Retraction (brace shoulders): trapezius, rhomboids.

NB: serratus anterior prevents 'winging' of the scapula as pressure is placed on the outstretched hand.

Recurrent dislocation of the shoulder (For initial dislocation, see p 696.) There are 2 types: *Atraumatic:* (5%) The patient is often a teenager with no history of trauma, and may have general joint laxity. Remember BRAA; *b*ilateral; treat by *r*ehabilitation; *a*traumatic; dislocates in *a*ll directions.

Traumatic: Dislocation is usually anterior (sometimes inferior, rarely posterior) and is secondary to trauma—which may be mild. Remember BUST: *B*ankart lesion (see below); *u*nidirectional; *s*urgical treatment; *t*raumatic. Abduction + lateral rotation of the arm (eg on putting on a coat) may cause dislocation. The capsule is attached to the neck of scapula but detached from the glenoid labrum (Bankart lesion). There may be a posterolateral 'dent' in the humeral head (seen on X-ray with arm medially rotated). Treatment is Bankart repair (sew capsule back to glenoid) or Putti-Platt operation (the subscapularis tendon is reefed, ie shortened).

With the rarer recurrent posterior dislocation, the capsule is torn from the back of the neck of scapula, the humeral dent is superomedial, and it is abduction and medial rotation which causes dislocation (eg seizure). *Treatment:* Reef the infraspinatus tendon.

Recurrent *subluxation* is also recognized (disabling and difficult to treat).

The painful shoulder

Remember that the neck may refer pain via C5 to the deltoid region and via C6, C7 and C8 to the superior border of the scapula. Diaphragmatic referral via C3 may also cause shoulder-tip pain.

Rotator cuff tears Tears in supraspinatus tendon (or adjacent subscapularis and infraspinatus) may be due to degeneration, or, less commonly, to a sudden jolt (eg a fall). Partial tears cause a painful arc (see below); complete tears limit shoulder abduction to the 45-60° provided for by scapular rotation. If the arm is passively abducted beyond 90° deltoid's contribution to abduction comes into play, which is then possible from this point. The full range of passive movement is present. Pain is felt at the shoulder-tip and upper arm; there is tenderness under the acromion. *Tests:* Ultrasound, MRI, or arthrography reveals communication between joint capsule and subacromial bursa. *Treatment:* Tendon repair may be successful, particularly if done early.

Painful arc syndrome This is pain on abduction between 45° and 160°. Causes of pain on abduction: 1 *Supraspinatus tendinitis* or partial rupture of supraspinatus tendon gives pain reproduced by pressure on the partially abducted arm. Typical age: 35-60. Only a proportion will have a painful arc (others have increasing pain up to full abduction), which is why the term *impingement syndrome* (as the greater tuberosity of the humerus catches under the acromion during abduction between eg 70 and 140°) is preferred.[1] *Treatment:* Active shoulder movement with physiotherapy; NSAIDs, eg naproxen 250mg/8h PO pc; subacromial bursa injection of steroid, eg triamcinolone acetonide 40mg with local anaesthetic; arthroscopic acromioplasty.
2 *Calcifying tendinitis:* Typical age: 25-55. There is acute inflammation of supraspinatus. Pain is maximal during the phase of resorption. *Treatment:* Physiotherapy; NSAIDs; steroid injection. Rarely, surgery to remove calcium may be needed.
3 *Acromio-clavicular joint osteoarthritis (AC joint OA):* This is common. *Treatment:* Steroid injection often helps. Excision of the lateral end of the clavical may be needed.

Tendinitis of long head of biceps Pain is felt in the anterior shoulder and is characteristically worse on forced contraction of biceps. *Treatment:* Anti-inflammatory drugs. Hydrocortisone injection to the tendon may give relief but risks tendon rupture. Technique: p 658.

Rupture of long head of biceps Discomfort occurs after 'something has gone' when lifting or pulling. A 'ball' appears in the muscle on elbow flexion. *Treatment:* Repair is rarely indicated as function remains.

Frozen shoulder (capsulitis) This may follow modest injury in older people. Pain may be severe. There is marked reduction of passive and active movement. Abduction to 90° is not possible. *Treatment:* NSAIDs, intra-articular steroid injections, physiotherapy to encourage movement and manipulation may all be used. Resolution may take 2-3yrs. It may be associated with cervical spondylosis, giving a more global restriction of movement.

Shoulder osteoarthritis This is not so common as hip or knee OA. Good success rates are being achieved by joint replacement.

1 JI Brox 1993 *BMJ* ii 899

The elbow

The normal range of flexion and extension at the elbow is 0°-150°. With the elbow flexed, supination and rotation of 90° should be possible. Pain felt at the elbow and in the middle outer aspect of the arm may radiate from the shoulder.

Etymology An ell is the distance from shoulder to wrist (or finger-tips), and the elbow (alino [ulna/ell=Old Teutonic for *arm*]-bogon [Old Teutonic for *bending*]) is strictly the *outer* part of the joint between fore and upper arm: but here we use the term to denote the entire joint.

Tennis elbow There is inflammation where the common extensor tendon inserts at the lateral epicondyle of the humerus (lateral epicondylitis) ± rupture of aponeurotic fibres. This is usually caused by strain.
Presentation: Pain is worst when the tendon is most stretched (flexion of the wrist and fingers with hand pronated). Pain is felt at the front the lateral condyle. Ask the patient to extend the wrist, and then pull on the hand to elicit pain. *Tests:* There are no X-ray findings.
Treatment: Pain often subsides in time, but injection of the tendon origin (p 658) gives more rapid pain relief. If this fails, physiotherapy may help, or an epicondylitis brace; with severe disability, operative stripping of the common extensor origin from the bone and then allowing it to fall back in place may provide relief.

Golfer's elbow This is medial humeral epicondylitis. It is less common than tennis elbow. Steroid injection may help, but be wary of the ulnar nerve; the brachial artery is also nearby.

Student's elbow This is a traumatic bursitis following pressure on the elbows, eg while engrossed in a long book. There is pain and swelling behind the olecranon. Other causes are septic and gouty bursitis (look for tophi). The bursa should be aspirated (send fluid for Gram stain and microscopy for crystals). Traumatic bursitis may then be injected with hydrocortisone. Septic bursitis should be formally drained.

Ulnar neuritis Osteoarthritic or rheumatoid narrowing of the ulnar groove and constriction of the ulnar nerve as it passes behind the medial epicondyle, or friction of the ulnar nerve due to cubitus valgus (a possible sequel to childhood supracondylar fractures) can cause fibrosis of the ulnar nerve and ulnar neuropathy. *Presentation:* Patients may experience clumsiness of the hand. Sensation is reduced over the little finger and medial half of ring finger. There is weakness of the small muscles of the hand innervated by the ulnar nerve (adductor pollicis, interossei, abductor digiti minimi and opponens digiti minimi).
Tests: Nerve conduction studies will confirm the site of the lesion.
Treatment: Operative release of the nerve to lie in a new channel in front of the elbow.

Deformities Cubitus valgus: The normal degree of valgus ('carrying angle') at the elbow is 10° in ♂, and 15° in ♀. Fractures at the lower end of humerus or interference with the lateral epiphyseal growth plate may cause the angle to be greater. As a result ulnar neuritis and osteoarthritis may occur. Treat if necessary. Association: Turner's.
Cubitus varus: Commonly after poorly reduced supracondylar fractures.

Osteoarthritis of the elbow Osteitis dissecans and fractures involving the joint are risk factors. *Tests:* Flexion and extension are usually impaired but rotation is full. *Treatment:* Operation is rarely indicated, but if pain is felt in the lateral compartment, the head of radius may be excised. Loose bodies may be removed if causing pain or locking of the joint.

631

Congenital dislocation of the hip (CDH)

1.3% of neonates have unstable hips, but only 0.1% have true dislocation. ♀/♂=6; left hips are affected 4 times as often as right (bilateral in ⅓). CDH is more common after breech delivery.

Diagnosis Examine hips of all babies in the 1st few days of life and at 6 weeks. Early detection hopes to result in early correction. ►Be alert to CDH throughout childhood surveillance (p 208). *Click test of Ortolani*[1]: With the baby supine, flex the hips to 90° and knees fully. The examiner places his middle finger over the greater trochanter, and thumb on inner thigh opposite the lesser trochanter. Diagnose a dislocated hip if slow hip abduction produces a palpable (often audible) jerk or jolt (ie rather more than a click)—as the femoral head slips back into the acetabulum. *The Barlow manoeuvre*[1]: With pelvis stabilized by one hand abduct each hip in turn 45°. Forward pressure by middle finger causes the femoral head to slip into acetabulum if hip is dislocated. If the femoral head slips over posterior lip of acetabulum and straight back again when pressure is exerted by the thumb it is 'unstable' (ie dislocatable not dislocated). Use both tests but avoid repeated tests as these may induce instability and dislocation.[2]

In older children signs may be delay in walking, abnormal waddling gait (affected leg is shorter), asymmetric thigh creases (extra crease on the affected side), and inability to fully abduct the affected hip. With bilateral involvement the perineum appears wide and lumbar lordosis is increased.

Imaging Ultrasound is the investigation of choice, because it is non-invasive and dynamic. It is unclear if ultrasound should be part of general screening. In one series of 4717 babies, 17 needed treatment, and in 5 of these clinical examination missed the diagnosis. 81 had ultrasound abnormalities which did not require treatment.[3]

Treatment If neonatal examination suggests dislocation/instability, use double nappies until reassessment at 3 weeks. If still a problem then, splint the hips in moderate abduction for 3 months (eg von Rosen splint). Excess abduction may case avascular necrosis of the head of femur. *From 6-18 months* closed reduction may be tried using a frame or gallows and traction to gradually abduct the hips to 80°. If this produces full reduction, the legs are then splinted, medially rotated and abducted; if it does not, various operations may be used *before 6yrs*—aiming to reduce excessive anteversion of the femoral neck (eg rotation osteotomy) and improve the acetabulum so the femoral head lies more deeply in it. *From 7-10yrs* similar operations may be appropriate (eg for high-lying femoral heads in false sockets). *From 11yrs*, operations (eg pelvic osteotomy) are principally for pain. Hip replacement may be needed in early adult life for painful osteoarthritis. Arthrodesis is an alternative.

Club foot

Neonatal club foot (talipes equinovarus; ♂/♀>1). The foot deformity consists of: 1 Inversion; 2 Adduction of forefoot relative to hindfoot; 3 Equinus (plantarflexion). The foot cannot be passively everted and dorsiflexed through the normal range. Treatment begins within 1 week of birth with weekly foot manipulations, holding it strapped or splinted in position between manipulations. The knee is flexed 90° in a splint to stop the baby drawing up his foot. If treatment has not corrected the foot by ~3 months, operative reduction is carried out. If this does not work, operations on soft tissues and/or bones of the foot may be carried out later (from 2 yrs).

1 T Barlow 1962 *J Bone & Joint Surg* 48 292   2 C Cheetham 1987 *Arch Dis Chil* 62 315
3 NM Clarke 1989 *J Bone & Joint Surg* 71 9

Knee history and examination

▶*Pain in the knee may be referred from the hip*. So examination must include the hip. Does internal hip rotation hurt?—so revealing hip pathology.

Ask about trauma, pain, swelling, mobility, locking, clicking and giving way (eg on exercise—a feature of cruciate ligament injury).

Examine the patient lying supine with legs fully exposed. Examine for swelling (*Causes:* Bone thickening, fluid, synovial thickening—this feels 'rubbery'). Look for wasting of the quadriceps. The presence of fluid can be confirmed by placing the palm of one hand above the patella over the suprapatellar pouch, and thumb and forefinger of the other hand below the patella. Fluid can be moved between the two (hydraulic pressure will be felt) by alternating the source of pressure. If 30–40ml fluid are present it may be possible to feel a 'patellar tap' (ballot patella against neighbouring bones). Patellar taps are absent with small or tense effusions (which may be up to 120ml).

Flexion and extension at the knee vary between persons. Flexion should be sufficient for the heel to touch the buttock. Compare extension with the 'good' side. Medial and lateral ligaments are examined with the knee flexed to 20–30° (to relax the posterior capsule), as one hand lifts the ankle off the couch, the other holds the knee just slightly flexed. The knee is stressed in abduction by abducting the ankle with one hand while pushing the knee medially with the hand behind the knee (tests medial ligament). Reverse the pressures to give adducting force to test lateral ligament. If these ligaments are torn the knee joint opens more widely when the relevant ligament is tested (compare knees against each other).

Test the cruciate ligaments with the knee 90° flexed, with the foot placed on couch with you sitting on it (anchor tibia). Place your fingers interlocking behind the knee clasping the sides of leg between the thumbs (each tip on a femoral condyle). With quadriceps relaxed, anteroposterior assess glide of tibia on femur (normal ~0.5cm). The anterior cruciate ligament prevents anterior glide; the posterior prevents posterior glide. Excessive glide in one direction (compare knees) suggests damage to the relevant ligament. Examination should also be performed in 20° of flexion (Lachman's test). A more sensitive test to determine if symptoms are really due to cruciate ligament damage (can be asymptomatic) is the 'pivot shift test': flex the knee, then put it in valgus; now extend it. If the anterior cruciate is ruptured, the knee jumps smartly forwards. (Often hard to elicit, unless very relaxed or under general anaesthesia.)

McMurray's rotation test is an unreliable way of detecting pedunculated tears of menisci. The knee is flexed, the tibia laterally rotated on the femur, then the knee is extended with tibia kept rotated. This is repeated with varying degrees of knee flexion, and then again with the tibia medially rotated on the femur. This manoeuvre is designed to jam the free end of a pedunculated meniscus in the joint—a click being felt and heard and pain experienced by the patient as the jammed tag is released as the knee is straightened. This test may not detect bucket-handle tears (p 710). Note: normal knees commonly produce patellar clicks. Simply eliciting joint-line tenderness may be a more valid test when combined with a history of mechanical locking.

Arthroscopy Arthroscopes enable internal structures of the knee to be seen and a definite diagnosis to be made without opening the joint. They also enable a wide range of operations to be done without opening the knee, so reducing convalescence time for those undergoing arthroscopic surgery. Most can be done as day cases under local anaesthesia. MRI, where available, is superseding diagnostic arthroscopy (OHCM p 794).

Pain in the knee

The common symptoms are anterior knee pain or pain and swelling. Anterior knee pain can be due to many causes (see below).

Chondromalacia patellae (This is a component of the 'anterior knee pain syndrome'.) It particularly affects young women and servicemen. Patellar aching is felt after prolonged sitting. Medial retropatellar tenderness and pain on patello-femoral compression occur. *Diagnosis* is clinical but if arthroscopy is performed softening and/or fibrillation of patellar articular cartilage is seen. *Treat* by vastus medialis strengthening exercises—lie on the back with the foot externally rotated. Lift the heel 10cm off the floor × 500/24h, relaxing muscles between lifts (exercises are boring but relieve pain in 80%). If symptoms persist despite a year's exercises, arthroscopic lateral retinacular release may be tried. If pain still persists, rarely consider patellectomy. Shaving the posterior surface of the patella gives rather uncertain results.

Excessive lateral pressure syndrome Retropatellar tenderness and pain are felt laterally. Exercise provokes pain. Vastus medialis exercises are less likely to give relief. Arthroscopy reveals a normal patella. Lateral retinacular release should relieve pain.

Bipartite patella This is usually an incidental X-ray finding but may give pain if the superolateral fragment is mobile with tenderness over the junction. Extra fragment excision relieves pain.

Recurrent subluxation of patella A tight lateral retinaculum causes the patella to sublux laterally giving medial pain and causing the knee to give way. It is commoner in girls and with valgus knees. It may be familial, or associated with joint laxity, a high-riding patella (*patella alta*), or a hypotrophic lateral femoral condyle. Examination reveals increased lateral patellar movement which may be accompanied by pain and the reflex contraction of quadriceps (ie a positive patellar apprehension test). If vastus medialis exercises fail to help, lateral retinacular release usually cures. Patellar tendon transfer is rarely needed.

Patella tendinitis This is usually initiated by a small tear in the patellar tendon, and can occur in any part of the patellar tendon. It is most commonly seen in sport players (insertional tendinitis of patellar tendon and patella called 'jumper's knee'). Tendinitis settles with rest and NSAIDs. For those unable to rest, steroid injection around (not into) the tendon may help. For Osgood–Schlatter's disease, see p 652.

Ilio-tibial tract syndrome Synovium deep to the ilio-tibial tract is inflamed where it rubs on the lateral femoral condyle. Common in runners, it settles with rest, NSAIDs or steroid injection.

Medial shelf syndrome The synovial fold above the medial meniscus is inflamed. Pain is superomedial. There may be brief locking of the knee (mimics a torn meniscus). *Diagnosis:* Arthroscopy. *Treatment:* Rest, NSAIDs, local steroid injection or division of the synovial fold arthroscopically.

Fat pad syndrome Tenderness deep to the patellar tendon may be caused by the fat pad being caught in the tibio-femoral joint. Pain resolves with rest.

Joint replacements

Joint replacement has been used for 30 years. Each year 40,000 hips and 15–20,000 knees are replaced in the UK (knees may soon overtake hips).

Hip replacements The usual indication for operation is pain (surgery for fixed flexion deformity affecting walking is less successful). Of the 30,000 operations carried out annually 75% are to replace osteoarthritic hips. These patients are usually elderly. Rheumatoid arthritis is the next most common indication (patients are often in their 30s or 40s). Other conditions which may result in replacement are: avascular necrosis of head of femur; congenitally dislocated hip; fractured neck of femur.

Many prostheses are available but most consist of a metal femoral component with an intramedullary stem usually held in place by bone cement, and a plastic acetabular component. Early success of operation occurs in 90% of cases. Early complications include: dislocation (0.5–3%) and deep infection (<1%). Deep infection is a disaster and although exchange prosthesis using antibiotic loaded cement may be possible, a Girdlestone excisional arthroplasty may be needed. A small proportion of implants fail to make a satisfactory interface with bone.

Later problems of loosening or infection are heralded by return of pain. If plain X-rays are inconclusive in the case of loosening, strontium or technetium scans may reveal increased bone activity. Suspected sepsis should be investigated by white cell count, ESR, and gallium scan. Scans are not reliable within 8 months of operation. Revision arthroplasty is more successful for loosening than for infection. By 9–10yrs post-op 11% of implants have been revised. Caution should be exercised in recommending replacement to those in their 60s who are likely to give extreme usage to the prosthesis. Excess weight should be lost as this also contributes to usage. Earlier replacement is used for rheumatoid as joints tend to be grossly affected younger—and excessive delay may result in operation upon very rarefied osteoporotic bone.

Knee replacement The knee is a more complicated joint than the hip, and designs for replacements have altered greatly, from long-stemmed hinge mechanisms, to short-stemmed articular replacements in favour currently. Indications for knee replacement: pain at rest, or disturbing sleep, or making housebound. Pain correlates poorly with radiological signs. Success rate: 95%. Joint survival: 90% last 15yrs (better than hips).[1] Revision rates are similar.

Other joints Joint spacers are used in finger joints for rheumatoid, with success. Elbow replacements are beginning to show some success. Shoulder replacement success rates are approaching those of knees.

Preventing thromboses DVT occurs in >⅔ of major orthopaedic events. Low-dose heparin halves DVT rate and lowers risk of fatal pulmonary embolus (PE) by ⅔. If low molecular weight heparin (5000 daltons) is used (dalteparin, enoxaparin, tinzaparin) rather than ordinary heparin (~13,000 daltons), the DVT and PE rate (fatal and non-fatal) falls further (2.3% down to 1.4% for PE). SE: bleeding (at operative site, intracranial, retroperitoneal). CI: uncontrolled bleeding/risk of bleeding (eg peptic ulcer); endocarditis; children. Dose example: dalteparin 2500U 2h pre-op and at 12h post-op, then 5000U once daily for 5 days.[2][3]

1 J Noble 1991 *BMJ* ii 262   2 *Drug Ther Bul* 1993 **31** 37   3 TF Imperiale 1994 *JAMA* **271** 1780

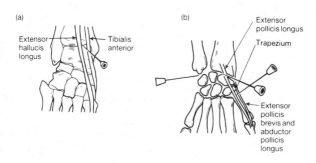

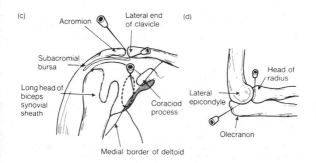

(a) Dorsal aspect of the right ankle indicating anatomical landmarks.
(b) Dorsal aspect of the right wrist indicating anatomical landmarks.
(c) Diagrammatic representation of the anterior aspect of the right shoulder region. The stippled areas indicate synovial membranes of the subacromial bursa and the glenohumeral joints. (d) The right elbow, flexed; the stippled area is the synovial membrane. (Reproduced by kind permission from M Crawley 1974 *Br J Hosp Med* **11** 747-55.)

Tendons

The tendon conditions which may cause orthopaedic problems are those of tendon rupture or tenosynovitis (which may be infective or inflammatory). The most common ruptures are the extensor tendons of the fingers, the Achilles tendon, the long head of biceps (p 612), supraspinatus (p 612) and the quadriceps expansion.

Ruptured extensor tendons of the fingers The long extensors inserting into the distal phalanx are particularly prone to rupture by trauma resulting in 'mallet' or 'baseball' fingers. The affected digit is splinted for 3–6 weeks (in extension). If the tendon does not reattach to bone spontaneously, the choice is between living with the deformity or operative repair. The long extensor of the thumb may rupture as a complication of fractures to the lower radius.
Treatment: Repair is effected using the lower end of the extensor indicis tendon re-routed to be sutured to the distal thumb tendon (as the thumb extensor is too frayed for direct repair). The extensors of the ring and little fingers are particularly prone to spontaneous rupture in rheumatoid arthritis (usually attrition rupture from the distal ulna).

Ruptured Achilles (calcaneal) tendon Sudden pain is felt at the back of the ankle during running or jumping as the tendon ruptures 5cm above the tendon insertion. Pain may be perceived as a kick. It is still possible to walk (with a limp), and some plantar flexion of the foot remains, but it is impossible to raise the heel from the floor when standing upon the affected leg. A gap may be palpated in the tendon course (particulary within 24h of the injury). On examination he cannot stand on tiptoe.
The squeeze test (Simmonds' test) is sensitive: ask the patient to kneel on a chair, while you squeeze both calves—if the Achilles is ruptured, there will be less plantar flexion on the affected side.
Treatment: Urgent tendon repair is often preferred to conservative treatment (immobilization of the lower leg in plaster with the foot slightly in equinus). Undiagnosed ruptures may repair, but tendon lengthening occurs.

Ruptured quadriceps expansion Injury may be direct (eg a blow) or indirect (stumbling causing sudden contraction of the apparatus). The quadriceps expansion encloses the patella and inserts into the tibial tubercle by the patellar tendon.
Treatment: Rupture can occur at the site of quadriceps insertion to patella, through the patella by fracture (if reduction is possible it is secured by wire or screws, if not, the patella is excised and tendon repair effected), or by avulsion of the patellar tendon from the tibial tubercle (in young patients the torn tendon may be resutured). After repairs, a plaster cylinder is worn for several weeks, then intensive physiotherapy is needed to regain knee function.

Tenosynovitis Tendons and their surrounding synovium may become locally inflamed (possibly due to strain) thereby causing pain (eg supraspinatus tendinitis and bicipital tendinitis p 612, De Quervain's p 616). Acute frictional synovitis at the wrist is another example, with swelling over the wrist and thumb extensors.
Treatment: In the latter case, the wrist is splinted for three weeks (leaving the fingers free) to allow inflammation to subside.

Nerve compression syndromes

For brachial plexus and peripheral nerve lesions, see p 724, 664 & 720.

Carpal tunnel syndrome ▶This is the commonest cause of hand pain at night. It is due to compression of the median nerve as it passes under the flexor retinaculum. Pregnancy, the Pill, myxoedema, lunate fracture (rare), rheumatoid arthritis, cardiac failure, or being premenstrual may all increase compression. *Clinical features* are of median nerve distribution:
●Tingling or pain are felt in the thumb, index finger, and middle finger.
●When the pain is at its worse, the patient characteristically flicks or shakes the wrist to bring about relief. Pain is especially common at night and after repetitive actions. Affected persons may experience clumsiness. Early on there may be no signs. Later there is:
●Wasting of the thenar eminence and decreased sensation over the palm.
●Lateral palmar sensation is spared as its supply (the palmar cutaneous branch of the median nerve) does not pass through the tunnel.
●Phalen's test: holding the wrist hyperflexed for 1–2mins reproduces the symptoms. (This is more reliable than Tinel's test—tapping over the tunnel to produce paraesthesiae. Note: Phalen's flexing, Tinnel's tapping.)

Tests: Nerve conduction studies are not usually needed (and may be -ve).[1]

Treatment: Diuretics may help. Carpal injection: see p 658 for technique. Wrist splints worn at night may relieve nocturnal pain. More permanent results are obtained from decompression by flexor retinaculum division.

Ulnar nerve compression at the elbow This is described on p 614.

Posterior interosseus nerve compression This radial nerve branch is sometimes compressed on passing through the supinator muscle (after forearm fracture). Patients experience weakness of the thumb and fingers. *Examination* may reveal weakness of long finger extensors, and short and long extensors of the thumb, but no sensory loss. *Treatment* after trauma is surgical decompression and springed splints to extend fingers.

Anterior interosseus nerve compression This median nerve branch may be compressed under the fibrous origin of flexor digitorum sublimis, causing weakness of pinch and pain along the forearm's radial border. Examination shows weakness of the long thumb flexor and flexor profundus to the index and middle fingers. *Treatment* is surgical decompression.

Ulnar nerve compression at the wrist Uncommon. Loss may be motor or sensory. *Diagnose* by electrophysiology. *Treat* by surgical decompression.

Meralgia paraesthetica If the lateral cutaneous nerve of the thigh is compressed (on leaving the pelvis just medial to the anterior superior iliac spine, eg by tight jeans), pain and paraesthesiae may be felt over the upper outer thigh. Sensation may be ↓ over this area. It is usually self-limiting, and can occur in pregnancy. *Treatment:* Cortisone and local anaesthetic injection at the anterior superior iliac spine gives unpredictable results.

Common peroneal compression Nerve compression against the head of fibula (eg plaster casts, thin patients lying unconscious, proximal fibula fracture) causes inability to dorsiflex the foot. Sensation may be ↓ over the dorsum of the foot. *Tests:* Electrophysiology. *Treatment:* Most recover spontaneously but surgical decompression may be needed. Physiotherapy and splint until foot-drop recovers.

1 M Shipley 1995 *BMJ* i 239

Soft tissue injuries

Correct management of soft tissue injuries reduces pain, recovery time and subsequent disability.

Treatment in the first 24 hours: 'RICE'

*R*est: A splint or plaster cast may help.

*I*ce: Cold is anaesthetic and causes vasoconstriction. Apply an ice pack (eg a packet of frozen peas wrapped in a cloth), a cold spray or cold compresses intermittently every 10 minutes.

*C*ompression: Strapping restricts swelling and further bleeding.

*E*levation: Ideally toes higher than bottom to improve drainage from the affected part and to reduce pain.

Rehabilitation Passive stretching to maintain joint mobility and muscle length, then progressive active exercise until the full range and strength of movement is restored, eg wobbleboards for ankles. Sportsmen must then retrain to full fitness.

Adjuvants: NSAIDs, rubefacients and ultrasound may aid recovery.

Sports injuries Many sports injuries can be prevented.
- Is the patient preparing the body for activity with a proper warm-up? Inadequate warm-up increases the risk of injury.
- Cooling-down is also important in reducing muscle soreness.
- Is protective equipment being worn?
- Many acute and chronic injuries are caused by unsuitable equipment, faulty technique or unwise training schedules—advice from a professional coach can be invaluable.

Common patterns of injury in sport: 'Shin splints'—shin soreness, common in unfit runners on hard surfaces and due to muscle tears, mild anterior compartment syndrome or stress fracture; knee pain—most causes on p 636; ligamentous ankle injury (p 712); plantar fasciitis (p 642); stress fractures such as march fracture—suspect with increasing bony pain despite normal X-ray; Achilles tendon problems; tennis elbow (p 614).

Overuse phenomena

Overuse phenomena at work (Work-related upper limb injury = Occupational overuse syndrome = Isometric contraction myopathy = Repetitive strain injury (RSI)) Activity requiring repetitive actions—particularly those associated with prolonged muscle contraction,[1] may lead to chronic symptoms. Employers have a duty to provide a safe working environment and well-designed chairs and tools, and frequent short breaks. Changes of posture and activity help to reduce work-related upper limb injury (this term is preferred to repetitive strain injury). The cost of these injuries in suffering, and hours lost from work, is considerable as treatment of established symptoms is often difficult and may necessitate change in employment—if one is available.

Compensation is a vexed issue, and recent Court judgments have gone in favour of the employers in some instances, and in favour of patients in others. Some people arguing that the condition does not exist as a separate medical entity[2] emphasize the lack of histopathology. It should be noted that this is not a pre-requisite for a disease (see *Sudden infant death*, p 286)—and in any case histopathology *is* sometimes demonstrable.[1] Treatments tried include splinting (may prolong the problem), physiotherapy, β-blockers for relaxation, and the Alexander technique for posture re-education.[3]

Those who use vacuum cleaners, assemble cars or play stringed instruments may all develop tennis elbow.

1 C Murray-Lesliw 1994 *Update* 48 587   2 A Mann 1994 *BMJ* i 269   3 P Brookes 1993 *BMJ* ii 1298

Testing peripheral nerves[1]

Nerve root	Muscle	Test—by asking the patient to:
C3,4	trapezius	Shrug shoulder, adduct scapula.
C4,5	rhomboids	Brace shoulder back.
C5,6,7	serratus anterior	Push forward against resistance.
C5,6,7,8	pectoralis major (clavicular head)	Adduct arm from above horizontal. and forward
C6,7,8 T1	pectoralis major (sternocostal head)	Adduct arm below horizontal.
C5	supraspinatus	Abduct arm the first 15°.
C5,6	infraspinatus	Externally rotate arm, elbow at side.
C6,7,8	latissimus dorsi	Adduct horizontal and lateral arm.
C5,6	biceps	Flex supinated forearm.
C5,6	deltoid	Abduct arm between 15° and 90°.
L4,5, S1	gluteus medius & minimum (superior gluteal nrv)	Internal rotation. at hip, hip abduction.
L5, S1,2	gluteus maximus (inferior gluteal nrv)	Extension at hip (lie prone).
L2,3,4	adductors (obturator nrv)	Adduct leg against resistance.

The radial nerve

C7,8	triceps	Extend elbow against resistance.
C5,6	brachioradialis	Flex elbow with forearm half way between pronation and supination.
C6,7	extensor carpi radialis longus	Extend wrist to radial side with fingers extended.
C5,6	supinator	Arm by side, resist hand pronation.
C7,8	extensor digitorum	Keep fingers extended at MCP joint.
C7,8	extensor carpi ulnaris	Extend wrist to ulnar side.
C7,8	abductor pollicis longus	Abduct thumb at 90° to palm.
C7,8	extensor pollicis brevis	Extend thumb at MCP joint.
C7,8	extensor pollicis longus	Resist thumb flexion at IP joint.

Median nerve

C6,7	pronator teres	Keep arm pronated against resistance.
C6,7,8	flexor carpi radialis	Flex wrist towards radial side.
C7,8 T1	flexor digitorum sublimis	Resist extension at PIP joint (while you fix his proximal phalanx).
C8 T1	flexor digitorum profundus I & II	Resist extension at the DIP joint of index finger.
C8 T1	flexor pollicis longus	Resist thumb extension at interphalangeal joint (fix proximal phalanx).
C8 T1	abductor pollicis brevis	Abduct thumb (nail at 90° to palm).
C8 T1	opponens pollicis	Thumb touches 5th finger-tip (nail parallel to palm).
C8 T1	1st & 2nd lumbricals	Extend PIP joint against resistance with MCP joint held hyperextended.

Rheumatoid arthritis

The central immunological event is said to be presentation of the culprit antigen (whatever this turns out to be) to T-helper cells, with subsequent cytokine-mediated synovial neutrophil exudate, which releases cartilage-degrading enzymes. There is a symmetrical synovitis, with arthralgia, stiffness, and swelling, and, if prolonged, destruction of peripheral joints.[1]

Prevalence 3% of females, 1% of males. Common in middle age (but any age affected). Association: HLA-DRW4. Incidence: 1/1000 adults/yr.[1]

Presentation Very variable. It may take time before RA can be distinguished from other inflammatory conditions. Onset is often insidious with malaise, transient pain, stiffness and weakness, but may be sudden—'One day I couldn't move'. Monoarthritis, palindromic arthritis—joints affected by turn—and systemic presentations occur.

Prognosis 20% have 1 attack, often acute, and recover completely. 30% are mildly disabled; 40% have prolonged, moderately disabling disease. 10% are severely disabled. Life expectancy↓, by 3 (♀) to 7yr (♂). Worse prognosis if: ♂ patient, palindromic onset, extra-articular features, nodules, erosions occurring within 3yrs of onset, rheumatoid factor ↑↑, HLA DR4 +ve.

Joint involvement Inflamed synovium (pannus) produces hot swollen tender joints and thickened tendons with associated soft-tissue swelling, hence the characteristic sausage-shaped fingers. Early morning stiffness may last for several hours. With time the pannus erodes the joints and tendons rupture and the irreversible changes typical of advanced RA are seen. *Fingers* develop 'swan neck' (hyperextended PIP joint but flexed DIP joint) or 'boutonnière' deformities (flexed PIP joint, extended MCP joint, hyperextended DIP joint); *thumbs* adopt Z-deformities. MCP joints and *wrists* sublux acquiring ulnar deviation: the ulnar styloid and radial head become prominent. *Knees* develop valgus or varus deformity. Popliteal 'Baker's' cysts occur and may rupture mimicking a DVT. In the *feet* subluxation of the metatarsal heads with associated hallux valgus, clawed toes and calloused causes pain—'walking on marbles'. ►*Atlanto-axial instability* may lead to cord compression; X-ray prior to intubation is wise.

Extra-articular involvement Weight loss, fever and malaise are common. Many systemic manifestations of RA are caused by vasculitis or nodules. 10% develop amyloid. Drug complications can affect most systems (p 672).
Skin: Nail fold infarcts, ulcers, rashes, palmar erythema.
Nodules: Firm, mobile, non-tender nodules are found on pressure areas eg forearm of 20% of sufferers. Nodules may also occur on tendon sheaths (causing triggering) or in lungs, myocardium, pericardium or sclera.
Eyes: Keratoconjunctivitis sicca (p 482), scleritis, episcleritis.
Cardiovascular system: Pericardial effusion, pericarditis, myocarditis.
Lungs: Pleurisy, pleural effusions, nodules, fibrosing alveolitis.
CNS: Entrapment neuropathy (eg carpal tunnel), compression syndromes.

Tests *Blood:* Normochromic anaemia, ESR↑. Both correlate with disease activity. Also: alk phos↑, platelets↑, lymphocytes↓, WCC↓ (in Felty's OHCM p 698). Rheumatoid factor: this anti-IgG autoantibody appears eventually in 80% of those with RA but is found in low titres in many chronic conditions. Antinuclear factor: +ve in 30% with RA (OHCM p 690).
Synovial fluid: Turbid; viscosity↓; clots, from high polymorph count.
X-ray changes: In order of appearance: soft-tissue swelling; juxta-articular osteoporosis; symmetrical loss of joint space; erosions at joint margins, subluxation and dislocation of joints.

1 R Madhok 1995 *Lancet* 346 481

Rheumatoid arthritis: treatment

The patient's attitude is crucial to wellbeing. A severely afflicted patient who does quads exercises daily and seeks ways around disability will be living an independent life long after a mildly affected patient who has 'given up'. Physiotherapy, occupational therapy and maintainance of morale are the mainstays of treatment, backed up by drugs and surgery.

Acute flare-ups initially require rest, analgesia and NSAIDs; then passive and active exercise can be introduced to maintain joint mobility and muscle tone. Chronic disease is characterized by pain, stiffness, malaise and permanent loss of function. Exercises, splints, and gadgets (eg kettle-tippers) protect joints. Special shoes and seat raises allow independence. Second-line drugs, intralesional steroids and synovectomy reduce disease activity. HRT may particularly help women patients. Skeletal collapse needs analgesia, surgery, appliances/home conversion, and social/financial help to mitigate isolation and income loss. Simple analgesia eg paracetamol is useful for pain relief. NSAIDs If no contraindication (asthma, active peptic ulcer) start an NSAID eg ibuprofen 400–800mg/8h PO pc (cheapest and least likely to cause GI bleeds) or naproxen 250–500mg/12h PO pc (twice a day dose, and is more anti-inflammatory than ibuprofen, and may cause less GI bleeding than other NSAIDs). Diclofenac 25–50mg/8h PO pc is similar to naproxen and can be combined with misoprostol (Arthrotec®)—an ulcer-protecting agent (OHCM p 490)—worth considering if history of dyspepsia and NSAIDs vital. One cannot predict which NSAID a patient will respond to. If 1–2 week trials of 3 NSAIDs do not control pain, or if synovitis for >6 months, consider a second-line drug.

Second-line drugs appear to modify the immunological derangement, hence reducing disease activity as measured by ESR and functional indices, eg grip. As they are slow to act and have a high incidence of adverse effects needing regular monitoring, their use is restricted to patients with severe, persistent, or progressive joint or extra-articular disease. Joint erosions are closely linked to long-term disability, and if developing in the first 2yrs suggest the need for a second-line drug.

Gold is given as sodium aurothiomalate 50mg/week IM or auranofin 3mg/12h PO. After 3 months on IM gold 80% of patients show an improvement and can be maintained on a lower dose but 60% stop due to side effects. Oral gold is less effective but less toxic. SE necessitating stopping treatment: mouth ulcers, rashes (risk of exfoliative dermatitis), persistent proteinuria, blood dyscrasias. ▶Monitor urine and FBC often.

Penicillamine up to 500mg/24h PO in use resembles IM gold.

Sulphasalazine ½g/6h PO (start with ½g/24h) is less effective than IM gold but works faster and is well tolerated, so may be used earlier. SE: D&V, sperm count↓, WCC↓. Monitor LFT & FBC. Warn to stop if sore throat (=WCC↓).

Hydroxychloroquine 200mg/24h PO is least effective. SE: irreversible retinopathy so ophthalmological monitoring required.

Immunosuppressants eg azathioprine are occasionally used.

Methotrexate consider if severe disease (do LFT, FBC, U&E, hepatitis B and C serology first; do liver biopsy if +ve, or if alcohol abuse). ▣

Steroids can reduce erosions if given early (prednisolone 7.5mg/day PO). It is not known if all should receive this. One problem is ↓bone density over long periods. Another problem is that it would be difficult to keep patients to <7.5mg/day, as, for the first months, there may be great symptomatic improvement, which then tails off, leaving patients wanting higher doses, and risking cataract, fluid retention, and peptic ulcers. ▣

Surgery may relieve pain and improve function eg arthroplasty (joint reconstruction/replacement), particularly for forefoot, hips and knees; arthrodesis (joint fusion) of ankle or foot. Synovectomy, tendon repair and transfer, joint debridement and nerve decompression are also useful. ▣

This page assumes that the resuscitator has an experienced person to help him and a well-equipped resuscitation room. On hearing that a major injury is coming in, summon experienced help (eg senior traumatologist). Remember ABC (*a*irway, *b*reathing, *c*irculation)

- Remove foreign bodies from the airway. Suck out any fluid.
- Is the patient breathing effetively? If not, assist ventilation first with bag-valve-mask + 100% O₂, then intubate the trachea. If upper airway obstruction prevents intubation do cricothyrotomy (OHCM p 753). Is there a tension pneumothorax?—relieve it now. If breathing spontaneously, give *all* patients O₂ at 15 litres/min through tight-fitting mask with reservoir. Cover and seal any open chest wound on 3 sides only.
- Is there a pulse? If not, start external cardiac massage. Summon help to bring a defibrillator.
- Apply electrode jelly, and position the paddles over the chest (over the sternum and in the midaxillary line). Get an ECG read-out, and defibrillate; continue as per 'cardiac arrest', p 310-11, for adults, see OHCM p 724. If the rhythm is acceptable, proceed as follows.
- Beware IPPV if there is surgical emphysema. This implies ruptured lung (eg from rib fracture) and requires prompt insertion of a chest drain.
- Set up a 0.9% saline IVI. Take blood: crossmatch and glucose ward test. Use 2 big cannulae—via a cut down (OHCM p 698), if IV access fails. Continue cardiopulmonary resuscitation.
- Take systolic BP. If <90mmHg and blood loss is the probable cause, give colloid (Haemaccel®) fast IV until BP↑, pulse↓, urine flows (>30 ml/h; catheterize at leisure—exclude urethral injury first) and crossmatched blood arrives (▶ask for group-compatible blood—only takes 5-10min).
- Remove clothing (eg with large scissors). What are the injuries? Circumferential neck or chest burns may require incising (escharotomy) to relieve pressure on the larynx.
- Assume that there is spinal instability. Ensure immobility by sand-bags, collars and tape, until radiology is carried out.
- Apply pressure and elevation to any actively bleeding part.
- Assess conscious level (AVPU, p 787).
- Examine all the peripheral pulses. 30% of fractures are missed in the resuscitation room.[1] X-ray skull, cervical spine (p 726), and pelvis. If intraperitoneal haemorrhage is suspected, see p 718.
- Blood gases; CXR—chest drain if pneumothorax, OHCM p 698.
- Chart pulse, BP, conscious level and pupil size every few mins. If pupils are unequal, summon expert neurological help. Give mannitol 1g/kg IVI to relieve cerebral oedema. See p 714.
- Full head-to-toe examination ('secondary survey'). If pneumothorax, then chest drain (OHCM p 698). Glasgow Coma Scale for coma level (p 680).
- If fits occur give 10mg diazepam IV followed by 100mg in 500ml 5% dextrose infused at 40ml/h if they continue (children: 0.3mg/kg IV, then 50µg/kg/h IVI)—beware apnoea. If fits continue, give 40-100ml of ready prepared 0.8% chlormethiazole IVI over 10mins (kept in the fridge)—adults only. The subsequent infusion rate is determined by the response.
- In lung contusion, near-drowning or smoke inhalation, consider early high-dose steroids (methylprednisolone 30mg/kg IV).
- Give tetanus toxoid booster ± human antitetanus immunoglobulin (p 686).
- Enlist experienced help in determining the priorities of treatment for each specific injury.

See also *Advanced Trauma Life Support* course manual, 1988, American College of Surgeons

The stitching of wounds

1 Remove foreign bodies and all necrotic material.

2 Clean with sterile gauze soaked in 1% cetrimide.

3 Control bleeding by pressure and elevation of the part. Deep absorbable stitches (eg Dexon®, Vicryl®, catgut) will aid apposition of skin edges (avoid in the hand, as may disrupt nerves).

4 Infiltrate skin to be sutured with 1% lignocaine (<3mg/kg; SE: seizures).
▶*Avoid adrenaline when anaesthetizing digits.*

5 If much tension is needed, delay suturing or use skin grafts.

6 Handle the tissues gently to prevent further injury.

7 Use interrupted stitches for closure. Start from the middle, or with any jagged edges. This aids accurate apposition.

8 Avoid eversion or inversion of the stitched skin edges.

9 Examples of which stitches to use when closing wounds: Adult's face: 5-0 monofilament nylon (remove after 5 days). Child's face: 6-0 monofilament nylon (remove after 3-5 days). Other parts of children: 5-0 catgut. The deep part will be absorbed and the top part will slough off after 10-14 days. Adult leg: 3-0 nylon (remove after 2 weeks). Shin skin flaps (following a fall on a hard object) may have a poor blood supply, especially if the length of the flap is more than twice as long as its breadth, and healing will be better with thin strips of elastoplast (eg Steristrip®), rather than stitches. Exact approximation of the edges is likely to be impossible, but skin grafts are rarely needed.
Adult arm, trunk, abdomen: 3-0 silk. Remove after ~9 days. Scalp: 3-0 silk. Remove after 5 days.

10 The method of stitching is best learned by demonstration by a surgeon. Practise stitching up a slashed banana skin in a sterile field.

11 Dress with sterile tulle to prevent further injury.

Alternatives to suturing Suturing requires anaesthesia and leaves scars. *Skin closure strips (Steristrips®)* are useful for minor lacerations, but not on hairy skin. Leave space between strips for ooze. They can also temporarily oppose edges of a larger wound while you suture. Wound infection is less of a risk than with stitches, particularly for animal bites.
Acrylic tissue adhesive (Histoacryl®) is quick, minimizes scarring and needs no follow-up. Dry the wound surface, and align edges carefully, as you cannot correct their position once you have glued the wound. Apply the glue thinly (use a capillary tube or disposable pipette), and press the edges together for 30sec. *Hair:* You may be able to close minor scalp wounds by tying strands of hair across the wound.

Prevention of tetanus Active immunization with tetanus toxoid is given as part of the triple vaccine during the first year of life (p 209). Boosters are given on starting school and in early adulthood. Once 5 injections have been given, revaccinate only at the time of significant injury.
Primary immunization of adults: 0.5ml tetanus toxoid IM repeated twice at monthly intervals.
Wounds: Any cut merits an extra dose of 0.5ml toxoid IM, unless already fully immune (a full course of toxoid or a booster in last 10 years). The non-immune will need 2 further injections (0.5ml IM) at monthly intervals. If partially immune (ie has had a toxoid booster or a full course >10 years previously), a single booster is all the toxoid that is needed.
Human tetanus immunoglobulin: This is required for the partially or non-immune patient (defined above) with dirty, old (>6h), infected, devitalized, or soil-contaminated wounds. Give 250-500 units IM, using a separate syringe and site to the toxoid injection.

▶If the immune status is unknown, assume he/she is non-immune.
Routine infant immunization started in 1961, so many adults are at risk. **687**

▶Hygiene educatoin and wound debridement are of vital importance.

Maximum doses of lignocaine—a rough guide[1]

0.25%	1.12ml/kg
0.5%	0.56ml/kg
1%	0.28ml/kg
2%	0.14ml/kg

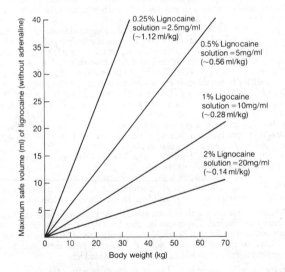

Maximum safe volume of plain lignocaine according to body weight.
(Maximum safe dose of lignocaine—without adrenaline 3mg/kg, with
adrenaline 7mg/kg; uses of different solutions—0.25-0.5% for infiltration
and intravenous regional anaesthesia, 1% for nerve blocks, epidural anaesthesia, and intravenous regional anaesthesia, 2% for nerve blocks.)

1 D Kelly 1983 *BMJ* i 1784

▶▶ Burns^{ATLS}

Most hospitals do not have their own burns unit. Resuscitate and organize expeditious transfer for all major burns. Always assess site, size, and depth of the burn; smaller burns in the young, the old, or at specific sites, may still require specialist help.

Assessment The size of the burn must be estimated to calculate fluid requirements. Ignore erythema. Consider transfer to a burns unit if >10% burn in children or elderly, or >20% in others. Use a Lund & Browder chart or the 'rule of nines':

Arm (all over) 9%	Front 18%	Head (all over) 9%	Palm 1%
Leg (all over) 18%	Back 18%	Genitals 1%	
	(NB: in children, head = 14%; leg = 14%)		

Estimating burn thickness: Partial thickness is painful, red, and blistered; full thickness is painless and white/grey.

Refer full thickness burns >5% of body area or full/partial thickness burns involving face, hands, eyes, genitalia to a burns unit.

Resuscitation ●*Airway:* Beware of upper airway obstruction developing if inhaled hot gases (only superheated steam will cause thermal damage distal to trachea)—suspect if singed nasal hairs or hoarse voice. Consider early intubation or surgical airway (OHCM p 753).

●*Breathing:* Give 100% O_2 if you suspect carbon monoxide poisoning (may have cherry-red skin, but often not) as $t_{1/2}$ carboxyhaemoglobin (COHb) falls from 250min to 40min (consider hyperbaric oxygen if pregnant; CNS signs; >20% COHb). SpO_2 measured by pulse oximeter is unreliable (falls eg 3% with up to 40% COHb). Do escharotomy bilaterally in anterior axillary line if thoracic burns impair chest excursion (p 689).

●*Circulation:* >10% burns in a child and >15% burns in adults require IV fluids. Put up 2 large-bore (14G or 16G) IV lines. Do not worry if you have to put these through burned skin. Secure them well: they are literally lifelines. It does not matter whether you give crystalloid or colloid: the volume is what is important. Use a 'Burns Calculator' flow chart[1] or a formula, eg:

●*Muir and Barclay formula* (popular in UK):
[weight (kg) × %burn]/2=ml colloid (eg Haemaccel®) per unit time
Time periods are 4h, 4h, 4h, 6h, 6h, 12h.

●*Parkland formula* (popular in USA):
4 × weight (kg) × %burn=ml Hartmann's solution in 24h, half given in 1st 8h (NB: unsatisfactory in children, so use a 'Burns Calculator').

▶NB: you must replace fluid from the time of burn, not from the time first seen in hospital. You must also give 1.5-2.0ml/kg/h 5% dextrose if Muir and Barclay formula is used. Formulae are only guides: adjust IVI rate according to clinical response and urine output (aim for >30ml/h in adults; >1ml/kg/h in children). Catheterize the bladder.

Treatment ▶Do *not* apply cold water to extensive burns: this may intensify shock. Do not burst blisters. Covering extensive partial thickness burns with sterile linen prior to transfer to a burns unit will deflect air currents and relieve pain. Use morphine in 1-2mg aliquots IV. Suitable dressings include Vaseline® gauze, or silver sulphadiazine under absorbent gauze; change every 1-2 days. Hands may be covered in silver sulphadiazine inside a plastic bag. Ensure tetanus immunity. Give 50ml whole blood for every 1% of full-thickness burn, half in the second 4h of IV therapy and half after 24h.

1 SM Milner 1993 *Lancet* **342** 1089

Smoke inhalation

In addition to thermal injury, the problems are carbon monoxide poisoning (see opposite), and cyanide poisoning.

Cyanide poisoning is common from smouldering plastics. It binds reversibly with ferric ions in enzymes—so stopping oxidative phosphorylation and causing dizziness, headaches and seizures. Tachycardia and dyspnoea soon give way to bradycardia and apnoea. The aim is to detatch cyanide from cytochrome oxidase by making attractive ferric opportunities by producing methaemoglobin (each molecule has 3 ferric haem groups) from Hb—using amyl nitrate 0.2-0.4ml via an Ambu bag.[1] Less safe alternatives are sodium nitrite solution (10ml of 3% IV over 3min; dimethylaminophenol (5ml of 5% IV over 1min). Also give sodium thiosulphate (25ml of 50% IV over 10min) to provide an extra source of sulphur to augment cyanide conversion to non-toxic thiocyanate. Alternatively, chelate cyanide with dicobalt edetate (20ml of 1.5% IV over 1min), followed by 50ml of glucose IV infusion 50%. (These are adult doses.)

Escharotomy

For chest escharotomy make incisions 1 and 2 first. If further expansion is needed make the others. Make incisions either side of trunk/limb or digit (one side only shown in diagrams). If it causes pain, burns are not full thickness and escharotomy is probably not needed. Substantial bleeding may necessitate covering with sterile dressings and blood transfusion.

Diagrams reproduced by kind permission of the *BMJ* and authors of reference 2.

1 R Langford 1989 *BMJ* ii 902   2 C Robertson 1990 *BMJ* ii 282

Fractures

Fractures may be *transverse*, *oblique*, or *spiral*. They are *comminuted* if there are >2 fragments. If broken ends protrude through the skin or into a body cavity, the fracture is *open* (an emergency). Young soft bones may bend and partially break (greenstick fractures). Compression forces yield *impacted* or *crush* fractures (in cancellous bone), and traction force causes tendons or ligaments to detach a piece of bone (*avulsion*). A *complicated* fracture causes significant damage to vessels, nerves or nearby organs. Separation of the parts is defined in terms of side-to-side *displacement*, *rotation* and *angulation* (see diagram). There may be dislocation or subluxation (partial loss of contact of joint surfaces) as a result of the injury. If it seems trivial, the bone may be diseased (*pathological* fracture eg in secondary deposits, osteomalacia, Paget's disease—p 676). Ask about the degree of force which produced the injury. If it seems trivial, the bone may be diseased (*pathological* fracture eg in secondary deposits, osteomalacia, Paget's disease—p 676).

Clinical features Pain; loss of function; tenderness; deformity; swelling; crepitus (grating of the broken ends—be gentle!).

X-rays Ask these questions: ●How many fractures are there?
●Can you see the whole bone, and the joint at each end?
●How much displacement is there? In what direction?
●Are the bones normal? Look for trapped air, bone pathology and any foreign body. Are more X-rays needed (IVU, arteriogram)?

Salter and Harris classification of epiphyseal injury (See diagram)
I   Seen in babies, or pathological conditions (eg scurvy).
II   The commonest.
III There is a displaced fragment.
IV   Union across the growth plate may interfere with bone growth.
V   Compression of the epiphysis causes deformity and stunting.

Treatment ●Correct shock (eg colloid IV, 500ml/½h, until systolic BP >100mmHg)—give blood if >1–1.5 litres lost, or continued bleeding. Monitor BP/15min, urine output (± CVP). Stop bleeding (may need open surgery).
●Relieve pain (eg morphine 10mg/4h IM or 1–2mg aliquots IV if shock).
●Sterile cover (eg Betadine®-soaked) over protruding bones and torn skin.
●Prevent infection: eg cloxacillin 500mg IM/IV + benzylpenicillin 600mg IV/IM; tetanus toxoid (0.5ml IM) if needed.

Methods of fracture management: Immobilize and reduce the part (if necessary, p 692), eg using a plaster of Paris cast. The problem with this method is that 'fracture disease' follows immobilization—muscle atrophy, stiff joints and osteoporosis. If possible, fractures involving joint articulations should be treated by open reduction, accurate reconstruction of the joint surfaces and fixation, such that immediate movement can occur. Otherwise, secondary osteoarthritis is inevitable. *Internal fixation* (eg with nails, screws and metal plates) of all fractures in those with polytrauma leads to large reductions in serious complications (fat embolism, acute respiratory distress syndrome, as well as lessening the time during which mechanical ventilation is needed).

External fixation uses screws into bone, a bar, and a means (cement or clamps) of attaching the bar to the screws to engage the fractured cortices. The screws may be driven through the bone and out of the far side of the limb, so that a second bar can add stability. Further rigidity is afforded by another set of screws at 90° to the first set. Because the intervention is away from the field of injury, this method is very useful when there are burns, and loss of skin and bone, or an open fracture. *Other methods:* plates (eg with 10 screw-holes), nails (Küntscher) and wires (Kirschner).

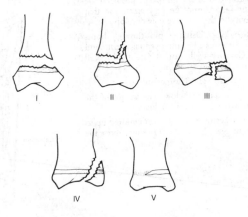

Fracture reduction and traction

A displaced fracture needs realignment (reduction) unless the function and appearance of the limb are satisfactory. For example, a small displacement acceptable in a fracture of the femoral shaft might not be acceptable in the radius, where it could cause difficulty in supination. Reduction also allows freeing of any structures trapped between bone ends. Accurate reduction helps revascularization (vital in subcapital fractures of the femur) and prevents later degenerative change if fractures involve joints. Internal and external fixation has removed the need for much traction in adults, but it is still used for children.

Methods of reduction *Manipulation* under anaesthesia (most cases).
Traction eg for subcapital femur fractures or spinal injury.
Open reduction aids accuracy, eg before internal fixation (p 690).

The next problem is to hold the reduction in place (using traction or fixation) during healing—which may take from 2 weeks (babies) to 12 weeks (a long bone in an adult).

Skin traction uses adhesive strapping to attach the load to the limb. The problems are that the load cannot be very great, and that sensitivity to the adhesive may develop.

Bone traction Using a pin through bone, greater forces can be employed (eg Steinmann pin, Gardner-Wells or Crutchfield skull calipers).

Fixed traction The Thomas splint.

Balanced traction (below) The weight of the patient balances against the load.

Gallows traction is suitable for children up to 2 years of age. The buttocks rise just above the bed.

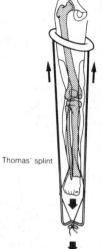

Thomas' splint

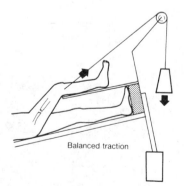

Balanced traction

Gallows traction

Algodystrophy

Synonyms[1]
Sudeck's atrophy
Reflex sympathetic dystrophy syndrome
Post-traumatic sympathetic atrophy
Shoulder-hand syndrome
Minor causalgia (causalgia means burning pain)
Post traumatic painful osteoporosis
Sudeck's post-traumatic osteodystrophy
Complex regional pain syndrome type I

Definitions 'A complex disorder of pain, sensory abnormalities, abnormal blood flow, sweating, and trophic changes in superficial or deep tissues.' The central event may be loss of vascular tone or supersensitivity to sympathetic neurotransmitters.[1,2]

Presentation This may follow weeks or months after an insult—which may be minor trauma, a fracture, zoster, or myocardial infarction.[2] Lancinating pain (which may have a trigger point) in a limb accompanies vasomotor instability. The limb may be cold and cyanosed, or hot and sweating (locally). Temperature sensitivity may be heightened. The skin of the affected part may be oedematous—and later becomes shiny and atrophic. Hyperreflexia, dystonic movements, and contractures may occur.[2]

Note that it is not the traumatized area where the symptoms occur: rather, it is some neighbouring area.

There are no systemic signs (no fever, tachycardia or lymphadenopathy). Timid, neurotic personalities are particularly affected—perhaps because of poor mobilization following the initial insult.[3]

Tests Patchy osteoporosis on X-ray, but no joint space narrowing or thinning of cartilage. Bone scintigraphy shows characteristic uniform uptake.

Treatment Refer to a pain clinic. Ordinary, standard pain-killers often have little effect. Consider physiotherapy and NSAIDs. Calcitonin and postganglionic sympathetic blockade (guanethidine or bretylium[4]) have advocates. Ultimately, the condition is self-limiting.

1 D Justins 1995 *BMJ* ii 812   2 KP Bhatia 1995 *BMJ* ii 811   3 RM Atkins 1987 in *OTM*, 2ed, OUP page 16.89 see also see 3ed 1996 page 3941)   4 AH Hord 1992 *Anaesthesia & Analgesia* 74 818   3

Trauma to the arm

Fracture of the clavicle This is common in the young, and often results from a fall on the outstretched arm (which in the elderly would be more likely to cause a Colles' fracture). A broad sling support for three weeks is usually all that is required. Fractures at the lateral end may need internal fixation.

Scapula and acromion fractures and acromio-clavicular dislocation rarely require anything more than sling support. Mobilize early.

Fractures of the humeral neck Often these are impacted, and stable; if there is displacement there is risk to the brachial plexus and axillary artery. Manipulation or open reduction may be needed if there is displacement of the upper humeral epiphysis (in children) or if there is coexisting shoulder dislocation.

Fractures of the humerus shaft This is often caused by a fall on an outstretched arm. Marked displacement often makes the diagnosis easy. Radial nerve injury may cause wrist drop. Splinting with metal gutter-shaped splints and support from the wrist to the neck (collar and cuff sling) usually gives satisfactory reduction. Immobilize for 8–12 weeks.

Anterior dislocation of the shoulder This may follow a fall on the arm or shoulder. Signs: loss of shoulder contour (flattening of the deltoid muscle), an anterior bulge from the head of the humerus, which may also be palpated in the axilla. Damage to the axillary nerve may cause deltoid paralysis and absent sensation on a patch below the shoulder. Before reduction, X-ray (Is there a fracture too?). Relieve pain (eg parenteral opioid, and Entonox® through the procedure). Methods of reduction:
Treatment by simple reduction: Apply longitudinal traction to the arm in abduction, and replace the head of the humerus by gentle pressure.
Kocher's method: A Flex the elbow and put traction on the arm.
B Apply forced external rotation and C adduction of the point of the elbow across the chest while the arm is flexed at the shoulder.
D Now apply internal rotation, so that the volar aspect of the forearm lies on the chest. Risk: fractures of the humerus.
The Hippocratic method:[1] With the patient lying on the floor, grasp the ipsilateral wrist, and exert downward traction on the arm, with counter-traction exerted by your shoeless foot in the axilla (this method provides your only chance to give a foot on your patient justifiably). Keep your knees and elbows straight, and lean backward. Your toes are now in a position to act as a fulcrum to relocate the head of the humerus as you slowly externally rotate the limb, and then adduct it. Offer pain relief before the manipulation. Some sources say that there is less danger to axillary vessels and the the humerus head with this method.

Surgery may be needed for recurrent dislocation (eg the Putti-Platt and Bankart reinforcement of the capsule and rotator cuff).

Posterior dislocation of the shoulder is rare, and may be caused by seizures or electroconvulsive therapy (if unparalysed). It may be hard to diagnose from an antero-posterior X-ray ('light-bulb' appearance of humeral head); lateral X-rays are essential.

1 D Fair 1993 Comments on *Pye's Surgical Handicraft* 16ed, 1950, *BMJ* ii 808

Fractures of the elbow and forearm

The humerus *Supracondylar fractures:* The patient is usually a child who has fallen on an outstretched hand. The elbow should be kept in extension after the injury to prevent damage to the brachial artery. Median and ulnar nerve palsies are rare complications. Reduce under anaesthesia, with X-ray guidance. Avoid flexing the elbow by more than 90°. Careful postoperative observations are required. The radial pulse may not return for 24h after the reduction, although limb perfusion is usually adequate to prevent Volkmann's ischaemic contracture (p 616). Dunlop traction may be required to maintain the reduction (the elbow is held semi-extended by axial traction while a second weight is suspended over the distal end of the upper arm). Internal fixation may be required if there is much instability.

Fractures of the medial condyle: These may require surgery if manipulation fails to bring about satisfactory reduction.

Fractures of the lateral condyle: Surgical fixation will be required. Complications: cubitus valgus and ulnar nerve palsy.

T-shaped intercondylar fractures of the humerus: This is a supracondylar fracture with a further break between the condyles. The presence of a fracture may be suggested by the hoisting of the sail-shaped fat pad on the anterior aspect of the distal humerus (the 'fat pad sign'), seen on the lateral elbow X-ray—which indicates an effusion. Compare with the other elbow. If no fracture is obvious, but an effusion is present, treat initially with a broad arm sling. Re-X-ray after 10 days (when fractures are more easily seen): if clear, start mobilization. For fractures, further immobilization (± internal fixation) may be needed; physiotherapy is vital in preventing prolonged stiffness.

Fractures of the head of the radius The elbow is swollen and tender, with the movements of pronation and supination particularly painful.
X-ray often shows an effusion, but minor fractures are commonly missed. Undisplaced fractures can be treated in a collar and cuff sling—if the fracture is displaced, internal fixation or excision of the radial head may be needed. Complications: radial nerve palsy (rare).

Pulled elbow The patient is a child (1–4yrs) who has been pulled up by the arms, while being lifted in play, causing the radial head to slip out of the annular ligament. Elbow rotation (forced supination with a thumb over the radial head) may be all that is required, producing a click on reduction. X-rays are not required.

Dislocation of the elbow The patient has usually fallen on a not yet fully outstretched hand, with elbow somewhat flexed, causing the ulna to be displaced backwards on the humerus, producing a swollen elbow, fixed in flexion. Brachial artery and median nerve damage are rare. Reduction (± anaesthesia): stand behind the patient, flex the elbow, and with your fingers around the epicondyles, push forwards on the olecranon with your thumbs. This may be aided by traction at the wrist from an assistant standing the other side of the trolley. Immobilize in a sling for 3 weeks.

Fractures of the olecranon are frequently part of a fracture-dislocation of the elbow. There may be wide displacement of the proximal fragment, caused by the action of triceps. Open reduction with internal fixation is often required.

Monteggia and **Galeazzi fractures** See p 754 and p 748.

Fractures and dislocations of the wrist

Colles' fracture This fracture at the distal radial head is common in osteoporotic post-menopausal ladies who fall on an outstretched hand. There is backward angulation and displacement producing a 'dinner-fork' wrist deformity when viewed in pronation (the fingers are the fork's prongs). Avulsion of the styloid process of the ulna may also occur. If there is much displacement reduction will be needed, particularly if there is backward and proximal shift of the distal fragment.

Treatment: Bier's block (regional anaesthesia)—method:
1 Place a double cuff around the arm.
2 Elevate arm for 1min, and then inflate the upper cuff to 250mmHg.
3 Inject 30–40ml 0.5% prilocaine into a vein on the back of the hand.
4 Manipulate the fracture 10min later.
5 20min after the injection deflate upper cuff, and inflate lower cuff.[1] NB: Sudden release of prilocaine into the circulation can cause fits. Other methods such as using smaller dose of anaesthetic infiltrated directly into the haematoma appear to be less effective.[1] Some centres rely on injecting the fracture haematoma with 10–20ml of bupivacaine 0.5%, with a wait of 30min before manipulation—complications: poor anaesthesia; sepsis. The alternative is general anaesthesia.

The manipulation: Prepare a plaster back slab up to the knuckles. Ask an assistant to hold the elbow. Apply traction to disimpact the fragment and push it forwards, and towards the ulnar side. Keeping the arm under traction, apply the slab, with wrist slightly flexed and hand in ulnar deviation. Support in a sling, once an X-ray has shown a good position. Do check X-ray in 5 days, when swelling has reduced. The plaster is then completed. Complications: median nerve symptoms (should resolve after good reduction); ruptured tendons; malunion; Sudeck's atrophy (p 695).

Smith's fracture and **Bennett's fracture** See p 758 and p 743.

Fractures of the scaphoid This fracture, which is notoriously easy to miss on X-rays, may result from a fall on the hand.
Signs: Swelling, pain on wrist movements and tenderness in the anatomical snuff box (between the tendons of extensor pollicis longus and abductor pollicis longus). *X-rays:* if the fracture is suspected request additional oblique 'scaphoid' views. If X-rays are –ve, and scaphoid fracture is a strong possibility, a bone scan is a good test.[2] If unavailable, put in plaster and X-ray at 2 weeks, by which time the fracture may be visible.
Treatment: Immobilize in plaster from below the elbow to beyond the knuckle (include the thumb up to the base of the nail) until union takes place (eg 8 weeks). If there is non-union, consider a bone graft or Herbert screw fixation. Complications: as the nutrient artery enters distally, there may be avascular necrosis of the proximal fragment. This may cause late degeneration in the wrist.

Dislocation of the carpus This may be anterior or posterior. Both require prompt manipulation or open reduction, and plaster immobilization for 6 weeks. Median nerve compression may occur.

1 A Cobb 1985 *BMJ* ii 1683   2 *Injury* 1992 **23** 77–9

Limb surgery in bloodless fields[1]

This may be achieved by using pneumatic tourniquets. The risk of inducing ischaemic changes in the limb is minimized by applying 'Bruner's rules'.

- Width of tourniquet: 10cm for the arm; 15cm for legs.
- Apply to the upper arm, or mid/upper thigh.
- Inflation–arm: 50-100mmHg above systolic BP; leg: double systolic.
- Deflation–must be within ~2 hours (3 hours is the absolute maximum).
- Avoid heating the limb (cooling is better, if feasible).
- Use ≥2 layers of orthopaedic wool to provide adequate padding (make sure it does not get wet with the skin preparation fluid, which should be aqueous, so that if wetting happens inadvertently, 'burns' do not occur).
- Apply only with the utmost caution to an unhealthy limb.
- Ensure the apparatus is calibrated weekly, and is well maintained.
- Document duration and pressure of tourniquet use.

1 I Braithwaite 1996 *Journal of the Medical Defence Union* 12 14

Fractures in the hand

Fractures of the metacarpals These only need manipulation if there is gross displacement. The fifth metacarpal is most commonly involved, often from a punching injury. Immobilization for up to 10 days in a wool and crêpe bandage is usually all that is required, with finger movement encouraged after the first 2–3 days. Longer periods of splinting in a plaster or 'boxing glove' bandage can result in a stiff hand. Refer any fractures with obvious rotational deformity (a clinical, not a radiological decision), as this can be disabling. Rotational fractures or fractures of ≥2 metacarpals usually require operative fixation with plate and screws.
►Beware wounds overlying metacarpophalangeal joints—these are often from the teeth of the punched victim, are contaminated bites, and may communicate with the joint.

Fractures of the proximal phalanx Spiral or oblique fractures occurring at this site are likely to be associated with a rotation deformity—and this must be corrected. Often, the only way to do this accurately is by open reduction and fixation.

Fractures of the intermediate phalanges These should be manipulated and splinted in flexion over a malleable metal splint, and the affected finger strapped to its neighbour. The aim is to control rotation, which may interfere with subsequent finger flexion.

Fractures of the terminal phalanges may be caused by crush injuries and are often compound. If closed, symptoms may be relieved by trephining the nail to reduce swellings (a lightly-held hypodermic needle can be used as a drill, or a paper clip heated in a paraffin flame will pierce the nail). Split skin grafts from the thenar eminence may be needed for partial amputations of the finger tip.

Mallet finger The tip of the finger droops because of avulsion of the extensor tendon's attachment to the terminal phalanx. If the avulsed tendon includes a piece of bone, union is made easier—using a plastic 'mallet finger splint' or a simple wooden spatula (with adhesive felt) which holds the terminal phalanx taped in extension. Use for 6 weeks. Interphalangeal arthrodesis may be needed if active extension remains limited. A poorer outcome is associated with delay in splinting, and in those >50yrs old.

Gamekeeper's thumb This is so-called because of the rupture of the ulnar collateral ligament of the metacarpophalangeal joint of the thumb during the forced thumb abduction that occurs when wringing a pheasant's neck. The same injury is described in dry ski-slope participants who fall and catch their thumb in the matting ('Hill-end thumb'). Diagnosis can be difficult as the thumb is so painful to examine, but to miss this injury may condemn the patient to a weak pincer grip—inject 1–2ml 1% plain lignocaine around the ligament to facilitate examination. Nearly always the ligament ends are turned into the joint, and are best dealt with by open operation and repair.

Fractures of the pelvis

▶*The pelvis is like a suit of armour: after damage there is much more concern about its contents than about the structure itself.*[1]

Because of the ring structure of the pelvis, single fractures are often stable and need no treatment other than a few weeks rest. In contrast, ≥2 fractures in the pelvis (with one above the level of the hip) renders the ring unstable and this constitutes a serious injury, with internal injuries in 25%. The force producing the fracture is usually a compression force, often from a traffic accident (60%).

Examples: Fractures of the ilium and pubic ramus (on one or both sides); fracture through a sacroiliac joint and a pubic ramus, as below.

Malgaigne's fracture (20% of all pelvic fractures, 60% of unstable ones): disruption of the pelvis anteriorly and posteriorly with displacement of a central fragment containing the hip joint.

Acetabular fractures Common sites: posterior lip or transverse.
Two 45° oblique X-rays (± CT scans) are needed to define injuries exactly.
Treatment: open reduction and reconstruction of the articular surface—to delay the onset of secondary osteoarthritis.

Complications ●Haemorrhage (eg internal iliac artery). Check foot pulses, BP, CVP + urine output often. Transfusion is often needed. ▶For rapid pelvis stabilizing in the resuscitation room, external fixation can be applied. This simple procedure dramatically reduces blood loss.
●Bladder rupture—may be intraperitoneal or extra-peritoneal.
●Vaginal and rectal perforation—look for bleeding. Both rare.
●Trapping of the sciatic nerve—there is persistent pain.
●Urethral rupture, often at the junction of the prostatic and membranous parts in males. The appearance of a drop of blood at the end of the urethra is suggestive. He may be unable to pass urine (avoid repeated tries). On rectal exam, the prostate may be elevated out of reach.

Treatment Relieve pain and replace blood. Suspend the patient in a pelvic sling promptly. The patient lies supine with the pelvis over the sling's webbing, which exerts upward and medial (thereby compressing) traction via weights and runners suspended above the bed. Traction is also applied to the legs. Surgical reconstruction may be undertaken.

If rupture of the urethra is suspected, insert a suprapubic catheter, and ask for expert urological help. If only a small volume of urine is found, the reason may be rupture of the bladder. Enlist expert help. A cystogram may be needed. Avoid urethral catheters: they may make a false passage.

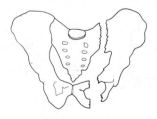

1 M Rang 1983 *Children's Fractures* p 233 JB Lippincott

Injury to the hip and femur

Incomplete or impacted femoral head fractures may present with coxa valgus; complete fracture with dislocation may present with coxa varus.

Intracapsular fractures occur just below the femoral head. This often causes external rotation and adduction of the leg. The injuring force may be trivial and the patient may be able to walk (but with difficulty). As the medial femoral circumflex artery supplies the head via the neck, ischaemic necrosis of the head may occur, particularly if there is much displacement. Femur fractures fill 20% of all UK orthopaedic beds. The incidence is rising (~1:100/yr in females aged 75–84). Mortality: 50%.

Treatment:
- Assess vital signs. Treat shock with Haemaccel®, but beware incipient heart failure. If present, monitor CVP.
- Relieve pain (eg morphine 0.2mg/kg IM+prochlorperazine 12.5mg IM).
- X-rays: a good quality lateral is essential to make the diagnosis if there is impaction or little displacement.
- Prepare for theatre: FBC, U&E, CXR, ECG, crossmatch 2U, consent. Tell anaesthetist about any drugs the patient is taking.
  Sort out any medical problems before embarking on surgery.
- In theatre: if displacement is minimal, either multiple screw fixation, or Garden screws. (Sliding nail plate units are not used for intracapsular fractures.)
  Alternatively, the head is excised and a prosthesis inserted.

Intertrochanteric–extracapsular fractures (between greater and lesser trochanters). They occur in a younger age group and, as blood supply is adequate, non-union is uncommon. *Treatment:* Traction and bed rest for 3 months—or internal fixation (eg with plate and nail or Richard's screw). The principle of the screw plate is to fix the fracture, but to allow compression by sliding.

Femoral shaft fracture ▶Is the femoral artery torn? (Look for swelling and check distal pulses.) Sciatic nerve injury may also occur. The proximal bone fragment will be flexed by iliopsoas, adducted by gluteus medius, and laterally rotated by gluteus maximus. The lower fragment will be pulled up by the hamstrings and adducted (with lateral rotation) by the adductors. *Treatment:* Typically, this is with a locked intramedullary nail. This is introduced proximally over a guide wire that can be manipulated across the fracture under X-ray control. This allows early mobilization. Alternative: manipulation under anaesthesia (exact reduction is not needed) with fixed traction with a Thomas knee splint, or skeletal traction, or sliding traction, with the thigh supported on a frame, or a cast brace with a hinge at the knee (this permits early weight bearing). Union takes 3–4 months. Intramedullary nail fixation allows early restoration of function.

Condylar fractures and tibial plateau fractures Being intra-articular, these demand accurate joint reconstruction to minimize secondary osteoarthritic changes.

Posterior hip dislocation These occur to front-seat passengers in car accidents, as the knee strikes the dashboard. The femoral head may be felt in the buttock. The leg is flexed, adducted and shortened. There may be damage to the popliteal division of the sciatic nerve. *Treatment:* Reduction under anaesthesia, by lifting the femoral head back into the joint. Traction (for about 3 weeks) enables the joint capsule to heal.

Fracture dislocation of the hip This is described on p 706.

Injuries of the lower leg

The patella Lateral dislocation may result from a blow to the side of the knee, causing a tear in the medial capsule and the quadriceps aponeurosis. It is an injury of young people. On examination, the knee is unable to flex. Recurrent dislocation may be related to under-development of the lateral femoral condyle. Sudden contraction of quadriceps may snap the patella transversely, whereas direct force may produce a multiple 'star-shaped' fracture (undisplaced within the quadriceps tendon).

The knee *Injury to the collateral ligaments* is common in sport, eg presenting with effusion ± tenderness over the ligament injured. Rest is needed, then firm support. If there is complete rupture, varus and valgus straining opens the joint if the knee is held in 10° flexion (and if the knee is held in extension, if a cruciate ligament is also torn).

Anterior cruciate tears: Typically, these follow posterior blows to the tibia or rotation injuries, when the foot is fixed to the ground. Signs: effusion, haemarthrosis, and a +ve 'draw' sign. (Immobilize the patient's leg by sitting on his foot, and then with the knee in 90° flexion, grip the upper tibia, and try to draw it towards you, away from the femur. Note: the reverse 'set back' suggests a posterior cruciate tear—eg after car accidents as the knee strikes the dashboard.) Do a 'pivot shift test' (p 634). Examination under anaesthesia may be needed. *Treatment:* This can be problematic. 3 weeks' rest in a plaster cylinder may help. If problems (knee instability /giving way) persist, consider reconstruction using part of the patella tendon. Physiotherapy is adequate for ~60%.

Semilunar cartilage tears: Medial cartilage tears (eg 'bucket-handle') follow forcible twists to a flexed knee (eg in football). Adduction with internal rotation causes lateral cartilage tears. Extension is limited (the knee 'locks') as the displaced segment lodges between femoral and tibial condyles. The patient has to stop what he is doing, and can only walk on tip-toe, if at all. The joint line is tender, and McMurray's test is +ve (p 634). If the 'handle' of the 'bucket' becomes free at one end ('parrot beak' tear), the knee suddenly gives way, rather than locking.

Tests: MRI; arthroscopy. *Treatment:* remove the damaged meniscus.

Fractures of tibia *Avulsion fractures* of the intercondylar region often occur with anterior cruciate injury. Sliding traction will be required, or surgery with open elevation of the joint surface.

Fractures of the tibial shaft are unlikely to be displaced, unless the fibula is also fractured. Compound fractures are common as there is so little anterior covering tissue. Tearing of the nutrient artery may make non-union a problem in fractures of the lower third of the tibia.

Treatment: Close compound fractures as quickly as possible. Reduce under anaesthesia. The treatment of choice for closed fracture is intramedullary nailing. Open fractures are treated with external fixation or unreamed intramedullary nails. Steinmann pins are now rarely used. A supportive dressing may be sufficient for isolated fibula fractures.

Pretibial lacerations The shin (particularly if elderly) has a poor blood supply. It is vulnerable to flap wounds, eg caused by steps of buses. *Treatment:* Try hard to iron out *all* the flap, and reposition it carefully: an ideal tool is the 'wrong' end of a Vacutaine® needle, sheathed in its rubber case. The important thing is to prevent tension—so skin closure with adhesive strips (eg Steristrip®), which can be loosened if the tissues swell, is better than sutures. Dress the wound and advise a support bandage, and leg elevation. Review to check for infection, wound tension, and necrosis.

Cord injury—physiotherapy and nursing

▶ *Turn every two hours to prevent pressure sores.* See p 726.

The chest Start regular physiotherapy early with coughing and breathing exercises, to prevent the sputum retention and pneumonia which are likely to follow diaphragmatic partial paralysis (eg C3–4 dislocation). If the lesion is above T10 segmental level, there is no effective coughing.

The straight lift (for transferring patients) One attendant supports the head with both hands under the neck so that the head lies on the arms. 3 lifters standing on the same side insert their arms under the patient, one at a time, starting at the top. After the lift withdraw in the reverse order.

The log roll 3 lifters stand on the same side of the patient. The one near the head has both arms over the patient's further arm; the middle one has one arm under the legs and the other arm holds the patient's iliac crest. The 3rd lifter supports the calves. A fourth person controls the head and neck, and gives the command to turn. The patient is then gently rolled laterally, with pillows to support the lumbar curve and to keep the position stable.

Posture Joints need to be placed in a full range of positions. Avoid hyper-extension of the knees. Aim to keep the feet flexed at 90° with a pillow between the soles and the foot of the bed. Avoid over-stretching of the soft tissues (deformity may result).

Bowels From the 2nd day of injury gentle manual evacuation using plenty of lubricant is needed. A flatus tube may be helpful in relieving distension once the ileus of spinal shock has passed.

Wheelchairs The patient should be kept sitting erect; adjust the footplates so that the thighs are supported on the wheelchair cushion and there is no undue pressure on the sacrum. Regular relief of pressure on the sacral and ischial areas is vital. Independence in transferring to bed or toilet will be a suitable aim for some patients with paraplegia. Expert skill is needed in assigning the correct wheelchair for any particular patient.

Standing and walking Using a 'tilt table', or the Oswestry standing frame with trunk support straps, the tetraplegic patient can gain the upright posture. If the level of injury is at L2–4, below-knee calipers and crutches will enable walking to take place. If the lesion is at T1–8, 'swing to gait' may be possible. The crutches are placed a short distance in front of the feet. By leaning on them and pushing down with the shoulders, both legs may be lifted and moved forwards together.

Sport Consider archery, darts, snooker, table tennis and swimming for those with paraplegia.

Personal qualities The personal qualities of the nurse, physiotherapist, and the patient are as the important as exact anatomic lesion. There may be great mood swings from euphoria to despair as the patient accustoms himself to his loss and his new body image.

10. Some unusual eponymous syndromes

Tests: FBC: Hb↑. CXR: 'wooden shoe' heart contour+RVH. ECG: RVH. *Prognosis:* 86% 32-yr survival after corrective surgery (JG Murphy 1993 *NEJM* **329** 593).

Fanconi anaemia[□] Aplastic anaemia associated with congenital absence of the radii, skin pigmentation, hypoplasia of the thumb, syndactyly, missing carpal bones, microsomy, microcephaly, strabismus, cryptgenitalism, cryptorchidism, IQ↓, deafness, short stature. Bone marrow transplant has been tried—as has umbilical cord blood (HUCB) haemopoietic progenitor cells. The disease may terminate in acute myeloid or monomyelocytic leukaemia.

Galeazzi fracture Radius shaft fracture with distal ulna subluxation.

Ganser syndrome This is a 'pseudodementia' in which the patient shows 'approximate answering' to the examiner's questions. In one patient the question 'What is the colour of the chair in the corner?' was answered thus: 'What corner? I don't know what a corner is. I don't see a chair. . .' Absurd remarks only occur as answers to questions. Disorientation is invariably present, but the intellectual deficit is inconstant (hence the 'pseudo'). Hysteria, hallucinations and a fluctuating level of consciousness are common. Often there is a preceding head injury.

Gaucher's syndrome This is the most prevalent lysosomal storage disorder. There is an autosomal recessive deficiency of lysosomal acid-beta-glucosidase, so glucosylceramide accumulates in macrophage lysosomes (producing so-called Gaucher's cells), causing progressive organomegaly, with marrow and CNS infiltration. There is pigmentation of the face and legs. *Diagnosis:* Skin fibroblast glucocerebrosidase assay. Death is often from pneumonia or bleeding. In infantile Gaucher's, there are marked CNS signs (rigid neck, dysphagia, catatonia, hyperreflexia, low IQ). *Treatment:* Growth, IQ, and psychomotor skills improve with enzyme therapy (human placental alglucerase = Ceredase®, metabolizes intracellular cerebrosides in spleen and liver; dose example 60u/kg/2 weeks IVI over 4h; cost to remission: ≈ £100,000, B Bembi 1994 *Lancet* ii 1679 & T Cox 1993 *Lancet* ii 694).

Hand–Schüller–Christian syndrome (Langerhans' cell histiocytosis) Other names include histiocytosis X, Letterer–Siwe disease, and eosinophilic granuloma of bone. Abnormal monoclonal Langerhans'-like cells are pathognomonic of the condition, which is putatively neoplastic. This is a destructive, infiltrative disease in which bone, liver, skin and spleen show necrotic (lytic) aggregates of eosinophils, plasma cells and vacuolated histiocytes. The lesions may show up on a technetium-99 labelled bone scan. The disease may occur in children or adults, and often starts with a polyp appearing in the external auditory meatus. Other features: dermatitis (erosions and pustules), exophthalmos, hepatosplenomegaly, lymphadenopathy (rarely suppurative with sinus formation), honeycomb lung, stomatitis, thrombocytopenia, anaemia, and a lethal 'leukaemia' most often seen in infants. *Treatment:* Bone surgery, radiotherapy, steroids or cytotoxics may induce remissions. F Cotter 1995 *BMJ* i 74

Hartnup's disease There is a congenital defect in tryptophan metabolism precipitated by light. The skin is thick and scaly, being hyperpigmented where exposed to light. Signs: nystagmus, ataxia, hyperreflexia. *Treatment:* give nicotinamide and vitamins from the B complex. Autosomal recessive.

Hunter's syndrome (Mucopolysaccharidosis II) Deficiency of iduronate sulphatase usually affects males (sex-linked recessive, but a new mutation in 33%), producing deafness, mental retardation, short stature, chronic diarrhoea, an unusual face, joint stiffness, corneal clouding and hepatosplenomegaly. *Diagnosis* is by measuring iduronate sulphatase in the serum. Even in its severe form it is milder than Hurler's syndrome.

Smith's fracture of the distal radius is known as the reversed Colles' fracture—there is forward (rather than backward) angulation of the radial fragment. Manipulation is needed, with the forearm held in full supination.

Still's disease (Juvenile rheumatoid arthritis) Typically presenting in a prepubertal girl who develops a mono- or polyarticular synovitis with erosion of cartilage ± fevers, pericarditis, iridocyclitis, pneumonitis, lymphadenopathy, splenomegaly. It accounts for 10% of all juvenile chronic arthritis, and has been hived off from this larger group as rheumatoid factor is present in the blood. Other subgroups are juvenile ankylosing spondylitis, psoriatic arthritis and ulcerative colitis-associated arthritis.
Tests: FBC: leucocytosis; ESR ↑; mild anaemia; ferritin↑; LFTs↑; albumin↓.
Treatment is supportive rather than curative. Mild exercise should be followed by one hour's rest per day. If the hips are affected, physiotherapy aims to prevent contractures by encouraging extension (eg lying prone on the floor to watch TV). Splinting, traction and non-weight-bearing exercises may be used. Hot baths relieve morning stiffness. Naproxen 5mg/kg/12h PO pc may be helpful. Give aspirin, up to 80mg/kg/day PO, to give levels <250mg/l (beware of liver and CNS toxicity). If there is severe systemic disturbance prednisolone 0.5mg/kg/day PO may be needed. For those with unremitting destructive disease, penicillamine, gold and hydroxychloroquine may help. Surgery may be needed to conserve joint function.

Sydenham's chorea (St Vitus' dance) Purposeless movement, exaggerated by tension, and disappearing on sleep, with clumsiness, grimacing, emotional lability, a darting lizard's tongue and unclear speech (OHCM p 40) complicating rheumatic fever (in <5%, p 194). Chorea may be the only feature, appearing up to 6 months after clinical and lab signs of streptococcal infection have abated. Differential diagnosis: Wilson's disease, juvenile Huntington's, thyrotoxicosis, SLE, polycythaemia, Na⁺↓, hypoparathyroidism, kernicterus, encephalitis lethargica, subdural haematoma, alcohol, phenytoin, neuroleptics, benign hereditary chorea, neuroacanthosis.▣ If needed, haloperidol, phenobarbitone, or diazepam may be tried.

Syme's amputation An amputation immediately proximal to the ankle.

Tay-Sachs disease This is an autosomal recessive gangliosidosis (Type I) which affects ~1:4000 Ashkenazic Jewish births. It is a disease of grey matter. There is reduced lysosomal hexosaminidase A. Low levels of enzyme are detectable in carriers. Children are normal until ~6 months old, when developmental delay, photophobia, myoclonic fits, hyperacusis and irritability occur. Ophthalmoscopy: cherry-red spot at macula. Death usually occurs at 3–5yrs of age. Prenatal diagnosis may be made by amniocentesis.

Tolosa-Hunt syndrome A lesion in the superior orbital fissure (or cavernous sinus) produces unilateral ophthalmoplegia with disordered sensation in the area of the first branch of the trigeminal nerve—a very convenient state of affairs for conducting cataract surgery, which may be temporarily induced by local anaesthetic agents and a longish needle.

Treacher-Collins' syndrome This is the association of lower eyelid notching, oblique palpebral fissures, flattening of the malar bones and an absent or hypoplastic zygoma. If these are associated with mandibular defects, ear defects and deafness, it is called Franceschetti's syndrome.

Turner's syndrome (XO) Prevalence: 1:2500 girls. Girls lack a sex chromosome, and this is associated with short stature (<130cm)—possibly the only sign, hyperconvex nails, broad chest, wide carrying angle (cubitus valgus), inverted nipples, broad chest, ptosis, nystagmus, webbed neck, coarctation of the aorta, left heart defects, leg lymphoedema. The gonads are rudimentary or absent. Mosaicism may occur (XO, XX). Treatment: Somatropin (human growth hormone) 0.14U/kg/24h SC helps prevent short stature (do not give if the epiphyses have fused). Association: Crohn's disease (BMJ 1994 ii 606).

Pre-operative care

Aims
1 To ensure that the right patient gets the right surgery. Have the symptoms and signs changed? If so, inform the surgeon.
2 To assess and balance the risks of anaesthesia.
3 To ensure that the patient is as fit as possible.
4 To decide on the type of anaesthesia and analgesia.
5 To allay all anxiety and pain.

The pre-operative visit Assess cardiovascular and respiratory systems, exercise tolerance, existing illnesses, drug therapy and allergies. Assess past history—of myocardial infarction, diabetes mellitus, asthma, hypertension, rheumatic fever, epilepsy, jaundice. Assess any specific risks—eg is the patient pregnant?
Have there been any anaesthetic problems (nausea, DVT)? Has he had a recent GA? (Do not repeat halothane within 6 months.)

Family history Ask questions about malignant hyperpyrexia (p 772); dystrophia myotonica (OHCM p 470); porphyria; cholinesterase problems; sickle-cell disease (test if needed). Does the patient have any specific worries?

Drugs Ask about allergy to any drug, antiseptic, plaster.
Anticoagulants: Epidural and spinal blocks are contraindicated. Beware regional anaesthesia. Check clotting.
Anticonvulsants: Give the usual dose up to 1h before surgery. Give drugs IV (or by NGT) post-op, until able to take oral drugs. Sodium valproate: an IV form is available (give the patient's usual dose). Phenytoin: give IV slowly (<50mg/min). IM phenytoin absorption is unreliable.
Antibiotics: Neomycin, streptomycin, kanamycin, polymixin and tetracyclines may prolong neuromuscular blockade.
Beta-blockers: Continue up to and including the day of surgery as this precludes a labile cardiovascular response.
Contraceptive steroids: Don't stop except perhaps before very major surgery.
Digoxin: Continue up to and including morning of surgery. Check for toxicity and do plasma K^+. Suxamethonium increases serum K^+ and can lead to ventricular arrhythmias in the fully digitalized. Potentiation of the vagotonic effects of digitalis by halothane and neostigmine may be precipitated by an increase in calcium and low potassium.
Diuretics: Beware hypokalaemia. Do U&E (and bicarbonate).
Hormone replacement therapy: As there is no increased risk of DVT or PE, there is no need to stop these agents.
Insulin: If on long acting preparations stabilize pre-operatively on short acting. IV 5% dextrose with KCl and insulin the morning of theatre allows accurate control. See OHCM p 108.
Levodopa: Possible arrhythmias when the patient is under GA.
Lithium: Stop 3 wks pre-op. It may potentiate neuromuscular blockade and cause arrhythmias. Beware post-op toxicity ± U&E imbalance; see p 354.
Monoamine oxidase inhibitors: Stop 3 weeks before surgery. Interaction with narcotics & anaesthetics may lead to hypotensive/hypertensive crisis.
Ophthalmic drugs: Anticholinesterases used to treat glaucoma (eg ecothiopate iodine) may cause sensitivity to, and prolong duration of, drugs metabolized by cholinesterases, eg suxamethonium.
Oral hypoglycaemics: No chlorpropamide 24h pre-op (long $t_{1/2}$).
Steroids: If the patient is on or has recently taken steroids give extra cover for the peri-operative period (OHCM p 110).
Tricyclics: These enhance adrenaline, exerting anticholinergic effects causing tachycardia, arrhythmias and low BP.

Pre-operative examination and tests

It the anaesthetist's duty to assess suitability for anaesthesia. The ward doctor assists by obtaining a good history and examination—and should also reassure, inform and and obtain written consent from the patient (eg remember to get consent for orchidectomy in orchidopexy procedures; also inform thyroidectomy patients of the risk of nerve damage).

Be alert to chronic lung disease, hypertension, arrhythmias, and murmurs (endocarditis prophylaxis needed?—see p 194). In rheumatoid arthritis do a lateral cervical spine X-ray to warn about difficult intubations.

Tests Be guided by the history and examination.
- U&E, FBC and ward test for blood glucose in most patients. If Hb <10g/dl tell anaesthetist. Investigate/treat as appropriate. U&E are particularly important if the patient is on diuretics, a diabetic, a burns patient, or has hepatic or renal diseases, or has starved, has an ileus, or is parenterally fed.
- Crossmatching: group and save for mastectomy, cholecystectomy. Crossmatch 2 units for Caesarean section; 4 units for a gastrectomy, and >6 units for abdominal aortic aneurysm surgery.
- Specific blood tests: LFT in jaundice, malignancy or alcohol abuse.
  Amylase in acute abdominal pain.
  Blood glucose in diabetic patients (OHCM p 108).
  Drug levels as appropriate (eg digoxin).
  Clotting studies in liver disease, DIC, massive blood loss, already on warfarin or heparin. Contact lab as special bottles are needed.
  HIV, HBsAg in high risk patients—after appropriate counselling.
  Sickle test in those from Africa, West Indies or Mediterranean area—and others whose origins are in malarial areas (including most of India).
  Thyroid function tests in those with thyroid disease.
- CXR: if known cardiorespiratory disease, pathology or symptoms.
- ECG: those with poor exercise tolerance, or history of myocardial ischaemia, hypertension or rheumatic fever.

Preparation ►Fast the patient.
- Check on any bowel or skin preparation needed.
- Start any DVT prophylaxis as indicated. One régime is heparin 5000U sc pre-op, then every 12h sc until ambulant. See p 654.
- Catheterize and insert nasogastric tube (before induction and its attendant risk of vomiting) as indicated.
- Book any pre-, intra-, or post-operative X-rays as needed. Make sure that the pathologist knows if there is to be a frozen section.
- Book post-operative physiotherapy.

ASA classification (American Society of Anesthesiologists)
1 Normally healthy.
2 Mild systemic disease.
3 Severe systemic disease that limits activity; not incapacitating.
4 Incapacitating systemic disease which poses a threat to life.
5 Moribund. Not expected to survive 24h even with operation.

You will see a space for an ASA number on most standard anaesthetic charts. It is an index of health at the time of surgery. The prefix E is used in emergencies.

Pre-medication

The patient should be aware of what will happen, where he will waken and how he will feel. Premedication aims to allay anxiety and contribute to ease of anaesthesia by decreasing secretions (much less important than when ether used), promoting amnesia, analgesia and decreasing vagal reflexes. *Timing:* 2h pre-op for oral drugs; 1h pre-op for IM drugs.

Examples for the 70kg man
- Lorazepam 2–3mg PO.
- Temazepam 10–20mg PO.
- Diazepam 5–20mg PO.

Some anaesthetists still use the traditional IM premeds. Examples:
- Morphine 10mg IM and atropine 0.6mg IM.
- Pethidine 50mg IM and promethazine 25mg IM.
- Papaveretum 15.4mg (=1ml 15.4mg/ml ampoule) with 0.4mg hyoscine hydrobromide IM. A combined ampoule is available. Avoid the obsolete term 'Om and Scop' which implies Omnopon® (now called papaveretum) given with hyoscine hydrobromide (=scopolamine hydrobromide). Noscapine, (implicated in polyploidy risk in those of childbearing age) is not contained in papaveretum manufactured since 1993.

Examples for children
- Trimeprazine 2mg/kg as a syrup.
- Always use oral premeds in children as first choice. Children over 1yr: if IM needed, give pethidine 0.5–2mg/kg or morphine 0.15mg/kg.
- Emla® local anaesthetic cream on at least 2 sites that may be used for IV access, and at the same time as the premed.

Specific pre-medication
- Antibiotics
- Bronchodilators, eg salbutamol nebulizer
- Nitrate patches, eg Transiderm Nitro®
- Steroids, eg minor operations 100mg of IM hydrocortisone 1h pre-op and 6h post-op. Major operations hydrocortisone 100mg/6h IM. Consider if adrenal insufficiency or adrenal surgery; steroid therapy for more than 2 weeks prior to surgery; or steroid therapy for more than 1 month in the year preceding surgery.

Common reasons for cancellation
- Cold or recent viral illness.
- Recent myocardial infarction (significant ↑ in mortality if GA is given within 6 months—believed to be related to degree of cardiac failure).
- Patient not adequately fasted.
- U&E imbalance (particularly K⁺); anaemia.
- Inadequate preparation (results not available, not crossmatched).
- Patient not in optimum condition—eg poor control of drug therapy (digoxin, thyroxine, phenytoin); exacerbation of illness.
- Undiagnosed or untreated hypertension.

Equipment

Careful checking of equipment is vital before any anaesthetic or sedative procedure. The essentials are:

- Tilting bed or trolley (in case of vomiting).
- High volume suction with rigid Yankaeur, and long suction catheters.
- Reliable oxygen supply, capable of delivering 15 litres/min.
- Self-inflating bag with oxygen reservoir, non-rebreathing valve and compatible mask (a 'bag-valve-mask' system).
- Oropharyngeal and nasopharyngeal airways.
- A range of anatomical face mask sizes with harness.
- Endotracheal tubes (range) and catheter mount.
- Anaesthetic circuit.
- Laryngoscope with range of blade sizes and spare batteries.
- Intravenous infusion cannulae and fluids.
- Anaesthetic and resuscitation drugs and anaesthetic gases.
- Defibrillator.
- Monitoring equipment (may include pulse oximeter and end tidal CO_2 monitor, p 772).

Inhalational agents

These are the vapours which in clinically useful concentrations help to maintain anaesthesia and decrease awareness. In Britain they are generally added to the fresh gas flow by passing a fraction of the carrier gas (generally N_2O/O_2) through a plenum vaporizer (in which the delivery of anaesthetic vapour does not depend on the patient's respiratory effort—as it does in 'draw over' systems).

Halothane This gas has little analgesic effect. It decreases cardiac output (vagal tone ↑, leading to bradycardia, vasodilation and hypotension). It sensitizes the myocardium to catecholamines, (beware in patients with arrhythmias; surgical infiltration with local anaesthetic and adrenaline). It bronchodilates so is useful in bronchospasm. It relaxes the uterus in deep anaesthesia and may lead to post-partum haemorrhage (hence 0.5% concentration used during Caesarean section). The 2 major but rare complications are malignant hyperpyrexia and hepatitis.
▶ Do not use halothane twice within a 6-month period.

Enflurane This is similar to the above, but less potent. Avoid in epileptics (EEG epileptiform activities seen during anaesthesia). It does not sensitize the heart to catecholamines as much as halothane.

Isoflurane This is an isomer of enflurane. Theoretically induction should be quick, but isoflurane is irritant, so coughing, laryngospasm or breath-holding may complicate the onset of anaesthesia.

Stopping inhalation reverses all the above effects—except for hepatitis resulting from drug metabolism.

The ideal IV anaesthetic agent

The ideal IV agent would be stable in solution, be water soluble and have a long shelf-life. It would be painless when given IV; non-irritant if injected extravascularly (with a low incidence of thrombosis) with some pain (as a warning) if given intra-arterially.

- It should act rapidly.
- Recovery should be quick and complete with no hangover effect.
- It should provoke no excitatory phenomena.
- Analgesic properties are advantageous.
- Respiratory and cardiovascular effects should be minimal.
- It should not interact with other anaesthetic agents.
- Hypersensitivity reactions are not ideal.
- There should be no post-op phenomena, eg nausea, or hallucinations.

This perfect drug does not exist.

Intravenous anaesthetic agents: Sodium thiopentone

Sodium thiopentone ($t_{1/2}$=11h) When mixed with water to give a 2.5% solution this barbiturate is stable for 24-48h, so make it up daily. It has a rapid onset of action (arm–brain circulation time about 30sec). Effects last 3-8mins. Awakening is due largely to redistribution, not metabolism. Some 30% of the injected dose is still present in the body after 24h, giving rise to the hangover effect. Patients must not drive, operate machinery etc. within 24h.

Dose: eg 4-6ml of 2.5% solution (3-5mg/kg, but less in the elderly and the premedicated, and more in children who generally require 6-7mg/kg, but some children need much less, eg 2mg/kg.) Subsequent doses are cumulative. Note thiopentone interacts with other drugs eg increase dose in those who consume much alcohol.

Uses: induction of GA; it is also a potent anticonvulsant.

Contraindications: ●Airway obstruction. ●Barbiturate allergy. ●Fixed cardiac output states. ●Hypovolaemia/shock. ●Porphyria.

▶Problems: Intra-arterial injection produces pain and blanching of the hand/limb below the level of injection due to arterial spasm, followed by ischaemic damage and gangrene—frequently following inadvertent brachial artery puncture in the antecubital fossa.

Treatment: 1 Leave the needle in the artery and inject an alpha-blocker eg tolazoline (5ml of a 1% solution, or procaine hydrochloride 10-20ml of a 0.5% solution). 2 Perform, or ask an experienced colleague to perform, brachial plexus or stellate ganglion block. (These measures should dilate vessels and reduce ischaemia.) 3 Heparin IV to stop thrombus forming. 4 Give pain relief. 5 Postpone surgery unless desperate.

Extra-vascular injection produces severe pain and local tissue necrosis. Infiltrate with hyaluronidase 1500iu.

Other IV anaesthetic agents (See p 772)

Methohexitone (t½=5h) This barbiturate is like thiopentone, but mixed with water it has a shelf life of 6 weeks. *Dose:* 1mg/kg at a rate of 2mg/sec. It often causes pain and involuntary movements on injection. Recovery and metabolism are swifter than for thiopentone so it is more suitable for day-case patients. Contraindications are the same as for thiopentone but in addition it should not be used in the epileptic as it may provoke fits in susceptible patients.

Etomidate (t½=3.5h) This is a carboxylated imidazole.
Dose: 0.1–0.3mg/kg. Histamine release is not a feature; rapidity of recovery and cardiovascular stability are; therefore it is suitable for day-case surgery, the elderly and those with compromised cardiovascular systems.

Ketamine Dose: 0.5–2mg/kg; t½=2.2h. This is a phencyclidine derivative. There may be a delay before the onset of sleep. Hypertonus and salivation are problems, but there is some maintenance of laryngeal reflexes (do not rely on this). Recovery is slow. Emergence phenomena are troublesome (delirium, hallucinations, nightmares; all made worse if the patient is disturbed during recovery). Cardiac output is unchanged or increased. It is a good 'on site' or 'in the field' agent, as it can be given IM and produces profound analgesia without compounding shock. ▶Avoid in the hypertensive patient, those with a history of stroke, or raised intracranial pressure (further ↑ produced), patients with a recent penetrating eye injury (risk of ↑ intraocular pressure), and psychiatric patients. Avoid infiltration with adrenaline.

Propofol This phenol derivative is available in soya oil and egg phosphatid. Dose: 2mg/kg. Rapid injection may cause hypotension. It acts and is metabolized quickly. It is used in day case surgery, short procedures, and IV as a sedative in ITU. CI: the extremities of age; egg allergy.

Neuromuscular blockers

These drugs act on the post-synaptic receptors at the NMJ (neuromuscular junction). There are two main groups:

1 Depolarizing agents eg suxamethonium (=succinylcholine, Scoline®). These drugs depolarize the post-synaptic membrane, causing paralysis by inhibiting the normal membrane polarity. They are partial agonists for acetylcholine and cause initial fasciculation, liberation of K^+ (beware in paraplegia and burns!), myoglobin and creatine kinase. 30% of patients get post-operative muscle pains. Suxamethonium is an ideal intubating agent: it has a rapid onset, a short duration of action (2–3min), and produces good relaxation. Note: a second dose, if required, should be preceded by atropine, as the vagotonic effects of suxamethonium can lead to profound bradycardia (notably in children). Beware suxamethonium (Scoline®) apnoea (p 772). Dose of suxamethonium 0.6mg/kg IV.

2 Non-depolarizing agents These drugs compete with acetylcholine at the NMJ—but without producing initial stimulation (see suxamethonium above). Repeated doses may be given without atropine. Their action can be reversed by anticholinesterases (neostigmine). They are used during balanced anaesthesia to facilitate IPPV and surgery. Length of action and side effects will govern anaesthetists' choices. Examples of the more common agents are:

Pancuronium Long acting (~1 hour), vagal blockade and sympatho-mimetic action. *Dose:* 50–100µg/kg IV then 10–20µg/kg IV as needed.

Vecuronium Lasting 20–30min, it is used if cardiovascular stability is important. No ganglion-blocking effect. *Dose:* 80–100µg/kg IV then 20–30 µg/kg IV as needed. Starting dose for infants ≤4 months: 10–20µg/kg. If over 5 months, as adult dose, but high intubation dose may not be needed.

Atracurium Lasting ~20 minutes this causes histamine release so avoid in asthma. Metabolism is by Hoffman Elimination (spontaneous molecular breakdown), so it is the drug of choice in renal and liver failure.
Dose: 300–600µg/kg IV then 100–200µg/kg IV as needed.

Practical conduct of anaesthesia: I

The practitioner administering the anaesthetic is responsible for the suitability of the surroundings, the adequacy of the available equipment, and his own competence to deal with potential complications. Equipment must be checked before even the shortest case.

All anaesthetic rooms should be treated like a church—respect the contents, keep quiet unless asked to participate and follow the lead given by the (ad)minister of the anaesthetic. They differ in that hats *must* be worn, and the only person allowed to go to sleep is the patient.

Induction May be gaseous or IV (IM possible with ketamine).

Gaseous:
- Start with nitrous oxide:oxygen 60%:40% mixture. In children, it is less frightening to start with a hand cupped from the end of the circuit onto the face than to apply the mask direct to face.
- Proceed with slow, careful introduction of the volatile agent in 0.5-1% incremental doses (rapid increase in concentration is unpleasant and leads to coughing and possible laryngospasm).
- Monitor vital signs closely, as the time factor separating a struggling child from a flaccid, apnoeic, overdosed child is short.
- Establish IV access when and where appropriate.

Indications for gaseous induction:
- Any patient with airway obstruction (actual or potential eg foreign body, tumour, or abscess).
- At the patient's request.

Intravenous:
- Establish IV access.
- A sleep-inducing dose of eg thiopentone is injected after a 2ml test dose to detect inadvertent arterial cannulation (p 764).
- Beware! Stimulation before anaesthesia is established can have drastic consequences (coughing, breath-holding, laryngospasm). Remember, noise is a stimulus too.

Airway control This is maintained either by holding a mask onto the face, or by intubation (p 771). To prevent airway obstruction the standard chin lift and/or jaw thrust manoeuvres are used. It may be facilitated by the use of an airway adjunct (eg oropharyngeal, or nasopharyngeal). Insertion of an airway adjunct may produce vomiting or laryngospasm.

Accidents and their prevention

Road accidents affect 1 in 4 people in their lifetime. They are the most common cause of death in those <45yrs.[1] 50% of deaths in the 15-19 age group are from road accidents. In the UK, road deaths have fallen since 1966 despite a 50% increase in numbers of licensed vehicles. There are now (1992) 4229 deaths/yr and 49,245 serious injuries/yr in the UK. Since seat belt wearing was made compulsory (1983), deaths and serious injury to car users have fallen by 10% and 18%. Casualty rates per 100 million vehicle km are similar for pedal cyclists (527-204 deaths/yr) and motorcyclists (596-469 deaths/yr).[2] Age is important: babies are rarely killed on roads compared with children, because their risk exposure is less. Deaths fall in middle age, rising again before falling in old age.

Alcohol is a major factor in at least 10% of all road accidents. 20% of drivers and 36% of pedestrians killed may be expected to have blood alcohol levels above the legal limit of 80mg/100ml (17.4 mmol/l). This rises to 50% of drivers and 73% of pedestrians killed between 22:00 and 04:00. There is some evidence that older drivers compensate to some extent for alcohol-impaired performance—by driving more carefully. Young drivers can compensate in this way, but may choose not to do so (peer group pressure or a delight in risk taking).

Benzodiazepines, antidepressants and antihistamines have also been implicated as contributing factors to road accidents, and it is important that patients are advised not to drive while taking these. If a patient has epilepsy or diabetes with hypoglycaemic attacks (or is otherwise unfit to drive, p 468) it is his duty to inform the licencing authorities, and the doctor's duty to request him to do so.

The prevention of accidents A vital cognitive shift occurs if the word 'accident' is replaced by 'preventable occurrence', which suggests that accidents are predictable and preventable. They happen because of laziness, haste, ignorance, bad design, false economies, and failure to apply existing knowledge—more often than because of truly random events. For example, over the years, thousands of aircraft passengers have been killed because of poor design of flight decks, which could have been prevented by using simple ergonomic principles.

Various schemes have been tried to help young drivers who abuse alcohol—eg the Driver Improvement System for Traffic Violators, and rehabilitation by re-education with driving instructors. Psychotherapy is more helpful than lectures. Health education posters which picture tragic consequences to a girl friend are successful.[1]

Legislation is an effective means of saving lives (eg seat belt and speed laws). Alcohol laws and those governing the road-worthiness of vehicles presumably prevent accidents, but their effectiveness has not been quantified. Another effective way of reducing alcohol-related road accidents is to provide adequate lighting, eg at difficult bends or junctions. Accidents in the home may be prevented by such simple measures as child-proof containers, putting holes in polythene bags, using toughened glass throughout the home, and using cooking pans with handles turned in away from grasping toddlers.

1 T Benjamin 1987 *Young Drivers Impaired by Alcohol and other Drugs*, Roy Soc Medicine 2 Dept of Transport 1993 *Road Accidents in Great Britain 1992*, HMSO

Accident statistics (UK)

Every year about 13,000 people are killed in accidents in Great Britain. This includes 1000 children.

Deaths and major injuries 1991[1] (Total deaths=13,014)

On the roads	4568 killed,[2] 336,000 injured
Home	5273
Work	615
Sport	268 (estimated)
Other transport (rail, air, water)	203
Other falls	979
Other (eg medical misadventure, lightning)	1108

Accidents in the home Old people are particularly at risk. Over half the males and three-quarters of females who die from accidents in their homes were 65yrs old or older. About half of those old ladies falling and fracturing their femur will be dead in 6 months (p 708). The commonest cause of accidental death in children is suffocation; in the 15–44 age group it is poisoning, and in the over-45s the chief cause is falls.

Hospital attendances for accidents 7.5 million people/year in the UK attend an accident and emergency department following an 'accident'. 350,000 (4.7%) of these are admitted. One-third of accidents occur in the home, and one-quarter are in those under 5yrs old.

Many more receive treatment from their general practitioner.

Accidents in children[3]
- ~ 10,000 children are permanently disabled by accidents each year.
- Accidents cause 1 child in 5 to attend the A&E department each year.
- Accidents are the commonest cause of death among children aged 1–14yrs, and they cause half of all deaths in those aged 10–14yrs.

Useful addresses: ●Scottish Chamber of Safety, Heriot-Watt University, Riccarton, Currie, Edinburgh EH14 4AS (UK).
- British Safety Council, 62 Challenors Road, London W6 9RS (UK).
- Construction health and safety group, St Ann's Road, Chertsey, Surrey KT16 9AT (UK)

---

1 Royal Society for the Prevention of Accidents (RoSPA), The Priory Queensway, Birmingham B4 6BS (tel. UK 0121 200 2461) 2 Department of Transport 1992 *Road Accidents in Great Britain: 1991*, HMSO 3 APLS group 1993 *Advanced Paediatric Life Support*, BMA

►►Basic life support (BLS)^{ATLS}

Synonyms Artificial Respiration; cardiopulmonary resuscitation (CPR).

Definition BLS is the provision of life support—expired air (your own) ventilation + external chest compression, without any equipment other than a simple airway adjunct, eg a mouth shield (to protect you from vomit) or Laerdal Pocket Mask® (to facilitate contact with the patient) which should be used if available.

SAFE approach ●As you approach the patient shout for help (pointing to an individual if possible, to activate him or her).
●Approach him with care—are there any hazards to you (p 790)?
●Free the patient from immediate dangers.
●Evaluate the patient's 'ABC' (see table below).

> The SAFE approach
> Shout for help
> Approach with care
> Free from danger
> Evaluate ABC (Airway, breathing, and circulation)

Establish unresponsiveness Shake gently by the shoulder while stabilising the forehead with the other hand. Ask "Are you all right?" If he responds, put in the recovery position (p 789)—if not, check for breathing.

Breathing Open and clear the airway (finger sweep; remove dental plate or *loose* false teeth—leave a well-fitting set, otherwise the mouth collapses making expired air ventilation difficult). If breathing, put in the recovery position; if not, check the pulse.

Circulation Feel the carotid pulse for 5 seconds. If felt, give 10 rescue breaths, then go for help. If no pulse, go for help *now*.

On return give 2 slow inflations (mouth-to-mouth/pocket mask), then 15 chest compressions (lower ⅓ of sternum, 2 fingers' breadth above xiphisternum; depress 4–5cm; at a rate of ~80/min). If there are 2 rescuers use a ratio of 1:5.

Children (1–8yrs) Same sequence, but use 1 hand 1 finger breadth above xiphisternum (rate 100/min; use ratio of 1:5 for one or two rescuers). Avoid blind finger sweep—may impact foreign body in conical upper airway; do look into the airway for easily removable foreign body. See p 310.

Infants (<1yr) If not breathing, give 5 rescue breaths, *then* check pulse (use brachial pulse as the neck is very short). If no pulse, give 20 cycles of 1:5 ventilations-to-compressions (compression rate 100/min) *then* go for help—take the baby with you and continue BLS while phoning for help. Again avoid blind finger sweeps. See p 310.

Helicopter transport[1]

▶It is often better to spend 30 minutes transporting a seriously injured person to a well-equipped trauma centre with consultants standing by, than to spend 10 minutes transporting such a person to a small A&E department where the most skilled help is not *immediately* available.

The importance of helicopters for casualty rescue and transport has been increasingly recognized since the Korean war. It is important to be aware of their limitations. Helicopters may be used for transporting casualties to hospital, or for interhospital transfer.

Advantages: Speed over long distances; access to remote areas; delivery of highly trained doctors and special equipment to the scene—eg ready to intubate, paralyse and ventilate, and give mannitol IV if head injury.[2]

Disadvantages: The most expensive method of casualty evacuation; noise and general stress, leading to anxiety and disorientation and hampering communication—reassure and provide with a headset; vibration exacerbating bleeding/pain from fracture sites; cold—beware in those rescued from sea/mountainside who may already be hypothermic; problems related to altitude; aircraft limitations, eg weather, suitable landing site, limited carriage space (especially if additional medical personnel); technical (police craft will not allow ECG monitor/oximeter due to magnetic radiation). NB: the gains of helicopter transfer depend on how many severe injuries occur. One UK study[2] concluded that at best only ~13 lives would be saved per year in the London area if it was reserved for the severest cases—trauma score >15; in lesser trauma, there is evidence that outcome is *less* good.[2]

Helicopter safety ●Always approach from the front of the aircraft, in full view of the pilot. Secure loose items, eg headgear.
●Do not enter/leave the rotor disc area without permission (thumbs up signal from pilot). Lower your head in the rotor disc area.
●Do not touch the winch strop/cable until the earthing lead has contacted the ground. Also, be sure to avoid the tail rotor.
●Make sure no-one is smoking within 50 metres of the aircraft.

Problems of altitude Hypoxia is unlikely unless there is cardiac or lung disease, anaemia, shock or chest trauma, as helicopters rarely fly high enough to produce a significant fall in P_aO_2.

Reduction in atmospheric pressure results in an expansion of enclosed gases on ascent. This produces pain in blocked sinuses, expansion of a pneumothorax, abdominal wound dehiscence (avoid flying for 10 days post surgery if possible) and renewed bleeding from a peptic ulcer. Remember drips may slow down.

On descent, beware of endotracheal tube cuffs and military antishock trousers (MAST) deflating significantly (particularly if applied at altitude eg on hillside). Rapid descent may induce barotrauma.

Specific problems ●Decompression sickness (p 802): if air is breathed under pressure (divers), nitrogen dissolves in blood and tissues. On rapid ascent after a dive the nitrogen will come out of solution as bubbles, producing joint pains ('the bends') ± urticaria, CNS defects and shortness of breath. ▶Do not fly if dived <30m within last 12h or >30m within 24h.
●Ischaemic chest pain or infarction is not a contraindication to flying and the fear that it may induce arrhythmias is unfounded.
●Psychiatric illness may preclude transport if the patient is considered a hazard to the aircraft's safety.
●Burns over 20% require preflight insertion of a nasogastric tube as a precaution against gaseous expansion of an ileus.

1 Brit Assoc for Immediate Care 1990 *Monographs on Immediate Care* (4)   2 J Nicholl 1995 *BMJ* ii 217

Drug index

The following is a drug index to therapeutic agents described in the text. The dosages included here are as an *aide memoire*, and should not substitute for the careful reading of the text before prescribing a drug with which you are not familiar. Paediatric doses are denoted 'paeds'.

Abidec® 0.6ml/24h po for infants... See *Cystic fibrosis* p 192
Acetazolamide 250mg/24h–500mg/12h po... See *Simple glaucoma* p 502
Acetazolamide 500mg po stat (im if vomiting) then 250mg/8h po... See *The red eye* p 494
Acyclovir 800mg five times a day po for 7 days... See *Ophthalmic shingles* p 484
Acyclovir as soon as the diagnosis is suspected. Give 5–10mg/kg/8h by ivi over 1h (250–500mg/m²)... See *Herpes simplex encephalitis* p 256
Acyclovir child dose: (20mg/kg/6h po), eg <2yrs: 200mg/6h po; 2–5yrs: 400mg/6h po. ≥6yrs: 800mg/6h... See *Varicella (herpes) zoster* p 216
Acyclovir ivi dose: 10mg/kg/8h (over 1h), with concentration <10mg/ml... See *Varicella (herpes) zoster* p 216
Acyclovir topically and 200mg/4h po for 5 days... See *Herpes simplex: vulval* p 30
Adrenaline in cardiac arrest (child)... See ▶▶*Cardiac arrest protocols* p 311
Adrenaline 10µg/kg (= 0.01ml/kg of 1:1000)... See ▶▶*im/iv adrenaline in anaphylaxis* p 303
Adrenaline 10µg/kg iv or 20µg/kg via et... See ▶▶*Resuscitation after delivery* p 230
Adrenaline dose by endotracheal tube: 100µg/kg... See ▶▶*im/iv adrenaline in anaphylaxis* p 303
Alflacalcidol 50 nanograms/kg/day... See *Renal failure (paeds)* p 280
Allopurinol 10–20mg/kg/24 po... See *Childhood leukaemia* p 190
Allopurinol: 100mg/24h po pc, ↑ as needed to 600mg/24h... See *The seronegative arthropathies* p 674
Almevax® 0.5ml sc if non-immune; delay until postnatal... See *The puerperium* p 148
Alverine 60–120mg/8h po gives unreliable results... See *Dysmenorrhoea* p 10
Aminophylline (0.15mg/kg/h ivi)... See *Neonatal intensive care* p 232
Aminophylline 5mg/kg by slow ivi... See ▶▶*Mendelson's syndrome* p 134
Aminophylline 5mg/kg iv over 20 min... See ▶▶*Childhood asthma* p 270
Amitriptyline (50mg/8–24h po. Starting dose example: 50mg at night)... See *Antidepressant drugs* p 340
Amitriptyline 25–50mg po at night for nerve pain... See *Terminal care* p 444
Amoxycillin 10–13mg/kg/8h po for 7 days is the drug of choice in under 5s... See *Fluid in the middle ear* p 538
Amoxycillin 250mg/8h po... See *Renal disease in pregnancy* p 160
Amoxycillin 250mg/8h po... See *Otitis media* p 536
Amoxycillin 250mg/8h po... See *Nasal injury and foreign bodies* p 550
Amoxycillin 500mg/8h po... See *Chlamydia during pregnancy* p 100
Amoxycillin 500mg/8h po or iv... See *The puerperium* p 148
Amoxycillin/clavulanic acid 1g/6h iv... See ▶▶*Obstetric shock* p 106
Ampicillin 30mg/kg im/iv... See *Infective endocarditis (paeds)* p 194
Ampicillin 30mg/kg im/iv 6h later... See *Infective endocarditis (paeds)* p 194
Ampicillin 30mg/kg/4h iv... See *Meningitis: the organisms (paeds)* p 260
Ampicillin 500mg iv... See *Cardiac disease in pregnancy* p 152
Ampicillin 500mg/6h iv... See *Prematurity* p 102
Ampicillin 500mg/6h iv... See *Post-partum haemorrhage (pph)* p 138
Ampicillin 500mg/6h iv... See *Renal disease in pregnancy* p 160
Amylobarbitone iv... See *The patient in disgrace* p 334
Anti-D 250u im to Rh–ve women... See *Amniocentesis* p 120
Antivaricella-zoster immunoglobulin (12.5u/kg im within 96h of exposure, max 625u)... See *Chickenpox* p 216
Arachis oil... See *The agents of terminal care* p 444
Aspirin (p 119) eg 75mg/24h po... See *Intrauterine growth retardation* p 104
Aspirin may be used up to 80mg/kg/day po... See *Juvenile arthritis* p 758
Atenolol 1–2mg/kg/day po... See *Blood pressure in childhood* p 304
Atracurium 300–600µg/kg iv then 100–200µg/kg iv as needed... See *Neuromuscular blockers* p 767

Drug index   ▶▶*Consult the pages referred to for further details about each drug.*

Atropine 0.3–0.6mg IV... See *Spinal cord injury—the first day* p 726
Atropine 0.6mg IM... See *Pre-medication* p 763
Auranofin 3mg/12h PO... See *Rheumatoid arthritis: treatment* p 672
Azathioprine... See *Rheumatoid arthritis: treatment* p 672
Azelaic acid cream... See *Acne vulgaris* p 592
Azithromycin (20mg/kg)... See *Tropical eye disease* p 512
Beclomethasone diproprionate 8×50µg puffs/24h... See *Allergic rhinitis* p 548
Benzhexol 1–4mg/6h PO... See *Schizophrenia: management* p 360
Benzhexol, eg 1mg/day PO, gradually increased up to 5mg/8h (tablets are 2mg or 5mg)... See *Blepharospasm* p 522
Benzylpenicillin (25mg/kg IM or IV)... See *Rheumatic fever (paeds)* p 194
Benzylpenicillin 25mg/kg/4h IV... See *Infective endocarditis (paeds)* p 194
Benzylpenicillin 25–50mg/kg/4h slow IV... See *Meningitis (paeds)* p 260
Benzylpenicillin 25–50mg/kg/4h slowly IV... See *Meningitis (paeds)* p 260
Benzylpenicillin 30mg/kg IM stat... See *Gonococcal conjunctivitis (paeds)* p 100
Benzylpenicillin 50mg/kg/4h IV... See *Meningitis: the organisms (paeds)* p 260
Benzylpenicillin 600mg IM (any age) given immediately... See *Meningitis: the organisms (paeds)* p 260
β-carotene (a provitamin) 1.2×10⁶ U PO... See *Xerophthalmia/keratomalacia* p 512
Betadine vaginal gel® for foul *rectal* or vaginal discharges... See *Agents of terminal care* p 444
Betahistine 8–16mg/8h PO... See *Menière's disease* p 546
Betamethasone sodium phosphate 0.1% nasal drops, instilled every 8h in the head-down-and-forward position (bent double at the hips, with nostrils pointing at the sky) for 48h... See *Nasal polyps* p548
Bicarbonate 1mmol/kg IV... See ►►*Infant cardiorespiratory arrest* p 310
Bicarbonate 1mmol/kg IV (4.2%... See ►►*Resuscitation after delivery* p 230
Bicarbonate 2.5ml/kg of 8.4% IV over 2h... See *Childhood ketoacidosis* p 264
Bicarbonate eg 1–2mmol/kg IV as a 4.2% solution... See *The ill neonate* p 238
Bicillin®, IM daily for 10 days... See *Intrauterine syphilis* p 98
Biperiden 2–5mg IM or IV slowly, up to 4 times daily.. See *Schizophrenia* p 360
Bisacodyl tablets (5mg), 1–2 at night help opiate-induced constipation... See *Terminal care* p 444
Bromocriptine 2.5 mg/12h PO... See *Sudden infant death syndrome* p 286
Bromocriptine 2.5mg PO on day 1, then 2.5mg/12h for 14 days... See *Stillbirth* p 136
Bromocriptine 2.5mg/12h PO days 10–26... See *The premenstrual syndrome* p 16
Budesonide 50–200µg/12h by large volume spacer... See ►►*Childhood asthma* p 270
Budesonide nebulized may also help croup—acute dose: 3 mths–12yrs: 0.5–1mg/12h (eg two 250mg Pulmicort Respules®; halve dose if chronic)... See *Croup* p 276
Bumetanide 1mg/24h PO... See *The agents of terminal care* p 444
Bupivacaine t½ = 3h. Slow onset, long duration. Max dose: 2mg/kg... See *Local anaesthesia* p 774
Buprenorphine (synthetic opiate; 'controlled' drug): 0.4mg/6h SL... See *Pain* p 778
Buprenorphine sublingual 0.2–0.4mg/8h... See *The agents of terminal care* p 444
Buserelin nasal spray 300µg/8h... See *Endometriosis* p 52
Calcipotriol ointment or cream... See *Psoriasis* p 586
Calcitonin 0.5–2u/12h SC... See *Paget's disease of bone* p 676
Calcitriol 6µg/day PO... See *Osteoporosis* p 676
Calcium chloride 10–30µg/kg IV... See ►►*Infant cardiorespiratory arrest* p 310
Calcium gluconate 10%, 0.5ml/kg IV over 5min... See *Renal failure (paeds)* p 280
Calcium gluconate 10%, 4.4ml/kg/day PO. IV treatment: 0.2ml/kg diluted in 4.8ml/kg of 0.9% saline over 10min... See *The ill neonate* p 238
Calcium lactate gluconate eg as Sandocal-1000®, 1 tablet per day PO... See *Osteoporosis* p 676
Captopril 0.5mg/kg/8h PO if >6 months old... See *BP in childhood* p 304
Carbamazepine... See *Epilepsy drugs: children* p 269
Carbimazole (OHCM p 542). Once control is achieved, aim to keep the dose at ≤10mg/24h PO... See *Thyroid disease in pregnancy* p 157
Carbimazole eg 15mg/24h PO for 18 months... See *Childhood hyperthyroidism* p 196
Carboprost 250µg (15-methyl prostaglandin F2α) eg as Hemabate® 1ml deep IM... See *Post-partum haemorrhage (PPH)* p 138

Drug index   ►►*Consult the pages referred to for further details about each drug.*

Nitrous oxide 50% in 50% O$_2$ (Entonox®) self-administered... See *Pain relief in labour* p 114

Nitrous oxide:oxygen 60%:40% mixture... See *Practical conduct of anaesthesia: I* p 768

Nitrous oxide/oxygen (Entonox®)... See *Pain* p 778

Norethisterone 5mg/8h PO... See *Menorrhagia* p 14

Norethisterone 5mg/bd PO from 4 days before period is due... See *Normal menstruation* p 8

Norethisterone 5–10mg/12h PO continuously... See *Endometriosis* p 52

Norethisterone enanthate 200µg... See *Depot contraception* p 68

Norgestrel 0.15mg PO from days 17–28... See *Metabolic bone disease—I: Osteoporosis* p 676

Norgestrel 150µg/24h, PO for 12 days out of 28... See *The menopause* p 18

Norplant®... See *Depot and implant hormonal contraception* p 67

Nortriptyline (25mg/6–24h PO. Start with 10mg/12h)... See *Antidepressants* p 340

Nystatin vaginal pessaries for 14 nights... See *Vaginal discharge* p 48

O$_2$ at 15 litres/min via mask with reservoir... See ▶▶*Antepartum haemorrhage* p 108

Oestradiol 1–2mg/24h PO... See *The menopause* p 18

Oestradiol implants (surgical) 25mg lasts ~36 weeks, 100mg lasts ~1yr... See *The menopause* p 18

Oestradiol patches supply 25–100µg/24h for 3–4 days... See *The menopause* p 18

Oestriol 0.1%... See *The menopause* p 18

Oestrogens 0.625mg/day PO continuously starting on the 1st day of the cycle... See *Metabolic bone disease—I: Osteoporosis* p 676

Orphenadrine 50 mg/8h PO... See *Schizophrenia: management* p 360

Oxybutynin 2.5mg/12h PO for 2 weeks; increasing to a maximum of 5mg/12h PO in elderly (5mg/8h non-elderly)... See *Urinary malfunction* p 70

Oxycodone 30mg suppositories (eg 30mg/8h, ≈30mg morphine)... See *The agents of terminal care* p 444

Oxytetracycline (250mg/12–24h PO 1h ac)... See *Rosacea and Rhinophyma* p 594

Oxytetracycline (500–250mg/12h PO 1h ac—start at higher dose... See *Acne* p 592

Oxytetracycline 250mg/12h 1h ac... See *The salivary glands* p 570

Oxytocin 5u IM... See *Normal labour* p 82

Oxytocin is given IV in 5% dextrose using a pump system (eg Ivac®). Infusions start at 2 milliunits (mu) per min, doubling every 20min until effective uterine contractions are produced (usually at a rate of 4–16mu/min: occasionally 32mu/min may be necessary)... See *Induction of labour* p 110

Pancrease®... See *Cystic fibrosis* p 192

Pancuronium 0.02–0.03mg/kg IV; then 0.03–0.09mg/kg/1.5–4h... See *Ventilating neonates* p 234

Pancuronium 50–100µg/kg IV then 10–20µg/kg IV as needed... See *Neuromuscular blockers* p 767

Papaveretum 7.7mg/ml: 1ml diluted with 9ml water. Dose: child up to 1 month: 0.15ml/kg IM; if 1–12 months, 0.15–0.2ml/kg IM of the above dilution. For older children (1–12yrs) use 0.02–0.03ml/kg IM of *undiluted* papaveretum 7.7mg/ml solution... See *Reducing pain in A&E (paeds)* p 678

Paracetamol 12mg/kg/6h PO... See *Fluid in the middle ear* p 538

Paracetamol 500mg–1g/6h PO... See *Osteoarthritis* p 668

Paracetamol syrup (120mg/5ml) for high fever or severe discomfort ever. Dose (per 6h PO): <3 mths old: 5–10mg/kg; 3 mths–1yr: 60–120mg; 1–5yrs: 120–250mg; 6–12yr 250–500mg. 60mg suppositories... See *Mumps (paeds)* p 214

Paracetamol, up to 4g/24h PO... See *Management of low back pain* p 624

Paraldehyde 0.4mg/kg PR in arachis oil, or IM (max 5ml/buttock to reduce risk of sterile abscess). Rule of thumb: 1ml/year to max 10ml... See *Epilepsy (paeds)* p 268

Paroxetine: 20mg each morning (↑in 10mg steps to 50mg—40mg if old)... See *Selective serotonin reuptake inhibitors* (SSRI) p 341

Pemoline 0.5–2mg/kg/24h PO... See *Behavioural questions (paeds)* p 206

Penicillamine up to 500mg/24h PO... See *Rheumatoid arthritis: treatment* p 672

Penicillin 10mg/kg/4h IV for the first few days, then PO for 3 months... See *Acute nephritis and nephrosis (paeds)* p 282

Drug index ▶▶*Consult the pages referred to for further details about each drug.*

Drug index ►►*Consult the pages referred to for further details about each drug.*

Pyrimethamine, 21 days of 0.25mg/kg/6h PO... See *Congeital toxoplasmosis* p 98

Ranitidine in elective sections, eg 150mg PO 2h pre-op... See *Caesarean section* p 132

Resonium A® 0.5g/kg PO... See *Renal failure (paeds)* p 280

Resperidone... See *Schizophrenia* p 360

Retinol palmitate 50,000u IM monthly... See *Xerophthalmia and keratomalacia* p 512

Retinyl palmitate 200,000u PO... See *Xerophthalmia and keratomalacia* p 512

Rifampicin 10mg/kg/12h PO for 48h for close contacts... See *Meningitis* p 260

Rifampicin 10–20mg/kg/24h PO (≤600mg/day)... See *Meningitis (paeds)* p 260

Rifampicin 15mg/kg/PO ac 3 times a week... See *Childhood tuberculosis* p 278

Ritodrine régime... See *Prematurity* p 102

Salbutamol 250µg IV... See *Complications of anaesthesia* p 772

Salbutamol 2mg nebulized... See ►►IM/IV *adrenaline in anaphylaxis* p 303

Salbutamol syrup 2mg/5ml. 5ml/8h (0.1mg/kg/8h if <2yr)... See *Childhood asthma* p 270

Salmeterol, 2 puffs/12h... See ►►*Childhood asthma* p 270

Sertraline: 50mg/24h PO, ↑ if needed by 50mg steps every few weeks to a maximum of 200mg/day; within 8 weeks reduce to a maintenance of 50–100mg/day PO... See *SSRI* p 341

Sodium aurothiomalate 50mg/week IM... See *Rheumatoid arthritis* p 672

Sodium bicarbonate... See *bicarbonate* (and pp 310, 230, 264, 238)

Sodium cromoglycate 2% eye drops/6h... See *Conjunctivitis* p 496

Sodium cromoglycate 2%, 2 × 2.6mg squeezes/4–6h... See *Aallergic rhinitis* p 548

Sodium nitrite solution 10ml of 3% IV over 3min... See *Smoke inhalation* p 689

Sodium nitroprusside 1–8µg/kg/min IVI by pump... See *Blood pressure in childhood* p 304

Sodium thiosulphate 25ml of 50% IV over 10min... See *Smoke inhalation* p 689

Sodium valproate sugar-free liquid is 200mg/5ml... See *Epilepsy (paeds)* p 268

Somatropin (human growth hormone) eg 0.14u/kg/day SC helps prevent short stature... See *Turner's syndrome* p 758

Spironolactone (1.2mg/kg/12h PO)... See *Acute nephritis/nephrosis (paeds)* p 282

Spironolactone 100mg/12h PO... See *The agents of terminal care* p 444

Spironolactone 25mg/6h PO days 18–26 of cycle... See *Premenstrual syndrome* p 16

Stesolid® give 5mg if <3yrs, or 10mg if >3yrs... See *Epilepsy (paeds)* p 268

Sulphadiazine 50mg/kg/12h PO... See *Congen. CMV/Toxoplasmosis* p 98

Sulphadimidine 250mg PO daily... See *Rhematic fever (paeds)* p 194

Sulphadimethoxine 1g stat, then 500mg/24h PO for 10 days... See *Trachoma* p 512

Sulphasalazine ½g/6h PO (start with ½g/24h); warn to stop if sore throat... See *Rheumatoid arthritis* p 672

Suxamethonium 0.6mg/kg IV... See *The practical conduct of anaesthesia: II* p 770

Temazepam 10–20mg PO... See *Pre-medication* p 763

Terbinafine is an expensive alternative (eg 250mg/24h PO for 1–3 months)... See *Superficial fungal infections* p 590

Terbutaline 10mg nebulized with O₂... See ►►*Childhood asthma* p 270

Terbutaline 250µg/4–6h (one puff)—eg with spacer and mask until 8yrs old... See ►►*Childhood asthma* p 270

Terfenadine 60mg/12h PO... See *Allergic rhinitis* p 548

Tetracycline 1% eye ointment is used 12-hourly for 5 days each month for 6 months... See *Tropical eye disease* p 512

Tetracycline 250mg/6h PO 1h ac... See *Pelvic infection* p 50

Tetracycline 250mg/6h PO 1h before food... See *Pelvic infection* p 50

Theophylline (as Slo-Phyllin®) dose: <12mg/kg/24h PO. Capsules are 60, 125, or 250mg... See ►►*Childhood asthma* p 270

Theophylline, eg 5mg/kg/8h PO syrup=60mg/5ml... See ►►*Childhood asthma* p 270

Thioridazine 25–100mg/6h No injection available... See *Schizophrenia* p 360

Thyroxine... See *Infant hypothyroidism* p 196

Tibolone 2.5mg/day PO... See *The menopause* p 18

Ticarcillin (50–60mg/kg/6h IV)... See *Cystic fibrosis* p 192

Tinzaparin... See *Joint replacements* p 654

Topical 1% clindamycin... See *Acne vulgaris* p 592

Drug index ►►*Consult the pages referred to for further details about each drug.*

Tranexamic acid 1–1.5g/6–8h PO... See *Menorrhagia* p 14

Trifluoperazine 5–10mg/12h PO, or 1–2mg/8h IM (child: 50µg/kg/day)... See *Schizophrenia* p 360

Trimeprazine 10–20mg/8h PO... See *Itch* p 578

Trimeprazine 2mg/kg as a syrup... See *Pre-medication (child)* p 763

Trimeprazine syrup (≤2mg/kg)... See *Common happenings in childhood* p 182

Trimethoprim 100mg PO prior to intercourse may help... See *Urinary malfunction* p 70

Trimethoprim 50mg/5ml, 50mg/12h PO for 5 d (halve if <6mths; double if >6yrs)... See *Childhood urinary tract infection (UTI)* p 188

Trimipramine (25–50mg/8 PO. Start with 50mg 2h before bed)... See *Antidepressant drugs* p 340

Triple (pertussis, tetanus, diphtheria) 0.5ml SC... See *Immunizations* p 209

Urea 40% in aqueous cream... See *Superficial fungal infections* p 590

Sodium valproate... See *Epilepsy drugs: children* p 269

Vecuronium 0.1mg/kg IV... See *The practical conduct of anaesthesia: II* p 770

Vigabatrin starting dose: 40mg/kg/day PO (sachets and tabs are 500mg). Max: 100mg/kg/day... See *Epilepsy: management (paeds)* p 268

Vit D (50 nanograms/kg of alfacalcidol/day PO... See *Renal failure (paeds)* p 280

Vit D 1000u/day... See *Breast feeding* p 178

Vit K 500µg PO to all babies within 24h of birth... See *The bleeding neonate* p 244

Zagreb antivenom (2 ampoules IVI at rate <1ml antivenom/min)—diluted in 2–3 volumes of 0.9% saline... See *Bites, stings & foreign bodies* p 736

Zidovudine 100mg 5 times a day PO antepartum, and, during labour (2mg/kg IVI over 1h, and then 1mg/kg IVI until delivery), and then 2mg/kg/6h PO for the baby for 6 weeks... See *Antenatal HIV infection* p 98

Zoster immune globulin 1000mg IM to the pregnant woman within 3 days of exposure to infection... See *Varicella (herpes) zoster* p 216

Zuclopenthixol acetate 50–100mg IM (buttock), 1 injection lasts 2–3 days... See *Mania* p 354

Drug index ▶▶*Consult the pages referred to for further details about each drug.*

Index

Bolder type denotes emergency topics Drugs: see Drug index, 807

New and updated topics for the 1996 reprinting

Blue chevrons (>>) indicate topics where more important changes have been made: the greater the number of chevrons, the more important the change. Consult the electronic version to ascertain the nature of the updates (*Mentor*, p 446—tel. 01132 582454).